New Frontiers in Parameterized Complexity and Algorithms

New Frontiers in Parameterized Complexity and Algorithms

Editors

Frances Rosamond
Neeldhara Misra
Meirav Zehavi

Basel • Beijing • Wuhan • Barcelona • Belgrade • Novi Sad • Cluj • Manchester

Editors
Frances Rosamond
Lebanese American University
Beirut
Lebanon

Neeldhara Misra
Indian Institute of Technology
Palaj
India

Meirav Zehavi
Ben-Gurion University
Beer-Sheva
Israel

Editorial Office
MDPI
St. Alban-Anlage 66
4052 Basel, Switzerland

This is a reprint of articles from the Special Issue published online in the open access journal *Algorithms* (ISSN 1999-4893) (available at: https://www.mdpi.com/journal/algorithms/special_issues/Parameterized_Complexity).

For citation purposes, cite each article independently as indicated on the article page online and as indicated below:

Lastname, A.A.; Lastname, B.B. Article Title. *Journal Name* **Year**, *Volume Number*, Page Range.

ISBN 978-3-7258-1301-8 (Hbk)
ISBN 978-3-7258-1302-5 (PDF)
doi.org/10.3390/books978-3-7258-1302-5

© 2024 by the authors. Articles in this book are Open Access and distributed under the Creative Commons Attribution (CC BY) license. The book as a whole is distributed by MDPI under the terms and conditions of the Creative Commons Attribution-NonCommercial-NoDerivs (CC BY-NC-ND) license.

Contents

Mugang Lin, Jianxin Wang, Qilong Feng and Bin Fu
Randomized Parameterized Algorithms for the Kidney Exchange Problem
Reprinted from: *Algorithms* 2019, 12, 50, doi:10.3390/a12020050 1

Max Bannach and Sebastian Berndt
Practical Access to Dynamic Programming on Tree Decompositions
Reprinted from: *Algorithms* 2019, 12, 172, doi:10.3390/a12080172 13

Ronald de Haan and Stefan Szeider
A Compendium of Parameterized Problems at Higher Levels of the Polynomial Hierarchy
Reprinted from: *Algorithms* 2019, 12, 188, doi:10.3390/a12090188 30

Nadia Creignou, Raïda Ktari, Arne Meier, Julian-Steffen Müller, Frédéric Olive and Heribert Vollmer
Parameterised Enumeration for Modification Problems
Reprinted from: *Algorithms* 2019, 12, 189, doi:10.3390/a12090189 58

Borislav Slavchev, Evelina Masliankova and Steven Kelk
A Machine Learning Approach to Algorithm Selection for Exact Computation of Treewidth
Reprinted from: *Algorithms* 2019, 12, 200, doi:10.3390/a12100200 74

Wenxing Lai
The Inapproximability of k-DOMINATINGSET for Parameterized AC^0 Circuits
Reprinted from: *Algorithms* 2019, 12, 230, doi:10.3390/a12110230 94

Robert Ganianand Sebastian Ordyniak
Solving Integer Linear Programs by Exploiting Variable-Constraint Interactions: A Survey
Reprinted from: *Algorithms* 2019, 12, 248, doi:10.3390/a12120248 108

Julien Baste, Lars Jaffke, Tomáš Masařík, Geevarghese Philip and Günter Rote
FPT Algorithms for Diverse Collections of Hitting Sets
Reprinted from: *Algorithms* 2019, 12, 254, doi:10.3390/a12120254 122

Laurent Bulteau and Mathias Weller
Parameterized Algorithms in Bioinformatics: An Overview
Reprinted from: *Algorithms* 2019, 12, 256, doi:10.3390/a12120256 140

N. S. Narayanaswamy and R. Vijayaragunathan
Parameterized Optimization in Uncertain Graphs—A Survey and Some Results
Reprinted from: *Algorithms* 2020, 13, 3, doi:10.3390/a13010003 173

Andreas Emil Feldmann, Karthik C. S., Euiwoong Lee and Pasin Manurangsi
A Survey on Approximation in Parameterized Complexity: Hardness and Algorithms
Reprinted from: *Algorithms* 2020, 13, 146, doi:10.3390/a13060146 200

Neeldhara Misra, Frances Rosamond and Meirav Zehavi
Special Issue "New Frontiers in Parameterized Complexity and Algorithms": Foreward by the Guest Editors
Reprinted from: *Algorithms* 2020, 13, 236, doi:10.3390/a13090236 260

Article
Randomized Parameterized Algorithms for the Kidney Exchange Problem

Mugang Lin [1,2,3], Jianxin Wang [1,*], Qilong Feng [1] and Bin Fu [4]

1. School of Computer Science and Engineering, Central South University, Changsha 410083, China; linmu718@csu.edu.cn (M.L.); csufeng@csu.edu.cn (Q.F.)
2. School of Computer Science and Technology, Hengyang Normal University, Hengyang 421002, China
3. Hunan Provincial Key Laboratory of Intelligent Information Processing and Application, Hengyang 421002, China
4. Department of Computer Science, The University of Texas Rio Grande Valley, Edinburg, TX 78539, USA; binfu@utrgv.edu
* Correspondence: jxwang@csu.edu.cn

Received: 28 December 2018; Accepted: 21 February 2019; Published: 25 February 2019

Abstract: In order to increase the potential kidney transplants between patients and their incompatible donors, kidney exchange programs have been created in many countries. In the programs, designing algorithms for the kidney exchange problem plays a critical role. The graph theory model of the kidney exchange problem is to find a maximum weight packing of vertex-disjoint cycles and chains for a given weighted digraph. In general, the length of cycles is not more than a given constant L (typically $2 \leq L \leq 5$), and the objective function corresponds to maximizing the number of possible kidney transplants. In this paper, we study the parameterized complexity and randomized algorithms for the kidney exchange problem without chains from theory. We construct two different parameterized models of the kidney exchange problem for two cases $L = 3$ and $L \geq 3$, and propose two randomized parameterized algorithms based on the random partitioning technique and the randomized algebraic technique, respectively.

Keywords: kidney exchange problem; randomized algorithm; parameterized algorithm; random partitioning; multilinear monomial detection

1. Introduction

Kidney transplantation is one of the best treatment methods for those who are suffering from end-stage renal failure. Unfortunately, it is quite difficult for some patients to find suitable kidneys because of a lack of deceased donors and incompatibility of their blood and tissue types with willing donors. Although some patients can buy living donor kidneys from marketplaces, it is commonly considered to be unethical to commercialize of human organs. Moreover, it is in fact illegal in most countries. Living donation is regarded as legal as long as it occurs as a gift instead of a commodity to make profit. But it often arises that a donor who in general is a relative or friend of a patient is willing to donate a kidney to the patient even though their kidneys are not physiologically compatible such that the kidney transplantation cannot be carried out. Therefore, it is a significant research topic to effectively utilize living-donor kidneys such that their incompatible intended recipients can obtain one compatible kidney from these living-donor kidneys by exchanging. In 1986, Rapaport [1] first presented the idea of kidney paired donation where two incompatible patient-donor pairs reciprocally exchange the donors' kidneys so that two patients respectively obtain a compatible kidney, where an incompatible patient-donor pair represents a patient and his willing but incompatible donor. Since then, kidney exchange programs have developed rapidly and have already been established in some countries, for example, the USA, the UK, Turkey, Switzerland, and South Korea [2].

The Kidney Exchange problem (KEP) is the central issue of kidney exchange programs from a computational point of view. In a set of incompatible patient-donor pairs, the KEP is to find the exchanges with the maximum collective benefit such that each patient-donor pair of the exchanges can donate a kidney to some pair and receive a compatible kidney from another pair. The collective benefit is usually measured by the total number of possible kidney transplants [3–8]. From a graph theory perspective, as a type of barter exchange, kidney exchange is implemented by cycles comprising patient-donor pairs, where each donor in a pair donates a kidney to the patient in the next pair. Theoretically speaking, long cycle exchanges are better, since they are beneficial in improving the chances that more incompatible pairs are contained into the exchanges. However, the length of cycles must be limited in a certain range for the following reasons. First, to ensure an exchange cycle is executed successfully, all operations should be performed simultaneously so as to reduce the risk that a donor might break his promise after his incompatible patient has obtained a kidney before he donates his kidney. Long cycle exchanges may bring substantial logistical difficulties because the number of personnel and facilities required for simultaneous operations is too large. Second, all exchanges of a cycle will be cancelled if the final crossmatch test before operations detects new incompatibilities. Hence, longer cycles may cause greater logistical complexity and greater exchange failure risk. As a result, the length of cycles is generally limited in the range $[2, L]$, where L is a given constant and usually $2 \leq L \leq 5$. Therefore, the KEP can be regarded as finding an optimal packing of vertex-disjoint cycles from a given directed graph such that the length of each cycle is at most L.

When considering altruistic donors, kidney exchange allows the inclusion of chains which start from altruistic donors and end with either a pair or a patient in a wait list. The KEP becomes more complex, and can be modelled as finding an optimal set of vertex-disjoint cycles and chains from the directed graph of given instance such that the length of each cycle is at most L [5]. In a chain exchange, if the donor of a patient-donor pair backs out of the chain after his patient has received a transplant, no pair in the remaining tail of the planned chain suffers irreparable harm. Hence, a chain exchange can be conducted non-simultaneously [9] and the length of a chain exchange can be much longer than that of a cycle exchange [10,11]. But a donor in a non-simultaneous exchange still has the risk of reneging, which can make some patient-donor pairs lose exchange opportunity, thus in some literatures the limitation of chain length is still considered, e.g., the same length constrained for chains and cycles in [12]. An altruistic donor can be seen as a patient-donor pair with a "dummy" patient which is compatible with each donor of other incompatible patient-donor pairs. Thus, in this way, chain exchange and cycle exchange can be treated similarly in optimization models. In this paper, we mainly address the KEP without chains.

Over the past decades, there has been a rich literature on models, market mechanisms, and optimal algorithms for the KEP. There are many different variations of the KEP when considering different scenarios and limitations. The reader is referred to surveys on KEP variants [2,13]. In this section, we mainly describe the previous work of the KEP from a computational point of view. The complexity of the KEP was investigated in [3,4], and [14]. When $L \geq 3$, the KEP is NP-hard, meaning that it is unlikely there exists a polynomial time exact algorithm.

For the KEP, most exact algorithms are based on Integer Programming (IP) approaches, and research work focuses mainly on IP formulations. Roth et al. [5] proposed two noncompact formulations of IP for the KEP: the edge formulation and the cycle formulation with exponential numbers of constraints and variables, respectively. Abraham et al. [3] compared the two formulations and showed that the cycle formulation is tighter than the edge formulation when the cycle size is limited. They presented an incremental formulation approach to solve the cycle formulation based on enhancing the column generation method and the branch-and-price search technique [15]. To reduce the number of constraints and variables of the two formulations from exponential to polynomial, Constantino et al. [2] gave two compact formulations for the KEP: the edge assignment formulation and the extended edge formulation, and showed by computational analysis that when the scale of KEP becomes large, the compact formulations have more merits than non-compact ones.

Klimentova et al. [16] presented the disaggregated cycle decomposition model which is regarded as a Dantzig-Wolfe decomposition [17] of the extended edge formulation proposed in [2], and solved the model by a branch-and-price method. Recently, Dickerson et al. [18] introduced models where chains are represented by only a polynomial number of position-indexed edge variables which bring about great improvements on running time for the KEP with chains. Mak-Hau [19] studied the edge formulations for the KEP with and without chains and presented some polyhedral results. For a detailed review of various IP approaches to solving the KEP, see a recent survey of Mak-Hau [20].

Approximation technique is also one of the main approaches of solving the KEP. Biró et al. [4] proved the APX-completeness of the maximum size KEP when $L = 3$, and gave an $(L - 1 + \varepsilon)$-approximation algorithm for the maximum cycle weight KEP and an $(L/2 + \varepsilon)$-approximation algorithm for the maximum size KEP (for any $\varepsilon > 0$). Luo et al. [21] proved three inapproximability results for the KEP in a weighted graph. It is NP-hard to approximate the KEP within $\frac{14}{13}$ under a fixed integer $L > 3$ and within $\frac{434}{433}$ when $L = 3$. In an unweighted graph, it is NP-hard to approximate the KEP within $\frac{698}{697}$ when $L = 3$. Jia et al. [22] presented near optimal approximation algorithms by combining heuristics and local search technique.

Recently, some researchers started studying exact exponential algorithms and parameterized algorithms for the KEP. For the maximum cycle weight KEP with $L = 3$, an $O^*(2^k)$ FPT algorithm [4] was proposed by turning it into a maximum weight matching problem, where k is the minimum size of an arc set S where there is at least one arc in S from each 3-cycle of the instance. However, the problem finding an arc set S with the minimum size is NP-hard. Recently, Xiao and Wang [8] studied the variant of the KEP which contains chain exchanges and allows some compatible patient-donor pairs to participate in exchanges to find better matched kidneys. They designed two $O(2^n n^3)$ exact algorithms based on dynamic programming and subset convolution for the KEP with and without length constraints, respectively. Dickerson et al. [23] contracted the representation of a compatibility graph by introducing a way to class the vertices in the graph into different types according to some actual meaning. Based on the number k of vertex types, Xiao and Wang [8] designed an FPT algorithm of running time $O(2^{O(k^2)}k^{2k^2} + n(n+m))$ for the KEP without length constraints by using decomposition technique.

Randomness is an essential resource for solving many computational problems, and randomization is a powerful tool for developing algorithms [24–29]. Recently, stochastic matching approaches have been applied to deal with the KEP [30,31]. In fact, stochastic methods can naturally describe many settings of barter exchange problems [32,33]. Thus, in the future stochastic approaches for the KEP will become an important research issue.

In this paper, we study randomized algorithms for the parameterized KEP without chains by the random partitioning technique [24,25] and the randomized algebraic technique [26–29]. For ease of description, in what following we use "KEP" to refer to the kidney exchange problem without chains. We consider the parameterized KEP involving only 2-cycle and 3-cycle exchanges (namely the case of $L = 3$), with respect to parameters that correspond to the numbers k_2 and k_3 of 2-cycle exchanges and 3-cycle exchanges, respectively. We note that the solution P of a given instance of the KEP involves only 2-cycle and 3-cycle exchanges. Suppose that the vertex set of a given instance can be correctly split into two subsets S_1 and S_2, where the subset S_1 involves only 2-cycle exchanges of P and the subset S_2 involves 3-cycle exchanges of P. Then we can easily find these 2-cycle exchanges in S_1 by transforming it into maximum matching problem and find these 3-cycle exchanges in S_2 by utilizing an existing 3-set-packing algorithm [34]. Hence, we obtain a parameterized algorithm of running time $O^*(4.6^{3k_3} \cdot 2^{2k_2})$ by the random partitioning technique. For the case of $L \geq 3$, we discuss the parameterized KEP with respect to the number k of possible kidney transplants. Applying the randomized algebraic technique, we reduce the KEP to the multilinear monomial detection problem, and obtain an $O^*(2^{(k+L)}(k+L)^2 L^3 n^L)$ algorithm based on the results of Koutis and Williams [27–29].

This paper is organized as follows. Section 2 covers the preliminaries and notations, and introduces the formal definitions of the KEP without chains. In Sections 3 and 4, we investigate the randomized algorithms for two parameterized models of the KEP, and we present two randomized FPT algorithms. Finally, in Section 5 we give a short summary of our works and some future directions.

2. Preliminaries and Problem Definitions

In this paper, all graphs are simple finite graphs. For an undirected graph G, $V(G)$ and $E(G)$ denote the sets of vertices and edges of G, respectively. When the graph is clear from the context, we write short notations V and E. For two vertices u and v, $\{u,v\}$ represents the edge between u and v. A matching of G is an edge set where there is no shared vertex between arbitrary two edges. A matching M is said to be maximum if for any other matching M_0, $|M| \geq |M_0|$. For a directed graph D, $V(D)$ and $A(G)$ denote the sets of vertices and arcs of D, respectively. We use short notations V and A when the graph is clear. For two vertices u and v, (u,v) denotes an arc from u to v. For a subset $U \subseteq V$, $D[U]$ represents the subgraph induced by U in graph D. Suppose that a graph W is a sequence of arcs $(v_1, v_2), (v_2, v_3), \cdots, (v_{k-1}, v_k)$ in a directed graph D. If the vertices $v_1, v_2, \cdots, v_k$ are distinct, then graph W is a directed path. If the vertices $v_1, v_2, \cdots, v_{k-1}$ are distinct, $v_1 = v_k$ and $k \geq 3$, then graph W is a directed cycle. A k-cycle is a cycle of length k.

2.1. Parameterized Complexity

Parameterized complexity, introduced by Downey and Fellows [35], is a two-dimensional framework to measure the complexity of NP-complete problems by a function depending on the input size n and a fixed parameter k. Problems with some fixed parameter k are called *parameterized problems*. A parameterized problem is called *fixed-parameter tractable* (FPT) [35,36] if there is an algorithm with running time of the form $f(k) \cdot n^{O(1)}$, where the $f(k)$ is a computable function independent of the input size n.

2.2. KEP Definitions

From the graph theory perspective, the KEP without chains may be modeled as finding a maximum weighted packing of vertex-disjoint cycle in a directed graph $D = (V, A)$, where each vertex corresponds to an incompatible patient-donor pair consisting of a patient and his incompatible donor, each arc (i, j) represents that the donor of i is compatible with the patient of j. The weight w_{ij} of each arc (i, j) represents the collective benefit of the kidney swap between the patient of j and the donor of i. In kidney exchange without altruistic donors, a donor is willing to donate his kidney if and only if his associated patient receives a kidney. The kidney transplants may proceed in the same cycle of the graph D because each vertex of the cycle is compatible with the donating kidney of the previous vertex. Thus, when a cycle consists of k patient-donor pairs, it is called a k-cycle exchange. Because the length of exchange cycles is restricted for logistical reasons and to avoid risk, for a k-cycle exchange, k does not exceed an integer L ($2 \leq L \leq 5$). The kidney exchange problem without chains is defined as follows:

Kidney Exchange Problem without Chains:
Input: Directed graph $D = (V, A)$, arc weight function $w : A \to R^+$, and positive integer L.
Question: Finding a maximum weighted set of vertex-disjoint cycles with length at most L.

When $L = 2$, the KEP becomes the maximum weighted matching problem for an undirected graph G, where $V(G) = V(D)$ [7]. For two vertices $v_i, v_j \in V(G)$ there is an edge $\{v_i, v_j\} \in E(G)$ in G if there are both arc $(v_i, v_j) \in A(D)$ and $(v_j, v_i) \in A(D)$ in D, and edge weight $w(\{v_i, v_j\}) = w(v_i, v_j) + w(v_j, v_i)$. Therefore, we can solve the KEP by utilizing the maximum matching algorithm which has running time $O(|V(G)|(|E(G)| + |V(G)| \log |V(G)|))$ [37]. If $L = \infty$ (i.e., there is no upper bound on the maximum length of a cycle), we can transform the KEP into the maximum weight perfect matching problem of a bipartite graph. The bipartite graph $G = (\{V_d, V_p\}, E)$ can be constructed in the following way: if vertex $v_i \in V(D)$ in digraph D, there are a vertex $v_{id} \in V_d$, a vertex $v_{ip} \in V_p$ and an edge $\{v_{id}, v_{ip}\} \in E$ with weight $w(\{v_{id}, v_{ip}\}) = 0$ in graph G; if arc $(v_i, v_j) \in A(D)$ ($i \neq j$) in digraph D, there is an edge

$\{v_{id}, v_{jp}\} \in E$ with weight $w(\{v_{id}, v_{jp}\}) = w(v_i, v_j)$ in graph G. If $\{v_{id}, v_{jp}\}$ is in the maximum weight perfect matching of graph G and $i \neq j$, then v_i and v_j are in the same exchange cycle of D. If $\{v_{id}, v_{ip}\}$ is in the maximum weight perfect matching of graph G, then v_i is not in any exchange cycle of D. Thus, if there is no upper bound on the length of cycles, the KEP can be transformed into the maximum weight perfect matching problem and solved in polynomial time. When $L \geq 3$, the KEP is NP-hard by reducing from the 3D-matching problem [3,4].

3. A Randomized Algorithm Based on Random Partitioning Technique

For the KEP, Roth [5] showed that in a large population, all the gains from exchange can be obtained by using 2-cycle, 3-cycle and 4-cycle exchanges. Especially, 2-cycle and 3-cycle exchanges capture almost all the gains from exchange, and their marginal contribution is significantly large. In practice, 2-cycle and 3-cycle exchanges are typically used, such as in the New England Program [38] and the kidney exchange program of the United Network for Organ Sharing [5]. In this section, we consider the maximum exchange size KEP whose exchanges is limited only 2-cycle and 3-cycle exchanges (namely $L = 3$). The model is defined as follows:

Parameterized $\text{KEP}_{L=3}$ ($p\text{-KEP}_{L=3}$):
Input: a simple directed graph $D = (V, A)$, and two positive integers k_2 and k_3;
Parameter: k_2 and k_3;
Question: Is there a packing of vertex-disjoint cycles in D, comprising at least k_2 2-cycles and at least k_3 3-cycles?

Applying random partitioning technique, we design a randomized algorithm for the $p\text{-KEP}_{L=3}$. Given an instance (D, k_2, k_3) of the $p\text{-KEP}_{L=3}$, we stochastically divide the vertex set V into two subsets V_2 and V_3 by putting v into V_2 or V_3 with probability $1/2$ for each vertex $v \in V$. When the instance is a Yes-instance, there exists a proper partition such that k_2 2-cycles are in $D[V_2]$ and k_3 3-cycles are in $D[V_3]$. We can easily find a packing with k_2 vertex-disjoint 2-cycles in $D[V_2]$ through transforming into the maximum matching problem of an undirected graph and invoking the corresponding algorithm, and find a packing with k_3 vertex-disjoint 3-cycles in $D[V_3]$ by transforming into the 3-set-packing problem and invoking the corresponding algorithm. Algorithm 1 RKEP(D, k_2, k_3) is the detailed randomized algorithm of the $p\text{-KEP}_{L=3}$ based on the random partitioning technique, where c is a constant which is applied to adjust the probability that Algorithm RKEP(D, k_2, k_3) fails.

Algorithm 1 RKEP(D, k_2, k_3)

Input: a directed graph $D = (V, A)$, positive integer c, and two positive integers k_2 and k_3;
Output: a packing of vertex-disjoint cycles in D, comprising at least k_2 2-cycles and at least k_3 3-cycles, otherwise return no solution exists.
1: $times=0$;
2: **while** $times \leq c \cdot 2^{3k_3+2k_2}$ **do**
3: $\quad times = times + 1$;
4: $\quad V_2 = V_3 = \emptyset$;
5: $\quad$ **for all** $v \in V$ **do**
6: $\quad\quad$ put v into V_2 or V_3 with probability $1/2$;
7: $\quad$ **end for**
8: $\quad$ for the induced subgraph $D[V_2]$, find a packing P_2 of at least k_2 vertex-disjoint 2-cycles;
9: $\quad$ **if** no P_2 exists **then**
10: $\quad\quad$ goto next loop iteration;
11: $\quad$ **end if**
12: $\quad$ for the induced subgraph $D[V_3]$, find a packing P_3 of at least k_3 vertex-disjoint 3-cycles;
13: $\quad$ **if** no P_3 exists **then**
14: $\quad\quad$ goto next loop iteration;
15: $\quad$ **end if**
16: $\quad$ **return** $P_2 \cup P_3$;
17: **end while**
18: **return** no solution exists;

Theorem 1. *For the p-KEP$_{L=3}$, there exists a randomized algorithm of running time $O^*(4.6^{3k_3} \cdot 2^{2k_2})$ which, for any given constant $c > 0$, produces the desired packing in a Yes-instance with probability at least $1 - \frac{1}{e^c}$.*

Proof. For the p-KEP$_{L=3}$, if an instance (D, k_2, k_3) is a No-instance, then no matter how the vertex set V is split, we cannot find a required packing in V_2 and V_3. Thus, no solution to the p-KEP$_{L=3}$ will be returned in line 18 of Algorithm RKEP(D, k_2, k_3).

Now assume that the given instance (D, k_2, k_3) is a Yes-instance and P is a solution to the p-KEP$_{L=3}$. Then there are k_2 vertex-disjoint 2-cycles and k_3 vertex disjoint 3-cycles in $D[P]$. Let P_2 and P_3 be the vertex sets of k_2 vertex-disjoint 2-cycles and k_3 vertex disjoint 3-cycles, respectively. Thus, $|P_2| = 2k_2$ and $|P_3| = 3k_3$. By random partitioning, if the vertex set V can be correctly divided into V_2 and V_3 such that $P_2 \subseteq V_2$ and $P_3 \subseteq V_3$, then we can obtain P_2 by maximum matching algorithm for V_2 in line 8 and obtain P_3 by 3-set-packing algorithm for V_3 in line 12. Thus in line 16, $P_2 \cup P_3$ can be correctly returned.

We analyze the probability that RKEP(D, k_2, k_3) fails to find a packing P in D. Since each vertex $v \in V$ is put into V_2 or V_3 with probability $1/2$ in each while loop, the vertices V_2 and V_3 can be correctly partitioned in line 5–7 with probability $(\frac{1}{2})^{2k_2} \cdot (\frac{1}{2})^{3k_3} = 2^{-2k_2-3k_3}$. Thus, the probability in each loop iteration is $1 - 2^{-2k_2-3k_3}$ such that there exists at least one vertex $v \in P_2$ which is put into V_3 or at least one vertex $u \in P_3$ which is put into V_2. Hence, after executing $c \cdot 2^{2k_2+3k_3}$ loops, the probability that P_2 and P_3 are not divided correctly is $(1 - 2^{-2k_2-3k_3})^{c \cdot 2^{2k_2+3k_3}} = ((1 - \frac{1}{2^{2k_2+3k_3}})^{2^{2k_2+3k_3}})^c \leq \frac{1}{e^c}$. Therefore, by the random partitioning operation, there exist $P_2 \subseteq V_2$ and $P_3 \subseteq V_3$ with a probability at least $1 - \frac{1}{e^c}$.

We analyze the running time of RKEP(D, k_2, k_3). In each loop iteration, lines 5–7 run in $O(|V(D)|)$ time. For line 8, we transform it into the maximum matching problem of an undirected graph G. Graph G can be constructed in time $O(|V_2|^2)$ in the following way: Let $V(G) = V_2$. For $v_i, v_j \in V_2$, if $(v_i, v_j) \in A(D[V_2])$ and $(v_j, v_i) \in A(D[V_2])$, then there is an edge $\{v_i, v_j\} \in E(G)$. Line 8 takes $O(|E(G)|\sqrt{|V_2|})$ time by applying Edmonds' algorithm [39] for the maximum matching problem of graph G. For line 12, we find a packing of k_3 vertex-disjoint 3-cycles in V_3 by transforming into the 3-set-packing problem and invoking the corresponding algorithm in $O^*(2.3^{3k_3})$ [34]. Thus, the running time of Algorithm 1 RKEP(D, k_2, k_3) is $O^*(4.6^{3k_3} \cdot 2^{2k_2})$. □

4. A Randomized Algorithm Based on Algebraic Technique

Efficient kidney exchange can be achieved by 2-cycle and 3-cycle exchange. However, with the progress of technology, cycle exchange with length more than 3 has proved useful in kidney exchange. Limitations on the maximum length of cycles may be weakened in the future. It is reported that a 9-cycle exchange was performed successfully in 2015 [20]. In this section, we discuss the KEP with $L \geq 3$ parameterized by the size of kidney transplants. The parameterized model is defined as follows:

Parameterized KEP (p-KEP):
Input: a simple directed graph $D = (V, A)$, and a positive integer k;
Parameter: k;
Question: In graph D, is there a packing S of vertex-disjoint cycles such that $|V(S)| \geq k$ and the length of each cycle is in the range $[2, L]$?

Here, L is a fixed integer at least 3. For an integer t and a packing S, if $|V(S)| = t$ and $t \geq k$, then S is called a solution set of size t for the p-KEP.

By the above definition, the p-KEP is a special class of the set packing problem. For the set packing problem, most work has focused on the case that each set has the same size, such as the deterministic time $O^*(2.3^{3k})$ algorithm for the 3-set k-packing problem [34] and the randomized algorithm with time $O^*(2^{mk})$ for the m-set k-packing problem [27]. However, so far relatively little research has been carried out on the set packing problem such as the p-KEP which the size of each set is limited in a range $[2, L]$. In the section, we develop a randomized algorithm for the p-KEP based on the algebraic technique, introduced by Koutis in [27]. Developed by Koutis and Williams [28,29], the randomized algebraic technique has been successfully applied to some graph problems and counting problems [26,40,41].

An outstanding example is the improved algorithm with running time $O(1.657^n)$ for the Hamiltonian path problem [26], which resolves the open problem for half of a century whether there exists an $O(c^n)$ time algorithm with some constant $c < 2$ for the Hamiltonian path problem. The main idea of the algebraic approach is as follows: By constructing an appropriate multivariate polynomial for a given problem instance, the problem is reduced to the multilinear monomial detection problem for the sum-product expansion of the polynomial represented by a monotone arithmetic circuit, where a monomial $x_1^{i_1} x_2^{i_2} \cdots x_k^{i_k}$ whose degree is $\sum_{j=1}^{k} i_j$ is called multilinear if each i_j ($1 \le j \le k$) is either 0 or 1, and the degree of a polynomial is the maximum degree of all monomials in its expansion. A monotone arithmetic circuit is a directed acyclic graph in which each vertex of in-degree 0 is an input gate labeled either with variable or a non-negative real constant, the other vertices have in-degree 2 and are labeled either "+" (addition gates) or "×" (multiplication gates), and there is exactly one vertex with out-degree 0 which is called output gate. A polynomial is represented by a monotone arithmetic circuit if it can be computed by the output gate of the circuit. The multilinear monomial detection problem is to decide whether a polynomial $P(X)$ which can be represented by an arithmetic circuit contains a multilinear monomial of degree at most k in its sum-product expansion. Koutis and Williams gave randomized algorithms for multilinear monomial detection problem in [27] and [29], respectively. In [28], they extended their methods and gave Lemma 1 which is rephrased as follows in terms of our notation.

Lemma 1 (see Lemma 1 in [28]). *Let X be a variable set, z be an extra variable, and $P(X, z)$ be a polynomial represented by a monotone arithmetic circuit of size $s(n)$. Assume that the sum-product expansion of $P(X, z)$ is expressed the form $\sum_{i \in \mathbb{N}} Q_i(X) z^i$. There is a randomized algorithm that on every $P(X, z)$ runs $O^*(2^k t^2 s(n))$ time, outputs "YES" with high probability if $Q_t(X)$ contains a multilinear monomial of degree at most k, and always outputs "NO" if there is no multilinear monomial of degree at most k in $Q_t(X)$.*

Our algorithm for the p-KEP relies on Lemma 1. For a given directed graph $D = (V, A)$ of the p-KEP, we construct a suitable polynomial $P(X, z)$ such that for a positive integer k, the coefficient of z^k is a polynomial $Q_k(X)$ in the sum-product expansion of $P(X, z)$, the degree of each monomial in $Q_k(X)$ is k, and each multilinear monomial in $Q_k(X)$ maps to a packing of vertex-disjoint cycle in $D = (V, A)$. Hence, we can prove that there is a solution set S of size k for the p-KEP if and only if $Q_k(X)$ contains a multilinear monomial of degree k (see Lemma 2 below).

According to the above idea, we first construct a polynomial from the given instance. In a given directed graph $D = (V, A)$, let vertex set $V = \{v_1, v_2, \cdots, v_n\}$, variable set $X = \{x_1, x_2, \cdots, x_n\}$ and z be an extra variable. There is a one-to-one mapping relationship between V and X which each vertex $v_i \in V$ is mapped to a product zx_i for $i = 1, \cdots, n$. If $v_{i_1} \cdots v_{i_l} v_{i_1}$ form a directed cycle C in graph D, then define a monomial $\pi(C) = zx_{i_1} \cdots zx_{i_l} = z^{|V(C)|} x_{i_1} \cdots x_{i_l}$. In graph D, let $\mathcal{C}$ denote the cycle set which contains all cycles of length in the range $[2, L]$, and $\mathcal{C}(v_i)$ denote a cycle set, each of which passes through vertex v_i. For graph D, we partition $\mathcal{C}$ into $\mathcal{C}(v_1), \mathcal{C}(v_2), \cdots, \mathcal{C}(v_n)$ by the following Partitioning Operation.

Partitioning Operation

1: $\mathcal{C} = \emptyset$;
2: **for** each $v_i \in V$ **do**
3:    In graph D, find cycle set $\mathcal{C}(v_i)$ such that $\mathcal{C}(v_i)$ contains all directed cycles of length in $[2, L]$
4:      and each cycle in $\mathcal{C}(v_i)$ passes through v_i;
5:    $\mathcal{C} = \mathcal{C} \cup \mathcal{C}(v_i)$;
6:    $D = D[V(D) - \{v_i\}]$;
7: **end for**

In the Partitioning Operation, we find cycle set $\mathcal{C}(v_i)$ by applying Tarjan's algorithm for strongly connected components [42]: find a strongly connected component by the depth-first search starting vertex v_i on graph D, where the search depth is at most L, and then find all cycles passing through v_i by the backtracking algorithm. It takes $O(|V| + |E|)$ time running Tarjan's algorithm. By the Partitioning Operation, we obtain $\mathcal{C}(v_1), \mathcal{C}(v_2), \cdots, \mathcal{C}(v_n)$ and $\mathcal{C}$. We define

$$p(\mathcal{C}(v_i)) = \begin{cases} 1, & \text{if } \mathcal{C}(v_i) = \emptyset; \\ 1 + \sum_{C \in \mathcal{C}(v_i)} \pi(C), & \text{if } \mathcal{C}(v_i) \neq \emptyset. \end{cases} \qquad (1)$$

$$P(\mathcal{C}) = \prod_{i=1}^{n} p(\mathcal{C}(v_i)). \qquad (2)$$

Let $P(X, z) = P(\mathcal{C})$. If we view z as an extra indeterminate, the sum-product expansion of $P(X, z)$ can be expressed as follows:

$$P(X, z) = 1 + \sum_{t=1}^{n} Q_t(X) z^t. \qquad (3)$$

Since each vertex $v_i \in V$ is mapped to a product zx_i (where $x_i \in X$). From expressions (1) and (2), we know that if in the expansion expression (3) of $P(X, z)$ there is a term which contains z^k, then $Q_k(X)$ is a polynomial, where the degree of each monomial is k. By the polynomial construction, we have the following lemma.

Lemma 2. *For a given graph D of the p-KEP, there is a solution set S of size k if and only if the polynomial $Q_k(X)$ contains a multilinear monomial of degree k.*

Proof. ($\Rightarrow$): For a given instance of the p-KEP, assume that S is a solution set of size k and it is comprised of the vertices in s mutually vertex-disjoint cycles of length in range $[2, L]$. We use $\mathcal{C}'$ to denote the cycle set and have $\mathcal{C}' \subseteq \mathcal{C}$. Let $\{C_i\} = \mathcal{C}' \cap \mathcal{C}(v_i)$ and $s_i = |V(\{C_i\})|, 1 \leq i \leq n$. By the Partitioning Operation, we know that $\mathcal{C}(v_i) \cap \mathcal{C}(v_j) = \emptyset$ when $i \neq j$. Since the cycles in $\mathcal{C}'$ are mutually vertex-disjoint, either $\{C_i\} = \emptyset$ or $\{C_i\}$ has only one cycle C_i of length in range $[2, L]$, we also have $s_i = 0$ when $\{C_i\} = \emptyset$ and $s_i = |V(C_i)|$ when C_i is a cycle. Thus, we have $\sum_{i=1}^{n} s_i = k$. For $1 \leq i \leq n$, let $p(C_i) = 1$ if $\{C_i\} = \emptyset$, or $p(C_i) = \pi(C_i)$ if C_i is a cycle in $\mathcal{C}(v_i)$. It follows from the above definition and expression (1) that $p(C_i)$ is a monomial term in $p(\mathcal{C}(v_i))$. Let $p(\mathcal{C}') = \prod_{i=1}^{n} p(C_i)$. Since the cycles in $\mathcal{C}'$ are mutually vertex-disjoint and $\sum_{i=1}^{n} s_i = k$, $p(\mathcal{C}') = z^k q(X)$ where $q(X)$ is a multilinear monomial of degree k. Hence, by expressions (1), (2) and (3), there is a degree k multilinear monomial $q(X)$ in the polynomial $Q_k(X)$.

($\Leftarrow$): Assume that there is a monomial of the form $z^k q(X)$ in the sum-product expansion of $P(X, z)$, where $q(X)$ is a multilinear monomial of degree k. Obviously, $q(X)$ is a monomial in $Q_k(X)$. It follows from expressions (1) and (2) that there exists an expression $z^k q(X) = \prod_{i=1}^{n} p_i$ (where p_i is corresponding to a term from $p(\mathcal{C}(v_i))$) such that either $p_i = 1$ or $p_i = \pi(C_i)$ for some cycle $C_i \in \mathcal{C}(v_i)$. Let $\mathcal{C}' = \{C_i | p_i \neq 1, p_i = \pi(C_i), 1 \leq i \leq n\}$. Since $q(X)$ is a multilinear monomial of degree k, the cycles in $\mathcal{C}'$ are mutually vertex-disjoint. Let s_i denote the degree of monomial p_i, then $s_i = 0$ when $p_i = 1$, or $s_i = |V(C_i)|$ when $p_i \neq 1$, and $\sum_{i=1}^{n} s_i = k$, namely, $\sum_{C \in \mathcal{C}'} |V(C)| = k$. By the Partitioning Operation, the length of each cycle in $\mathcal{C}'$ is in the range $[2, L]$. Therefore, the vertex set of $\mathcal{C}'$ is a solution of size k for the p-KEP. □

For the p-KEP, we require to decide whether there is a packing S of vertex-disjoint cycles of lengths in the range $[2, L]$ such that $|V(S)| \geq k$. We can prove the following Lemma 3.

Lemma 3. *For a given graph D of the p-KEP, if each polynomial $Q_i(X)$ for $k \leq i \leq k + L - 1$ does not contain any multilinear monomial of degree i, then there does not exist a solution S with $|V(S)| \geq k$ for the given instance.*

Proof. By Lemma 2, we know that there is no solution of size in range $[k, k+L-1]$ if each polynomial $Q_i(X)$ for $k \leq i \leq k+L-1$ does not contain any multilinear monomial of degree i. Thus, we only prove there is no solution of size at least $k+L$ by contradiction. Assume that there is a solution S with $|V(S)| \geq k+L$ when each polynomial $Q_i(X)$ for $k \leq i \leq k+L-1$ does not contain any multilinear monomial of degree i. Suppose that S is comprised of mutually vertex-disjoint cycles $C_1, C_2, \cdots, C_l$ of length in range $[2, L]$ and $S = \{C_1, C_2, \cdots, C_l\}$. Let $S_{max<k} \subset S$ be the cycle set whose $|V(S_{max<k})|$ is maximum and $|V(S_{max<k})| < k$. Thus, there is a vertex-disjoint cycle set $S' = S_{max<k} \cup \{C_i\}$ where $C_i \in S - S_{max<k}$ and $|V(S')| \in [k, k+L-1]$. Otherwise, $|V(C_i)| > L$ and it contradicts that $|V(C_i)| \in [2, L]$ for $1 \leq i \leq l$. Thus, S' is also a solution of size $|V(S')| \in [k, k+L-1]$. By Lemma 2, we can get that there is a polynomial $Q_{|V(S')|}(X)$ containing a multilinear monomial of degree $|V(S')|$. It contradicts the fact that each polynomial $Q_i(X)$ for $k \leq i \leq k+L-1$ dose not contain any multilinear monomial of degree i. □

Now, we represent the polynomial $P(X, z)$ by a monotone arithmetic circuit. First, we require $n+2$ input gates for the variables set X, variable z and constant 1. For a directed cycle C of length in the range $[2, L]$, we represent monomial $\pi(C)$ by at most $(L-1)$ "×" gates. Since there are at most $\sum_{i=2}^{L} \binom{n}{i}$ cycles in $\mathcal{C}$, we need at most $(L-1) \sum_{i=2}^{L} \binom{n}{i}$ "×" gates and at most $n + \sum_{i=2}^{L} \binom{n}{i}$ "+" gates to represent all of $p(\mathcal{C}(v_j))$ for $1 \leq j \leq n$. At last, we need at most n "×" gates to represent the products of all $p(\mathcal{C}(v_j))$ for $1 \leq j \leq n$. Since $\sum_{i=2}^{L} \binom{n}{i} = O(Ln^L)$, the size $s(n)$ of the whole circuit is $O(L^2 n^L)$. Based on Lemmas 1, 2 and 3, we can call the algorithm in [28] to decide whether in $P(X, z)$ there is polynomial $Q_i(X)$ for $k \leq i \leq k+L-1$ which contains a multilinear monomial of degree i. If it exists, then graph D has a solution for the p-KEP and return "YES". Otherwise, it returns "NO". By Lemma 1, it takes time $O^*(2^i i^2 L^2 n^L)$ to decide whether there exists a polynomial $Q_i(X)$ containing a multilinear monomial of degree i, and we need do this at most L times. Hence, the total time of the algorithm is $T(n) = O^*(2^{(k+L)}(k+L)^2 L^3 n^L)$.

Note that parameter k and constant L are at most n. Therefore, by Lemmas 1, 2, and 3, we obtain clearly the following Theorem 2.

Theorem 2. *There is a randomized algorithm of running time $O^*(2^{(k+L)}(k+L)^2 L^3 n^L)$ for the p-KEP which produces the desired packing in a Yes-instance with high probability.*

5. Conclusions

We develop two randomized algorithms for the p-KEP$_{L=3}$ and the p-KEP. By stochastically partitioning the vertex set V into two subsets V_2 and V_3 which respectively contains 2-cycle or 3-cycle exchanges of the solution for a given instance, we propose a randomized algorithm of running time $O^*(4.6^{3k_3} \cdot 2^{2k_2})$ for the p-KEP$_{L=3}$. By applying the randomized algebraic technique, we transform the p-KEP into the multilinear monomial detection problem, and obtain the other randomized algorithm which can be executed in time $O^*(2^{(k+L)}(k+L)^2 L^3 n^L)$. Both algorithms improve previous theoretical running-time bounds for the KEP. Our algorithms can be extended to the case where the KEP involves chain exchanges of length bounded by a fixed constant by viewing each altruistic donor as a patient-donor pair whose patient is compatible with each donor of other incompatible patient-donor pairs, but not compatible with other altruistic donor. They will be interesting to further extend the algebraic technique to maximum weight variants of the KEP with and without chains, and study the effectiveness of the two algorithms on realistic instances.

In fact, patients' autonomy and their individual optimality criteria are critical factors which we should consider in the kidney exchange problem. The game-theoretical approaches were introduced in [43,44] and [6] to study the kidney exchange problem. In these kidney exchange game models, there also exist a lot of NP-complete problems [14,45–47]. Developing randomized methods for those NP-complete problems is also an interesting research topic for the future research.

Author Contributions: Conceptualization, M.L. and J.W.; methodology, M.L., J.W., Q.F. and B.F.; formal analysis, M.L. and B.F.; writing–original draft preparation, M.L. and B.F.; writing–review and editing, M.L., J.W. and Q.F.; project administration, J.W.; funding acquisition, J.W., Q.F. and B.F.

Funding: This work was supported in part by the National Natural Science Foundation of China under Grants 61772179, 61672536, 61872450, 61420106009, and 61503128, in part by Hunan Provincial Natural Science Foundation of China under Grant 2019JJ40005, in part by the Scientific Research Fund of Hunan Provincial Education Department under Grant 17C0222, in part by Application-oriented Special Disciplines, Double First-Class University Project of Hunan Province(Xiangjiaotong [2018] 469), and in part by the Science and Technology Plan Project of Hunan Province under Grant 2016TP1020.

Acknowledgments: We are grateful to the reviewers for their valuable and helpful comments which lead to the improvement of this paper.

Conflicts of Interest: The authors declare no conflict of interest.

Abbreviations

The following abbreviations are used in this manuscript:

KEP   Kidney Exchange Problem
IP   Integer Programming
FPT   Fixed-Parameter Tractable

References

1. Rapaport, F.T. The case for a living emotionally related international kidney donor exchange registry. *Transpl. Proc.* **1986**, *18*, 5–9.
2. Constantino, M.; Klimentova, X.; Viana, A.; Rais, A. New insights on integer-programming models for the kidney exchange problem. *Eur. J. Oper. Res.* **2013**, *231*, 57–68. [CrossRef]
3. Abraham, D.J.; Blum, A.; Sandholm, T. Clearing algorithms for barter exchange markets: enabling nationwide kidney exchanges. In Proceedings of the 8th ACM Conference on Electronic Commerce, San Diego, CA, USA, 11–15 June 2007; pp. 295–304.
4. Biró, P.; Manlove, D.F.; Rizzi, R. Maximum weight cycle packing in directed graphs, with application to kidney exchange programs. *Discrete Math. Algorithms Appl.* **2009**, *1*, 499–517. [CrossRef]
5. Roth, A.E.; Sönmez, T.; Ünver, M.U. Efficient kidney exchange: Coincidence of wants in markets with compatibility-based preferences. *Am. Econ. Rev.* **2007**, *97*, 828–851. [CrossRef] [PubMed]
6. Roth, A.E.; Sönmez, T.; Ünver, M.U. Kidney exchange. *Q. J. Econ.* **2004**, *119*, 457–488. [CrossRef]
7. Segev, D.L.; Gentry, S.E.; Warren, D.S.; Reeb, B.; Montgomery, R.A. Kidney paired donation and optimizing the use of live donor organs. *JAMA* **2005**, *293*, 1883–1890. [CrossRef] [PubMed]
8. Xiao, M.; Wang, X. Exact algorithms and complexity of kidney exchange. In Proceedings of the Twenty-Seventh International Joint Conference on Artificial Intelligence, Stockholm, Sweden, 13–19 July 2018; pp. 555–561.
9. Ashlagi, I.; Gilchrist, D.S.; Roth, A.E.; Rees, M.A. Nonsimultaneous chains and dominos in kidney-paired donation–Revisited. *Am. J. Transpl.* **2011**, *11*, 984–994. [CrossRef] [PubMed]
10. Glorie, K.M.; van de Klundert, J.J.; Wagelmans, A.P.M. Kidney exchange with long chains: an efficient pricing algorithm for clearing barter exchanges with branch-and-price. *Manuf. Serv. Oper. Manag.* **2014**, *16*, 498–512. [CrossRef]
11. Anderson, R.; Ashlagi, I.; Gamarnik, D.; Roth, A.E. Finding long chains in kidney exchange using the traveling salesman problem. *Proc. Natl. Acad. Sci. USA* **2015**, *112*, 663–668. [CrossRef] [PubMed]
12. Manlove, D.F.; O'malley, G. Paired and altruistic kidney donation in the UK: algorithms and experimentation. *J. Exp. Algorithmics* **2015**, *19*, 6:1–6:21. [CrossRef]
13. Gentry, S.E.; Montgomery, R.A.; Segev, D.L. Kidney paired donation: Fundamentals, limitations, and expansions. *Am. J. Kidney Dis.* **2011**, *57*, 144–151. [CrossRef] [PubMed]
14. Huang, C. Circular stable matching and 3-way kidney transplant. *Algorithmica* **2010**, *58*, 137–150. [CrossRef]
15. Barnhart, C.; Johnson, L.E.; Nemhauser, G.L.; Savelsbergh, M.W.; Vance, P.H. Branch-and-price: column generation for solving huge integer programs. *Oper. Res.* **1998**, *46*, 316–329. [CrossRef]

16. Klimentova, X.; Alvelos, F.; Viana, A. A new branch-and-price approach for the kidney exchange problem. In Proceedings of the 14th International Conference on Computational Science and Its Applications, Guimarães, Portugal, 30 June–3 July 2014; pp. 237–252.
17. Dantzig, G.; Wolfe, P. Decomposition principle for linear programs. *Oper. Res.* **1960**, *8*, 101–111. [CrossRef]
18. Dickerson, J.P.; Manlove, D.F.; Plaut, B.; Sandholm, T.; Trimble, J. Position-indexed formulations for kidney exchange. In Proceedings of the 2016 ACM Conference on Economics and Computation, Maastricht, The Netherlands, 24–28 July 2016; pp. 25–42.
19. Mak-Hau, V. A polyhedral study of the cardinality constrained multi-cycle and multi-chain problem on directed graphs. *Comput. Oper. Res.* **2018**, *99*, 13–26. [CrossRef]
20. Mak-Hau, V. On the kidney exchange problem: Cardinality constrained cycle and chain problems on directed graphs: A survey of integer programming approaches. *J. Comb. Optim.* **2017**, *33*, 35–59. [CrossRef]
21. Luo, S.; Tang, P.; Wu, C.; Zeng, J. Approximation of barter exchanges with cycle length constraints. *arXiv* **2018**, arXiv:1605.08863.
22. Jia, Z.; Tang, P.; Wang, R.; Zhang, H. Efficient near-optimal algorithms for barter exchange. In Proceedings of the 16th Conference on Autonomous Agents and MultiAgent Systems, São Paulo, Brazil, 8–12 May 2017; pp. 362–370.
23. Dickerson, J.P.; Kazachkov, A.M.; Procaccia, A.D.; Sandholm, T. Small representations of big kidney exchange graphs. In Proceedings of the 31st AAAI Conference on Artificial Intelligence, San Francisco, CA, USA, 4–9 February 2017; pp. 487–493.
24. Chen, J.; Liu, Y.; Lu, S.; Sze, S.H.; Zhang, F. Iterative expansion and color coding: An improved algorithm for 3D-matching. *ACM T. Algorithms* **2012**, *8*, 6. [CrossRef]
25. Feng, Q.; Wang, J.; Li, S.; Chen, J. Randomized parameterized algorithms for P_2-Packing and Co-Path Packing problems. *J. Comb. Optim.* **2015**, *29*, 125–140. [CrossRef]
26. Björklund, A. Determinant sums for undirected hamiltonicity. In Proceedings of the 51th Annual IEEE Symposium on Foundations of Computer Science, Las Vegas, NV, USA, 23–26 October 2010; pp. 173–182.
27. Koutis, I. Faster algebraic algorithms for path and packing problems. In Proceedings of the 35th International Colloquium on Automata, Languages and Programming, Reykjavik, Iceland, 7–11 July 2008; pp. 575–586.
28. Koutis I.; Williams, R. Limits and applications of group algebras for parameterized problems. In Proceedings of the 36th International Colloquium on Automata, Languages and Programming, Rhodos, Greece, 5–12 July 2009; pp. 653–664.
29. Williams, R. Finding paths of length k in $O^*(2^k)$ time. *Inform. Process. Lett.* **2009**, *109*, 315–318. [CrossRef]
30. Assadi, S.; Khanna, S.; Li, Y. The stochastic matching problem with (very) few queries. In Proceedings of the 2016 ACM Conference on Economics and Computation, Maastricht, Netherlands, 24–28 July 2016; pp. 43–60.
31. Chen, N.; Immorlica, N.; Karlin, A.R.; Mahdian, M.; Rudra, A. Approximating matches made in heaven. In Proceedings of 36th International Colloquium on Automata, Languages and Programming, Rhodes, Greece, 5–12 July 2009; pp. 266–278.
32. Awasthi, P.; Sandholm, T. Online stochastic optimization in the large: Application to kidney exchange. In Proceedings of the 21st International Joint Conference on Artificial Intelligence, Pasadena, CA, USA, 11–17 July 2009; pp. 405–411.
33. Fang, W.; Filos-Ratsikas, A.; Frederiksen, S.K.S.; Tang, P.; Zuo, S. Randomized assignments for barter exchanges: Fairness vs. efficiency. In Proceedings of the 4th International Conference on Algorithmic Decision Theory, Lexington, KY, USA, 27–30 September 2015; pp. 537–552.
34. Goyal, P.; Misra, N.; Panolan, F.; Zehavi, M. Deterministic algorithms for matching and packing problems based on representative sets. *SIAM J. Discrete Math.* **2015**, *29*, 1815–1836. [CrossRef]
35. Downey, R.G.; Fellows, M.R. *Parameterized Complexity*; Springer: Berlin, Germany, 1999.
36. Cygan, M.; Fomin, F.V.; Kowalik, Ł.; Lokshtanov, D.; Marx, D.; Pilipczuk, M.; Pilipczuk, M.; Saurabh, S. *Parameterized Algorithms*; Springer: Berlin, Germany, 2015.
37. Gabow, H.N. Data structures for weighted matching and nearest common ancestors with linking. In Proceedings of the 1st Annual ACM-SIAM Symposium on Discrete Algorithms, San Francisco, CA, USA, 22–24 January 1990; pp. 434–443.
38. Hanto, R.L.; Saidman, S.; Roth, A.E.; Delmonico, F. The evolution of a successful kidney paired donation program. *Transplantation* **2010**, *90*, 940. [CrossRef]

39. Edmonds, J. Maximum matching and a polyhedron with 0, 1-vertices. *J. Res. Nat. Bur. Stand. Sec. B* **1965**, *69*, 125–130. [CrossRef]
40. Björklund, A.; Husfeldt, T.; Kaski, P.; Koivisto, M. Narrow sieves for parameterized paths and packings. *J. Comput. Syst. Sci.* **2017**, *87*, 119–139. [CrossRef]
41. Fomin, F.V.; Lokshtanov, D.; Raman, V.; Saurabh, S.; Rao, B.V.R. Faster algorithms for finding and counting subgraphs. *J. Comput. Syst. Sci.* **2012**, *78*, 698–706. [CrossRef]
42. Tarjan R. Depth-First Search and Linear Graph Algorithms. *SIAM J. Comput.* **1972**, *1*, 146–160. [CrossRef]
43. Cechlárová, K.; Fleiner, T.; Manlove, D.F. The kidney exchange game. In Proceedings of the 8th International Symposium on Operational Research, Nova Gorica, Slovenia, Balkans, 28–30 September 2005; pp. 77–83.
44. Cechlárová, K.; Biró, P. Inapproximability of the kidney exchange problem. *Inform. Process. Lett.* **2007**, *101*, 199–202.
45. Biró, P.; McDermid, E. Three-sided stable matchings with cyclic preferences. *Algorithmica* **2010**, *58*, 5–18. [CrossRef]
46. Cechlárová, K.; Lacko, V. The kidney exchange problem: How hard is it to find a donor? *Ann. Oper. Res.* **2012**, *193*, 255–271. [CrossRef]
47. Mészáros-Karkus, Z. Hardness results for stable exchange problems. *Theor. Comput. Sci.* **2017**, *670*, 68–78. [CrossRef]

© 2019 by the authors. Licensee MDPI, Basel, Switzerland. This article is an open access article distributed under the terms and conditions of the Creative Commons Attribution (CC BY) license (http://creativecommons.org/licenses/by/4.0/).

Article

Practical Access to Dynamic Programming on Tree Decompositions [†]

Max Bannach [1] and Sebastian Berndt [2],*

[1] Institute for Theoretical Computer Science, Universität zu Lübeck, 23562 Lübeck, Germany
[2] Department of Computer Science, Kiel University, 24103 Kiel, Germany
* Correspondence: seb@informatik.uni-kiel.de
[†] This paper is an extended version of our paper published in ESA 2018.

Received: 18 July 2019; Accepted: 13 August 2019; Published: 16 August 2019

Abstract: Parameterized complexity theory has led to a wide range of algorithmic breakthroughs within the last few decades, but the practicability of these methods for real-world problems is still not well understood. We investigate the practicability of one of the fundamental approaches of this field: dynamic programming on tree decompositions. Indisputably, this is a key technique in parameterized algorithms and modern algorithm design. Despite the enormous impact of this approach in theory, it still has very little influence on practical implementations. The reasons for this phenomenon are manifold. One of them is the simple fact that such an implementation requires a long chain of non-trivial tasks (as computing the decomposition, preparing it, ...). We provide an easy way to implement such dynamic programs that only requires the definition of the update rules. With this interface, dynamic programs for various problems, such as 3-COLORING, can be implemented easily in about 100 lines of structured Java code. The theoretical foundation of the success of dynamic programming on tree decompositions is well understood due to Courcelle's celebrated theorem, which states that every MSO-definable problem can be efficiently solved if a tree decomposition of small width is given. We seek to provide practical access to this theorem as well, by presenting a lightweight model checker for a small fragment of MSO_1 (that is, we do not consider "edge-set-based" problems). This fragment is powerful enough to describe many natural problems, and our model checker turns out to be very competitive against similar state-of-the-art tools.

Keywords: fixed-parameter tractability; treewidth; model checking

1. Introduction

Parameterized algorithms aim to solve intractable problems on instances where the value of some parameter tied to the complexity of the instance is small. This line of research has seen enormous growth in the last few decades and produced a wide range of algorithms—see, for instance, [1]. More formally, a problem is *fixed-parameter tractable* (in FPT), if every instance I can be solved in time $f(\kappa(I)) \cdot \text{poly}(|I|)$ for a computable function $f \colon \mathbb{N} \to \mathbb{N}$, where $\kappa(I)$ is the *parameter* of I. While the impact of parameterized complexity to the theory of algorithms and complexity cannot be overstated, its practical component is much less understood. Very recently, the investigation of the practicability of fixed-parameter tractable algorithms for real-world problems has started to become an important subfield (see e. g., [2,3]). We investigate the practicability of dynamic programming on tree decompositions—one of the most fundamental techniques of parameterized algorithms. A general result explaining the usefulness of tree decompositions was given by Courcelle [4], who showed that *every* property that can be expressed in monadic second-order logic (MSO) is fixed-parameter tractable if it is parameterized by treewidth of the input graph. By combining this result (known as Courcelle's Theorem) with the $f(\text{tw}(G)) \cdot |G|$ algorithm of Bodlaender [5] to compute an optimal tree decomposition in FPT-time, a wide range of graph-theoretic problems is known to be solvable on these tree-like graphs. Unfortunately, both ingredients of this approach are very expensive in practice.

One of the major achievements concerning practical parameterized algorithms was the discovery of a practically fast algorithm for treewidth due to Tamaki [6]. Concerning Courcelle's Theorem, there are currently two contenders concerning efficient implementations of it: D-Flat, an Answer Set Programming (ASP) solver for problems on tree decompositions [7]; and Sequoia, an MSO solver based on model checking games [8]. Both solvers allow to solve very general problems and the corresponding overhead might, thus, be large compared to a straightforward implementation of the dynamic programs for specific problems.

Our Contributions

In order to study the practicability of dynamic programs on tree decompositions, we expand our tree decomposition library Jdrasil with an easy to use interface for such programs: the user only needs to specify the *update rules* for the different kind of nodes within the tree decomposition. The remaining work—computing a suitable optimized tree decomposition and performing the actual run of the dynamic program—is done by Jdrasil. This allows users to implement a wide range of algorithms within very few lines of code and, thus, gives the opportunity to test the practicability of these algorithms quickly. This interface is presented in Section 3.

In order to balance the generality of MSO solvers and the speed of direct implementations, we introduce an MSO fragment (actually, a MSO_1 fragment), which avoids quantifier alternation, in Section 4. By concentrating on this fragment, we are able to build a model checker, called Jatatosk, that runs nearly as fast as direct implementations of the dynamic programs. To show the feasibility of our approach, we compare the running times of D-Flat, Sequoia, and Jatatosk for various problems. It turns out that Jatatosk is competitive against the other solvers and, furthermore, its behaviour is much more consistent (that is, it does not fluctuate greatly on similar instances). We conclude that concentrating on a small fragment of MSO gives rise to practically fast solvers, which are still able to solve a large class of problems on graphs of bounded treewidth.

2. Preliminaries

All graphs considered in this paper are undirected, that is, they consist of a set of vertices V and of a symmetric edge-relation $E \subseteq V \times V$. We assume the reader to be familiar with basic graph theoretic terminology—see, for instance, [9]. A *tree decomposition* of a graph $G = (V, E)$ is a tuple (T, ι) consisting of a rooted tree T and a mapping ι from nodes of T to sets of vertices of G (which we call *bags*) such that (1) for all $v \in V$ there is a non-empty, connected set $\{ x \in V(T) \mid v \in \iota(x) \}$, and (2) for every edge $\{v, w\} \in E$ there is a node y in T with $\{v, w\} \subseteq \iota(y)$. The *width* of a tree decomposition is the maximum size of one of its bags minus one, and the *treewidth* of G, denoted by $\text{tw}(G)$, is the minimum width any tree decomposition of G must have.

In order to describe dynamic programs over tree decompositions, it turns out be helpful to transform a tree decomposition into a more structured one. A *nice tree decomposition* is a triple (T, ι, η) where (T, ι) is a tree decomposition and $\eta \colon V(T) \to \{\text{leaf}, \text{introduce}, \text{forget}, \text{join}\}$ is a labeling such that (1) nodes labeled "leaf" are exactly the leaves of T, and the bags of these nodes are empty; (2) nodes n labeled "introduce" or "forget" have exactly one child m such that there is exactly one vertex $v \in V(G)$ with either $v \notin \iota(m)$ and $\iota(n) = \iota(m) \cup \{v\}$ or $v \in \iota(m)$ and $\iota(n) = \iota(m) \setminus \{v\}$, respectively; (3) nodes n labeled "join" have exactly two children x, y with $\iota(n) = \iota(x) = \iota(y)$. A *very nice tree decomposition* is a nice tree decomposition that also has exactly one node labeled "edge" for every $e \in E(G)$, which virtually introduces the edge e to the bag—that is, whenever we introduce a vertex, we assume it to be "isolated" in the bag until its incident edges are introduced. It is well known that any tree decomposition can efficiently be transformed into a very nice one without increasing its width (essentially, traverse through the tree and "pull apart" bags) [1]. Whenever we talk about tree decompositions in the rest of the paper, we actually mean very nice tree decompositions. Note that this increases the space needed to store the decomposition but only by a small factor of $O(\operatorname{tw}(G))$. Due to the exponential (in $\operatorname{tw}(G)$) space requirement typical for the dynamic programs, this overhead is usually negligible. However, we want to stress that all our interfaces also support "just" nice tree decompositions.

We assume the reader to be familiar with basic logic terminology and give just a brief overview over the syntax and semantic of monadic second-order logic (MSO)—see, for instance, [10] for a detailed introduction. A *vocabulary* (or *signature*) $\tau = (R_1^{a_1}, \ldots, R_n^{a_n})$ is a set of *relational symbols* R_i of arity $a_i \geq 1$. A τ-*structure* is a set U—called *universe*—together with an *interpretation* $R_i^U \subseteq R^{a_i}$ of the relational symbols in U. Let $x_1, x_2, \ldots$ be a sequence of *first-order variables* and $X_1, X_2, \ldots$ be a sequence of *second-order variables* X_i of arity $\operatorname{ar}(X_i)$. The atomic τ-formulas are $x_i = x_j$ for two first-order variables and $R(x_{i_1}, \ldots, x_{i_k})$, where R is either a relational symbol or a second-order variable of arity k. The set of τ-formulas is inductively defined by (1) the set of atomic τ-formulas; (2) Boolean connections $\neg \phi$, $(\phi \vee \psi)$, and $(\phi \wedge \psi)$ of τ-formulas ϕ and ψ; (3) quantified formulas $\exists x \phi$ and $\forall x \phi$ for a first-order variable x and a τ-formula ϕ; (4) quantified formulas $\exists X \phi$ and $\forall X \phi$ for a second-order variable X of arity 1 and a τ-formula ϕ. The set of *free variables* of a formula ϕ consists of the variables that appear in ϕ but are not bounded by a quantifier. We denote a formula ϕ with free variables $x_1, \ldots, x_k, X_1, \ldots, X_\ell$ as $\phi(x_1, \ldots, x_k, X_1, \ldots, X_\ell)$. Finally, we say a τ-structure $\mathcal{S}$ with an universe U is a *model* of an τ-formula $\phi(x_1, \ldots, x_k, X_1, \ldots, X_\ell)$ if there are elements $u_1, \ldots, u_k \in U$ and relations $U_1, \ldots, U_\ell$ with $U_i \subseteq U^{\operatorname{ar}(X_i)}$ with $\phi(u_1, \ldots, u_k, U_1, \ldots, U_\ell)$ being true in $\mathcal{S}$. We write $\mathcal{S} \models \phi(u_1, \ldots, u_k, U_1, \ldots, U_\ell)$ in this case.

Example 1. *Graphs can be modeled as $\{E^2\}$-structures with universe $U = V(G)$ and a symmetric interpretation of E. Properties such as "is 3-colorable" can then be described by formulas as:*

$$\tilde{\phi}_{3\text{col}} = \exists R \exists G \exists B \ (\forall x \ R(x) \vee G(x) \vee B(x)) \wedge (\forall x \forall y \ E(x,y) \rightarrow \bigwedge_{C \in \{R,G,B\}} \neg C(x) \vee \neg C(y)).$$

For instance, we have ⊠ $\models \tilde{\phi}_{3\text{col}}$ and ⊠ $\not\models \tilde{\phi}_{3\text{col}}$. Here, and in the rest of the paper, we indicate with the tilde (as in $\tilde{\phi}$) that we will present a more refined version of ϕ later on.

The *model checking* problem asks, given a logical structure $\mathcal{S}$ and a formula ϕ, whether $\mathcal{S} \models \phi$ holds. A *model checker* is a program that solves this problem. To be useful in practice, we also require that a model checker outputs an assignment of the free and existential bounded variables of ϕ in the case of $\mathcal{S} \models \phi$.

3. An Interface for Dynamic Programming on Tree Decompositions

It will be convenient to recall a classical viewpoint of dynamic programming on tree decompositions to illustrate why our interface is designed the way it is. We will do so by the guiding example of 3-COLORING: Is it possible to color the vertices of a given graph with three colors such that adjacent vertices never share the same color? Intuitively, a dynamic program for 3-COLORING will work bottom-up on a very nice tree decomposition and manages a set of possible colorings per node. Whenever a vertex is introduced, the program "guesses" a color for this vertex; if a vertex is forgotten, we have to remove it from the bag and identify configurations that become eventually equal; for join-bags, we just have to take the configurations that are present in both children; and, for edge bags, we have to reject colorings in which both endpoints of the introduced edge have the same color. To formalize this vague algorithmic description, we view it from the perspective of automata theory.

3.1. The Tree Automaton Perspective

Classically, dynamic programs on tree decompositions are described in terms of tree automata [10]. Recall that, in a very nice tree decomposition, the tree T is rooted and binary; we assume that the children of T are ordered. The mapping ι can then be seen as a function that maps the nodes of T to symbols from some alphabet Σ. A naïve approach to manage ι would yield a huge alphabet (depending on the size of the graph). We thus define the so-called *tree-index*, which is a map idx: $V(G) \to \{0, \ldots, \text{tw}(G)\}$ such that no two vertices that appear in the same bag share a common tree-index. The existence of such an index follows directly from the property that every vertex is forgotten exactly once: We can simply traverse T from the root to the leaves and assign a free index to a vertex V when it is forgotten, and release the used index once we reach an introduce bag for v. The symbols of Σ then only contain the information for which tree-index there is a vertex in the bag. From a theoretician's perspective, this means that $|\Sigma|$ depends only on the treewidth; from a programmer's perspective, the tree-index makes it much easier to manage data structures that are used by the dynamic program.

Definition 1 (Tree Automaton). *A nondeterministic bottom-up* tree automaton *is a tuple* $A = (Q, \Sigma, \Delta, F)$, *where Q is the set of* states *of the automaton, $F \subseteq Q$ is the set of* accepting states, Σ *is an* alphabet, *and* $\Delta \subseteq (Q \cup \{\bot\}) \times (Q \cup \{\bot\}) \times \Sigma \times Q$ *is a* transition relation *in which* $\bot \notin Q$ *is a special symbol to treat nodes with less than two children. The automaton is* deterministic *if, for every $x, y \in Q \cup \{\bot\}$ and every $\sigma \in \Sigma$, there is exactly one $q \in Q$ with $(x, y, \sigma, q) \in \Delta$.*

Definition 2 (Computation of a Tree Automaton). *The* computation *of a nondeterministic bottom-up tree automaton $A = (Q, \Sigma, \Delta, F)$ on a labeled tree (T, ι) with $\iota: V(T) \to \Sigma$ and root $r \in V(T)$ is an assignment $q: V(T) \to Q$ such that, for all $n \in V(T)$, we have (1) $(q(x), q(y), \iota(n), q(n)) \in \Delta$ if n has two children x, y; (2) $(q(x), \bot, \iota(n), q(n)) \in \Delta$ or $(\bot, q(x), \iota(n), q(n)) \in \Delta$ if n has one child x; (3) $(\bot, \bot, \iota(n), q(n)) \in \Delta$ if n is a leaf. The computation is* accepting *if $q(r) \in F$.*

Simulating Tree Automata

A dynamic program for a decision problem can be formulated as a nondeterministic tree automaton that works on the decomposition—see the left side of Figure 1 for a detailed example. Observe that a nondeterministic tree automaton A will process a labeled tree (T, ι) with n nodes in time $O(n)$. When we simulate such an automaton deterministically, one might think that a running time of the form $O(|Q| \cdot n)$ is sufficient, as the automaton could be in any potential subset of the Q states at some node of the tree. However, there is a pitfall: for every node, we have to compute the set of potential states of the automaton depending on the sets of potential states of the children of that node, leading to a quadratic dependency on $|Q|$. This can be avoided for transitions of the form $(\bot, \bot, \iota(x), p)$, $(q, \bot, \iota(x), p)$, and $(\bot, q, \iota(x), p)$, as we can collect potential successors of every state of the child and compute the new set of states in linear time with respect to the cardinality of the set. However, transitions of the form $(q_i, q_j, \iota(x), p)$ are difficult, as we now have to merge two sets of states. In detail, let x be a node with children y and z and let Q_y and Q_z be the set of potential states in which the automaton eventually is in at these nodes. To determine Q_x, we have to check for every $q_i \in Q_y$ and every $q_j \in Q_z$ if there is a $p \in Q$ such that $(q_i, q_j, \iota(x), p) \in \Delta$. Note that the number of states $|Q|$ can be quite large, as for tree decompositions with bags of size k the set Q is typically of cardinality $2^{\Omega(k)}$, and we will thus try to avoid this quadratic blow-up.

Observation 1. *A tree automaton can be simulated in time $O(|Q|^2 \cdot n)$.*

Unfortunately, the quadratic factor in the simulation cannot be avoided in general, as the automaton may very well contain a transition for all possible pairs of states. However, there are some special cases in which we can circumnavigate the increase in the running time.

Definition 3 (Symmetric Tree Automaton). *A* symmetric nondeterministic bottom-up tree automaton *is a nondeterministic bottom-up tree automaton $A = (Q, \Sigma, \Delta, F)$ in which all transitions $(l, r, \sigma, q) \in \Delta$ satisfy either $l = \bot$, $r = \bot$, or $l = r$.*

Assume as before that we wish to compute the set of potential states for a node x with children y and z. Observe that, in a symmetric tree automaton, it is sufficient to consider the set $Q_y \cap Q_z$ and that the intersection of two sets can be computed in linear time if we take some care in the design of the underlying data structures.

Observation 2. *A symmetric tree automaton can be simulated in time $O(|Q| \cdot n)$.*

The right side of Figure 1 illustrates the deterministic simulation of a symmetric tree automaton. The massive time difference in the simulation of tree automata and symmetric tree automata significantly influenced the design of the algorithms in Section 4, in which we try to construct an automaton that is (1) "as symmetric as possible" and (2) allows for taking advantage of the "symmetric parts" even if the automaton is not completely symmetric.

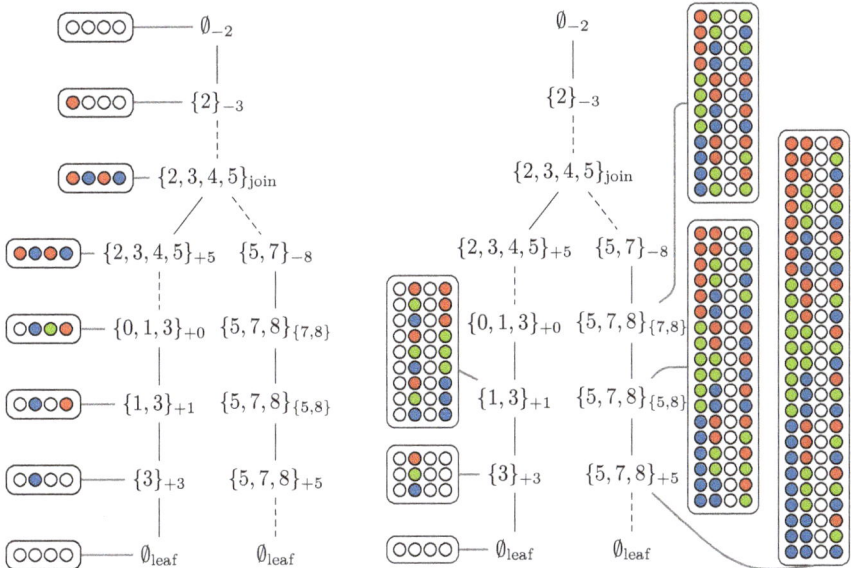

Figure 1. The *left* picture shows a part of a tree decomposition of the grid graph with vertices $\{0,\ldots,8\}$ (i.e., with rows $\{i, i+1, i+2\}$ for $i \in \{0,3,6\}$ and columns $\{i, i+3, i+6\}$ for $i \in \{0,1,2\}$). The index of a bag shows the type of the bag: a positive sign means "introduce", a negative one "forget", a pair represents an "edge"-bag, and the text is self explanatory. Solid lines represent real edges of the decomposition, while dashed lines illustrate a path (that is, some bags are skipped). On the left branch of the decomposition, a run of a nondeterministic tree automaton with tree-index $\begin{pmatrix} 0 & 1 & 2 & 3 & 4 & 5 & 6 & 7 & 8 \\ 2 & 3 & 0 & 1 & 2 & 3 & 0 & 1 & 0 \end{pmatrix}$ for 3-COLORING is illustrated. To increase readability, states of the automaton are connected to the corresponding bags with gray lines, and, for some nodes, the states are omitted. In the *right* picture, the same automaton is simulated deterministically.

3.2. The Interface

We introduce a simple Java-interface to our library Jdrasil, which originally was developed for the computation of tree decompositions only. The interface is built up from two classes: StateVectorFactory and StateVector. The only job of the factory is to generate StateVector objects for the leaves of the tree decomposition, or with the terms of the previous section: "to define the initial states of the tree automaton." The StateVector class is meant to model a vector of potential states in which the nondeterministic tree automaton is at a specific node of the tree decomposition. Our interface does not define what a "state" is, or how a collection of states is managed. The only thing the interface requires a user to implement is the behaviour of the tree automaton when it reaches a node of the tree decomposition, i.e., given a StateVector for some node x of the tree decomposition and the information that the next node y reached by the automaton is of a certain type, the user has to compute the `StateVector` for y. To this end, the interface contains the methods shown in Listing 1.

Listing 1: The four methods of the interface describe the behaviour of the tree automaton. Here, "T" is a generic type for vertices. Each function obtains as a parameter the current bag and a tree-index "idx". Other parameters correspond to bag-type specifics, e. g., the introduced or forgotten vertex v.

```
StateVector<T> introduce(Bag<T> b, T v, Map<T, Integer> idx);
StateVector<T> forget(Bag<T> b, T v, Map<T, Integer> idx);
StateVector<T> join(Bag<T> b, StateVector<T> o, Map<T, Integer> idx);
StateVector<T> edge(Bag<T> b, T v, T w, Map<T, Integer> idx);
```

This already rounds up the description of the interface, everything else is done by Jdrasil. In detail, given a graph and an implementation of the interface, Jdrasil will compute a tree decomposition (see [11] for the concrete algorithms used by Jdrasil), transform this decomposition into a very nice tree decomposition, potentially optimize the tree decomposition for the following dynamic program, and finally traverse through the tree decomposition and simulate the tree automaton described by the implementation of the interface. The result of this procedure is the `StateVector` object assigned to the root of the tree decomposition.

3.3. Example: 3-Coloring

Let us illustrate the usage of the interface with our running example of 3-COLORING. A `State` of the automaton can be modeled as a simple integer array that stores a color (an integer) for every vertex in the bag. A `StateVector` stores a set of `State` objects, that is, essentially a set of integer arrays. Introducing a vertex v to a `StateVector` therefore means that three duplicates of each stored state have to be created, and, for every duplicate, a different color has to be assigned to v. Listing 2 illustrates how this operation could be realized in Java.

Listing 2: Exemplary implementation of the `introduce` method for 3-COLORING. In the listing, the variable `states` is stored by the `StateVector` object and represents all currently possible states.

```
StateVector<T> introduce(Bag<T> b, T v, Map<T, Integer> idx) {
Set<State> newStates = new HashSet<>();
for (State state : states) { // 'states' is the set of states
for (int color = 1; color <= 3; color++) {
State newState = new State(state); // copy the state
newState.colors[idx.get(v)] = color;
newStates.add(newState);
}
}
states = newStates;
return this;
}
```

The three other methods can be implemented in a very similar fashion: in the `forget`-method, we set the color of v to 0; in the `edge`-method, we remove states in which both endpoints of the edge have the same color; and, in the `join`-method, we compute the intersection of the state sets of both `StateVector` objects. Note that, when we forget a vertex v, multiple states may become identical, which is handled here by the implementation of the Java `Set`-class, which takes care of duplicates automatically.

A reference implementation of this 3-COLORING solver is publicly available [12], and a detailed description of it can be found in the manual of Jdrasil [13]. Note that this implementation is only meant to illustrate the interface and that we did not make any effort to optimize it. Nevertheless, this very simple implementation (the part of the program that is responsible for the dynamic program only contains about 120 lines of structured Java-code) performs surprisingly well.

4. A Lightweight Model Checker for an MSO-Fragment

Experiments with the coloring solver of the previous section have shown a huge difference in the performance of general solvers as D-Flat and Sequoia against a concrete implementation of a tree automaton for a specific problem (see Section 5). This is not necessarily surprising, as a general solver needs to keep track of way more information. In fact, an MSO model checker on formula ϕ can probably (unless P = NP) not run in time $f(|\phi| + \text{tw}) \cdot \text{poly}(n)$ for any elementary function f [14]. On the other hand, it is not clear (in general) what the concrete running time of such a solver is for a concrete formula or problem (see e.g., [15] for a sophisticated analysis of some running times in Sequoia). We seek to close this gap between (slow) general solvers and (fast) concrete algorithms. Our approach is to concentrate only on a fragment of MSO, which is powerful enough to express many natural problems, but which is restricted enough to allow model checking in time that matches or is close to the running time of a concrete algorithm for the problem. As a bonus, we will be able to derive upper bounds on the running time of the model checker directly from the syntax of the input formula.

Based on the interface of Jdrasil, we have implemented a publicly available prototype called Jatatosk [16]. In Section 5, we describe various experiments on different problems on multiple sets of graphs. It turns out that Jatatosk is competitive against the state-of-the-art solvers D-Flat and Sequoia. Arguably, these two programs solve a more general problem and a direct comparison is not entirely fair. However, the experiments do reveal that it seems very promising to focus on smaller fragments of MSO (or perhaps any other description language) in the design of treewidth based solvers.

4.1. Description of the Used MSO-Fragment

We only consider vocabularies τ that contain the binary relation E^2, and we only consider τ-structures with a symmetric interpretation of E^2, that is, we only consider structures that contain an undirected graph (but may also contain further relations). The fragment of MSO that we consider is constituted by formulas of the form $\phi = \exists X_1 \ldots \exists X_k \bigwedge_{i=1}^n \psi_i$, where the X_j are second-order variables and the ψ_i are first-order formulas of the form

$$\psi_i \in \{ \forall x \forall y\ E(x,y) \to \chi_i,\ \forall x \exists y\ E(x,y) \land \chi_i,\ \exists x \forall y\ E(x,y) \to \chi_i,$$
$$\exists x \exists y\ E(x,y) \land \chi_i,\ \forall x\ \chi_i,\ \exists x\ \chi_i\ \}.$$

Here, the χ_i are quantifier-free first-order formulas in canonical normal form. It is easy to see that this fragment is already powerful enough to encode many classical problems as 3-COLORING ($\tilde{\phi}_{3\text{col}}$ from the introduction is part of the fragment), or VERTEX-COVER (note that, in this form, the formula is not very useful—every graph is a model of the formula if we choose $S = V(G)$; we will discuss how to handle optimization in Section 4.4):

$$\tilde{\phi}_{\text{vc}} = \exists S \forall x \forall y\ E(x,y) \to S(x) \lor S(y).$$

4.2. A Syntactic Extension of the Fragment

Many interesting properties, such as connectivity, can easily be expressed in MSO, but not directly in the fragment that we study. Nevertheless, a lot of these properties can directly be checked by a model checker if it "knows" what kind of properties it actually checks. We present a *syntactic extension* of our MSO-fragment which captures such properties. The extension consists of three new second order quantifiers that can be used instead of $\exists X_i$.

The first extension is a *k-partition quantifier*, which quantifies over partitions of the universe:

$$\exists^{\text{partition}} X_1, \ldots, X_k \equiv \exists X_1 \exists X_2 \ldots \exists X_k \left(\forall x \bigvee_{i=1}^{k} X_i(x) \right) \wedge \left(\forall x \bigwedge_{i=1}^{k} \bigwedge_{j \neq i} \neg X_i(x) \wedge \neg X_j(x) \right).$$

This quantifier has two advantages. First, formulas like $\tilde{\phi}_{3\text{col}}$ can be simplified to

$$\phi_{3\text{col}} = \exists^{\text{partition}} R, G, B \; \forall x \forall y \; E(x,y) \rightarrow \bigwedge_{C \in \{R,G,B\}} \neg C(x) \vee \neg C(y),$$

and, second, the model checking problem for them can be solved more efficiently: the solver directly "knows" that a vertex must be added to exactly one of the sets.

We further introduce two quantifiers that work with respect to the symmetric relation E^2 (recall that we only consider structures that contain such a relation). The $\exists^{\text{connected}} X$ quantifier guesses an $X \subseteq U$ that is connected with respect to E (in graph theoretic terms), that is, it quantifies over connected subgraphs. The $\exists^{\text{forest}} F$ quantifier guesses a $F \subseteq U$ that is acyclic with respect to E (again in graph theoretic terms), that is, it quantifies over subgraphs that are forests. These quantifiers are quite powerful and allow, for instance, expressing that the graph induced by E^2 contains a triangle as minor:

$$\phi_{\text{triangle-minor}} = \exists^{\text{connected}} R \; \exists^{\text{connected}} G \; \exists^{\text{connected}} B \; \cdot$$
$$\left(\forall x \; (\neg R(x) \vee \neg G(x)) \wedge (\neg G(x) \vee \neg B(x)) \wedge (\neg B(x) \vee \neg R(x)) \right)$$
$$\wedge \left(\exists x \exists y \; E(x,y) \wedge R(x) \wedge G(y) \right) \wedge \left(\exists x \exists y \; E(x,y) \wedge G(x) \wedge B(y) \right)$$
$$\wedge \left(\exists x \exists y \; E(x,y) \wedge B(x) \wedge R(y) \right).$$

We can also express problems that usually require more involved formulas in a very natural way. For instance, the FEEDBACK-VERTEX-SET problem can be described by the following formula (again, optimization will be handled in Section 4.4):

$$\tilde{\phi}_{\text{fvs}} = \exists S \; \exists^{\text{forest}} F \; \forall x \; S(x) \vee F(x).$$

4.3. Description of the Model Checker

We describe our model checker in terms of a nondeterministic tree automaton that works on a tree decomposition of the graph induced by E^2 (note that, in contrast to other approaches in the literature (see, for instance, [10]), we do not work on the Gaifman graph). We define any state of the automaton as a bit-vector, and we stipulate that the initial state at every leaf is the zero-vector. For any quantifier or subformula, there will be some area in the bit-vector reserved for that quantifier or subformula and we describe how state transitions effect these bits. The "algorithmic idea" behind the implementation of these transitions is not new, and a reader familiar with folklore dynamic programs on tree decompositions (for instance for VERTEX-COVER or STEINER-TREE) will probably recognize them. An overview over common techniques can be found in the standard textbooks [1,10].

4.3.1. The Partition Quantifier

We start with a detailed description of the k-partition quantifier $\exists^{\text{partition}} X_1, \ldots, X_q$ (in fact, we do not implement an additional $\exists X$ quantifier, as we can always state $\exists X \equiv \exists^{\text{partition}} X, \bar{X}$): Let k be the maximum bag-size of the tree decomposition. We reserve $k \cdot \log_2 q$ bit in the state description (Here, and in the following, $\log(x)$ denotes the number of bits to store x, i. e., $\log(x) = \lfloor \log_2(x) \rfloor + 1$), where each block of length $\log q$ indicates in which set X_i the corresponding element of the bag is. On an introduce-bag (e. g., for $v \in U$), the nondeterministic automaton guesses an index $i \in \{1, \ldots, q\}$ and sets the $\log q$ bits that are associated with the tree-index of v to i. Equivalently, the corresponding bits are cleared when the automaton reaches a forget-bag. As the partition is independent of any edges, an edge-bag does not change any of the bits reserved for the k-partition quantifier. Finally, on join-bags, we may only join states that are identical on the bits describing the partition (as otherwise the vertices of the bag would be in different partitions)—meaning this transition is symmetric with respect to these bits (in terms of Section 3.1).

4.3.2. The Connected Quantifier

The next quantifier we describe is $\exists^{\text{connected}} X$, which has to overcome the difficulty that an introduced vertex may not be connected to the rest of the bag in the moment it got introduced, but may be connected to it when further vertices "arrive." The solution to this dilemma is to manage a partition of the bag into $k' \leq k$ connected components $P_1, \ldots, P_{k'}$, for which we reserve $k \cdot \log k$ bit in the state description. Whenever a vertex v is introduced, the automaton either guesses that it is not contained in X and clears the corresponding bits, or it guesses that $v \in X$ and assigns some P_i to v. Since v is isolated in the bag in the moment of its introduction (recall that we work on a very nice tree decomposition), it requires its own component and is therefore assigned to the smallest empty partition P_i. When a vertex v is forgotten, there are four possible scenarios:

1. $v \notin X$, then the corresponding bits are already cleared and nothing happens;
2. $v \in X$ and $v \in P_i$ with $|P_i| > 1$, then v is just removed and the corresponding bits are cleared;
3. $v \in X$ and $v \in P_i$ with $|P_i| = 1$ and there are other vertices w in the bag with $w \in X$, then the automaton rejects the configuration, as v is the last vertex of P_i and may not be connected to any other partition anymore;
4. $v \in X$ is the last vertex of the bag that is contained in X, then the connected component is "done", the corresponding bits are cleared and one additional bit is set to indicate that the connected component cannot be extended anymore.

When an edge $\{u, v\}$ is introduced, components might need to be merged. Assume $u, v \in X$, $u \in P_i$, and $v \in P_j$ with $i < j$ (otherwise, an edge-bag does not change the state), then we essentially perform a classical union-operation from the well-known union-find data structure. Hence, we assign all vertices that are assigned to P_j to P_i. Finally, at a join-bag, we may join two states that agree locally on the vertices that are in X (that is, they have assigned the same vertices to some P_i); however, they do not have to agree in the way the different vertices are assigned to P_i (in fact, there does not have to be an isomorphism between these assignments). Therefore, the transition at a join-bag has to connect the corresponding components analogous to the edge-bags—in terms of Section 3.1, this transition is not symmetric.

The description of the remaining quantifiers and subformulas is very similar and summarized in the following overview:

$\exists^{\text{forest}} X$	
#Bit:	$k \cdot \log k$
Introduce:	As for $\exists^{\text{connected}} X$.
Forget:	Just clear the corresponding bits.
Edge:	As for $\exists^{\text{connected}} X$, but reject if two vertices of the same component are connected.
Join:	As for $\exists^{\text{connected}} X$, but track if the join introduces a cycle.

$\forall x \forall y \, E(x,y) \to \chi_i$	
#Bit:	0
Introduce:	-
Forget:	-
Edge:	Reject if χ_i is not satisfied for the vertices of the edge.
Join:	-

$\forall x \exists y\ E(x,y) \wedge \chi_i$	
#Bit:	k
Introduce:	-
Forget:	Reject if the bit corresponding to v is not set.
Edge:	Set the bit of v if χ_i is satisfied.
Join:	Compute the logical-or of the bits of both states.

$\exists x \forall y\ E(x,y) \rightarrow \chi_i$	
#Bit:	$k+1$
Introduce:	Set the corresponding bit.
Forget:	If the corresponding bit is set, set the additional bit.
Edge:	If χ_i is not satisfied, clear the corresponding bit.
Join:	Compute the logical-and of all but the last bit, for the last bit use a logical-or.

$\exists x \exists y\ E(x,y) \wedge \chi_i$	
#Bit:	1
Introduce:	-
Forget:	-
Edge:	Set the bit if χ_i is satisfied.
Join:	Compute logical-or of the bit in both states.

$\forall x\ \chi_i\ (\exists x\ \chi_i)$	
#Bit:	0 (1)
Introduce:	Test if χ_i is satisfied and reject if not (set the bit if so).
Forget:	-
Edge:	-
Join:	- (Compute logical-or of the bit in both states.)

4.4. Extending the Model Checker to Optimization Problems

As the example formulas from the previous section already indicate, performing model checking alone will not suffice to express many natural problems. In fact, every graph is a model of the formula $\tilde{\phi}_{vc}$ if S simply contains all vertices. It is therefore a natural extension to consider an optimization version of the model checking problem, which is usually formulated as follows [1,10]: Given a logical structure $\mathcal{S}$, a formula $\phi(X_1, \ldots, X_p)$ of the MSO-fragment defined in the previous section with free unary second-order variables $X_1, \ldots, X_p$, and weight functions $\omega_1, \ldots, \omega_p$ with $\omega_i \colon U \rightarrow \mathbb{Z}$; find $S_1, \ldots, S_p$ with $S_i \subseteq U$ such that $\sum_{i=1}^{p} \sum_{s \in S_i} \omega_i(s)$ is *minimized* under $\mathcal{S} \models \phi(S_1, \ldots, S_p)$, or conclude that $\mathcal{S}$ is not a model for ϕ for any assignment of the free variables. We can now correctly express the (actually *weighted*) optimization version of VERTEX-COVER as follows:

$$\phi_{vc}(S) = \forall x \forall y\ E(x,y) \rightarrow \big(S(x) \vee S(y)\big).$$

Similarly, we can describe the optimization version of DOMINATING-SET if we assume the input does not have isolated vertices (or is reflexive), and we can also fix the formula $\tilde{\phi}_{fvs}$:

$$\phi_{ds}(S) = \forall x \exists y\ E(x,y) \wedge \big(S(x) \vee S(y)\big),$$
$$\phi_{fvs}(S) = \exists^{forest} F\ \forall x\ \big(S(x) \vee F(x)\big).$$

We can also *maximize* the term $\sum_{i=1}^{p} \sum_{s \in S_i} \omega_i(s)$ by multiplying all weights with -1 and, thus, express problems such as INDEPENDENT-SET:

$$\phi_{is}(S) = \forall x \forall y\ E(x,y) \rightarrow \big(\neg S(x) \vee \neg S(y)\big).$$

The implementation of such an optimization is straightforward: there is a 2-partition quantifier for every free variable X_i that partitions the universe into X_i and $\bar{X}_i$. We assign a current value of $\sum_{i=1}^{p} \sum_{s \in S_i} \omega_i(s)$ to every state of the automaton, which is adapted if elements are "added" to some of the free variables at introduce nodes. Note that, since we optimize an affine function, this does not increase the state space: even if multiple computational paths lead to the same state with different values at some node of the tree, it is well defined which of these values is the optimal one. Therefore, the cost of optimization only lies in the partition quantifier, that is, we pay with k bits in the state

4.5. Handling Symmetric and Non-Symmetric Joins

In Section 4.3, we have defined the states of our automaton with respect to a formula, Table 1 gives an overview of the number of bits we require for the different parts of the formula. Let $\text{bit}(\phi, k)$ be the number of bits that we have to reserve for a formula ϕ and a tree decomposition of maximum bag size k that is, the sum over the required bits of each part of the formula. By Observation 1, this implies that we can simulate the automaton (and hence solve the model checking problem) in time $O^*((2^{\text{bit}(\phi,k)})^2)$, or by Observation 2 in time $O^*(2^{\text{bit}(\phi,k)})$, if the automaton is symmetric (The notation O^* supresses polynomial factors). Unfortunately, this is not always the case, in fact, only the quantifier $\exists^{\text{partition}} X_1, \ldots, X_q$, the bits needed to optimize over free variables, as well as the formulas that do not require any bits, yield a symmetric tree automaton. This means that the simulation is wasteful if we consider a mixed formula (for instance, one that contains a partition and a connected quantifier). To overcome this issue, we partition the bits of the state description into two parts: first, the "symmetric" bits of the quantifiers $\exists^{\text{partition}} X_1, \ldots, X_q$ and the bits required for optimization, and in the "asymmetric" ones of all other elements of the formula. Let $\text{symmetric}(\phi, k)$ and $\text{asymmetric}(\phi, k)$ be defined analogously to $\text{bit}(\phi, k)$. We implement the join of states as in the following lemma, allowing us to deduce the running time of the model checker for concrete formulas. Table 2 provides an overview for formulas presented here.

Lemma 1. *Let x be a join node of T with children y and z, and let Q_y and Q_z be sets of states in which the automaton may be at y and z. Then, the set Q_x of states in which the automaton may be at node x can be computed in time $O^*\left(2^{\text{symmetric}(\phi,k) + 2 \cdot \text{asymmetric}(\phi,k)}\right)$.*

Proof. To compute Q_x, we first split Q_y into $B_1, \ldots, B_q$ such that all elements in one B_i share the same "symmetric bits." This can be done in time $|Q_y|$ using bucket-sort. Note that we have $q \leq 2^{\text{symmetric}(\phi,k)}$ and $|B_i| \leq 2^{\text{asymmetric}(\phi,k)}$. With the same technique, we identify for every element v in Q_z its corresponding partition B_i. Finally, we compare v with the elements in B_i to identify those for which there is a transition in the automaton. This yields an overall running time of the form $|Q_z| \cdot \max_{i=1}^{q} |B_i| \leq 2^{\text{bit}(\phi,k)} \cdot 2^{\text{asymmetric}(\phi,k)} = 2^{\text{symmetric}(\phi,k) + 2 \cdot \text{asymmetric}(\phi,k)}$. □

Table 1. The table shows the precise number of bits we reserve in the description of a state of the tree automaton for different quantifiers and formulas. The values are with respect to a tree decomposition with maximum bag size k.

Quantifier/Formula	Number of Bits
free variables $X_1, \ldots, X_q$	$q \cdot k$
$\exists^{\text{partition}} X_1, \ldots, X_q$	$k \cdot \log q$
$\exists^{\text{connected}} X$	$k \cdot \log k + 1$
$\exists^{\text{forest}} X$	$k \cdot \log k$
$\forall x \forall y\ E(x,y) \to \chi_i$	0
$\forall x \exists y\ E(x,y) \wedge \chi_i$	k
$\exists x \forall y\ E(x,y) \to \chi_i$	$k+1$
$\exists x \exists y\ E(x,y) \wedge \chi_i$	1
$\forall x\ \chi_i$	0
$\exists x\ \chi_i$	1

Table 2. The table gives an overview of formulas ϕ used within this paper, together with the values symmetric(ϕ, k) and asymmetric(ϕ, k), as well as the precise time our algorithm will require to model check an instance for that particular formula.

ϕ	symmetric(ϕ,k)	asymmetric(ϕ,k)	Time
ϕ_{3col}	$k \cdot \log(3)$	0	$O^*(3^k)$
$\phi_{\text{vc}}(S)$	k	0	$O^*(2^k)$
$\phi_{\text{ds}}(S)$	k	k	$O^*(8^k)$
$\phi_{\text{triangle-minor}}$	0	$3k \cdot \log(k) + 3$	$O^*(k^{6k})$
$\phi_{\text{fvs}}(S)$	k	$k \cdot \log(k)$	$O^*(2^k k^{2k})$

5. Applications and Experiments

To show the feasibility of our approach, we have performed experiments for widely investigated graph problems: 3-COLORING (Figure 2), VERTEX-COVER (Figure 3), DOMINATING-SET (Figure 4), INDEPENDENT-SET (Figure 5), and FEEDBACK-VERTEX-SET (Figure 6). All experiments were performed on an Intel Core processor containing four cores of 3.2 GHz each and 8 Gigabyte RAM. Jdrasil was used with Java 8.1 and both Sequoia and D-Flat were compiled with gcc 7.2. All compilations were performed with the default optimization settings. The implementation of Jatatosk uses hashing to realize Lemma 1, which works well in practice. We use a data set assembled from different sources containing graphs with 18 to 956 vertices and treewidth 3 to 13. The first source is a collection of transit graphs from GTFS-transit feeds [17] that was also used for experiments in [18], the second source is real-world instances collected in [19], and the last one is that of the PACE challenge [2] with treewidth at most 11. In each of the experiments, the left picture always shows the difference of Jatatosk against D-Flat and Sequoia. A positive bar means that Jatatosk is faster by this amount in seconds, and a negative bar means that either D-Flat or Sequoia is faster by that amount. The bars are capped at 100 seconds. On every instance, Jatatosk was compared against the solver that was faster on this particular instance. The image also shows for every instance the treewidth of the input. The right image always shows a cactus plot that visualizes the number of instances that can be solved by each of the solvers in x seconds, that is, faster growing functions are better. For each experiment, there is a table showing the average, standard deviation, and median of the time (in seconds) each solver needed to solve the problem. The best values are highlighted.

The experiments reveal that Jatatosk is faster than its competitors on many instances. However, there are also formulas such as the one for the vertex cover problem on which one of the other solvers performs better on some of the instances. For an overall picture, the cactus plot in Figure 7 is the *sum* of all cactus plots from the experiments. It reveals that overall Jatatosk in fact outperforms its competitors. However, we stress once more that the comparison is not completely fair, as both Sequoia and D-Flat are powerful enough to model check the whole of MSO (and actually also MSO$_2$), while Jatatosk can only handle a fragment of MSO$_1$.

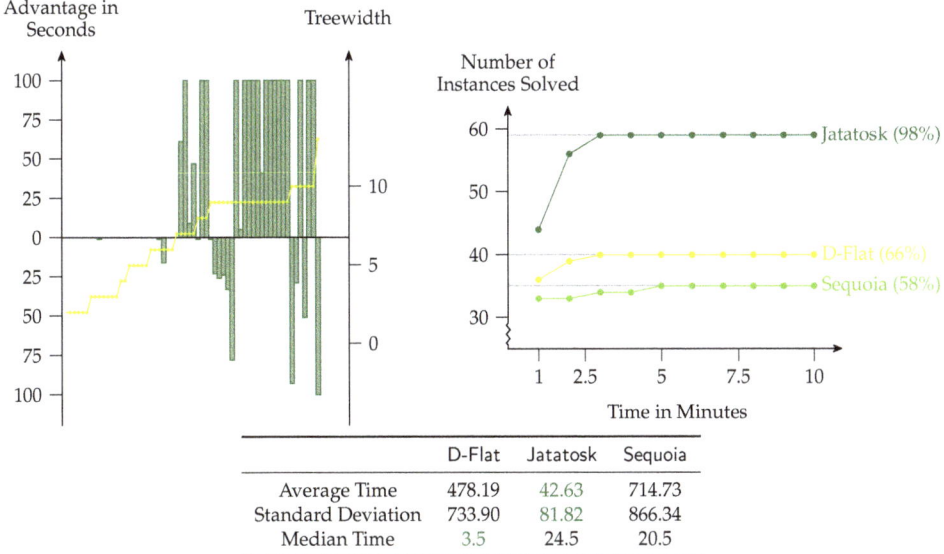

Figure 2. Experiment for the 3-COLORING problem.

As we can see, Jatatosk outperforms Sequoia and D-Flat if we consider the graph coloring problem. Its average time is more than a factor of 10 smaller than the average time of its competitors and Jatatosk solves about 30% of the instances more. This result is not surprising, as the fragment used by Jatatosk is directly tailored towards coloring and, thus, Jatatosk has a natural advantage.

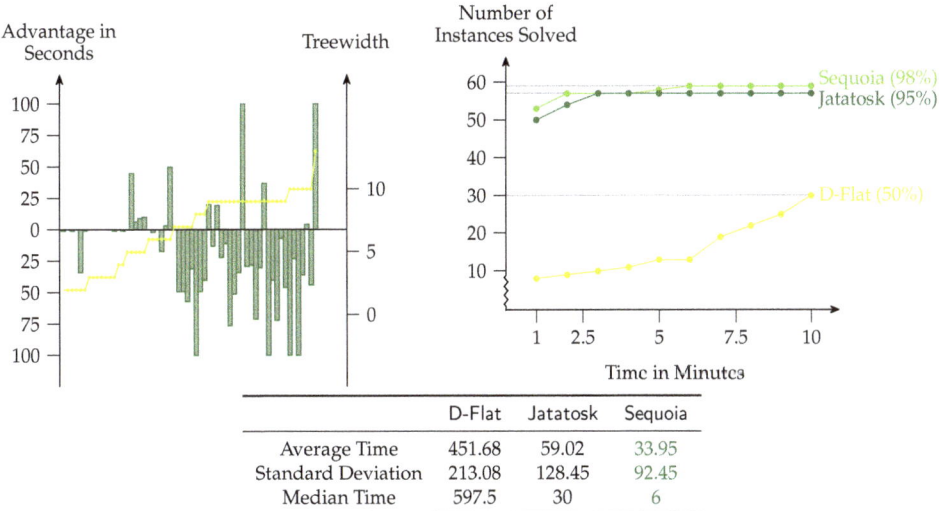

Figure 3. Experiment for the VERTEX-COVER problem.

The VERTEX-COVER problem is solved best by Sequoia, which becomes apparent if we consider the difference plot. Furthermore, the average time used by Sequoia is better than the time used by Jatatosk. However, considering the cactus plot, the difference between Jatatosk and Sequoia with respect to solved instances is small, while D-Flat falls behind a bit. We assume that the similarity between Jatatosk and Sequoia is because both compile internally a similar algorithm.

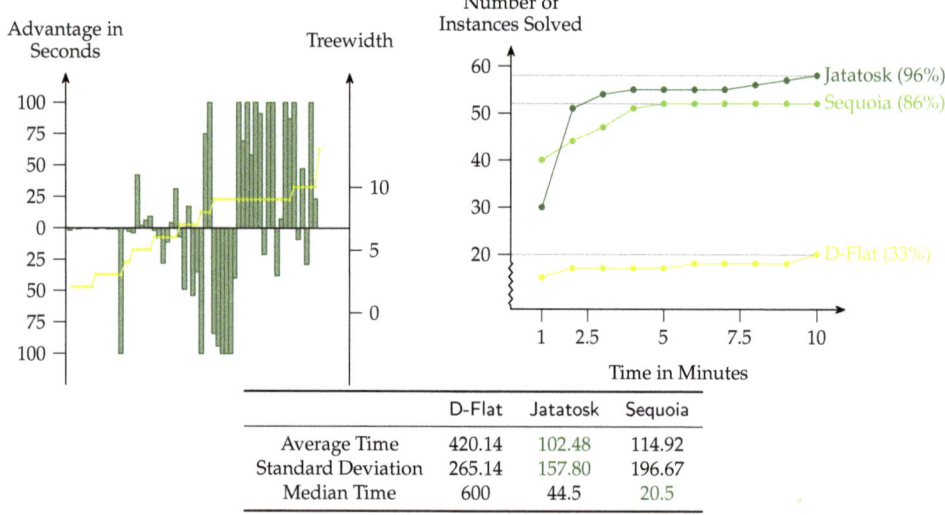

	D-Flat	Jatatosk	Sequoia
Average Time	420.14	102.48	114.92
Standard Deviation	265.14	157.80	196.67
Median Time	600	44.5	20.5

Figure 4. Experiment for the DOMINATING-SET problem.

In this experiment Jatatosk performs best with respect to all: the difference plot, the cactus plot, and the used average time. However, the difference to Sequoia is small and there are in fact a couple of instances that are solved faster by Sequoia. We are surprised by this result, as the worst-case running time of Jatatosk for DOMINATING-SET is $O^*(8^k)$ and, thus, far from optimal. Furthermore, Sequoia promises in theory a better performance with a running time of the form $O^*(5^k)$ [15].

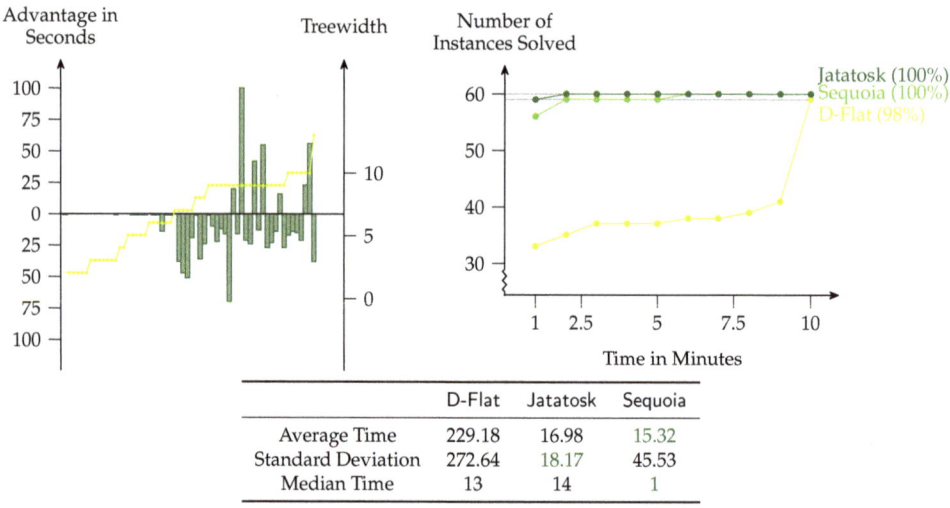

	D-Flat	Jatatosk	Sequoia
Average Time	229.18	16.98	15.32
Standard Deviation	272.64	18.17	45.53
Median Time	13	14	1

Figure 5. Experiment for the INDEPENDENT-SET problem.

This is the simplest test set that we consider within this paper, which is reflected by the fact that all solvers are able to solve almost the whole set. The difference between Jatatosk and Sequoia is minor: while Jatatosk has a slightly better peek-performance, there are more instances that are solved faster by Sequoia than the other way around.

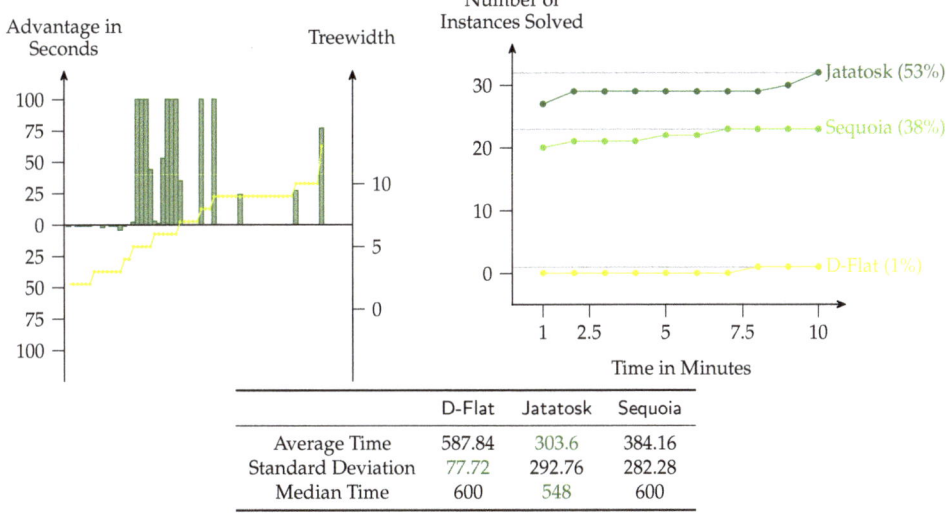

Figure 6. Experiment for the FEEDBACK-VERTEX-SET problem.

This is the hardest test set that we consider within this paper, which is reflected by the fact that no solver is able to solve much more than 50% of the instances. Jatatosk outperforms its competitors, which is reflected in the difference plot, the cactus plot, and the used average time. We assume this is because Jatatosk uses the dedicated forest quantifier directly, while the other tools have to infer the algorithmic strategy from a more general formula.

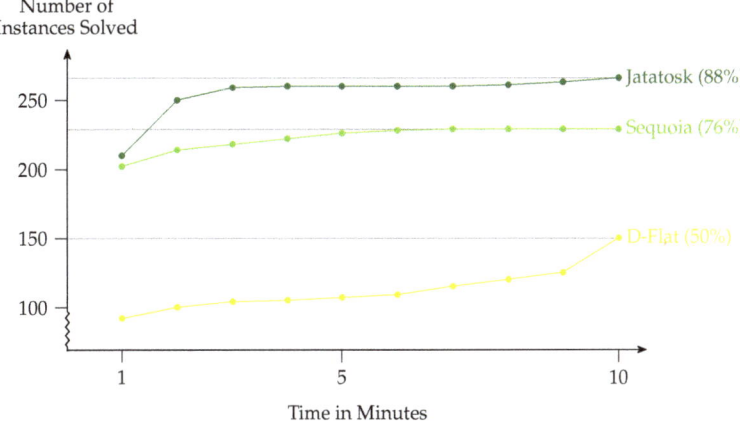

Figure 7. The cactus plot is the sum of the plots from Figures 2–6. Over all experiments, there are 300 instances in total that were tested. Jatatosk solves 88% of these, while Sequoia only manages to solve 76% and D-Flat 50% of the instances.

6. Conclusions and Outlook

We investigated the practicability of dynamic programming on tree decompositions, which is arguably one of the cornerstones of parameterized complexity theory. We implemented a simple interface for such programs and used it to build a competitive graph coloring solver with just a few lines of code. We hope that this interface allows others to implement and explore various dynamic programs. The whole power of these algorithms is well captured by Courcelle's Theorem, which states that there is an efficient version of such a program for every problem definable in monadic second-order logic. We took a step towards practice by implementing a "lightweight" version of a model checker for a small fragment of the logic. This fragment turns out to be powerful enough to express many natural problems such as 3-COLORING, FEEDBACK-VERTEX-SET, and more. Various experiments showed that the model checker is competitive against the state-of-the-art solvers D-Flat and Sequoia. It therefore seems promising, from a practical perspective, to study smaller fragments of MSO.

Author Contributions: Conceptualization, M.B. and S.B.; methodology, M.B. and S.B.; software, M.B.; validation, S.B.; formal analysis, M.B. and S.B.; investigation, S.B.; resources, S.B.; data curation, M.B. and S.B.; writing—original draft preparation, M.B. and S.B.; writing—review and editing, M.B. and S.B.; visualization, M.B.; supervision, M.B. and S.B.; project administration, M.B. and S.B.; funding acquisition, M.B. and S.B.

Funding: This research received no external funding.

Conflicts of Interest: The authors declare no conflict of interest.

References

1. Cygan, M.; Fomin, F.V.; Kowalik, Ł.; Lokshtanov, D.; Marx, D.; Pilipczuk, M.; Pilipczuk, M.; Saurabh, S. *Parameterized Algorithms*; Springer: Berlin, Germany, 2015. [CrossRef]
2. Dell, H.; Husfeldt, T.; Jansen, B.M.P.; Kaski, P.; Komusiewicz, C.; Rosamond, F.A. The first parameterized algorithms and computational experiments challenge. In Proceedings of the 11th International Symposium on Parameterized and Exact Computation (IPEC), Aarhus, Denmark, 24–26 August 2016; pp. 30:1–30:9. [CrossRef]
3. Dell, H.; Komusiewicz, C.; Talmon, N.; Weller, M. The PACE 2017 parameterized algorithms and computational experiments challenge: The second iteration. In Proceedings of the 12th International Symposium on Parameterized and Exact Computation (IPEC), Vienna, Austria, 6–8 September 2017; pp. 30:1–30:12. [CrossRef]
4. Courcelle, B. The monadic second-order logic of graphs. I. Recognizable sets of finite graphs. *Inf. Comput.* **1990**, *85*, 12–75. [CrossRef]
5. Bodlaender, H.L. A linear-time algorithm for finding tree-decompositions of small treewidth. *SIAM. J. Comput.* **1996**, *25*, 1305–1317. [CrossRef]
6. Tamaki, H. Positive-Instance driven dynamic programming for treewidth. In Proceedings of the 25th Annual European Symposium on Algorithms (ESA), Vienna, Austria, 4–6 September 2017; pp. 68:1–68:13. [CrossRef]
7. Abseher, M.; Bliem, B.; Charwat, G.; Dusberger, F.; Hecher, M.; Woltran, S. D-FLAT: Progress Report. Available online: https://www.dbai.tuwien.ac.at/research/report/dbai-tr-2014-86.pdf (accessed on 5 June 2019).
8. Langer, A.J. Fast Algorithms for Decomposable Graphs. Ph.D. Thesis, The RWTH Aachen University, Aachen, Germany, 2013.
9. Diestel, R. *Graph Theory: Springer Graduate Text Gtm 173*, 4th ed.; Springer: Berlin, Germany, 2012; Volume 173.
10. Flum, J.; Grohe, M. *Parameterized Complexity Theory*; Springer: Berlin, Germany, 2006. [CrossRef]
11. Bannach, M.; Berndt, S.; Ehlers, T. Jdrasil: A modular library for computing tree decompositions. In Proceedings of the 16th International Symposium on Experimental Algorithms (SEA), London, UK, 21–23 June 2017; pp. 28:1–28:21. [CrossRef]
12. Bannach, M. Jdrasil for Graph Coloring. Available online: https://github.com/maxbannach/Jdrasil-for-GraphColoring (accessed on 23 January 2019). Commit: a5e52a8.
13. Bannach, M.; Berndt, S.; Ehlers, T. Jdrasil. Available online: http://www.github.com/maxbannach/jdrasil (accessed on 5 June 2019). Commit: dfa1eee.
14. Frick, M.; Grohe, M. The complexity of first-order and monadic second-order logic revisited. *Ann. Pure Appl. Logic* **2004**, *130*, 3–31. [CrossRef]
15. Kneis, J.; Langer, A.; Rossmanith, P. Courcelle's theorem—A game-theoretic approach. *Discret. Optim.* **2011**, *8*, 568–594. [CrossRef]
16. Bannach, M.; Berndt, S. Jatatosk. Available online: https://github.com/maxbannach/Jatatosk/commit/45e306cfac5a273416870ec0bd9cd2c7f39a6932 (accessed on 8 April 2019).

17. Fichte, J.K. gtfs2graphs—A Transit Feed to Graph Format Converter. Available online: https://github.com/daajoe/gtfs2graphs/commit/219944893f874b365de1ed87fc265fd5d19d5972 (accessed on 20 April 2018).
18. Fichte, J.K.; Lodha, N.; Szeider, S. SAT-Based local improvement for finding tree decompositions of small width. In Proceedings of the International Conference on Theory and Applications of Satisfiability Testing (SAT), Melbourne, Australia, 28 August–1 September 2017; Springer: Cham, Switzerland, 2017; pp. 401–411.
19. Abseher, M.; Dusberger, F.; Musliu, N.; Woltran, S. Improving the efficiency of dynamic programming on tree decompositions via machine learning. *J. Artif. Intell. Res.* **2015**, *58*, 275–282. [CrossRef]

 © 2019 by the authors. Licensee MDPI, Basel, Switzerland. This article is an open access article distributed under the terms and conditions of the Creative Commons Attribution (CC BY) license (http://creativecommons.org/licenses/by/4.0/).

Article

A Compendium of Parameterized Problems at Higher Levels of the Polynomial Hierarchy

Ronald de Haan [1,*] and Stefan Szeider [2]

[1] Institute for Logic, Language and Computation (ILLC), University of Amsterdam, 1000–1183 Amsterdam, The Netherlands
[2] Algorithms and Complexity Group, Institute for Logic and Computation, Faculty of Informatics, Technische Universität Wien, 1040 Vienna, Austria
* Correspondence: me@ronalddehaan.eu

Received: 19 July 2019; Accepted: 3 September 2019; Published: 9 September 2019

Abstract: We present a list of parameterized problems together with a complexity classification of whether they allow a fixed-parameter tractable reduction to SAT or not. These problems are parameterized versions of problems whose complexity lies at the second level of the Polynomial Hierarchy or higher.

Keywords: parameterized complexity; polynomial hierarchy; fpt-reductions

1. Introduction

The remarkable performance of today's SAT solvers (see, e.g., [1]) offers a practically successful strategy for solving NP-complete combinatorial search problems by reducing them in polynomial time to the propositional satisfiability problem—also known as SAT. In order to apply this strategy to problems that are harder than NP, one needs to employ reductions that are more powerful than polynomial-time reductions. A compelling option for this strategy is to use fixed-parameter tractable reductions (or fpt-reductions)—i.e., reductions that are computable in time $f(k)n^c$ for some computable function f and some constant c—as they can exploit some structural aspects of the problem instances in terms of a problem parameter k.

To illustrate this approach, let us consider as an example the problem of deciding the truth of quantified Boolean formulas (QBFs) of the form $\varphi = \exists X.\forall Y.\psi$, where ψ is a DNF formula over the variables in $X \cup Y$. (We define and consider this example problem—as well as the parameterized variants of this problem that we consider in this section—in more detail in Section 3.3.) This problem is Σ_2^P-complete [2,3], so there is no polynomial-time reduction from this problem to SAT, unless the Polynomial Hierarchy collapses. In other words, the strategy of first reducing the problem to SAT, and then using a SAT solver to solve the problem seems not viable.

Another strategy would be to consider the problem from a parameterized point of view—i.e., investigating in which cases structural properties of the input allow the problem to be solved efficiently (in fpt-time). For example, if the incidence treewidth of the DNF formula ψ is bounded, we can solve the problem in fpt-time [4,5]. However, this is the case only in very restricted settings, and so this strategy is also only viable to a limited extent.

The approach of using fpt-reductions to SAT combines the merits of the above two strategies. It uses the remarkable performance of today's SAT solvers, and it can exploit structural properties of the problem inputs. As a result, this approach has the potential of being useful in a much wider range of settings than either of the two separate strategies. It can work for problems that lie at higher levels of the Polynomial Hierarchy. Moreover, it can work for choices of problem parameters that are much less restrictive than those needed to get traditional fixed-parameter tractability results.

As a demonstration of this, let us consider two other parameterized variants of our example problem of deciding the truth of formulas of the form $\varphi = \exists X.\forall Y.\psi$. In the first variant, the parameter is the treewidth of the incidence graph of ψ where we delete all nodes corresponding to variables in X—in other words, the incidence graph of ψ with respect to the universally quantified variables. This variant is para-NP-complete [6,7], which means that it allows an fpt-reduction to SAT. Thus, for this variant, we can exploit the problem parameter to efficiently reduce the problem to SAT, and subsequently use a SAT solver to solve the problem. In the second variant that we consider, the parameter is the treewidth of the incidence graph of ψ where we delete all nodes corresponding to variables in Y—in other words, the incidence graph of ψ with respect to the existentially quantified variables. This variant is para-Σ_2^P-complete [6,7], indicating that fpt-reductions to SAT are not possible (unless the Polynomial Hierarchy collapses).

For both of these variants of our example problem, both the approach of polynomial-time reducing the problem to SAT and the approach of finding (traditional) fixed-parameter tractable algorithms do not work. However, for only one of them, an fpt-reduction to SAT is possible (under common complexity-theoretic assumptions). This indicates that, to establish whether an fpt-reduction to SAT is possible, a suitable parameterized complexity analysis at higher levels of the Polynomial Hierarchy is needed. Various parameterized complexity classes have been developed for this purpose [7,8] that, together with previously studied classes, enable such analyses. This includes both classes that correspond to various types of fpt-reductions to SAT—e.g., para-NP, para-co-NP, and FPT^{NP}[few]—and classes that indicate that no fpt-reductions to SAT are possible, under suitable complexity-theoretic assumptions—e.g., $\Sigma_2^P[k*]$, $\Sigma_2^P[*k,1]$, $\Sigma_2^P[*k,P]$, $\Pi_2^P[k*]$, $\Pi_2^P[*k,1]$, $\Pi_2^P[*k,P]$, para-Σ_2^P, para-Π_2^P, PH[level], and para-PSPACE. (We define all of these parameterized complexity classes in Sections 2.2 and 2.3.)

Outline

In this compendium, we give a list of parameterized problems that are based on problems at higher levels of the Polynomial Hierarchy, together with a complexity classification indicating whether they allow a (many-to-one or Turing) fpt-reduction to SAT or not.

The compendium that we provide is similar in concept to the compendia by Schaefer and Umans [9] and Cesati [10] that also list problems along with their computational complexity. We group the list by the type of problems. A list of problems grouped by their complexity can be found at the end of this paper.

The remainder of the paper is structured as follows. In Section 2, we give an overview of the parameterized complexity classes involved in the classification of whether problems allow an fpt-reduction to SAT, as well as some other notions and definitions that are used throughout the paper. Then, in Sections 3–6, we present (i) problems related to (quantified variants of) propositional logic, (ii) problems from the area of Knowledge Representation and Reasoning, (iii) graph problems, and (iv) other problems, respectively. Finally, we conclude in Section 7.

2. Preliminaries

2.1. Computational Complexity

We assume that the reader is familiar with basic notions from the theory of computational complexity, such as the complexity classes P and NP. For more details, we refer to textbooks on the topic (see, e.g., [11,12]).

There are many natural decision problems that are not contained in the classical complexity classes P and NP (under some common complexity-theoretic assumptions). The *Polynomial Hierarchy* [2,3,12,13] contains a hierarchy of increasing complexity classes Σ_i^P, for all $i \geq 0$. We give a characterization of these classes based on the satisfiability problem of various classes of quantified Boolean formulas. A *quantified Boolean formula* is a formula of the form $Q_1 X_1 Q_2 X_2 \ldots Q_m X_m \psi$, where each Q_i is either $\forall$ or $\exists$, the X_i are disjoint sets of propositional variables, and ψ is a Boolean formula over the variables in $\bigcup_{i=1}^{m} X_i$. The quantifier-free part of such formulas is called the *matrix* of the formula. Truth of such formulas is defined in the usual way. Let $\gamma = \{x_1 \mapsto d_1, \ldots, x_n \mapsto d_n\}$ be a function that maps some variables of a formula φ to other variables or to truth values. We let $\varphi[\gamma]$ denote the application of such a substitution γ to the formula φ. We also write $\varphi[x_1 \mapsto d_1, \ldots, x_n \mapsto d_n]$ to denote $\varphi[\gamma]$. For each $i \geq 1$, the decision problem QSAT$_i$ is defined as follows.

QSAT$_i$
Instance: A quantified Boolean formula $\varphi = \exists X_1 \forall X_2 \exists X_3 \ldots Q_i X_i \psi$, where Q_i is a universal quantifier if i is even and an existential quantifier if i is odd.
Question: Is φ true?

Input formulas to the problem QSAT$_i$ are called Σ_i^P-formulas. For each nonnegative integer $i \geq 0$, the complexity class Σ_i^P can be characterized as the closure of the problem QSAT$_i$ under polynomial-time reductions [2,3]—that is, all decision problems that are polynomial-time reducible to QSAT$_i$. The Σ_i^P-hardness of QSAT$_i$ holds already when the matrix of the input formula is restricted to 3CNF for odd i, and restricted to 3DNF for even i. Note that the class Σ_0^P coincides with P, and the class Σ_1^P coincides with NP. For each $i \geq 1$, the class Π_i^P is defined as co-Σ_i^P—that is, $\Pi_i^P = \{ \{0,1\}^* \setminus L : L \in \Sigma_i^P \}$.

The classes Σ_i^P and Π_i^P can also be defined by means of nondeterministic Turing machines with an oracle. Intuitively, oracles are black-box machines that can solve a problem in a single time step—for more details, see, e.g., Chapter 3 of [11]. For any complexity class C, we let NP^C be the set of decision problems that are decided in polynomial time by a nondeterministic Turing machine with an oracle for a problem that is in the class C. Then, the classes Σ_i^P and Π_i^P, for $i \geq 0$, can be equivalently defined by letting $\Sigma_0^P = \Pi_0^P = P$, and for each $i \geq 1$ letting $\Sigma_i^P = \text{NP}^{\Sigma_{i-1}^P}$ and $\Pi_i^P = \text{co-NP}^{\Sigma_{i-1}^P}$.

The Polynomial Hierarchy also includes complexity classes between Σ_i^P and Σ_{i+1}^P—such as the classes Δ_{i+1}^P and Θ_{i+1}^P. The class Δ_{i+1}^P consists of all decision problems that are decided in polynomial time by a deterministic Turing machine with an oracle for a problem that is in the class Σ_i^P. Similarly, the class Θ_{i+1}^P consists of all decision problems that are decided in polynomial time by a deterministic Turing machine with an oracle for a problem that is in the class Σ_i^P, with the restriction that the Turing machine is only allowed to make $O(\log n)$ oracle queries, where n denotes the input size [14,15]. It holds that $\Sigma_i^P \cup \Pi_i^P \subseteq \Theta_{i+1}^P \subseteq \Delta_{i+1}^P \subseteq \Sigma_{i+1}^P \cap \Pi_{i+1}^P$.

There are also natural decision problems that are located between NP and Θ_2^P. The *Boolean Hierarchy* (BH) [16–18] consists of a hierarchy of complexity classes BH$_i$, for each $i \geq 1$, that can be used to classify the complexity of decision problems between NP and Θ_2^P. Each class BH$_i$ can be characterized as the class of problems that can be reduced to the problem BH$_i$-SAT, which is defined inductively as follows. The problem BH$_1$-SAT consists of all sequences (φ) of length 1, where φ is a satisfiable propositional formula. For even $i \geq 2$, the problem BH$_i$-SAT consists of all sequences $(\varphi_1, \ldots, \varphi_i)$ of propositional formulas such that both $(\varphi_1, \ldots, \varphi_{i-1}) \in$ BH$_{(i-1)}$-SAT and φ_i is unsatisfiable. For odd $i \geq 2$, the problem BH$_i$-SAT consists of all sequences $(\varphi_1, \ldots, \varphi_i)$ of propositional formulas such that $(\varphi_1, \ldots, \varphi_{i-1}) \in$ BH$_{(i-1)}$-SAT or φ_i is satisfiable. The class BH$_2$ is also denoted by DP, and the problem BH$_2$-SAT is also denoted by SAT-UNSAT. The class BH is defined as the union of all BH$_i$, for $i \geq 1$. It holds that NP $\cup$ co-NP $\subseteq$ BH$_2 \subseteq$ BH$_3 \subseteq \cdots \subseteq$ BH $\subseteq \Theta_2^P$.

2.2. Parameterized Complexity

We introduce some core notions from parameterized complexity theory. For an in-depth treatment, we refer to other sources [19–23]. A *parameterized problem* L is a subset of $\Sigma^* \times \mathbb{N}$ for some finite alphabet Σ. For an instance $(x, k) \in \Sigma^* \times \mathbb{N}$, we call x the *main part* and k the *parameter*. The following generalization of polynomial-time computability is commonly regarded as the main tractability notion of parameterized complexity theory. A parameterized problem L is *fixed-parameter tractable* if there exists a computable function f and a constant c such that there exists an algorithm that decides whether $(x,k) \in L$ in time $f(k) \cdot |x|^c$, where $|x|$ denotes the size of x. Such an algorithm is called an *fpt-algorithm*, and this amount of time is called *fpt-time*. FPT is the class of all parameterized problems that are fixed-parameter tractable. If the parameter is constant, then fpt-algorithms run in polynomial time where the order of the polynomial is independent of the parameter. This provides a good scalability in the parameter in contrast to running times of the form $|x|^k$, which are also polynomial for fixed k, but are already impractical for, say, $k > 5$.

Parameterized complexity also generalizes the notion of polynomial-time reductions. Let $L \subseteq \Sigma^* \times \mathbb{N}$ and $L' \subseteq (\Sigma')^* \times \mathbb{N}$ be two parameterized problems. A *(many-one) fpt-reduction* from L to L' is a mapping $R : \Sigma^* \times \mathbb{N} \to (\Sigma')^* \times \mathbb{N}$ from instances of L to instances of L' for which there exist some computable function $g : \mathbb{N} \to \mathbb{N}$ such that for all $(x,k) \in \Sigma^* \times \mathbb{N}$: (i) (x,k) is a yes-instance of L if and only if $(x',k') = R(x,k)$ is a yes-instance of L', (ii) $k' \leq g(k)$, and (iii) R is computable in fpt-time. Let K be a parameterized complexity class. A parameterized problem L is K-*hard* if for every $L' \in K$ there is an fpt-reduction from L' to L. A problem L is K-*complete* if it is both in K and K-hard. Reductions that satisfy properties (i) and (ii) but that are computable in time $O(|x|^{f(k)})$, for some fixed computable function f, we call *xp-reductions*.

The parameterized complexity classes W[t], for $t \geq 1$, W[SAT], and W[P] can be used to give evidence that a given parameterized problem is not fixed-parameter tractable. These classes are based on the satisfiability problems of Boolean circuits and formulas. We consider *Boolean circuits* with a single output gate. We call input nodes *variables*. We distinguish between *small* gates, with fan-in ≤ 2, and *large* gates, with fan-in > 2. The *depth* of a circuit is the length of a longest path from any variable to the output gate. The *weft* of a circuit is the largest number of large gates on any path from a variable to the output gate. We say that a circuit C is in *negation normal form* if all negation nodes in C have variables as inputs. A *Boolean formula* can be considered as a Boolean circuit where all gates have fan-out ≤ 1. We adopt the usual notions of truth assignments and satisfiability of a Boolean circuit. We say that a truth assignment for a Boolean circuit has *weight k* if it sets exactly k of the variables of the circuit to true. We denote the class of Boolean circuits with depth u and weft t by CIRC$_{t,u}$. We denote the class of all Boolean circuits by CIRC, and the class of all Boolean formulas by FORM. For any class $\mathcal{C}$ of Boolean circuits, we define the following parameterized problem:

p-WSAT$[\mathcal{C}]$
Instance: A Boolean circuit $C \in \mathcal{C}$, and an integer k.
Parameter: k.
Question: Does there exist an assignment of weight k that satisfies C?

We denote closure under fpt-reductions by $[\,\cdot\,]^{\text{fpt}}$—that is, for any set $\mathcal{L}$ of parameterized problems, $[\,\mathcal{L}\,]^{\text{fpt}}$ is the set of all parameterized problems L' that are fpt-reducible to some problem $L \in \mathcal{L}$. The classes W[t] are defined by letting W[t] $= [\,\{\,p\text{-WSAT}[\text{CIRC}_{t,u}] : u \geq 1\,\}\,]^{\text{fpt}}$ for all $t \geq 1$. The classes W[SAT] and W[P] are defined by letting W[SAT] $= [\,p\text{-WSAT}[\text{FORM}]\,]^{\text{fpt}}$ and W[P] $= [\,p\text{-WSAT}[\text{CIRC}]\,]^{\text{fpt}}$.

Let K be a classical complexity class, e.g., NP. The parameterized complexity class para-K is defined as the class of all parameterized problems $L \subseteq \Sigma^* \times \mathbb{N}$, for some finite alphabet Σ, for which there exist a computable function $f : \mathbb{N} \to \Sigma^*$, and a problem $P \subseteq \Sigma^*$ such that $P \in K$ and for all instances $(x,k) \in \Sigma^* \times \mathbb{N}$ of L we have that $(x,k) \in L$ if and only if $(x, f(k)) \in P$. (Here, we implicitly use a representation of pairs of strings in $\Sigma^* \times \Sigma^*$ as strings in Σ^*.) Intuitively, the class para-K consists of all problems that are in K after a precomputation that only involves the parameter. The class para-NP can also be defined via nondeterministic fpt-algorithms [24]. The class para-K can be seen as a direct analogue of the class K in parameterized complexity. In particular, for the case of K = P, we have FPT = para-P.

We consider the following (trivial) parameterization of SAT, the satisfiability problem for propositional logic. We let SAT$_1 = \{\,(\varphi, 1) : \varphi \in \text{SAT}\,\}$. In other words, SAT$_1$ is the parameterized variant of SAT where the parameter is the constant value 1. Similarly, we let UNSAT$_1 = \{\,(\varphi, 1) : \varphi \in$

UNSAT }. The problem SAT$_1$ is para-NP-complete, and the problem UNSAT$_1$ is para-co-NP-complete. In other words, the class para-NP consists of all parameterized problems that can be fpt-reduced to SAT$_1$, and para-co-NP consists of all parameterized problems that can be fpt-reduced to UNSAT$_1$.

Another analogue to the classical complexity class K is the parameterized complexity class XK^{nu}, that is defined as the class of those parameterized problems Q whose slices Q_k are in K, i.e., for each positive integer k the classical problem $Q_k = \{\, x : (x,k) \in Q \,\}$ is in K [20]. For instance, the class XP^{nu} consists of those parameterized problems whose slices are decidable in polynomial time. Note that this definition is non-uniform, that is, for each positive integer k, there might be a completely different polynomial-time algorithm that witnesses that Q_k is polynomial-time solvable. There are also uniform variants XK of these classes XK^{nu}. We define XP to be the class of parameterized problems Q for which there exists a computable function f and an algorithm A that decides whether $(x,k) \in Q$ in time $|x|^{f(k)}$ [20,22,24]. Similarly, we define XNP to be the class of parameterized problems that are decidable in nondeterministic time $|x|^{f(k)}$. Its dual class we denote by Xco-NP. Alternatively, we can view XNP as the class of parameterized problems for which there exists an xp-reduction to SAT$_1$ and Xco-NP as the class of parameterized problems for which there exists an xp-reduction to UNSAT$_1$. (For any $L \in$ XNP, we know that L can be xp-reduced to SAT$_1$ by following a suitable variant of the proof of the Cook-Levin Theorem [25,26]. Conversely, any parameterized problem L that can be xp-reduced to SAT$_1$ we can solve in nondeterministic time $|x|^{f(k)}$ by first carrying out the xp-reduction, and then solving the resulting instance of SAT$_1$. The case for Xco-NP and UNSAT$_1$ is entirely analogous.)

2.3. Fpt-Reductions to SAT and Parameterized Complexity Classes at Higher Levels of the PH

Problems in NP and co-NP can be encoded into SAT in such a way that the time required to produce the encoding and consequently also the size of the resulting SAT instance are polynomial in the input (the encoding is a polynomial-time many-one reduction). Typically, the SAT encodings of problems proposed for practical use are of this kind (see, e.g., [27]). For problems that are "beyond NP", say for problems on the second level of the PH, such polynomial SAT encodings do not exist, unless the PH collapses. However, for such problems, there still could exist SAT encodings which can be produced in fpt-time with respect to some parameter associated with the problem. In fact, such fpt-time SAT encodings have been obtained for various problems on the second level of the PH [28–31]. The classes para-NP and para-co-NP contain exactly those parameterized problems that admit such a many-one fpt-reduction to SAT$_1$ and UNSAT$_1$, respectively. Thus, with fpt-time encodings, one can go significantly beyond what is possible by conventional polynomial-time SAT encodings.

Fpt-time encodings to SAT also have their limits. Clearly, para-Σ_2^P-hard and para-Π_2^P-hard parameterized problems do not admit fpt-time encodings to SAT, even when the parameter is a fixed constant, unless the PH collapses. There are problems that apparently do not admit fpt-time encodings to SAT, but seem not to be para-Σ_2^P-hard nor para-Π_2^P-hard either. Recently, several complexity classes have been introduced to classify such intermediate problems [7,8,30]. These parameterized complexity classes are dubbed the k-$*$ class and the $*$-k hierarchy, inspired by their definition, which is based on the following weighted variants of the quantified Boolean satisfiability problem that is canonical for the second level of the PH. The problem $\Sigma_2^P[k*]$-WSAT$(\mathcal{C})$ provides the foundation for the k-$*$ class.

$\Sigma_2^P[k*]$-WSAT
Instance: A quantified Boolean formula $\exists X.\forall Y.\psi$, and an integer k.
Parameter: k.
Question: Does there exist a truth assignment α to X with weight k such that for all truth assignments β to Y the assignment $\alpha \cup \beta$ satisfies ψ?

Similarly, the problem $\Sigma_2^P[*k]$-WSAT$(\mathcal{C})$ provides the foundation for the $*$-k hierarchy—where $\mathcal{C}$ is a class of Boolean circuits. (The parameterized problems $\Sigma_2^P[*k]$-WSAT$(\mathcal{C})$ seem not to be fpt-reducible to each other for various classes $\mathcal{C}$ of Boolean circuits—similarly to the problems p-WSAT$[\mathcal{C}]$ that are used to define the classes W[t], W[SAT], and W[P]. This is in contrast to the case of $\Sigma_2^P[k*]$-WSAT, where we can use a Tseitin transformation [32] to reduce arbitrary Boolean circuits to equisatisfiable 3CNF formulas.)

$\Sigma_2^P[*k]$-WSAT($\mathcal{C}$)
Instance: A Boolean circuit $C \in \mathcal{C}$ over two disjoint sets X and Y of variables, and an integer k.
Parameter: k.
Question: Does there exist a truth assignment α to X such that for all truth assignments β to Y with weight k the assignment $\alpha \cup \beta$ satisfies C?

The parameterized complexity class $\Sigma_2^P[k*]$ (also called the k-* class) is then defined as follows:

$$\Sigma_2^P[k*] = [\ \Sigma_2^P[k*]\text{-WSAT}\ ]^{\text{fpt}}.$$

Similarly, the classes of the *-k hierarchy are defined as follows:

$$\Sigma_2^P[*k, t] = [\ \{\ \Sigma_2^P[*k]\text{-WSAT}(\text{CIRC}_{t,u}) : u \geq 1\ \}\ ]^{\text{fpt}},$$
$$\Sigma_2^P[*k, \text{SAT}] = [\ \Sigma_2^P[*k]\text{-WSAT}(\text{FORM})\ ]^{\text{fpt}}, \text{ and}$$
$$\Sigma_2^P[*k, P] = [\ \Sigma_2^P[*k]\text{-WSAT}(\text{CIRC})\ ]^{\text{fpt}}.$$

Note that these definitions are entirely analogous to those of the parameterized complexity classes of the W-hierarchy [20]. The following inclusion relations hold between the classes of the *-k hierarchy:

$$\Sigma_2^P[*k, 1] \subseteq \Sigma_2^P[*k, 2] \subseteq \cdots \subseteq \Sigma_2^P[*k, \text{SAT}] \subseteq \Sigma_2^P[*k, P].$$

(See also Figure 1 for a visual overview of these inclusion relations.)

Dual to the classical complexity class Σ_2^P is its co-class Π_2^P, whose canonical complete problem is complementary to the problem QSAT$_2$. Similarly, we can define dual classes for the k-* class and for each of the parameterized complexity classes in the *-k hierarchy. These co-classes are based on problems complementary to the problems $\Sigma_2^P[k*]$-WSAT and $\Sigma_2^P[*k]$-WSAT—i.e., these problems have as yes-instances exactly the no-instances of $\Sigma_2^P[k*]$-WSAT and $\Sigma_2^P[*k]$-WSAT, respectively. Equivalently, these complementary problems can be considered as variants of $\Sigma_2^P[k*]$-WSAT and $\Sigma_2^P[*k]$-WSAT where the existential and universal quantifiers are swapped, and are therefore denoted with $\Pi_2^P[k*]$-WSAT and $\Pi_2^P[*k]$-WSAT. We use a similar notation for the dual complexity classes, e.g., we denote co-$\Sigma_2^P[*k, t]$ by $\Pi_2^P[*k, t]$.

The class $\Sigma_2^P[k*]$ includes the class para-co-NP as a subset, and is contained in the class Xco-NP as a subset. Similarly, each of the classes $\Sigma_2^P[*k, t]$ include the the class para-NP as a subset, and is contained in the class XNP. Under some common complexity-theoretic assumptions, the class $\Sigma_2^P[k*]$ can be separated from para-NP on the one hand, and para-Σ_2^P on the other hand. In particular, assuming that NP $\neq$ co-NP, it holds that $\Sigma_2^P[k*] \not\subseteq$ para-NP, that para-NP $\not\subseteq \Sigma_2^P[k*]$ and that $\Sigma_2^P[k*] \subsetneq$ para-Σ_2^P [7,8]. Similarly, the classes $\Sigma_2^P[*k, t]$ can be separated from para-co-NP and para-Σ_2^P. Assuming that NP $\neq$ co-NP, it holds that $\Sigma_2^P[*k, 1] \not\subseteq$ para-co-NP, that para-co-NP $\not\subseteq \Sigma_2^P[*k, P]$ and thus in particular that para-co-NP $\not\subseteq \Sigma_2^P[*k, 1]$, and that $\Sigma_2^P[*k, P] \subsetneq$ para-Σ_2^P [7,8].

One can also enhance the power of polynomial-time SAT encodings by considering polynomial-time algorithms that can query a SAT solver multiple times—that is, polynomial-time Turing reductions. Such an approach has been shown to be quite effective in practice (see, e.g., [33–35]) and extends the scope of SAT solvers to problems in the class Δ_2^p, but not to problems that are Σ_2^p-hard or Π_2^p-hard. In addition, here, switching from polynomial-time to fpt-time provides a significant increase in power. The class para-Δ_2^P contains all parameterized problems that can be decided by an fpt-algorithm that can query a SAT oracle multiple times—i.e., by an fpt-time Turing reduction to SAT. (One can prove this by following the proof of Theorem 4 in [24] that FPT = para-P, with the modification that the algorithms are given access to a SAT oracle.) In addition, one could restrict the number of queries that the algorithm is allowed to make. The class para-Θ_2^P consists of all parameterized problems that can be decided by an fpt-algorithm that can query a SAT oracle at most $f(k)\log n$ many times, where k is the parameter value, n is the input size, and f is some computable function. (This statement one can prove by following the proof of Theorem 4 in [24] that FPT = para-P, with the modification that the algorithms can query a SAT oracle an amount of times that depends logarithmically on the input size.) Restricting the number of queries even further, we define the parameterized complexity class FPT^{NP}[few] as the class of all parameterized problems that can be decided by an fpt-algorithm that can query a SAT oracle at most $f(k)$ times, where k is the parameter value and f is some computable function [7,8].

We get the parameterized analogue para-PSPACE of the class PSPACE by using the definition of para-K for K = PSPACE. Similarly, we can define the parameterized complexity class XPSPACE, consisting of all parameterized problems Q for which there exists a computable function f and an algorithm A that decides whether $(x,k) \in Q$ in space $|x|^{f(k)}$. We also consider another parameterized variant of PSPACE, which is based on parameterizing the number of quantifier alternations in QSAT. An unbounded number of quantifier alternations in this problem results in the class PSPACE, and bounding the number of quantifier alternations by a constant leads to some fixed level of the PH. The parameterized complexity class PH[level] is based on bounding the number of quantifier alternations by the problem parameter [7,36]. Formally, we consider the following parameterized problem QSAT(level).

> QSAT(level)
> *Instance:* A quantified Boolean formula $\varphi = \exists X_1 \forall X_2 \exists X_3 \ldots Q_k X_k \psi$, where Q_k is a universal quantifier if k is even and an existential quantifier if k is odd, and where ψ is quantifier-free.
> *Parameter:* k.
> *Question:* Is φ true?

The parameterized complexity class PH[level] is defined to be the class of all parameterized problems that can be fpt-reduced to QSAT(level). We have that para-$\Sigma_2^P \cup$ para-$\Pi_2^P \subseteq$ PH[level] $\subseteq$ para-PSPACE.

An overview of the parameterized complexity classes relevant for this paper can be found in Figure 1.

In the early literature on this topic [6,28,30,37–41], the class $\Sigma_2^P[k*]$ appeared under the names $\exists^k \forall$ and $\exists^k \forall^*$. Similarly, the classes $\Sigma_2^P[*k,t]$ appeared under the names $\exists \forall^k$-W[t] and $\exists^* \forall^k$-W[t]. In addition, the class FPT^{NP}[few] appeared under the name FPT^{NP}[$f(k)$].

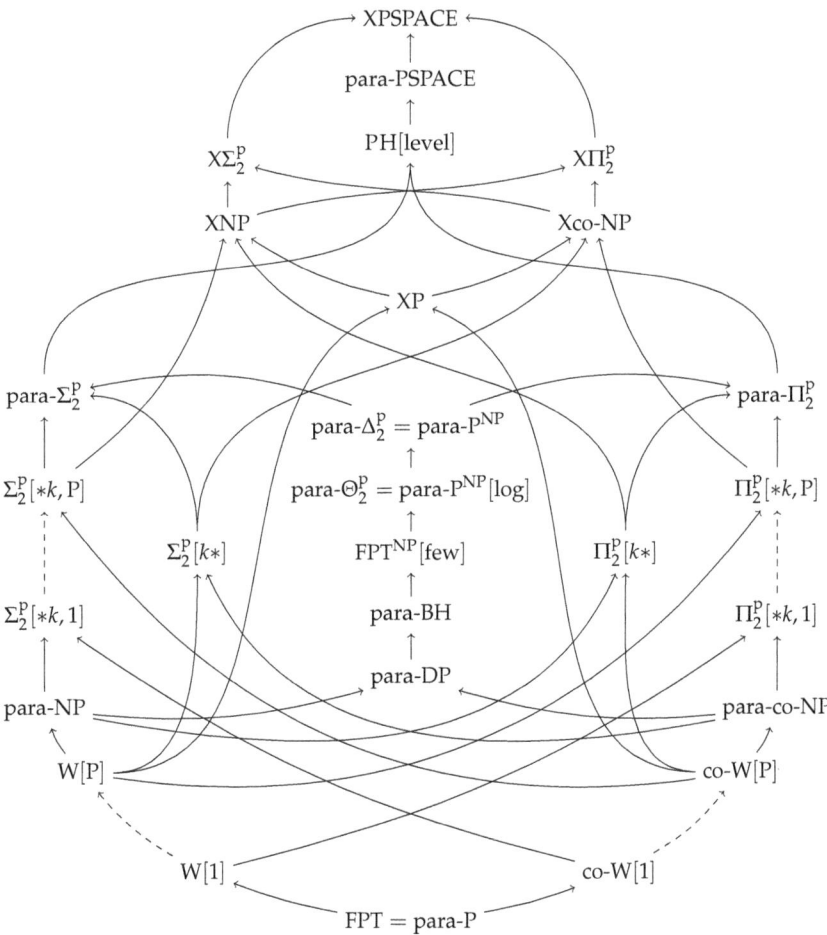

Figure 1. An overview of parameterized complexity classes up to the second level of the Polynomial Hierarchy, and higher. Dashed lines indicate a hierarchy of classes—e.g., between W[1] and W[P] lies the hierarchy W[1] ⊆ W[2] ⊆ ⋯ ⊆ W[P].

2.4. Treewidth and Tree Decompositions

We conclude this section with explaining the notions of tree decompositions and treewidth that we use in various places through the paper. For more details on these notions, we refer to textbooks—e.g., [19–23,42].

A tree decomposition of a graph $G = (V, E)$ is a pair $(\mathcal{T}, (B_t)_{t \in T})$ where $\mathcal{T} = (T, F)$ is a rooted tree and $(B_t)_{t \in T}$ is a family of subsets of V such that:

- for every $v \in V$, the set $B^{-1}(v) = \{\, t \in T : v \in B_t \,\}$ is nonempty and connected in $\mathcal{T}$; and
- for every edge $\{v, w\} \in E$, there is a $t \in T$ such that $v, w \in B_t$.

The *width* of the decomposition $(\mathcal{T}, (B_t)_{t \in T})$ is the number $\max\{\, |B_t| : t \in T \,\} - 1$. The *treewidth* of G is the minimum of the widths of all tree decompositions of G. Let G be a graph and k a nonnegative integer. There is an fpt-algorithm that computes a tree decomposition of G of width k if it exists, and fails otherwise [43]. We call a tree decomposition $(\mathcal{T}, (B_t)_{t \in T})$ *nice* if every node $t \in T$ is of one of the following four types:

- *leaf node*: t has no children and $|B_t| = 1$;
- *introduce node*: t has one child t' and $B_t = B_{t'} \cup \{v\}$ for some vertex $v \notin B_{t'}$;
- *forget node*: t has one child t' and $B_t = B_{t'} \setminus \{v\}$ for some vertex $v \in B_{t'}$; or
- *join node*: t has two children t_1, t_2 and $B_t = B_{t_1} = B_{t_2}$.

Given any graph G and a tree decomposition of G of width k, a nice tree decomposition of G of width k can be computed in polynomial time [42].

3. Propositional Logic Problems

We start with the quantified circuit satisfiability problems on which the k-$*$ and $*$-k hierarchies are based. We present two canonical forms of the problems in the k-$*$ hierarchy. For problems in the $*$-k hierarchy, we let $\mathcal{C}$ range over classes of Boolean circuits.

$\Sigma_2^P[k*]$-WSAT($\mathcal{C}$)
Instance: A Boolean circuit $C \in \mathcal{C}$ over two disjoint sets X and Y of variables, and an integer k.
Parameter: k.
Question: Does there exist a truth assignment α to X of weight k, such that for all truth assignments β to Y the assignment $\alpha \cup \beta$ satisfies C?

Complexity: $\Sigma_2^P[k*]$-complete [7,8].

$\Sigma_2^P[k*]$-WSAT
Instance: A quantified Boolean formula $\phi = \exists X. \forall Y. \psi$, and an integer k.
Parameter: k.
Question: Does there exist a truth assignment α to X with weight k, such that $\forall Y. \psi[\alpha]$ evaluates to true?

Complexity: $\Sigma_2^P[k*]$-complete [7,8].

$\Sigma_2^P[*k]$-WSAT($\mathcal{C}$)
Instance: A Boolean circuit $C \in \mathcal{C}$ over two disjoint sets X and Y of variables, and an integer k.
Parameter: k.
Question: Does there exist a truth assignment α to X, such that for all truth assignments β to Y of weight k the assignment $\alpha \cup \beta$ satisfies C?

Complexity:
$\Sigma_2^P[*k, t]$-complete when restricted to circuits of weft t, for any $t \geq 1$ (by definition);
$\Sigma_2^P[*k, \text{SAT}]$-complete if $\mathcal{C} = \text{FORM}$ (by definition);
$\Sigma_2^P[*k, P]$-complete if $\mathcal{C} = \text{CIRC}$ (by definition).

3.1. Weighted Quantified Boolean Satisfiability for the k-$*$ Classes

Consider the following variants of $\Sigma_2^P[k*]$-WSAT, most of which are $\Sigma_2^P[k*]$-complete.

$\Sigma_2^P[k*]$-WSAT(3DNF)
Instance: A quantified Boolean formula $\phi = \exists X. \forall Y. \psi$ with $\psi \in $ 3DNF, and an integer k.
Parameter: k.
Question: Does there exist a truth assignment α to X with weight k, such that $\forall Y. \psi[\alpha]$ evaluates to true?

Complexity: $\Sigma_2^P[k*]$-complete [7,8].

$\Sigma_2^P[k*]$-WSAT$^{\leq k}$
Instance: A quantified Boolean formula $\phi = \exists X.\forall Y.\psi$, and an integer k.
Parameter: k.
Question: Does there exist an assignment α to X with weight at most k, such that $\forall Y.\psi[\alpha]$ evaluates to true?
Complexity: $\Sigma_2^P[k*]$-complete [7,8].

$\Sigma_2^P[k*]$-WSAT$^{\geq k}$
Instance: A quantified Boolean formula $\phi = \exists X.\forall Y.\psi$, and an integer k.
Parameter: k.
Question: Does there exist an assignment α to X with weight at least k, such that $\forall Y.\psi[\alpha]$ evaluates to true?
Complexity: para-Σ_2^P-complete [7].

$\Sigma_2^P[k*]$-WSAT^{n-k}
Instance: A quantified Boolean formula $\phi = \exists X.\forall Y.\psi$, and an integer k.
Parameter: k.
Question: Does there exist an assignment α to X with weight $|X| - k$, such that $\forall Y.\psi[\alpha]$ evaluates to true?
Complexity: $\Sigma_2^P[k*]$-complete [7].

3.2. Weighted Quantified Boolean Satisfiability in the *-k Hierarchy

Let $d \geq 2$ be an arbitrary constant. Then, the following problem is also $\Sigma_2^P[*k, 1]$-complete.

$\Sigma_2^P[*k]$-WSAT$(d\text{-DNF})$
Instance: A quantified Boolean formula $\varphi = \exists X.\forall Y.\psi$ with $\psi \in d\text{-DNF}$, and an integer k
Parameter: k.
Question: Does there exist an assignment α to X, such that for all assignments β to Y of weight k the assignment $\alpha \cup \beta$ satisfies ψ?
Complexity: $\Sigma_2^P[*k, 1]$-complete for any $d \geq 2$ [7,8].

The problem $\Sigma_2^P[*k]$-WSAT(2-DNF) is $\Sigma_2^P[*k, 1]$-hard, even when we restrict the input formula to be anti-monotone in the universal variables, i.e., the universal variables occur only in negative literals [7,8].

Let C be a Boolean circuit with input nodes Z that is in negation normal form, and let $Y \subseteq Z$ be a subset of the input nodes. We say that C is *monotone in Y* if the only negation nodes that occur in the circuit C act on input nodes in $Z \setminus Y$, i.e., input nodes in Y can appear only positively in the circuit. Similarly, we say that C is *anti-monotone in Y* if the only nodes that have nodes in Y as input are negation nodes, i.e., all input nodes in Y appear only negatively in the circuit. The following problems are $\Sigma_2^P[*k, P]$-complete.

$\Sigma_2^P[*k]$-WSAT($\forall$-monotone)
Instance: A Boolean circuit $C \in \text{CIRC}$ over two disjoint sets X and Y of variables that is in negation normal form and that is monotone in Y, and an integer k.
Parameter: k.
Question: Does there exist a truth assignment α to X, such that for all truth assignments β to Y of weight k the assignment $\alpha \cup \beta$ satisfies C?
Complexity: $\Sigma_2^P[*k, P]$-complete [7,8].

$\Sigma_2^P[*k]$-WSAT($\forall$-anti-monotone)
Instance: A Boolean circuit $C \in \text{CIRC}$ over two disjoint sets X and Y of variables that is in negation normal form and that is anti-monotone in Y, and an integer k.
Parameter: k.
Question: Does there exist a truth assignment α to X, such that for all truth assignments β to Y of weight k the assignment $\alpha \cup \beta$ satisfies C?
Complexity: $\Sigma_2^P[*k, P]$-complete [7].

It remains open what the exact parameterized complexity is of $\exists$-monotone and $\exists$-anti-monotone variants of $\Sigma_2^P[*k]$-WSAT—i.e., variants of $\Sigma_2^P[*k]$-WSAT that are based on circuits $C \in \text{CIRC}$ over two disjoint sets X and Y of variables that are in negation normal form and that are (anti-)monotone in X. The proofs used to show $\Sigma_2^P[*k, P]$-completeness of the $\forall$-monotone and $\forall$-anti-monotone variants [7,8] do not immediately carry over to this case.

3.3. Quantified Boolean Satisfiability with Bounded Treewidth

Let $\psi = \delta_1 \vee \cdots \vee \delta_u$ be a DNF formula. For any subset $Z \subseteq \text{Var}(\psi)$ of variables, we define *the incidence graph* $\text{IG}(Z, \psi)$ *of* ψ *with respect to* Z to be the graph $\text{IG}(Z, \psi) = (V, E)$, where $V = Z \cup \{\delta_1, \ldots, \delta_u\}$ and $E = \{\, \{\delta_j, z\} : 1 \leq j \leq u$ and $z \in Z$ and z occurs in the clause $\delta_j \,\}$. If ψ is a DNF formula, $Z \subseteq \text{Var}(\psi)$ is a subset of variables, and $(\mathcal{T}, (B_t)_{t \in T})$ is a tree decomposition of $\text{IG}(Z, \psi)$, we let $\text{Var}(t)$ denote $B_t \cap Z$, for any $t \in T$.

The following parameterized decision problems are variants of QSAT_2, where the treewidth of the incidence graph graph for certain subsets of variables is bounded.

$\text{QSAT}_2(\text{itw})$
Instance: A quantified Boolean formula $\varphi = \exists X.\forall Y.\psi$, with ψ in DNF.
Parameter: The treewidth of the incidence graph $\text{IG}(X \cup Y, \psi)$ of ψ with respect to $X \cup Y$.
Question: Is φ satisfiable?
Complexity: fixed-parameter tractable [4,5].

$\text{QSAT}_2(\exists\text{-itw})$
Instance: A quantified Boolean formula $\varphi = \exists X.\forall Y.\psi$, with ψ in DNF.
Parameter: The treewidth of the incidence graph $\text{IG}(X, \psi)$ of ψ with respect to X.
Question: Is φ satisfiable?
Complexity: para-Σ_2^P-complete [6,7].

$\text{QSAT}_2(\forall\text{-itw})$
Instance: A quantified Boolean formula $\varphi = \exists X.\forall Y.\psi$, with ψ in DNF.
Parameter: The treewidth of the incidence graph $\text{IG}(Y, \psi)$ of ψ with respect to Y.
Question: Is φ satisfiable?
Complexity: para-NP-complete [6,7].

The above problems are parameterized by the treewidth of the incidence graph of the formula ψ (with respect to different subsets of variables). Since computing the treewidth of a given graph is NP-hard, it is unlikely that the parameter value can be computed in polynomial time for these problems. Alternatively, one could consider a variant of the problem where a tree decomposition of width k is given as part of the input.

3.4. Other Quantified Boolean Satisfiability

The following parameterized quantified Boolean satisfiability problem is para-NP-complete.

QSAT(#∀-vars)
Instance: A quantified Boolean formula φ.
Parameter: The number of universally quantified variables of φ.
Question: Is φ true?

Complexity: para-NP-complete [6,44,45].

3.5. Minimization for DNF Formulas

Let φ be a propositional formula in DNF. We say that a set C of literals is an *implicant of* φ if all assignments that satisfy $\bigwedge_{l \in C} l$ also satisfy φ. Moreover, we say that a DNF formula φ' is a *term-wise subformula* of φ' if for all terms $t' \in \varphi'$ there exists a term $t \in \varphi$ such that $t' \subseteq t$. The following parameterized problems are natural parameterizations of problems shown to be Σ_2^P-complete by Umans [46].

SHORTEST-IMPLICANT-CORE(core size)
Instance: A DNF formula φ, an implicant C of φ, and an integer k.
Parameter: k.
Question: Does there exist an implicant $C' \subseteq C$ of φ of size k?

Complexity: $\Sigma_2^P[k*]$-complete [6,7].

SHORTEST-IMPLICANT-CORE(reduction size)
Instance: A DNF formula φ, an implicant C of φ of size n, and an integer k.
Parameter: k.
Question: Does there exist an implicant $C' \subseteq C$ of φ of size $n - k$?

Complexity: $\Sigma_2^P[k*]$-complete [6,7].

DNF-MINIMIZATION(reduction size)
Instance: A DNF formula φ of size n, and an integer k.
Parameter: k.
Question: Does there exist a term-wise subformula φ' of φ of size $n - k$ such that $\varphi \equiv \varphi'$?

Complexity: $\Sigma_2^P[k*]$-complete [6,7].

DNF-MINIMIZATION(core size)
Instance: A DNF formula φ of size n, and an integer k.
Parameter: k.
Question: Does there exist an DNF formula φ' of size k, such that $\varphi \equiv \varphi'$?

Complexity: para-co-NP-hard, in $\text{FPT}^{\text{NP}}[\text{few}]$, and in $\Sigma_2^P[k*]$ [6,7].

3.6. Sequences of Propositional Formulas

The following problem is related to a Boolean combination of satisfiability checks on a sequence of propositional formulas. This is a parameterized version of the problem BH_i-SAT, which is canonical for the different levels of the Boolean Hierarchy (see Section 1). The problem is complete for the class

$\text{FPT}^{\text{NP}}[\text{few}]$.

BH-SAT(level)
Instance: A positive integer k and a sequence $(\varphi_1, \ldots, \varphi_k)$ of propositional formulas.
Parameter: k.
Question: Is it the case that $(\varphi_1, \ldots, \varphi_k) \in \text{BH}_k\text{-SAT}$?
Complexity: $\text{FPT}^{\text{NP}}[\text{few}]$-complete [7,28,37].

The above problem is used to show the following lower bound result for $\text{FPT}^{\text{NP}}[\text{few}]$-complete problems. No $\text{FPT}^{\text{NP}}[\text{few}]$-hard problem can be decided by an fpt-algorithm that uses only $O(1)$ many queries to an NP oracle, unless the Polynomial Hierarchy collapses to the third level [18,37].

The next $\text{FPT}^{\text{NP}}[\text{few}]$-complete problem is based on the problem SAT-UNSAT $= \{ (\varphi_1, \varphi_2) : \varphi_1 \in \text{SAT}, \varphi_2 \in \text{UNSAT} \}$.

BOUNDED-SAT-UNSAT-DISJUNCTION
Instance: A family $(\varphi_i, \varphi'_i)_{i \in [k]}$ of pairs of propositional formulas.
Parameter: k.
Question: Is there some $\ell \in [k]$ such that $(\varphi_\ell, \varphi'_\ell) \in \text{SAT-UNSAT}$?
Complexity: $\text{FPT}^{\text{NP}}[\text{few}]$-complete [7,28,37].

3.7. Maximal Models

The following $\text{FPT}^{\text{NP}}[\text{few}]$-complete problems are based on various notions of (local) maximality for models of propositional formulas. Let φ be a (satisfiable) propositional formula, and let $X \subseteq \text{Var}(\varphi)$ be a subset of variables. Then, a truth assignment $\alpha : \text{Var}(\varphi) \to \{0,1\}$ is an *X-maximal model* of φ if (i) α satisfies φ and (ii) there is no truth assignment $\beta : \text{Var}(\varphi) \to \{0,1\}$ that satisfies φ and that sets more variables among X to true than α. Moreover, take an arbitrary ordering over the variables $X \subseteq \text{Var}(\varphi)$—for the sake of presentation, let $X = \{x_1, \ldots, x_k\}$ and let the ordering $<$ specify $x_1 < \cdots < x_k$. We say that a truth assignment $\alpha : \text{Var}(\varphi) \to \{0,1\}$ is the *lexicographically X-maximal model* of φ (with respect to the given ordering) if (i) α satisfies φ and (ii) each truth assignment $\beta : \text{Var}(\varphi) \to \{0,1\}$ that satisfies φ (with $\alpha \neq \beta$) has the property that there is some $1 \leq \ell \leq k$ such that $\alpha(x_\ell) = 1$ and $\beta(x_\ell) = 0$ and for all $1 \leq \ell' < \ell$ it holds that $\alpha(x_{\ell'}) = \beta(x_{\ell'})$.

LOCAL-MAX-MODEL
Instance: A satisfiable propositional formula φ, a subset $X \subseteq \text{Var}(\varphi)$ of variables, and a variable $w \in X$.
Parameter: $|X|$.
Question: Is there a X-maximal model of φ that sets w to true?
Complexity: $\text{FPT}^{\text{NP}}[\text{few}]$-complete [7,47].

LOCAL-LEX-MAX-MODEL
Instance: A satisfiable propositional formula φ, a subset $X \subseteq \text{Var}(\varphi)$ of variables, and a variable $w \in X$.
Parameter: $|X|$.
Question: Is there a lexicographically X-maximal model of φ that sets w to true?
Complexity: $\text{FPT}^{\text{NP}}[\text{few}]$-complete [7].
(The problem LOCAL-LEX-MAX-MODEL isn't considered explicitly in the literature. $\text{FPT}^{\text{NP}}[\text{few}]$-completeness for this problem follows immediately from proofs in the literature—i.e., Propositions 73 and 74 in [7]).

> ODD-LOCAL-MAX-MODEL
> *Instance:* A propositional formula φ, and a subset $X \subseteq \text{Var}(\varphi)$ of variables.
> *Parameter:* $|X|$.
> *Question:* Do the X-maximal models of φ set an odd number of variables in X to true?
>
> **Complexity:** $\text{FPT}^{\text{NP}}[\text{few}]$-complete [7].

> ODD-LOCAL-LEX-MAX-MODEL
> *Instance:* A propositional formula φ, a subset $X \subseteq \text{Var}(\varphi)$ of variables, and an ordering $<$ over the variables in X.
> *Parameter:* $|X|$.
> *Question:* Does the lexicographically X-maximal model of φ (w.r.t. $<$) set an odd number of variables in X to true?
>
> **Complexity:** $\text{FPT}^{\text{NP}}[\text{few}]$-complete [7].

4. Knowledge Representation and Reasoning Problems

4.1. Disjunctive Answer Set Programming

The following problems from the setting of disjunctive answer set programming (ASP) are based on the notions of disjunctive logic programs and answer sets for such programs (cf. [48,49]). A *disjunctive logic program* P is a finite set of rules of the form $r = (a_1 \vee \cdots \vee a_k \leftarrow b_1, \ldots, b_m, \text{not } c_1, \ldots, \text{not } c_n)$, for $k, m, n \geq 0$, where all a_i, b_j and c_l are atoms. For each such rule, $a_1 \vee \cdots \vee a_k$ is called the *head* of the rule, and $b_1, \ldots, b_m, \text{not } c_1, \ldots, \text{not } c_n$ is called the *body* of the rule. A rule is called *disjunctive* if $k > 1$, and it is called *normal* if $k \leq 1$ (note that we only call rules with strictly more than one disjunct in the head disjunctive). A rule is called *dual-normal* if $m \leq 1$. A program is called normal if all its rules are normal, it is called negation-free if all its rules are negation-free, and it is called dual-normal if all its rules are dual-normal. We let $\text{At}(P)$ denote the set of all atoms occurring in P. By *literals*, we mean atoms a or their negations $\text{not } a$. The *(Gelfond–Lifschitz) reduct* of a program P with respect to a set M of atoms, denoted P^M, is the program obtained from P by: (i) removing rules with $\text{not } a$ in the body, for each $a \in M$, and (ii) removing literals $\text{not } a$ from all other rules [50]. An *answer set* A of a program P is a subset-minimal model of the reduct P^A. One important decision problem is to decide, given a disjunctive logic program P, whether P has an answer set.

We consider various parameterizations of this problem. Two of these are related to atoms that must be part of any answer set of a program P. We identify a subset $\text{Comp}(P)$ of *compulsory atoms*, that any answer set must include. Given a program P, we let $\text{Comp}(P)$ be the smallest set such that: (i) if $(w \leftarrow \text{not } w)$ is a rule of P, then $w \in \text{Comp}(P)$; and (ii) if $(b \leftarrow a_1, \ldots, a_n)$ is a rule of P, and $a_1, \ldots, a_n \in \text{Comp}(P)$, then $b \in \text{Comp}(P)$. We then let the set $\text{Cont}(P)$ of *contingent atoms* be those atoms that occur in P but are not in $\text{Comp}(P)$. We call a rule *contingent* if all the atoms that appear in the head are contingent. Another of the parameterizations that we consider is based on the notion of backdoors to normality for disjunctive logic programs. A set B of atoms is a *normality-backdoor* for a program P if deleting the atoms $b \in B$ from the rules of P results in a normal program. Deciding if a program P has a normality-backdoor of size at most k can be decided in fixed-parameter tractable time [29].

> ASP CONSISTENCY(#cont.atoms)
> *Instance:* A disjunctive logic program P.
> *Parameter:* The number of contingent atoms of P.
> *Question:* Does P have an answer set?
>
> **Complexity:** para-co-NP-complete [7,30].

ASP-CONSISTENCY(#cont.rules)
Instance: A disjunctive logic program P.
Parameter: The number of contingent rules of P.
Question: Does P have an answer set?
Complexity: $\Sigma_2^P[k*]$-complete [7,30].

ASP-CONSISTENCY(#disj.rules)
Instance: A disjunctive logic program P.
Parameter: The number of disjunctive rules of P.
Question: Does P have an answer set?
Complexity: $\Sigma_2^P[*k, P]$-complete [7,30].

ASP-CONSISTENCY(#dual-normal.rules)
Instance: A disjunctive logic program P.
Parameter: The number of rules of P that are dual-normal.
Question: Does P have an answer set?
Complexity: $\Sigma_2^P[*k, P]$-complete [7].

ASP-CONSISTENCY(str.norm.bd-size)
Instance: A disjunctive logic program P.
Parameter: The size of the smallest normality-backdoor for P.
Question: Does P have an answer set?
Complexity: para-NP-complete [29].

ASP-CONSISTENCY(max.atom.occ.)
Instance: A disjunctive logic program P.
Parameter: The maximum number of times that any atom occurs in P.
Question: Does P have an answer set?
Complexity: para-Σ_2^P-complete [7,30].

4.2. Robust Constraint Satisfaction

The following problem is based on the class of robust constraint satisfaction problems introduced by Gottlob [51] and Abramsky, Gottlob and Kolaitis [52]. These problems are concerned with the question of whether every partial assignment of a particular size can be extended to a full solution, in the setting of constraint satisfaction problems.

A *CSP instance* N is a triple (X, D, C), where X is a finite set of *variables*, the *domain* D is a finite set of *values*, and C is a finite set of *constraints*. Each constraint $c \in C$ is a pair (S, R), where $S = \text{Var}(c)$, the *constraint scope*, is a finite sequence of distinct variables from X, and R, the *constraint relation*, is a relation over D whose arity matches the length of S, i.e., $R \subseteq D^r$, where r is the length of S.

Let $N = (X, D, C)$ be a CSP instance. A *partial instantiation* of N is a mapping $\alpha : X' \to D$ defined on some subset $X' \subseteq X$. We say that α *satisfies* a constraint $c = ((x_1, \ldots, x_r), R) \in C$ if $\text{Var}(c) \subseteq X'$ and $(\alpha(x_1), \ldots, \alpha(x_r)) \in R$. If α satisfies all constraints of N then it is a *solution* of N. We say that α *violates* a constraint $c = ((x_1, \ldots, x_r), R) \in C$ if there is no extension β of α defined on $X' \cup \text{Var}(c)$ such that $(\beta(x_1), \ldots, \beta(x_r)) \in R$.

Let k be a positive integer. We say that a CSP instance $N = (X, D, C)$ is *k-robustly satisfiable* if for each instantiation $\alpha : X' \to D$ defined on some subset $X' \subseteq X$ of k many variables (i.e., $|X'| = k$)

that does not violate any constraint in C, it holds that α can be extended to a solution for the CSP instance (X, D, C).

ROBUST-CSP-SAT
Instance: A CSP instance (X, D, C), and an integer k.
Parameter: k.
Question: Is (X, D, C) k-robustly satisfiable?

Complexity: $\Pi_2^P[k*]$-complete [7,30].

4.3. Abductive Reasoning

The setting of (propositional) abductive reasoning can be formalized as follows. An *abduction instance* $\mathcal{P}$ consists of a tuple (V, H, M, T), where V is the set of *variables*, $H \subseteq V$ is the set of *hypotheses*, $M \subseteq V$ is the set of *manifestations*, and T is the theory, a formula in CNF over V. It is required that $M \cap H = \emptyset$. A set $S \subseteq H$ is a *solution* (or *explanation*) of $\mathcal{P}$ if (i) $T \cup S$ is consistent and (ii) $T \cup S \models M$. One central problem is to decide, given an abduction instance $\mathcal{P}$ and an integer m, whether there exists a solution S of $\mathcal{P}$ of size at most m. This problem is Σ_2^P-complete in general [53].

ABDUCTION(Krom-bd-size):
Input: an abduction instance $\mathcal{P} = (V, H, M, T)$, and a positive integer m.
Parameter: The size of the smallest strong 2CNF-backdoor for T.
Question: Does there exist a solution S of $\mathcal{P}$ of size at most m?

Complexity: para-NP-complete [31].

ABDUCTION(#non-Krom-clauses):
Input: an abduction instance $\mathcal{P} = (V, H, M, T)$, and a positive integer m.
Parameter: The number of clauses in T that contains more than 2 literals.
Question: Does there exist a solution S of $\mathcal{P}$ of size at most m?

Complexity: $\Sigma_2^P[*k, 1]$-complete [7].

ABDUCTION(Horn-bd-size):
Input: an abduction instance $\mathcal{P} = (V, H, M, T)$, and a positive integer m.
Parameter: The size of the smallest strong Horn-backdoor for T.
Question: Does there exist a solution S of $\mathcal{P}$ of size at most m?

Complexity: para-NP-complete [31].

ABDUCTION(#non-Horn-clauses):
Input: an abduction instance $\mathcal{P} = (V, H, M, T)$, and a positive integer m.
Parameter: The number of clauses in T that are not Horn.
Question: Does there exist a solution S of $\mathcal{P}$ of size at most m?

Complexity: $\Sigma_2^P[*k, P]$-complete [7].

---

5. Graph Problems

5.1. Clique Extensions

Let $G = (V, E)$ be a graph. A clique $C \subseteq V$ of G is a subset of vertices that induces a complete subgraph of G, i.e., $\{v, v'\} \in E$ for all $v, v' \in C$ such that $v \neq v'$. The W[1]-complete problem of determining whether a graph has a clique of size k is an important problem in the W-hierarchy, and is used in many W[1]-hardness proofs. We consider a related problem that is complete for $\Pi_2^p[*k,1]$.

SMALL-CLIQUE-EXTENSION
Instance: A graph $G = (V, E)$, a subset $V' \subseteq V$, and an integer k.
Parameter: k.
Question: Is it the case that for each clique $C \subseteq V'$, there is some k-clique D of G such that $C \cup D$ is a $(|C| + k)$-clique?

Complexity: $\Pi_2^p[*k,1]$-complete [7].

5.2. Graph Coloring Extensions

The following problem related to extending colorings to the leaves of a graph to a coloring on the entire graph, is Π_2^p-complete in the most general setting [54].

Let $G = (V, E)$ be a graph. We will denote those vertices v that have degree 1 by *leaves*. We call a (partial) function $c: V \to \{1, 2, 3\}$ a 3-*coloring (of G)*. Moreover, we say that a 3-coloring c is *proper* if c assigns a color to every vertex $v \in V$, and if for each edge $e = \{v_1, v_2\} \in E$ it holds that $c(v_1) \neq c(v_2)$. The problem of deciding, given a graph $G = (V, E)$ with n many leaves and an integer m, whether any 3-coloring that assigns a color to exactly m leaves of G (and to no other vertices) can be extended to a proper 3-coloring of G, is Π_2^p-complete [54]. We consider several parameterizations.

3-COLORING-EXTENSION(degree)
Instance: a graph $G = (V, E)$ with n many leaves, and an integer m.
Parameter: the degree of G.
Question: can any 3-coloring that assigns a color to exactly m leaves of G (and to no other vertices) be extended to a proper 3-coloring of G?

Complexity: para-Π_2^p-complete [7,41].

3-COLORING-EXTENSION(#leaves)
Instance: a graph $G = (V, E)$ with n many leaves, and an integer m.
Parameter: n.
Question: can any 3-coloring that assigns a color to exactly m leaves of G (and to no other vertices) be extended to a proper 3-coloring of G?

Complexity: para-NP-complete [7,41].

3-COLORING-EXTENSION(#col.leaves)
Instance: a graph $G = (V, E)$ with n many leaves, and an integer m.
Parameter: m.
Question: can any 3-coloring that assigns a color to exactly m leaves of G (and to no other vertices) be extended to a proper 3-coloring of G?

Complexity: $\Pi_2^p[k*]$-complete [7,41].

> 3-COLORING-EXTENSION(#uncol.leaves)
> *Instance:* a graph $G = (V, E)$ with n many leaves, and an integer m.
> *Parameter:* $n - m$.
> *Question:* can any 3-coloring that assigns a color to exactly m leaves of G (and to no other vertices) be extended to a proper 3-coloring of G?
>
> **Complexity:** para-Π_2^P-complete [7,41].

6. Other Problems

6.1. First-Order Logic Model Checking

First-order logic model checking is at the basis of a well-known hardness theory in parameterized complexity theory [22]. The following problem, also based on first-order logic model checking, offers another characterization of the parameterized complexity class $\Sigma_2^P[k*]$. We introduce a few notions that we need for defining the model checking perspective on $\Sigma_2^P[k*]$. A *(relational) vocabulary* τ is a finite set of relation symbols. Each relation symbol R has an *arity* $\text{arity}(R) \geq 1$. A *structure* $\mathcal{A}$ of vocabulary τ, or τ-*structure* (or simply *structure*), consists of a set A called the *domain* and an interpretation $R^\mathcal{A} \subseteq A^{\text{arity}(R)}$ for each relation symbol $R \in \tau$. We use the usual definition of truth of a first-order logic sentence φ over the vocubulary τ in a τ-structure $\mathcal{A}$. We let $\mathcal{A} \models \varphi$ denote that the sentence φ is true in structure $\mathcal{A}$.

> $\Sigma_2^P[k*]$-MC
> *Instance:* A first-order logic sentence $\varphi = \exists x_1, \ldots, x_k. \forall y_1, \ldots, y_n. \psi$ over a vocabulary τ, where ψ is quantifier-free, and a finite τ-structure $\mathcal{A}$.
> *Parameter:* k.
> *Question:* Is it the case that $\mathcal{A} \models \varphi$?
>
> **Complexity:** $\Sigma_2^P[k*]$-complete [41].

6.2. Quantified Fagin Definability

The W-hierarchy can also be defined by means of Fagin-definable parameterized problems [22], which are based on Fagin's characterization of NP. We provide an additional characterization of the class $\Pi_2^P[k*]$ by means of some parameterized problems that are quantified analogues of Fagin-defined problems.

Let τ be an arbitrary vocabulary, and let $\tau' \subseteq \tau$ be a subvocabulary of τ. We say that a τ-structure $\mathcal{A}$ *extends* a τ'-structure $\mathcal{B}$ if (i) $\mathcal{A}$ and $\mathcal{B}$ have the same domain, and (ii) $\mathcal{A}$ and $\mathcal{B}$ coincide on the interpretation of all relational symbols in τ', i.e., $R^\mathcal{A} = R^\mathcal{B}$ for all $R \in \tau'$. We say that $\mathcal{A}$ *extends $\mathcal{B}$ with weight k* if $\sum_{R \in \tau \setminus \tau'} |R^\mathcal{A}| = k$. Let φ be a first-order formula over τ with a free relation variable X of arity s.

We let Π_2 denote the class of all first-order logic formulas of the form $\forall y_1, \ldots, y_n. \exists x_1, \ldots, x_m. \psi$, where ψ is quantifier-free. Let $\varphi(X)$ be a first-order logic formula over τ, with a free relation variable X with arity s. Consider the following parameterized problem.

> $\Pi_2^P[k*]$-$\text{FD}_\varphi^{(\tau,\tau')}$
> *Instance:* A τ'-structure $\mathcal{B}$, and an integer k.
> *Parameter:* k.
> *Question:* Is it the case that for each τ-structure $\mathcal{A}$ extending $\mathcal{B}$ with weight k, there exists some relation $S \subseteq A^s$ such that $\mathcal{A} \models \varphi(S)$?
>
> **Complexity:** in $\Pi_2^P[k*]$ for each $\varphi(X)$, τ' and τ; $\Pi_2^P[k*]$-hard for some $\varphi(X) \in \Pi_2$, τ' and τ [7].

Note that this means the following. We let T denote the set of all relational vocabularies, and for any $\tau \in T$ we let FO_τ^X denote the set of all first-order logic formulas over the vocabulary τ with a free relation variable X. We then get the following characterization of $\Pi_2^p[k*]$ [7]:

$$\Pi_2^p[k*] = [\, \{\, \Pi_2^p[k*]\text{-FD}_\varphi^{(\tau',\tau)} : \tau \in T, \tau' \subseteq \tau, \varphi \in \text{FO}_\tau^X \,\} \,]^{\text{fpt}}.$$

Additionally, the following parameterized problem is hard for $\Pi_2^p[*k, 1]$.

$\Pi_2^p[*k]\text{-FD}_\varphi^{(\tau,\tau')}$
Instance: A τ'-structure $\mathcal{B}$, and an integer k.
Parameter: k.
Question: Is it the case that for each τ-structure $\mathcal{A}$ extending $\mathcal{B}$, there exists some relation $S \subseteq A^s$ with $|S| = k$ such that $\mathcal{A} \models \varphi(S)$?
Complexity: $\Pi_2^p[*k, 1]$-hard for some $\varphi(X) \in \Pi_2$ [7].

6.3. Symbolic Model Checking for Temporal Logics

The following problems are concerned with the task of verifying whether a temporal logic formula is true in a Kripke structure. This task is of importance in the area of software and hardware verification (see, e.g., [55,56]). We consider three different temporal logics: *linear-time temporal logic* (LTL), *computation tree logic* (CTL), and CTL*, which is a superset of both LTL and CTL.

Truth of formulas φ in these logics is defined over Kripke structures $\mathcal{M}$—we write $\mathcal{M} \models \varphi$ if φ is true in $\mathcal{M}$. We consider Kripke structures that are represented symbolically using propositional formulas. Moreover, for each of the logics $\mathcal{L} \in \{\text{LTL}, \text{CTL}, \text{CTL}^*\}$, we consider various restricted fragments $\mathcal{L}\backslash X$, $\mathcal{L}\backslash U$ and $\mathcal{L}\backslash U,X$, where certain operators are disallowed. For a full description of the syntax and semantics of these logics, how Kripke structures can be represented symbolically and the restricted fragments that we consider, we refer to Appendix A.

In general, the problem of deciding whether a given temporal logic formula φ is true in a given symbolically represented Kripke structure $\mathcal{M}$ is PSPACE-hard, even when restricted to formulas φ of constant size [7,36]. We consider a variant of this problem where the Kripke structures have limited recurrence diameter. The *recurrence diameter $rd(\mathcal{M})$* of a Kripke structure $\mathcal{M}$ is the length of the longest simple (non-repeating) path in $\mathcal{M}$.

SYMBOLIC-MC*[$\mathcal{L}$]
Input: A symbolically represented Kripke structure $\mathcal{M}$, $rd(\mathcal{M})$ in unary, and an $\mathcal{L}$ formula φ.
Parameter: $|\varphi|$.
Question: $\mathcal{M} \models \varphi$?
Complexity:
- para-co-NP-complete if $\mathcal{L} = \text{LTL}\backslash U,X$ [7,36],
- PH[level]-complete if $\mathcal{L} \in \{\text{CTL}, \text{CTL}\backslash X, \text{CTL}\backslash U, \text{CTL}\backslash U,X, \text{CTL}^*\backslash U,X\}$ [7,36], and
- para-PSPACE-complete if $\mathcal{L} \in \{\text{CTL}^*, \text{CTL}^*\backslash X, \text{CTL}^*\backslash U\}$ [7,36].

6.4. Judgment Aggregation

The following problems are related to judgment aggregation, in the domain of computational social choice. Judgment aggregation studies procedures that combine individuals' opinions into a collective group opinion.

An *agenda* is a finite nonempty set $\Phi = \{\varphi_1, \ldots, \varphi_n, \neg\varphi_1, \ldots, \neg\varphi_n\}$ of formulas that is closed under complementation. A *judgment set* J for an agenda Φ is a subset $J \subseteq \Phi$. We call a judgment set J

complete if $\varphi_i \in J$ or $\neg \varphi_i \in J$ for all formulas φ_i, and we call it *consistent* if there exists an assignment that makes all formulas in J true. Let $\mathcal{J}(\Phi)$ denote the set of all complete and consistent subsets of Φ. We call a sequence $J \in \mathcal{J}(\Phi)^n$ of complete and consistent subsets a *profile*. A (resolute) *judgment aggregation procedure* for the agenda Φ and n individuals is a function $F : \mathcal{J}(\Phi)^n \to \mathcal{P}(\Phi \backslash \varnothing) \backslash \varnothing$ that returns for each profile J a non-empty set $F(J)$ of non-empty judgment sets. An example is the *majority rule* F^{maj}, where $F^{\mathrm{maj}}(J) = \{J^*\}$ and where $\varphi \in J^*$ if and only if φ occurs in the majority of judgment sets in J, for each $\varphi \in \Phi$. We call F *complete* and *consistent*, if each $J^* \in F(J)$ is complete and consistent, respectively, for every $J \in \mathcal{J}(\Phi)^n$. For instance, the majority rule F^{maj} is complete, whenever the number n of individuals is odd. An agenda Φ is *safe* with respect to an aggregation procedure F, if F is consistent when applied to profiles of judgment sets over Φ. We say that an agenda Φ satisfies the *median property (MP)* if every inconsistent subset of Φ has itself an inconsistent subset of size at most 2. Safety for the majority rule can be characterized in terms of the median property as follows: an agenda Φ is safe for the majority rule if and only if Φ satisfies the MP [57,58]. The problem of deciding whether an agenda satisfies the MP is Π_2^p-complete [57].

MAJ-AGENDA-SAFETY(formula size)
Instance: an agenda Φ.
Parameter: $\ell = \max\{\,|\varphi| : \varphi \in \Phi\,\}$.
Question: Is Φ safe for the majority rule?
Complexity: para-Π_2^p-complete [7,37].

MAJ-AGENDA-SAFETY(degree)
Instance: an agenda Φ containing only CNF formulas.
Parameter: The degree d of Φ.
Question: Is Φ safe for the majority rule?
Complexity: para-Π_2^p-complete [7,37].

MAJ-AGENDA-SAFETY(degree + formula size)
Instance: an agenda Φ containing only CNF formulas, where $\ell = \max\{\,|\varphi| : \varphi \in B(\Phi)\,\}$, and where d is the degree of Φ.
Parameter: $\ell + d$.
Question: Is Φ safe for the majority rule?
Complexity: para-Π_2^p-complete [7,37].

The above three parameterized problems remain para-Π_2^p-hard even when restricted to agendas based on formulas that are Horn formulas containing only clauses of size at most 2 [7,37].

MAJ-AGENDA-SAFETY(agenda size)
Instance: an agenda Φ.
Parameter: $|\Phi|$.
Question: Is Φ safe for the majority rule?
Complexity: FPT^{NP}[few]-complete [7,37].

Moreover, the following upper and lower bounds on the number of oracle queries are known for the above problem. MAJ-AGENDA-SAFETY(agenda size) can be decided in fpt-time using $2^{O(k)}$ queries to an NP oracle, where $k = |\Phi|$ [7,37]. In addition, there is no fpt-algorithm that decides MAJ-AGENDA-SAFETY(agenda size) using $o(\log k)$ queries to an NP oracle, unless the Polynomial

Hierarchy collapses [7,37].

MAJ-AGENDA-SAFETY(counterexample size)
Instance: an agenda Φ, and an integer k.
Parameter: k.
Question: Does every inconsistent subset Φ' of Φ of size k have itself an inconsistent subset of size at most 2?
Complexity: $\Pi_2^P[k*]$-hard [7,37].

Let $\Phi = \{\varphi_1, \ldots, \varphi_m, \neg\varphi_1, \ldots, \neg\varphi_m\}$ be an agenda, where each φ_i is a CNF formula. We define the following graphs that are intended to capture the interaction between formulas in Φ. The *formula primal graph* of Φ has as vertices the variables $\text{Var}(\Phi)$ occurring in the agenda, and two variables are connected by an edge if there exists a formula φ_i in which they both occur. The *formula incidence graph* of Φ is a bipartite graph whose vertices consist of (1) the variables $\text{Var}(\Phi)$ occurring in the agenda and (2) the formulas $\varphi_i \in \Phi$. A variable $x \in \text{Var}(\Phi)$ is connected by an edge with a formula $\varphi_i \in \Phi$ if x occurs in φ_i, i.e., $x \in \text{Var}(\varphi_i)$. The *clausal primal graph* of Φ has as vertices the variables $\text{Var}(\Phi)$ occurring in the agenda, and two variables are connected by an edge if there exists a formula φ_i and a clause $c \in \varphi_i$ in which they both occur. The *clausal incidence graph* of Φ is a bipartite graph whose vertices consist of (1) the variables $\text{Var}(\Phi)$ occurring in the agenda and (2) the clauses c occurring in formulas $\varphi_i \in \Phi$. A variable $x \in \text{Var}(\Phi)$ is connected by an edge with a clause c of the formula $\varphi_i \in \Phi$ if x occurs in c, i.e., $x \in \text{Var}(c)$. Consider the following parameterizations of the problem MAJ-AGENDA-SAFETY.

MAJ-AGENDA-SAFETY(f-tw)
Instance: an agenda Φ containing only CNF formulas.
Parameter: The treewidth of the formula primal graph of Φ.
Question: Is Φ safe for the majority rule?
Complexity: fixed-parameter tractable [7,37].

MAJ-AGENDA-SAFETY(f-tw*)
Instance: an agenda Φ containing only CNF formulas.
Parameter: The treewidth of the formula incidence graph of Φ.
Question: Is Φ safe for the majority rule?
Complexity: para-Π_2^P-complete [7,37].

MAJ-AGENDA-SAFETY(c-tw)
Instance: an agenda Φ containing only CNF formulas.
Parameter: The treewidth of the clausal primal graph of Φ.
Question: Is Φ safe for the majority rule?
Complexity: para-co-NP-complete [7,37].

MAJ-AGENDA-SAFETY(c-tw*)
Instance: an agenda Φ containing only CNF formulas.
Parameter: The treewidth of the clausal incidence graph of Φ.
Question: Is Φ safe for the majority rule?
Complexity: para-co-NP-complete [7,37].

6.5. Planning

The following parameterized problems are related to planning in the face of uncertainty. We begin by describing the framework of SAS$^+$ planning (see, e.g., [59]). Let $V = \{v_1, \ldots, v_n\}$ be a finite set of *variables* over a finite *domain* D. Furthermore, let $D^+ = D \cup \{\mathbf{u}\}$, where $\mathbf{u}$ is a special *undefined* value not present in D. Then D^n is the set of *total states* and $(D^+)^n$ is the set of *partial states* over V and D. Intuitively, a state $(d_1, \ldots, d_n) \in D^n$ corresponds to an assignment that assigns to each variable $v_i \in V$ the value $d_i \in D$, and a partial state corresponds to a partial assignment that assigns a value to some variables $v_i \in V$. Clearly, $D^n \subseteq (D^+)^n$—that is, each total state is also a partial state. Let $(d_1, \ldots, d_n) = s \in (D^+)^n$ be a state. Then, the value of a variable v_i in state s is denoted by $s[v_i] = d_i$.

An *SAS$^+$ instance* is a tuple $\mathbb{P} = (V, D, A, I, G)$, where V is a set of variables, D is a domain, A is a set of *actions*, $I \in D^n$ is the *initial state* and $G \in (D^+)^n$ is the (partial) *goal state*. Each action $a \in A$ has a *precondition* $\mathrm{pre}(a) \in (D^+)^n$ and an *effect* $\mathrm{eff}(a) \in (D^+)^n$.

We will frequently use the convention that a variable has the value $\mathbf{u}$ in a precondition/effect unless a value is explicitly specified. Furthermore, by a slight abuse of notation, we denote actions and partial states such as preconditions, effects, and goals as follows. Let $a \in A$ be an action, and let $\{p_1, \ldots, p_m\} \subseteq V$ be the set of variables that are not assigned by $\mathrm{pre}(a)$ to the value $\mathbf{u}$—that is, $\{v \in V : \mathrm{pre}(a)[v] \neq \mathbf{u}\} = \{p_1, \ldots, p_m\}$. Moreover, suppose that $\mathrm{pre}(a)[p_1] = d_1, \ldots, \mathrm{pre}(a)[p_m] = d_m$. Then, we denote the precondition $\mathrm{pre}(a)$ by $\mathrm{pre}(a) = \{p_1 \mapsto d_1, \ldots, p_m \mapsto d_m\}$. In particular, if $\mathrm{pre}(a)$ is the partial state such that $\mathrm{pre}(a)[v] = \mathbf{u}$ for each $v \in V$, we denote $\mathrm{pre}(a)$ by $\emptyset$. We use a similar notation for effects. Let a be the action with $\mathrm{pre}(a) = \{p_1 \mapsto d_1, \ldots, p_m \mapsto d_m\}$ and $\mathrm{eff}(a) = \{e_1 \mapsto d'_1, \ldots, e_\ell \mapsto d'_\ell\}$. We then use the notation $a : \{p_1 \mapsto d_1, \ldots, p_m \mapsto d_m\} \to \{e_1 \mapsto d'_1, \ldots, e_{m'} \mapsto d'_{m'}\}$ to describe the action a.

Let $a \in A$ be an action and $s \in D^n$ be a state. Then, a is *valid in s* if for all $v \in V$, either $\mathrm{pre}(a)[v] = s[v]$ or $\mathrm{pre}(a)[v] = \mathbf{u}$. The *result of a in s* is the state $t \in D^n$ defined as follows. For all $v \in V$, $t[v] = \mathrm{eff}(a)[v]$ if $\mathrm{eff}(a)[v] \neq \mathbf{u}$ and $t[v] = s[v]$ otherwise. Let $s_0, s_\ell \in D^n$ and let $\omega = (a_1, \ldots, a_\ell)$ be a sequence of actions (of length ℓ). We say that ω is a *plan from s_0 to s_ℓ* if either (i) ω is the empty sequence (and $\ell = 0$, and thus $s_0 = s_\ell$), or (ii) there are states $s_1, \ldots, s_{\ell-1} \in D^n$ such that for each $i \in [\ell]$, a_i is valid in s_{i-1} and s_i is the result of a_i in s_{i-1}. A state $s \in D^n$ is a *goal state* if for all $v \in V$, either $G[v] = s[v]$ or $G[v] = \mathbf{u}$. An action sequence ω is a *plan for $\mathbb{P}$* if ω is a plan from I to a goal state.

We also allow actions to have *conditional effects* (see, e.g., [60]). A conditional effect is of the form $s \triangleright t$, where $s, t \in (D^+)^n$ are partial states. Intuitively, such a conditional effect ensures that the variable assignment t is only applied if the condition s is satisfied. When allowing conditional effects, the effect of an action is not a partial state $\mathrm{eff}(a) \in (D^+)^n$, but a set $\mathrm{eff}(a) = \{s_1 \triangleright t_1, \ldots, s_\ell \triangleright t_\ell\}$ of conditional effects. (For the sake of simplicity, we assume that the partial states $t_1, \ldots, t_\ell$ are non-conflicting—that is, there exist no $v \in V$ and no $i_1, i_2 \in [\ell]$ with $i_1 < i_2$ such that $\mathbf{u} \neq t_{i_1}[v] \neq t_{i_2}[v] \neq \mathbf{u}$.) The result of an action a with $\mathrm{eff}(a) = \{s_1 \triangleright t_1, \ldots, s_\ell \triangleright t_\ell\}$ in a state s (in which a is valid) is the state $t \in D^n$ that is defined as follows. For all $v \in V$, $t[v] = t_i[v]$ if there exists some $i \in [\ell]$ such that s_i is satisfied in s and $t_i[v] \neq \mathbf{u}$, and $t[v] = s[v]$ otherwise.

For the parameterized planning problems that we consider, in addition to the variables V, we consider a set V_u of n' variables—constituting an uncertain planning instance $\mathbb{P} = (V, V_u, D, A, I, G)$. Intuitively, the value of the variables in V_u in the initial state is unknown. The question is whether there exists an action sequence ω such that, for each state $I_0 \in D^{n+n'}$ that extends I—that is, $I_0[v] = I[v]$ for each $v \in V$—it holds that ω is a plan for the SAS$^+$ instance $(V \cup V_u, D, A, I_0, G)$. If there exists such an action sequence ω, we say that ω is a plan that *works* for each complete initial state I_0 that extends I.

> BOUNDED-UNCERTAIN-PLANNING
> *Instance:* an uncertain planning instance $\mathbb{P} = (V, V_u, D, A, I, G)$, and an integer k.
> *Parameter:* k.
> *Question:* Is there a plan of length k for $\mathbb{P}$ that works for each complete initial state I_0 that extends I?
> **Complexity:** $\Sigma_2^P[k*]$-complete [7,39].

> POLYNOMIAL-PLANNING(bounded-deviation)
> *Instance:* an uncertain planning instance $\mathbb{P} = (V, V_u, D, A, I, G)$, an integer d, and an integer u given in unary.
> *Parameter:* d.
> *Question:* Is there a plan for $\mathbb{P}$ of length $\leq u$ that works for each complete initial state I_0 that extends I and for which at most d unknown variables deviate from the base value?
> **Complexity:** $\Sigma_2^P[*k, P]$-complete [7,39].

6.6. Turing Machine Halting

The following problems are related to alternating Turing machines (ATMs), possibly with multiple tapes. ATMs are nondeterministic Turing machines where the states are divided into existential and universal states, and where each configuration of the machine is called existential or universal according to the state that the machine is in. A run ρ of the ATM $\mathbb{M}$ on an input x is a tree whose nodes correspond to configurations of $\mathbb{M}$ in such a way that (1) for each non-root node v of the tree with parent node v', the machine $\mathbb{M}$ can transition from the configuration corresponding to v' to the configuration corresponding to v, (2) the root node corresponds to the initial configuration of $\mathbb{M}$, and (3) each leaf node corresponds to a halting configuration. A computation path in a run of $\mathbb{M}$ is a root-to-leaf path in the run. Moreover, the nodes of a run ρ are labelled accepting or rejecting, according to the following definition. A leaf of ρ is labelled accepting if the configuration corresponding to it is an accepting configuration, and the leaf is labelled rejecting if it is a rejecting configuration. A non-leaf node of ρ that corresponds to an existential configuration is labelled accepting if at least one of its children is labelled accepting. A non-leaf node of ρ that corresponds to a universal configuration is labelled accepting if all of its children are labelled accepting. An ATM $\mathbb{M}$ is 2-alternating if for each input x, each computation path in the run of $\mathbb{M}$ on input x switches at most once from an existential configuration to a universal configuration, or vice versa. For more details on the terminology, we refer to the work of De Haan and Szeider [8,41] and to the work of Flum and Grohe—Appendix A.1 in [22].

We consider the following restrictions on ATMs. An $\exists\forall$-*Turing machine* (or simply $\exists\forall$-*machine*) is a 2-alternating ATM, where the initial state is an existential state. Let $\ell, t \geq 1$ be positive integers. We say that an $\exists\forall$-machine $\mathbb{M}$ *halts (on the empty string) with existential cost ℓ and universal cost t* if: (1) there is an accepting run of $\mathbb{M}$ with the empty input ϵ, and (2) each computation path of $\mathbb{M}$ contains at most ℓ existential configurations and at most t universal configurations. The following problem, where the number of Turing machine tapes is given as part of the input, is $\Sigma_2^P[k*]$-complete.

> $\Sigma_2^P[k*]$-TM-HALT*.
> *Instance:* Positive integers $m, k, t \geq 1$, and an $\exists\forall$-machine $\mathbb{M}$ with m tapes.
> *Parameter:* k.
> *Question:* Does $\mathbb{M}$ halt on the empty string with existential cost k and universal cost t?
> **Complexity:** $\Sigma_2^P[k*]$-complete [7,8].

Let $m \geq 1$ be a constant integer. Then, the following parameterized decision problem, where the number of Turing machine tapes is fixed, is also $\Sigma_2^P[k*]$-complete.

> $\Sigma_2^P[k*]$-TM-HALT^m.
> *Instance:* Positive integers $k, t \geq 1$, and an $\exists\forall$-machine $\mathbb{M}$ with m tapes.
> *Parameter:* k.
> *Question:* Does $\mathbb{M}$ halt on the empty string with existential cost k and universal cost t?
>
> **Complexity:** $\Sigma_2^P[k*]$-complete [7,8].

In addition, the parameterized complexity class $\Sigma_2^P[k*]$ can also be characterized by means of alternating Turing machines in the following way. Let P be a parameterized problem. An $\Sigma_2^P[k*]$-*machine for* P is a $\exists\forall$-machine $\mathbb{M}$ such that there exists a computable function f and a polynomial p such that: (1) $\mathbb{M}$ decides P in time $f(k) \cdot p(|x|)$; and (2) for all instances (x, k) of P and each computation path R of $\mathbb{M}$ with input (x, k), at most $f(k) \cdot \log|x|$ of the existential configurations of R are nondeterministic. We say that a parameterized problem P *is decided by some* $\Sigma_2^P[k*]$-*machine* if there exists a $\Sigma_2^P[k*]$-machine for P. Then, $\Sigma_2^P[k*]$ is exactly the class of parameterized decision problems that are decided by some $\Sigma_2^P[k*]$-machine [7,8].

7. Conclusions

In this paper, we provided a list of parameterized problems that are based on problems at higher levels of the Polynomial Hierarchy, together with a complexity classification indicating whether they allow a (many-to-one or Turing) fpt-reduction to SAT or not. These complexity classifications are based in part on recently developed parameterized complexity classes—e.g., $\Sigma_2^P[k*]$ and $\Sigma_2^P[*k, P]$ [7,8], FPT^{NP}[few] [6,7] and PH[level] [7,36]. The problems that we considered are related to propositional logic, quantified Boolean satisfiability, disjunctive answer set programming, constraint satisfaction, (propositional) abductive reasoning, cliques, graph coloring, first-order logic model checking, temporal logic model checking, judgment aggregation, planning, and (alternating) Turing machines.

Author Contributions: Conceptualization, Methodology, Formal analysis, Investigation, Project administration, R.d.H. and S.S; Resources, Funding acquisition, S.S.; Writing—original draft preparation, R.d.H., Writing—review and editing, R.d.H. and S.S.; Supervision, S.S.

Funding: This research was funded by the European Research Council (ERC), project 239962 (COMPLEX REASON), and the Austrian Science Fund (FWF), project P26200 (Parameterized Compilation).

Acknowledgments: Open Access Funding by the Austrian Science Fund (FWF).

Conflicts of Interest: The authors declare no conflict of interest. The funders had no role in the design of the study; in the collection, analyses, or interpretation of data; in the writing of the manuscript, or in the decision to publish the results.

Appendix A

We begin by defining the syntax of the logic LTL. LTL formulas φ are formed according to the following grammar (here p ranges over a fixed set P of propositional variables), given by:

$$\varphi ::= p \mid \neg\varphi \mid (\varphi \wedge \varphi) \mid X\varphi \mid F\varphi \mid (\varphi U \varphi).$$

(Further temporal operators that are considered in the literature can be defined in terms of the operators X and U.)

The semantics of LTL is defined along paths of Kripke structures. A *Kripke structure* is a tuple $\mathcal{M} = (S, R, V, s_0)$, where S is a finite set of *states*, where $R \subseteq S \times S$ is a binary relation on the set of states called the *transition relation*, where $V : S \to 2^P$ is a *valuation function* that assigns each state to a set of propositions, and where $s_0 \in S$ is the *initial state*. We say that a finite sequence $s_1 \ldots s_\ell$ of states $s_i \in S$ is a *finite path* in $\mathcal{M}$ if $(s_i, s_{i+1}) \in R$ for each $i \in [\ell-1]$. Similarly, we say that an infinite sequence $s_1 s_2 s_3 \ldots$ of states $s_i \in S$ is an *infinite path* in $\mathcal{M}$ if $(s_i, s_{i+1}) \in R$ for each $i \geq 1$.

Let $\mathcal{M} = (S, R, V, s_0)$ be a Kripke structure, and $\bar{s}_1 = s_1 s_2 s_3 \ldots$ be a path in $\mathcal{M}$. Moreover, let $\bar{s}_i = s_i s_{i+1} s_{i+2} \ldots$ for each $i \geq 2$. Truth of LTL formulas φ on paths $\bar{s}$ (denoted $\bar{s} \models \varphi$) is defined inductively as follows:

$\bar{s}_i \models p$ if $p \in V(s_i)$,
$\bar{s}_i \models \varphi_1 \wedge \varphi_2$,   if $\bar{s}_i \models \varphi_1$ and $\bar{s}_i \models \varphi_2$,
$\bar{s}_i \models \neg \varphi$,   if $\bar{s}_i \not\models \varphi$,
$\bar{s}_i \models X\varphi$,   if $\bar{s}_{i+1} \models \varphi$,
$\bar{s}_i \models F\varphi$,   if   for some $j \geq 0, \bar{s}_{i+j} \models \varphi$,
$\bar{s}_i \models \varphi_1 U \varphi_2$ if   there is some $j \geq 0$ such that $\bar{s}_{i+j} \models \varphi_2$ and $\bar{s}_{i+j'} \models \varphi$ for each $j' \in [0, j-1]$.

Then, we say that an LTL formula φ is true in the Kripke structure $\mathcal{M}$ (denoted $\mathcal{M} \models \varphi$) if for all infinite paths $\bar{s}$ starting in s_0 it holds that $\bar{s} \models \varphi$.

Next, we define the syntax of the logic CTL*, which consists of two different types of formulas: state formulas and path formulas. When we refer to CTL* formulas without specifying the type, we refer to state formulas. Given the set P of atomic propositions, the syntax of CTL* formulas is defined by the following grammar (here Φ denotes CTL* *state formulas*, φ denotes CTL* *path formulas*, and p ranges over P), given by:

$$\Phi ::= p \mid \neg \Phi \mid (\Phi \wedge \Phi) \mid \exists \varphi,$$

$$\varphi ::= \Phi \mid \neg \varphi \mid (\varphi \wedge \varphi) \mid X\varphi \mid F\varphi \mid (\varphi U \varphi).$$

Path formulas have the same intended meaning as LTL formulas. State formulas, in addition, allow explicit quantification over paths, which is not possible in LTL.

Formally, the semantics of CTL* formulas are defined inductively as follows. Let $\mathcal{M} = (S, R, V, s_0)$ be a Kripke structure, $s \in S$ be a state in $\mathcal{M}$ and $\bar{s}_1 = s_1 s_2 s_3 \ldots$ be a path in $\mathcal{M}$. Again, let $\bar{s}_i = s_i s_{i+1} s_{i+2} \ldots$ for each $i \geq 2$. The truth of CTL* state formulas Φ on states $s \in S$ (denoted $s \models \Phi$) is defined as follows:

$s \models p$ if $p \in V(s)$,
$s \models \Phi_1 \wedge \Phi_2$,   if $s \models \Phi_1$ and $s \models \Phi_2$,
$s \models \neg \Phi$,   if $s \not\models \Phi$,
$s \models \exists \varphi$,   if   there is some path $\bar{s}$ in $\mathcal{M}$ starting in s such that $\bar{s} \models \varphi$.

The truth of CTL* path formulas φ on paths $\bar{s}$ (denoted $\bar{s} \models \varphi$) is defined as follows:

$\bar{s}_i \models \Phi$ if $s_i \models \Phi$,
$\bar{s}_i \models \varphi_1 \wedge \varphi_2$,   if $\bar{s}_i \models \varphi_1$ and $\bar{s}_i \models \varphi_2$,
$\bar{s}_i \models \neg \varphi$,   if $\bar{s}_i \not\models \varphi$,
$\bar{s}_i \models X\varphi$,   if $\bar{s}_{i+1} \models \varphi$,
$\bar{s}_i \models F\varphi$,   if   for some $j \geq 0, \bar{s}_{i+j} \models \varphi$,
$\bar{s}_i \models \varphi_1 U \varphi_2$,   if   there is some $j \geq 0$ such that $\bar{s}_{i+j} \models \varphi_2$ and $\bar{s}_{i+j'} \models \varphi$ for each $j' \in [0, j-1]$.

Then, we say that a CTL* formula Φ is true in the Kripke structure $\mathcal{M}$ (denoted $\mathcal{M} \models \Phi$) if $s_0 \models \Phi$.

The syntax of the logic CTL is defined similarly to the syntax of CTL*. Only the grammar for path formulas φ differs, namely:

$$\varphi ::= X\Phi \mid F\Phi \mid (\Phi U \Phi).$$

In particular, this means that every CTL state formula, (CTL formula for short) is also a CTL* formula. The semantics for CTL formulas is defined as for their CTL* counterparts. Moreover, we say that a CTL formula Φ is true in the Kripke structure $\mathcal{M}$ (denoted $\mathcal{M} \models \Phi$) if $s_0 \models \Phi$.

For each of the logics $\mathcal{L} \in \{\text{LTL}, \text{CTL}, \text{CTL}^*\}$, we consider the fragments $\mathcal{L}\backslash X$, $\mathcal{L}\backslash U$ and $\mathcal{L}\backslash U,X$. In the fragment $\mathcal{L}\backslash X$, the X-operator is disallowed. Similarly, in the fragment $\mathcal{L}\backslash U$, the U-operator is disallowed. In the fragment $\mathcal{L}\backslash U,X$, neither the X-operator nor the U-operator is allowed.

Next, we define how Kripke structures can be represented symbolically using propositional formulas. Let $P = \{p_1, \ldots, p_m\}$ be a finite set of propositional variables. A *symbolically represented Kripke* structure over P is a tuple $\mathcal{M} = (\varphi_R, \alpha_0)$, where $\varphi_R(x_1, \ldots, x_m, x'_1, \ldots, x'_m)$ is a propositional formula over the variables $x_1, \ldots, x_m, x'_1, \ldots, x'_m$, and where $\alpha_0 \in \{0,1\}^m$ is a truth assignment to the variables in P. The Kripke structure associated with $\mathcal{M}$ is (S, R, V, α_0), where $S = \{0,1\}^m$ consists of all truth assignments to P, where $(\alpha, \alpha') \in R$ if and only if $\varphi_R[\alpha, \alpha']$ is true, and where $V(\alpha) = \{\, p_i : \alpha(p_i) = 1\,\}$.

References

1. Vardi, M.Y. Boolean satisfiability: Theory and engineering. *Commun. ACM* **2014**, *57*, 5.
2. Stockmeyer, L.J. The polynomial-time hierarchy. *Theor. Comput. Sci.* **1976**, *3*, 1–22.
3. Wrathall, C. Complete Sets and the Polynomial-Time Hierarchy. *Theor. Comput. Sci.* **1976**, *3*, 23–33.
4. Chen, H. Quantified Constraint Satisfaction and Bounded Treewidth. In Proceedings of the 16th European Conference on Artificial Intelligence (ECAI 2004), Valencia, Spain, 22–27 August 2004; pp. 161–165.
5. Feder, T.; Kolaitis, P.G. Closures and dichotomies for quantified constraints. In *Electronic Colloquium on Computational Complexity (ECCC)*; Technical Report TR06-160; Weizmann Institute of Science: Rehovot, Israel, 2006.
6. De Haan, R.; Szeider, S. Fixed-parameter tractable reductions to SAT. In Proceedings of the 17th International Symposium on the Theory and Applications of Satisfiability Testing (SAT 2014), Vienna, Austria, 14–17 July 2014; Egly, U., Sinz, C., Eds.; Springer: Berlin, Germany, 2014; Volume 8561, pp. 85–102.
7. De Haan, R. Parameterized Complexity in the Polynomial Hierarchy. Ph.D. Thesis, Technische Universität Wien, Vienna, Austria, 2016.
8. De Haan, R.; Szeider, S. Parameterized complexity classes beyond para-NP. *J. Comput. Syst. Sci.* **2017**, *87*, 16–57.
9. Schaefer, M.; Umans, C. Completeness in the Polynomial-Time hierarchy: A Compendium. *SIGACT News* **2002**, *33*, 32–49.
10. Cesati, M. Compendium of Parameterized Problems. Available online: http://cesati.sprg.uniroma2.it/research/compendium/ (accessed on 4 September 2019).
11. Arora, S.; Barak, B. *Computational Complexity—A Modern Approach*; Cambridge University Press: Cambridge, UK, 2009; pp. I–XXIV, 1–579.
12. Papadimitriou, C.H. *Computational Complexity*; Addison-Wesley: Boston, MA, USA, 1994.
13. Meyer, A.R.; Stockmeyer, L.J. The Equivalence Problem for Regular Expressions with Squaring Requires Exponential Space. In Proceedings of the 13th Annual Symposium on Switching & Automata Theory (SWAT), College Park, MD, USA, 25–27 October 1972; pp. 125–129.
14. Hemachandra, L.A. The strong exponential hierarchy collapses. *J. Comput. Syst. Sci.* **1989**, *39*, 299–322.
15. Buss, S.R.; Hay, L. On truth-table reducibility to SAT. *Inf. Comput.* **1991**, *91*, 86–102.
16. Cai, J.; Gundermann, T.; Hartmanis, J.; Hemachandra, L.A.; Sewelson, V.; Wagner, K.W.; Wechsung, G. The Boolean Hierarchy I: Structural Properties. *SIAM J. Comput.* **1988**, *17*, 1232–1252.
17. Chang, R.; Kadin, J. The Boolean Hierarchy and the Polynomial Hierarchy: A Closer Connection. *SIAM J. Comput.* **1993**, *25*, 169–178.
18. Kadin, J. The Polynomial Time Hierarchy Collapses if the Boolean Hierarchy Collapses. *SIAM J. Comput.* **1988**, *17*, 1263–1282.
19. Cygan, M.; Fomin, F.V.; Kowalik, L.; Lokshtanov, D.; Marx, D.; Pilipczuk, M.; Pilipczuk, M.; Saurabh, S. *Parameterized Algorithms*; Springer: Berlin, Germany, 2015.
20. Downey, R.G.; Fellows, M.R. Parameterized Complexity. In *Monographs in Computer Science*; Springer: New York, NY, USA, 1999.
21. Downey, R.G.; Fellows, M.R. Texts in Computer Science. *Fundamentals of Parameterized Complexity*; Springer: Berlin, Germany, 2013.
22. Flum, J.; Grohe, M. Parameterized Complexity Theory. In *Texts in Theoretical Computer Science. An EATCS Series*; Springer: Berlin, Germany, 2006; Volume XIV.
23. Niedermeier, R. Invitation to Fixed-Parameter Algorithms. In *Oxford Lecture Series in Mathematics and Its Applications*; Oxford University Press: Oxford, UK, 2006.
24. Flum, J.; Grohe, M. Describing parameterized complexity classes. *Inf. Comput.* **2003**, *187*, 291–319.
25. Cook, S.A. The Complexity of Theorem-Proving Procedures. In Proceedings of the Third Annual ACM Symposium on Theory of Computing, Shaker Heights, OH, USA, 3–5 May 1971; pp. 151–158.

26. Levin, L. Universal sequential search problems. *Probl. Inf. Transm.* **1973**, *9*, 265–266.
27. Prestwich, S.D. CNF Encodings. In *Handbook of Satisfiability*; Biere, A., Heule, M., van Maaren, H., Walsh, T., Eds.; IOS Press: Amsterdam, The Netherlands, 2009; pp. 75–97.
28. Endriss, U.; De Haan, R.; Szeider, S. Parameterized Complexity Results for Agenda Safety in Judgment Aggregation. In Proceedings of the 14th International Conference on Autonomous Agents and Multiagent Systems (AAMAS 2015), Istanbul, Turkey, 4–8 May 2015.
29. Fichte, J.K.; Szeider, S. Backdoors to Normality for Disjunctive Logic Programs. *ACM Trans. Comput. Log.* **2015**, *17*, 7:1–7:23.
30. De Haan, R.; Szeider, S. The Parameterized Complexity of Reasoning Problems Beyond NP. In Proceedings of the Fourteenth International Conference on the Principles of Knowledge Representation and Reasoning (KR 2014), Vienna, Austria, 20–24 July 2014; Baral, C., De Giacomo, G., Eiter, T., Eds.; AAAI Press: Palo Alto, CA, USA, 2014.
31. Pfandler, A.; Rümmele, S.; Szeider, S. Backdoors to Abduction. In Proceedings of the 23rd International Joint Conference on Artificial Intelligence (IJCAI 2013), Beijing, China, 3–9 August 2013; Rossi, F., Ed.; AAAI Press: Palo Alto, CA, USA, 2013.
32. Tseitin, G.S. Complexity of a Derivation in the Propositional Calculus. *Zap. Nauchn. Sem. Leningrad Otd. Mat. Inst. Akad. Nauk SSSR* **1968**, *8*, 23–41.
33. Belov, A.; Lynce, I.; Marques-Silva, J. Towards efficient MUS extraction. *AI Commun.* **2012**, *25*, 97–116.
34. Dvořák, W.; Järvisalo, M.; Wallner, J.P.; Woltran, S. Complexity-sensitive decision procedures for abstract argumentation. *Artif. Intell.* **2014**, *206*, 53–78.
35. Marques-Silva, J.; Janota, M.; Belov, A. Minimal Sets over Monotone Predicates in Boolean Formulae. In Proceedings of the 25th International Conference Computer Aided Verification (CAV 2013), Saint Petersburg, Russia, 13–19 July 2013; Sharygina, N., Veith, H., Eds.; Springer: Berlin, Germany, 2013; Volume 8044, pp. 592–607.
36. De Haan, R.; Szeider, S. Parameterized Complexity Results for Symbolic Model Checking of Temporal Logics. In Proceedings of the Fifteenth International Conference on the Principles of Knowledge Representation and Reasoning (KR 2016), Cape Town, South Africa, 25–29 April 2016; pp. 453–462.
37. Endriss, U.; De Haan, R.; Szeider, S. Parameterized Complexity Results for Agenda Safety in Judgment Aggregation. In Proceedings of the 5th International Workshop on Computational Social Choice (COMSOC-2014), Istanbul, Turkey, 4–8 May 2014.
38. De Haan, R. An Overview of Non-Uniform Parameterized Complexity. In *Electronic Colloquium on Computational Complexity (ECCC)*; Technical Report TR15-130; Weizmann Institute of Science: Rehovot, Israel, 2015.
39. De Haan, R.; Kronegger, M.; Pfandler, A. Fixed-parameter Tractable Reductions to SAT for Planning. In Proceedings of the 24th International Joint Conference on Artificial Intelligence (IJCAI 2015), Buenos Aires, Argentina, 25–31 July 2015.
40. De Haan, R.; Szeider, S. Compendium of Parameterized Problems at Higher Levels of the Polynomial Hierarchy. In *Electronic Colloquium on Computational Complexity (ECCC)*; Technical Report TR14–143; Weizmann Institute of Science: Rehovot, Israel, 2014.
41. De Haan, R.; Szeider, S. Machine Characterizations for Parameterized Complexity Classes beyond para-NP. In Proceedings of the 41st Conference on Current Trends in Theory and Practice of Computer Science (SOFSEM 2015), Pec pod Sněžkou, Czech Republic, 24–29 January 2015; Springer: Berlin, Germany, 2015; Volume 8939.
42. Kloks, T. *Treewidth: Computations and Approximations*; Springer: Berlin, Germany, 1994.
43. Bodlaender, H.L. A linear-time algorithm for finding tree-decompositions of small treewidth. *SIAM J. Comput.* **1996**, *25*, 1305–1317.
44. Ayari, A.; Basin, D.A. QUBOS: Deciding Quantified Boolean Logic Using Propositional Satisfiability Solvers. In Proceedings of the 4th International Conference on Formal Methods in Computer-Aided Design (FMCAD 2002), Portland, OR, USA, 6–8 November 2002; Aagaard, M., O'Leary, J.W., Eds.; Springer: Berlin, Germany, 2002; Volume 2517, pp. 187–201.
45. Biere, A. Resolve and Expand. In Proceedings of the Seventh International Conference on Theory and Applications of Satisfiability Testing (SAT 2004), Vancouver, BC, Canada, 10–13 May 2004; pp. 59–70.

46. Umans, C. Approximability and Completeness in the Polynomial Hierarchy. Ph.D. Thesis, University of California, Berkeley, CA, USA, 2000.
47. De Haan, R. Parameterized Complexity Results for the Kemeny Rule in Judgment Aggregation. In Proceedings of the 22nd European Conference on Artificial Intelligence (ECAI 2016), The Hague, The Netherlands, 29 August–2 September 2016; Kaminka, G.A., Fox, M., Bouquet, P., Hüllermeier, E., Dignum, V., Dignum, F., van Harmelen, F., Eds.; IOS Press: Amsterdam, The Netherlands, 2016; Volume 285, pp. 1502–1510.
48. Brewka, G.; Eiter, T.; Truszczynski, M. Answer set programming at a glance. *Commun. ACM* **2011**, *54*, 92–103.
49. Marek, V.W.; Truszczynski, M. Stable models and an alternative logic programming paradigm. In *The Logic Programming Paradigm: A 25-Year Perspective*; Springer: Berlin, Germany, 1999; pp. 169–181.
50. Gelfond, M.; Lifschitz, V. Classical Negation in Logic Programs and Disjunctive Databases. *New Gener. Comput.* **1991**, *9*, 365–386.
51. Gottlob, G. On minimal constraint networks. *Artif. Intell.* **2012**, *191–192*, 42–60.
52. Abramsky, S.; Gottlob, G.; Kolaitis, P.G. Robust Constraint Satisfaction and Local Hidden Variables in Quantum Mechanics. In Proceedings of the 23rd International Joint Conference on Artificial Intelligence (IJCAI 2013), Beijing, China, 3–9 August 2013; Rossi, F., Ed.; AAAI Press: Palo Alto, CA, USA, 2013.
53. Eiter, T.; Gottlob, G. The complexity of logic-based abduction. *J. ACM* **1995**, *42*, 3–42.
54. Ajtai, M.; Fagin, R.; Stockmeyer, L.J. The Closure of Monadic NP. *J. Comput. Syst. Sci.* **2000**, *60*, 660–716.
55. Baier, C.; Katoen, J.P. *Principles of Model Checking*; MIT Press: Cambridge, MA, USA, 2008.
56. Clarke, E.M.; Grumber, O.; Peled, D.A. *Model Checking*; MIT Press: Cambridge, MA, USA, 1999.
57. Endriss, U.; Grandi, U.; Porello, D. Complexity of Judgment Aggregation. *J. Artif. Intell. Res.* **2012**, *45*, 481–514.
58. Nehring, K.; Puppe, C. The structure of strategy-proof social choice—Part I: General characterization and possibility results on median spaces. *J. Econ. Theory* **2007**, *135*, 269–305.
59. Bäckström, C.; Nebel, B. Complexity Results for SAS+ Planning. *Comput. Intell.* **1995**, *11*, 625–656.
60. Pednault, E.P.D. ADL: Exploring the Middle Ground Between STRIPS and the Situation Calculus. In Proceedings of the 1st International Conference on Principles of Knowledge Representation and Reasoning (KR 1989), Toronto, ON, Canada, 15–18 May 1989; pp. 324–332.

© 2019 by the authors. Licensee MDPI, Basel, Switzerland. This article is an open access article distributed under the terms and conditions of the Creative Commons Attribution (CC BY) license (http://creativecommons.org/licenses/by/4.0/).

Article

Parameterised Enumeration for Modification Problems †

Nadia Creignou [1], Raïda Ktari [2], Arne Meier [3,*], Julian-Steffen Müller [3], Frédéric Olive [1] and Heribert Vollmer [3]

1. CNRS, LIS, Aix-Marseille Université, 13003 Marseille, France; nadia.creignou@univ-amu.fr (N.C.); frederic.olive@lif.univ-mrs.fr (F.O.)
2. Pôle technologique de Sfax, Université de Sfax, Sfax 3000, Tunisia; raida.ktari@isims.usf.tn
3. Institut für Theoretische Informatik, Leibniz Universität Hannover, 30167 Hannover, Germany; mueller@thi.uni-hannover.de (J.-S.M.); vollmer@thi.uni-hannover.de (H.V.)
* Correspondence: meier@thi.uni-hannover.de; Tel.: +49-(0)511-762-19768
† This paper is an extended version of our paper published in Parameterized Enumeration for Modification Problems. In Proceedings of the Language and Automata Theory and Applications—9th International Conference, Nice, France, 2–6 March 2015.

Received: 14 July 2019; Accepted: 5 September 2019; Published: 9 September 2019

Abstract: Recently, Creignou et al. (Theory Comput. Syst. 2017), introduced the class DelayFPT into parameterised complexity theory in order to capture the notion of efficiently solvable parameterised enumeration problems. In this paper, we propose a framework for parameterised ordered enumeration and will show how to obtain enumeration algorithms running with an FPT delay in the context of general modification problems. We study these problems considering two different orders of solutions, namely, lexicographic order and order by size. Furthermore, we present two generic algorithmic strategies. The first one is based on the well-known principle of self-reducibility and is used in the context of lexicographic order. The second one shows that the existence of a neighbourhood structure among the solutions implies the existence of an algorithm running with FPT delay which outputs all solutions ordered non-decreasingly by their size.

Keywords: parameterised complexity; enumeration; bounded search tree; parameterised enumeration; ordering

1. Introduction

Given a computational problem one often is interested in generating all solutions. For instance, one wants to list all answers to a query to a database [1] or is interested in all hits that a web search engine produces [2]. Even in bioinformatics [3] or computational linguistics [4] such enumeration problems play a crucial role. In this setting, one is more interested in the delay between output solutions rather than in the overall runtime of such algorithms. Here, a uniform stream of solutions is highly desired. Johnson et al. [5] explain in their seminal paper that the notion of the complexity class DelayP, which consists of problems whose delay is bounded by a polynomial in the input length, is very important.

A view on studied enumeration problems fuels the observation that often a specific order in the output solutions is very central: Many applications benefit from printing "cheap" solutions first. Moreover, enumerating all solutions in non-decreasing order allows to determine not only the smallest solution, but also the kth-smallest one. Such a generating algorithm allows finding the smallest solution obeying further constraints (at each generation step one verifies which constraints match). Unfortunately, this technique cannot guarantee efficient enumeration because a long prefix of candidates may not satisfy them. Yet, this technique is very versatile due to its applicability to any additional decidable constraint [6]. Now, we want to exemplify this observation.

Creignou and Hébrard [7] studied, within the well-known Schaefer framework for Boolean constraint satisfaction problems [8], which classes of propositional CNF formulas enumerating all satisfying solutions is possible in DelayP. They showed that for the classes of Horn, anti-Horn, affine or bijunctive formulas, such an algorithm exists. However, for any other class of formulas, having a DelayP algorithm implies P = NP. Interestingly, their proof builds on the self-reducibility of the propositional satisfiability problem. By the approach of a flashlight search, that is, trying first an assignment 0 and then 1, they observed that their enumeration algorithm obeys lexicographic order.

Later, Creignou et al. [9] studied enumerating satisfying assignments for propositional formulas in non-decreasing weight. Surprisingly, now, efficiently enumerating is only possible for Horn formulas and width-2 affine formulas (that is, affine formulas with at most two literals per clause). To achieve their result, the authors exploited priority queues to ensure enumeration in order (as was observed already by Johnson et al. [5]).

While parameterised enumeration had already been considered before (see, e.g., the works of Fernau, Damaschke and Fomin et al. [10–12]), the notion of fixed-parameter tractable delay was novel, leading to the complexity class DelayFPT [13]. Intuitively, the "polynomial time" in the definition of DelayP here is substituted by a fixed-parameter runtime-bound of the form $n^{O(1)} \cdot f(k)$, where n denotes the input length, k is the input parameter and f is a computable function. This introduces the notion of efficiency in the context of the parameterised world, that is, fixed-parameter tractability (FPT), to the enumeration framework. Creignou et al. [13] investigated a wealth of problems from propositional logic and developed enumeration algorithms based on self-reducibility and on the technique of kernelisation. Particularly, the membership of an enumeration problem in DelayFPT can be characterised by a specifically tailored form of kernelisability, very much as in the context of usual decision problems.

As this area of parameterised enumeration is rather young and has received less attention, we want to further support this topic with this paper. Here, we study ordered enumeration in the context of parameterised complexity. First, we introduce arbitrary orders to the parameterised enumeration framework. Then we consider the special context of graph modification problems where we are interested in ordered enumeration for the two mostly studied orders, namely by lexicographic and by non-decreasing size (where the size is the number of modifications that have to be made). We use two algorithmic strategies, depending on the respective order as follows. Based on the principle of self-reducibility we obtain DelayFPT (and polynomial-space) enumeration algorithms for lexicographic order, as soon as the decision problem is efficiently solvable. Secondly, we present a DelayFPT enumeration algorithm for order by size as soon as a certain FPT-computable neighbourhood function on the solutions set exists (see Theorem 1). Notice that the presented algorithms do not enumerate the set of minimal solutions but the set of solutions of bounded size. Extending to such solutions from minimal ones in the enumeration process is not generally trivial. To cope with the order, we use a priority queue that may require exponential space in the input length (as there exist potentially that many solutions).

Eventually, we show that the observed principles and algorithmic strategies can be applied to general modification problems as well. For instance, a general modification problem could allow to flip bits of a string. It is a rather rare situation that a general algorithmic scheme is developed. Usually algorithms are devised on a very individual basis. We prove a wide scope of applicability of our method by presenting new FPT delay ordered enumeration algorithms for a large variety of problems, such as cluster editing [14], triangulation [15], triangle deletion [16], closest-string [17] and backdoor sets [18]. Furthermore, there already exists work which adopts the introduced framework of Creignou et al. [13] in the area of conjunctive query enumeration [19], triangle enumeration [20], combinatorial optimisation [21], abstract argumentation [22] and global constraints [23].

2. Preliminaries

We start by defining parameterised enumeration problems with a specific ordering and their corresponding enumeration algorithms. Most definitions in this section transfer those of Johnson et al. and Schmidt [5,24] from the context of enumeration and those of Creignou et al. [13] from the context of parameterised enumeration to the context of parameterised ordered enumeration.

The studied orderings of enumeration problems in this paper are quasi-orders which will be defined in the following.

Definition 1 (Quasi-Order). *Let R be a set and $\preceq$ a binary relation on R. Then $\preceq$ is a preorder (or quasi-order) if we have for all elements $a, b, c \in R$:*

- $a \preceq a$ and
- *if $a \preceq b$ and $b \preceq c$ then $a \preceq c$.*

We will write $z \not\preceq y$ whenever $z \preceq y$ is not true.

Now, we proceed by introducing parameterised enumeration problems with ordering. Intuitively, the corresponding enumeration algorithm for such problems has to obey the given ordering, that is, it has to produce solutions without violating that ordering.

Definition 2. *A parameterised enumeration problem with ordering is a quadruple $E = (I, \kappa, \text{Sol}, \preceq)$ such that the following holds:*

- *I is the set of instances.*
- *$\kappa \colon I \to \mathbb{N}$ is the parameterisation function; κ is required to be polynomial time computable.*
- *Sol is a function such that for all $x \in I$, $\text{Sol}(x)$ is a finite set, the set of solutions of x. Further we write $\mathcal{S} = \bigcup_{x \in I} \text{Sol}(x)$.*
- *$\preceq$ is a quasi-order on $\mathcal{S}$.*

Notice that this order on all solutions is only a short way of simultaneously giving an order for each instance. Furthermore, we will write an index E letter, e.g., I_E, κ_E, to denote that we are talking about an instance set, parameterisation function, etc., of a given enumeration problem E. In the next step, we fix the notion of enumeration algorithms in our setting.

Definition 3 (Enumeration Algorithm). *Let $E = (I, \kappa, \text{Sol}, \preceq)$ be a parameterised enumeration problem with ordering. Then an algorithm $\mathcal{A}$ is an enumeration algorithm for E if the following holds:*

- *For every $x \in I$, $\mathcal{A}(x)$ terminates after a finite number of steps.*
- *For every $x \in I$, $\mathcal{A}(x)$ outputs exactly the elements of $\text{Sol}(x)$ without duplicates.*
- *For every $x \in I$ and $y, z \in \text{Sol}(x)$, if $y \preceq z$ and $z \not\preceq y$ then $\mathcal{A}(x)$ outputs solution y before solution z.*

Before we define complexity classes for parameterised enumeration, we need the notion of delay for enumeration algorithms.

Definition 4 (Delay). *Let $E = (I, \kappa, \text{Sol}, \preceq)$ be a parameterised enumeration problem with ordering and $\mathcal{A}$ be an enumeration algorithm for E. Let $x \in I$ be an instance. The ith delay of $\mathcal{A}$ is the elapsed runtime with respect to $|x|$ of $\mathcal{A}$ between outputting the ith and $(i+1)$th solution in $\text{Sol}(x)$. The 0th delay is the precomputation time which is the elapsed runtime with respect to $|x|$ of $\mathcal{A}$ from the start of the computation to the first output statement. Analogously, the nth delay, for $n = |\text{Sol}(x)|$, is the postcomputation time, which is the elapsed runtime with respect to $|x|$ of $\mathcal{A}$ after the last output statement until $\mathcal{A}$ terminates. Then, the delay of $\mathcal{A}$ is the maximum over all $0 \leq i \leq n$ of the ith delay of $\mathcal{A}$.*

Now we are able to define two different complexity classes for parameterised enumeration following the notion of Creignou et al. [13].

Definition 5. *Let $E = (I, \kappa, \mathrm{Sol}, \preceq)$ be a parameterised enumeration problem. We say that E is* FPT *enumerable if there exists an enumeration algorithm $\mathcal{A}$, a computable function $f \colon \mathbb{N} \to \mathbb{N}$ and a polynomial p such that for every $x \in I$, $\mathcal{A}$ outputs all solutions of $\mathrm{Sol}(x)$ in time $f(\kappa(x)) \cdot p(|x|)$.*

An enumeration algorithm $\mathcal{A}$ is a DelayFPT *algorithm if there exists a computable function $f \colon \mathbb{N} \to \mathbb{N}$ and a polynomial p such that for every $x \in I$, $\mathcal{A}$ outputs all solutions of $\mathrm{Sol}(x)$ with delay of at most $f(\kappa(x)) \cdot p(|x|)$.*

The class DelayFPT *consists of all parameterised enumeration problems that admit a DelayFPT-enumeration algorithm.*

Some of our enumeration algorithms will make use of priority queues to enumerate all solutions in the correct order and to avoid duplicates. We will follow the approach of Johnson et al. [5]. For an instance x of a parameterised enumeration problem whose sizes of solutions are polynomially bounded in $|x|$; we use a priority queue Q to store a subset of $\mathrm{Sol}(x)$ of cardinality potentially exponential in $|x|$. The insert operation of Q requires $O(|x| \cdot \log |\mathrm{Sol}(x)|)$ time. The extract minimum operation requires $O(|x| \cdot \log |\mathrm{Sol}(x)|)$ time, too. It is important, however, that the computation of the order between two elements takes at most $O(|x|)$ time. As pointed out by Johnson et al., the required queue can be implemented with the help of standard balanced tree schemes [25].

2.1. Graph Modification Problems

Graph modification problems have been studied for a long time in computational complexity theory [26]. Already in the monograph by Garey and Johnson [27], among the graph-theoretic problems considered, many fall into this problem class. To the best of our knowledge, graph modification problems were studied in the context of parameterised complexity for the first time in [28].

In this paper, we consider only undirected graphs. Let $\mathcal{G}$ denote the set of all undirected graphs. A graph property $\mathcal{P} \subseteq \mathcal{G}$ is a set of graphs. Given a graph property $\mathcal{P}$ and an undirected graph G, we write $G \models \mathcal{P}$ if the graph G obeys the property $\mathcal{P}$, that is, $G \in \mathcal{P}$.

Definition 6 (Graph Operations). *A graph operation for G is either of the following:*

- *removing a vertex: A function $\mathrm{rem}_v \colon \mathcal{G} \to \mathcal{G}$ such that $\mathrm{rem}_v(G)$ is the graph obtained by removing the vertex v from G (if v is present; otherwise rem_v is the identity) and deleting all incident edges to v,*
- *adding/removing an edge: A function $\mathrm{add}_{\{u,v\}}, \mathrm{rem}_{\{u,v\}} \colon \mathcal{G} \to \mathcal{G}$ such that $\mathrm{add}_{\{u,v\}}(G), \mathrm{rem}_{\{u,v\}}(G)$ is the graph obtained by adding/removing the edge $\{u, v\}$ to G if u and v are present in G; otherwise both functions are the identity*

Two operations o, o' are dependent if

- *$o = \mathrm{rem}_v$ and $o' = \mathrm{rem}_{\{u,v\}}$ (o removes the vertex v and o' removes an edge incident to v) or*
- *$o = \mathrm{rem}_{\{u,v\}}$ and $o' = \mathrm{add}_{\{u,v\}}$ (o removes the edge $\{u, v\}$ and o' adds the same edge $\{u, v\}$ again).*

A set of operations is consistent if it does not contain two dependent operations. Given such a consistent set of operations S, the graph obtained from G by applying the operations in S on G is denoted by $S(G)$.

Now, we turn towards the definition of solutions and will define minimality in terms of being inclusion-minimal.

Definition 7 (Solutions). *Given a graph property $\mathcal{P}$, a graph G, $k \in \mathbb{N}$ and a set of operations O, we say that S is a solution for (G, k, O) with respect to $\mathcal{P}$ if the following three properties hold:*

1. *$S \subseteq O$ is a consistent set of operations,*
2. *$|S| \leq k$ and*
3. *$S(G) \models \mathcal{P}$.*

A solution S is minimal if there is no solution S' such that $S' \subsetneq S$.

Cai [28] was interested in the following parameterised graph modification decision problem with respect to a given graph property $\mathcal{P}$:

Problem:	$\mathcal{M}_\mathcal{P}$
Input:	(G, k, O), G undirected graph, $k \in \mathbb{N}$, O set of operations on G.
Parameter:	The integer k.
Question:	Does there exist a solution for (G, k, O) with respect to $\mathcal{P}$?

Some of the most important examples of graph modification problems are presented now. A chord in a graph $G = (V, E)$ is an edge between two vertices of a cycle C in G which is not part of C. A given graph $G = (V, E)$ is triangular (or chordal) if each of its induced cycles of four or more nodes has a chord. The problem TRIANGULATION then asks, given an undirected graph G and $k \in \mathbb{N}$, whether there exists a set of at most k edges such that adding this set of edges to G makes it triangular. Yannakakis showed that this problem is NP complete [15]. Kaplan et al. [29] and independently Cai [28] have shown that the parameterised problem is in FPT. For this problem, a solution is a set of edges which have to be added to the graph to make the graph triangular. Observe that, in this special case of the modification problem, the underlying property $\mathcal{P}$, "to be triangular", does not have a finite forbidden set characterisation (since cycles of any length are problematic). Nevertheless, we will see later, that one can efficiently enumerate all minimal solutions as well.

A cluster is a graph such that all its connected components are cliques. In order to transform (or modify) a graph G we allow here only two kinds of operations: Adding or removing an edge. CLUSTER-EDITING asks, given a graph G and a parameter k, whether there exists a consistent set of operations of cardinality at most k such that $S(G)$ is cluster. It was shown by Shamir et al., that the problem is NP complete [14].

The problem TRIANGLE-DELETION asks whether a given graph can be transformed into a triangle-free graph by deletion of at most k vertices. Yannakakis has shown that the problem is NP complete [16].

Analogous problems can be defined for many other classes of graphs, e.g., line graphs, claw-free graphs, Helly circular-arc graphs, etc., see [30].

Now, we turn towards the main focus of the paper. Here, we are interested in corresponding enumeration problems with ordering. In particular, we will focus on two well-known preorders, lexicographic ordering and ordering by size. Since our solutions are subsets of an ordered set of operations, they can be encoded as binary strings in which the ith bit from right indicates whether the ith operation is in the subset. We define the lexicographic ordering of solutions as the lexicographic ordering of these strings. Then, the size of a solution simply is its cardinality.

Problem:	ENUM-$\mathcal{M}_\mathcal{P}^{\text{LEX}}$
Input:	(G, k, O), G undirected graph, $k \in \mathbb{N}$, O ordered set of operations on G.
Parameter:	The integer k.
Output:	All solutions of (G, k, O) with respect to $\mathcal{P}$ in lexicographic order.

Problem:	ENUM-$\mathcal{M}_\mathcal{P}^{\text{SIZE}}$
Input:	(G, k, O), G undirected graph, $k \in \mathbb{N}$, O set of operations on G.
Parameter:	The integer k.
Output:	All solutions of (G, k, O) with respect to $\mathcal{P}$ in non-decreasing size.

If the context is clear, we omit the subscript $\mathcal{P}$ for the graph modification problem and simply write $\mathcal{M}$. Furthermore, we write $\text{Sol}_\mathcal{M}(x)$ for the function associating solutions to a given instance, and also $\mathcal{S}_\mathcal{M}$ for the set of all solutions of $\mathcal{M}$.

3. Enumeration of Graph Modification Problems with Ordering

In this section, we study the two previously introduced parameterised enumeration problems with ordering (lexicographic and size ordering).

3.1. Lexicographic Ordering

We first prove that, for any graph property $\mathcal{P}$, if the decision problem $\mathcal{M}_\mathcal{P}$ is in FPT then there is an efficient enumeration algorithm for ENUM-$\mathcal{M}_\mathcal{P}^{\text{LEX}}$.

Lemma 1. *Let $\mathcal{M}_\mathcal{P}$ be a graph modification problem. If $\mathcal{M}_\mathcal{P}$ is in* FPT *then* ENUM-$\mathcal{M}_\mathcal{P}^{\text{LEX}} \in$ DelayFPT *with polynomial space.*

Proof. Algorithm 1 enumerates all solutions of an instance of a given modification problem $\mathcal{M}_\mathcal{P}$ by the method of self-reducibility (it is an extension of the flash light search of Creignou and Hébrard [7]). The algorithm uses a function ExistsSol(G, k, O) that tests if the instance (G, k, O) of the modification problem $\mathcal{M}_\mathcal{P}$ has a solution. By the assumption of the lemma, $\mathcal{M}_\mathcal{P} \in$ FPT so this function runs in FPT time. We use calls to this function to avoid exploration of branches of the recursion tree that do not lead to any output. Moreover, we ensure that the solutions using o_p have to be consistent. This consistency check runs in polynomial time for graph operations. The rest yields a search tree of depth at most k. From this it follows that, for any instance of length n, the time beween the output of any two solutions is bounded by $f(k) \cdot p(n)$ for some polynomial p and a computable function f. □

Algorithm 1: Enumerate all solutions of $\mathcal{M}_\mathcal{P}$ in lexicographic order

Input: (G, k, O): A graph G, $k \in \mathbb{N}$, an ordered set of operations $O = \{o_1, \ldots, o_n\}$
Output: all consistent sets $S \subseteq O$ s.t. $|S| \leq k$, $S(G) \models \mathcal{P}$ in lexicographic order

1 **if** ExistsSol(G, k, O) **then** Generate$(G, k, O, \emptyset)$;

Procedure Generate(G, k, O, S):
1 **if** $O = \emptyset$ or $k = 0$ **then return** S;
2 **else**
3   let o_p be the lexicographically last operation in O, $O' := O \setminus \{o_p\}$;
4   **if** ExistsSol$(S(G), k, O')$ **then** Generate$(S(G), k, O', S)$;
5   **if** $S \cup \{o_p\}$ is consistent and ExistsSol$((S \cup \{o_p\})(G), k-1, O')$ **then**
6   Generate$((S \cup \{o_p\})(G), k-1, O', S \cup \{o_p\})$.

Corollary 1. ENUM-TRIANGULATION$^{\text{LEX}} \in$ DelayFPT *with polynomial space.*

Proof. Kaplan et al. [29] and Cai [28] showed that TRIANGULATION $\in$ FPT. Now, by applying Lemma 1, we get the result. □

Cai [28] identified a class of graph properties whose associated modification problems belong to FPT. Let us introduce some terminology.

Definition 8. *Given two graphs $G = (V, E)$ and $H = (V', E')$, we write $H \trianglelefteq G$ if H is an induced subgraph of G, i.e., $V' \subseteq V$ and $E' = E \cap (V' \times V')$. Let $\mathcal{F}$ be a set of graphs and $\mathcal{P}$ be a graph property. We say that $\mathcal{F}$ is a forbidden set characterisation of $\mathcal{P}$ if for any graph G it holds that: $G \models \mathcal{P}$ iff for all $H \in \mathcal{F}, H \ntrianglelefteq G$.*

Among the problems presented in the previous section (see page 5), TRIANGLE-DELETION and CLUSTER-EDITING have a finite forbidden set characterisation, namely by triangles and paths of length two. In contrast to that, TRIANGULATION has a forbidden set characterisation which is infinite, since cycles of arbitrary length are problematic. Actually, for properties having a finite forbidden set characterisation, the corresponding modification problem is fixed-parameter tractable. Together with Lemma 1, this provides a positive result in terms of enumeration.

Proposition 1 ([28]). *If a property $\mathcal{P}$ has a finite forbidden set characterisation then $\mathcal{M}_\mathcal{P}$ is in FPT.*

Corollary 2. *For any graph modification problem, if $\mathcal{P}$ has a finite forbidden set characterisation then* ENUM-$\mathcal{M}_\mathcal{P}^{\text{LEX}} \in$ DelayFPT *with polynomial space.*

Proof. This result follows by combining Proposition 1 with Lemma 1. □

3.2. Size Ordering

A common strategy in the enumeration context consists of defining a notion of a neighbourhood that allows to compute a new solution from a previous one with small amounts of computation time (see, e.g., the work of Avis and Fukuda [31]). We introduce the notion of a neighbourhood function, which, roughly speaking, generates some initial solutions from which all solutions can be produced. A priority queue then takes care of the ordering and avoids duplicates, which may require exponential space. For the graph modification problems of interest, we show that if the inclusion-minimal solutions can be generated in FPT, then such a neighbourhood function exists, accordingly providing a DelayFPT-enumeration algorithm. In the following, $\mathbb{O}$ (the "seed") is a technical symbol that will be used to generate the initial solutions.

Definition 9. *Let $\mathcal{M}$ be a graph modification problem. A neighbourhood function for $\mathcal{M}$ is a (partial) function $\mathcal{N}_\mathcal{M}: I_\mathcal{M} \times (\mathcal{S}_\mathcal{M} \cup \{\mathbb{O}\}) \to 2^{\mathcal{S}_\mathcal{M}}$ such that the following holds:*

1. *For all $x = (G, k, O) \in I_\mathcal{M}$ and $S \in \text{Sol}_\mathcal{M}(x) \cup \{\mathbb{O}\}$, $\mathcal{N}_\mathcal{M}(x, S)$ is defined.*
2. *For all $x \in I_\mathcal{M}$, $\mathcal{N}_\mathcal{M}(x, \mathbb{O}) = \emptyset$ if $\text{Sol}_\mathcal{M}(x) = \emptyset$, and $\mathcal{N}_\mathcal{M}(x, \mathbb{O})$ is an arbitrary set of solutions otherwise.*
3. *For all $x \in I_\mathcal{M}$ and $S \in \text{Sol}_\mathcal{M}(x)$, if $S' \in \mathcal{N}_\mathcal{M}(x, S)$ then $|S| < |S'|$.*
4. *For all $x \in I_\mathcal{M}$ and all $S \in \text{Sol}_\mathcal{M}(x)$, there exists $p > 0$ and $S_1, \ldots, S_p \in \text{Sol}_\mathcal{M}(x)$ such that (i) $S_1 \in \mathcal{N}_\mathcal{M}(x, \mathbb{O})$, (ii) $S_{i+1} \in \mathcal{N}_\mathcal{M}(x, S_i)$ for $1 \leq i < p$ and (iii) $S_p = S$.*

Furthermore, we say that $\mathcal{N}_\mathcal{M}$ is FPT computable, when $\mathcal{N}_\mathcal{M}(x, S)$ is computable in time $f(\kappa(x)) \cdot \text{poly}(|x|)$ for any $x \in I_\mathcal{M}$ and $S \in \text{Sol}_\mathcal{M}(x)$.

As a result, a neighbourhood function for a problem $\mathcal{M}$ is a function that in a first phase computes from scratch an initial set of solutions (see Definition 9(2)). In many of our applications below, $\mathcal{N}_\mathcal{M}(x, \mathbb{O})$ will be the set of all minimal solutions for x. In a second phase these solutions are iteratively extended (see condition (3)), where condition (4) guarantees that we do not miss any solution, as we will see in the next theorem.

Theorem 1. *Let $\mathcal{M}$ be a graph modification problem. If $\mathcal{M}$ admits a neighbourhood function $\mathcal{N}_\mathcal{M}$ that is FPT-computable, then* ENUM-$\mathcal{M}^{\text{SIZE}} \in$ DelayFPT.

Proof. Algorithm 2 outputs all solutions in DelayFPT time. By the definition of the priority queue (recall in particular that insertion of an element is done only if the element is not yet present in the queue) and by the fact that all elements of $\mathcal{N}_\mathcal{M}((G, k, O), S)$ are of bigger size than S by Definition 9(3), it is easily seen that the solutions are output in the right order and that no solution is output twice.

Besides, no solution is omitted. Indeed, given $S \in \text{Sol}_\mathcal{M}(G, k, O)$ and $S_1, \ldots, S_p$ associated with S by Definition 9(4), we prove by induction that each S_i is inserted in Q during the run of the algorithm:

$i = 1$: This proceeds from line 2 of the algorithm.
$i > 1$: The solution S_{i-1} is inserted in Q by induction hypothesis and hence all elements of $\mathcal{N}_\mathcal{M}((G, k, O), S_{i-1})$, including S_i, are inserted in Q (line 5 of Algorithm 2). Consequently, each S_i is inserted in Q and then output during the run. In particular, this is true for $S = S_p$.

Finally, we claim that Algorithm 2 (see Figure 1 for a graphical representation) runs in DelayFPT time. Indeed, the delay between the output of two consecutive solutions is bounded by the time required to compute a neighbourhood of the form $\mathcal{N}_{\mathcal{M}}((G,k,O), \mathbb{O})$ or $\mathcal{N}_{\mathcal{M}}((G,k,O), S)$ and to insert all its elements in the priority queue. This is in FPT due to the assumption on $\mathcal{N}_{\mathcal{M}}$ being FPT computable and as there is only a single extraction and many FPT insertion operations in the queue. □

Algorithm 2: DelayFPT algorithm for ENUM-$\mathcal{M}$

Input : (G, k, O) : G is an undirected graph, $k \in \mathbb{N}$, and O is a set of operations.
1 compute $\mathcal{N}_{\mathcal{M}}((G, k, O), \mathbb{O})$;
2 insert all elements of $\mathcal{N}_{\mathcal{M}}((G, k, O), \mathbb{O})$ into priority queue Q (ordered by size);
3 **while** Q is not empty **do**
4   **extract** the minimum solution S of Q and output it;
5   **insert** all elements of $\mathcal{N}_{\mathcal{M}}((G, k, O), S)$ into Q;

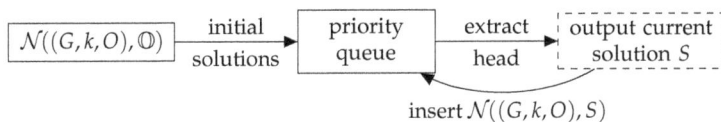

Figure 1. Structure of Algorithm 2.

A natural way to provide a neighbourhood function for a graph modification problem $\mathcal{M}$ is to consider the inclusion minimal solutions of $\mathcal{M}$. Let us denote by MIN-$\mathcal{M}$ the problem of enumerating all inclusion minimal solutions of $\mathcal{M}$.

Theorem 2. *Let $\mathcal{M}$ be a graph modification problem. If MIN-$\mathcal{M}$ is FPT enumerable then ENUM-$\mathcal{M}^{SIZE} \in$ DelayFPT.*

Proof. Let $\mathcal{A}$ be an FPT algorithm for MIN-$\mathcal{M}$. Because of Theorem 1, it is sufficient to build an FPT-neighbourhood function for $\mathcal{M}$. For an instance (G, k, O) of $\mathcal{M}$ and for $S \in \text{Sol}_{\mathcal{M}}(G, k, O) \cup \{\mathbb{O}\}$, we define $\mathcal{N}_{\mathcal{M}}((G, k, O), S)$ as the result of Algorithm 3.

Algorithm 3: Procedure for computing $\mathcal{N}_{\mathcal{M}}((G, k, O), S)$

Input : $(G, k, O), S$: G is an undirected graph, $k \in \mathbb{N}$, O and S are sets of operations.
1 **if** $S = \mathbb{O}$ **then return** $\mathcal{A}(G, k, O)$;
2 res := $\emptyset$;
3 **for all** $t \in O$ **do**
4   **for all** $S' \in \mathcal{A}((S \cup \{t\})(G), k - |S| - 1, O \setminus \{t\})$ **do**
5   **if** $S \cup S' \cup \{t\}$ is consistent **then** res := res $\cup \{S \cup S' \cup \{t\}\}$;
6 **return** res;

Accordingly, the function $\mathcal{N}_{\mathcal{M}}$ clearly fulfils Conditions 2 and 3 of Definition 9. We prove by induction that it also satisfies Condition 4 (that is, each solution T of size k comes with a sequence $T_1, \ldots, T_p = T$ such that $T_1 \in \mathcal{N}_{\mathcal{M}}((G, k, O), \mathbb{O})$ and $T_{i+1} \in \mathcal{N}_{\mathcal{M}}((G, k, O), T_i)$ for each i). If T is a minimal solution for (G, k, O), then $T \in \mathcal{N}_{\mathcal{M}}((G, k, O), \mathbb{O})$ and the expected sequence $(T_i)_{1 \leq i \leq p}$ reduces to $T_1 = T$. Otherwise, there exists an $S \in \text{Sol}_{\mathcal{M}}(G, k, O)$ and a non-empty set of transformations, say $S' \cup \{t\}$, such that $T = S \cup S' \cup \{t\}$ and there is no solution for G between S and $S \cup S' \cup \{t\}$. This entails that S' is a minimal solution for $((S \cup \{t\})(G), k - |S| - 1)$ and, as a consequence, $T \in \mathcal{N}_{\mathcal{M}}((G, k, O), S)$ (see lines 4–5 of Algorithm 3). The conclusion follows from the induction hypothesis that guarantees the existence of solutions $S_1, \ldots, S_q$ such that $S_1 \in \mathcal{N}_{\mathcal{M}}((G, k, O), \mathbb{O})$,

$S_{i+1} \in \mathcal{N}_\mathcal{M}((G,k,O), S_i)$ and $S_q = S$. The expected sequence $T_1, \ldots, T_p$ for T is nothing but $S_1, \ldots, S_q, T$. To conclude, it remains to show that Algorithm 3 is FPT. This follows from the fact that $\mathcal{A}$ is an FPT algorithm (Lines 1 and 4 of Algorithm 3). □

Corollary 3. ENUM-TRIANGULATION$^{\text{SIZE}} \in$ DelayFPT.

Proof. All inclusion-minimal k-triangulations can be output in time $O(2^{4k} \cdot |E|)$ for a given graph G and $k \in \mathbb{N}$ as shown by Kaplan et al. [29] (Theorem 2.4). This immediately yields the expected result via Theorem 2. □

Corollary 4. *For any property $\mathcal{P}$ that has a finite forbidden set characterisation, the problem* ENUM-$\mathcal{M}_\mathcal{P}^{\text{SIZE}}$ *is in* DelayFPT.

Proof. The algorithm developed by Cai [28] for the decision problem is based on a bounded search tree, whose exhaustive examination provides all size minimal solutions in FPT. Theorem 2 yields the conclusion. □

Corollary 5. ENUM-CLUSTER-EDITING^{SIZE} *and* ENUM-TRIANGLE-DELETION^{SIZE} *are in* DelayFPT.

Proof. Both problems have a finite forbidden set characterisation. For the cluster editing problem, paths of length two are the forbidden pattern, and, regarding ENUM-TRIANGLE-DELETION^{SIZE}, the forbidden patterns are triangles. The result then follows from Corollary 4. □

4. Generalisation to Modification Problems

In this section, we will show how the algorithmic strategy that has been defined and formalised in the context of graph modification can be of use for many other problems, coming from various combinatorial frameworks.

Definition 10 (General Operations). *Let $Q \subseteq \Sigma^*$ be a language defined over an alphabet and let $x \in \Sigma^*$ be an input. A set of operations $\Omega(Q) = \{\omega_n \colon \Sigma^* \to \Sigma^* \mid n \in \mathbb{N}\}$ is an infinite set of operations on instances of Q. We say that an operation ω is valid with respect to an instance $x \in Q$, if $\omega(x) \in Q$. We write Ω/x for the set of possible (valid) operations on an instance x.*

Two operations ω, ω' are dependent with respect to an instance $x \in Q$ if

- $\omega(\omega'(x)) = x$ or (intuitively, ω and ω' cancel out)
- $\omega(\omega'(x)) = \omega'(x)$ or $\omega(\omega'(x)) = \omega(x)$. (intuitively, the order of ω and ω' does not matter)

A set of operations $O \subseteq \Omega/x$ is consistent with respect to x if it does not contain two dependent operations. Similarly, we say that an operation ω is consistent with a set S if and only if $S \cup \{\omega\}$ is consistent.

For instance, the set Ω could contain operations that add edges or, in another case, flip bits. How exactly Ω is defined highly depends on the corresponding language Q.

Example 1. *Let $\mathcal{G} \subseteq \{0,1\}^*$ be the language of all undirected graphs encoded by adjacency matrices. Then $\Omega(\mathcal{G})$ is the set of all graph operations in the sense of Definition 6: Removing vertices or edges, adding edges. Note that $\Omega(\mathcal{G})$ contains all operations of the kind*

$$\text{rem}_i \colon \mathcal{G} \to \mathcal{G}, \quad \text{rem}_{\{i,j\}} \colon \mathcal{G} \to \mathcal{G}, \quad \text{add}_{\{i,j\}} \colon \mathcal{G} \to \mathcal{G}$$

for all $i, j \in \mathbb{N}$. Furthermore, let $G = (V, E) \in \{0,1\}^*$ be a concrete input graph. As a result, Ω/G then is the restriction of Ω to those $i, j \in \mathbb{N}$ such that $v_i, v_j \in V$ encode vertices in G.

Similarly to how it was defined in Section 2.1, a property is just a set. In the following context, it is a subset of a considered language Q. Intuitively, in the concept of graph modification problems, one may think of Q as $\mathcal{G}$. Then a graph property $\mathcal{P}$ is just a subset of $\mathcal{G}$.

Definition 11 (General Solutions). *Let $Q \subseteq \Sigma^*$ be a language defined over an alphabet, $S \subseteq \Omega/x$ be a finite set of operations on $x \in Q$ and $\mathcal{P} \subseteq Q$ be a property. We say that S is a solution (of x) if S is a consistent set of operations and $S(x) \in \mathcal{P}$. Furthermore, we denote by $\mathcal{S}_Q := \bigcup_{x \in Q} \{ S \mid S \text{ is a solution of } x \}$ the set of all solutions for every instance $x \in Q$. In addition, $\mathrm{Sol}(x)$ is the set of solutions for every instance $x \in Q$.*

Example 2. *Continuing the previous example, if the property $\mathcal{P}$ is "to be a cluster" then a consistent solution S to a given graph is a sequence of removing vertices, adding and deleting of edges where*

- there is no edge (i, j) added or deleted such that vertex i or j is removed,
- there is no edge (i, j) added and removed and
- $S(G) \models \mathcal{P}$.

Similarly, adding edge (i, j) together with removing vertex i or j or removing edge (i, j) is an inconsistent set of operations.

Now we want to define the corresponding decision and enumeration tasks. On that account, let $\mathcal{P}$ be a property, $\Pi = (Q, \kappa)$ be a parametrised problem with $Q \subseteq \Sigma^*$ and Ω be a set of operations.

Problem:	$\Pi_\mathcal{P}$—parametrised modification problem Π w.r.t. a property $\mathcal{P}$ over Σ
Input:	$x \in \Sigma^*, k \in \mathbb{N}, \Omega/x$ set of operations.
Parameter:	The integer k.
Question:	Is there a solution $S \subseteq \Omega/x$ and $|S| \leq k$?

Problem:	ENUM-MIN-$\Pi_\mathcal{P}$—parametrised minimum enumeration modification problem w.r.t. a property $\mathcal{P}$ over Σ
Input:	$x \in \Sigma^*, k \in \mathbb{N}, \Omega/x$ set of operations.
Parameter:	The integer k.
Output:	All minimal (w.r.t. some order) solutions $S \subseteq \Omega/x$ with $|S| \leq k$.

The enumeration modification problem where we want to output all possible sets of transformations on a given instance x (and not only the minimum ones) then is ENUM-$\Pi_\mathcal{P}$.

In the following, we show how the notion of neighbourhood functions can be generalised as well. This will in turn yield generalisations of the results for graph modification problems afterwards.

Definition 12. *Let Σ be an alphabet, $\mathcal{P} \subseteq \Sigma^*$ be a property and $\Pi_\mathcal{P}$ be a parametrised modification problem over Σ. A neighbourhood function for $\Pi_\mathcal{P}$ is a (partial) function $\mathcal{N}_{\Pi_\mathcal{P}} \colon \Sigma^* \times (\mathcal{S}_{\Pi_\mathcal{P}} \cup \{\mathbb{O}\}) \to 2^{\mathcal{S}_{\Pi_\mathcal{P}}}$ such that the following holds:*

1. *For all $x \in \Sigma^*$ and $S \in \mathrm{Sol}_{\Pi_\mathcal{P}}(x) \cup \{\mathbb{O}\}$, $\mathcal{N}_{\Pi_\mathcal{P}}(x, S)$ is defined.*
2. *For all $x \in \Sigma^*$, $\mathcal{N}_{\Pi_\mathcal{P}}(x, \mathbb{O}) = \emptyset$ if $\mathrm{Sol}_{\Pi_\mathcal{P}}(x) = \emptyset$, and $\mathcal{N}_{\Pi_\mathcal{P}}(x, \mathbb{O})$ is an arbitrary set of solutions otherwise.*
3. *For all $x \in \Sigma^*$ and $S \in \mathrm{Sol}_{\Pi_\mathcal{P}}(x)$, if $S' \in \mathcal{N}_{\Pi_\mathcal{P}}(x, S)$ then $|S| < |S'|$.*
4. *For all $x \in \Sigma^*$ and all $S \in \mathrm{Sol}_{\Pi_\mathcal{P}}(x)$, there exists $p > 0$ and $S_1, \ldots, S_p \in \mathrm{Sol}_{\Pi_\mathcal{P}}(x)$ such that (i) $S_1 \in \mathcal{N}_{\Pi_\mathcal{P}}(x, \mathbb{O})$, (ii) $S_{i+1} \in \mathcal{N}_{\Pi_\mathcal{P}}(x, S_i)$ for $1 < i < p$ and (iii) $S_p = S$.*

Furthermore, we say that $\mathcal{N}_{\Pi_\mathcal{P}}$ is FPT computable when $\mathcal{N}_{\Pi_\mathcal{P}}(x, S)$ is computable in time $f(k) \cdot \mathrm{poly}(|x|)$ for any $x \in \Sigma^*$ and $S \in \mathrm{Sol}_{\Pi_\mathcal{P}}(x)$.

As already announced before, we are able to state generalised versions of Theorems 1 and 2 which can be proven in a similar way. However, one has to replace the graph modification problems by general modification problems.

Corollary 6. *Let $\mathcal{P}$ be a property, $\Pi \subseteq \Sigma^* \times \mathbb{N}$ be a parameterised modification problem and Ω be a set of operations such that Ω/x is finite for all $x \in \Sigma^*$. If $\Pi_{\mathcal{P}}$ admits a neighbourhood function that is FPT computable then* ENUM-$\Pi_{\mathcal{P}} \in$ DelayFPT *using*

- *polynomial space for lexicographic order and*
- *exponential space for size order.*

Corollary 7. *Let $\mathcal{P}$ be a property, $\Pi \subseteq \Sigma^* \times \mathbb{N}$ be a parameterised modification problem and Ω be a set of operations such that Ω/x is finite for all $x \in \Sigma^*$. If* ENUM-MIN-$\Pi_{\mathcal{P}}$ *is FPT enumerable and the consistency of solutions can be checked in FPT then* ENUM-$\Pi_{\mathcal{P}} \in$ DelayFPT *and using*

- *polynomial space for lexicographic order and*
- *exponential space for size order.*

4.1. Closest String

In the following, we consider a central NP complete problem in coding theory [32]. Given a set of binary strings I, we want to find a string s with maximum Hamming distance $\max\{d_H(s, s') \mid s' \in I\} \leq d$ for a $d \in \mathbb{N}$, where $d_H(s, s')$ is the Hamming distance between two strings.

Definition 13 (Bit-Flip operation). *Given a string $w = w_1 \cdots w_n$ with $w_i \in \{0, 1\}, n \in \mathbb{N}$ and a set $S \subseteq \{1, \ldots, n\}$, $S(w)$ denotes the string obtained from w by flipping the bits indicated by S, more formally $S(w) := S(w_1) \cdots S(w_n)$, where $S(w_i) = 1 - w_i$ if $i \in S$ and $S(w_i) = w_i$ otherwise.*

The corresponding parametrised version is the following.

Problem:	CLOSEST-STRING
Input:	A sequence $(s_1, s_2, \ldots, s_k)$ of k strings over $\{0, 1\}$ each of given length $n \in N$ and an integer $d \in N$.
Parameter:	The integer d.
Question:	Does there exist $S \subseteq \{1, \ldots, n\}$ such that $d_H(S(s_1), s_i) \leq d$ for all $1 \leq i \leq k$?

Proposition 2 ([17]). CLOSEST-STRING *is in* FPT.

Moreover, an exhaustive examination of a bounded search tree constructed from the idea of Gramm et al. [17] (Figure 1), allows to produce all minimal solutions of this problem in FPT. Accordingly, we get the following result for the corresponding enumeration problems.

Theorem 3.
- ENUM-CLOSEST-STRING$^{\text{LEX}} \in$ DelayFPT *with polynomial space.*
- ENUM-CLOSEST-STRING$^{\text{SIZE}} \in$ DelayFPT *with exponential space.*

Proof. Ω is just the set of operations which flip the ith bit of a string for every $i \in \mathbb{N}$. By combining Proposition 2 with Corollary 7 we get the desired result. □

4.2. Backdoors

In this section, we will consider the concept of backdoors. Let $\mathcal{C}$ be a class of propositional formulas. Intuitively, a $\mathcal{C}$ backdoor is a set of variables of a given propositional formula with the following property. Applying assignments over these variables to the formula always yields a formula in the class $\mathcal{C}$. Of course, one aims for formula classes for which satisfiability can be decided efficiently. Informally speaking, with the parameter backdoor size of a formula one tries to describe a distance to tractability. This definition was first introduced by Golmes, Williams and Selman [18] to model short distances to efficient subclasses. Until today, backdoors gained copious attention in many different areas: abduction [33], answer set programming [34,35], argumentation [36], default logic [37], temporal logic [38], planning [39] and constraint satisfaction [40,41].

Consider a formula ϕ in conjunctive normal form. Denote by $\phi[\tau]$ for a partial truth assignment τ the result of removing all clauses from ϕ which contain a literal ℓ with $\tau(\ell) = 1$ and removing literals ℓ with $\tau(\ell) = 0$ from the remaining clauses.

Definition 14. *Let $\mathcal{C}$ be a class of CNF formulas and ϕ be a CNF formula. A set $V \subseteq Vars(\phi)$ of variables of ϕ is a strong $\mathcal{C}$ backdoor set of ϕ if for all truth assignments $\tau \colon V \to \{0,1\}$ we have that $\phi[\tau] \in \mathcal{C}$.*

Definition 15 ([42,43]). *Let $\mathcal{C}$ be a class of CNF formulas and ϕ be a CNF formula. A set $V \subseteq Vars(\phi)$ of variables of ϕ is a $\mathcal{C}$-deletion backdoor set of ϕ if $\phi[V]$ is in $\mathcal{C}$, where $\phi[V]$ denotes the formula obtained from ϕ by deleting in ϕ all occurrences of variables from V.*

Definition 16 (Weak Backdoor Sets). *Let $\mathcal{C}$ be a class of CNF formulas and ϕ be a propositional CNF formula. A set $V \subseteq Vars(\phi)$ of variables from ϕ is a weak $\mathcal{C}$ backdoor set of ϕ if there exists an assignment $\theta \in \Theta(V)$ such that $\phi[\theta] \in \mathcal{C}$ and $\phi[\theta]$ is satisfiable.*

Now let us consider the following enumeration problem.

Problem:	ENUM-WEAK-BACKDOORSET($\mathcal{C}$)
Input:	A formula ϕ in CNF, $k \in \mathbb{N}$.
Parameter:	The integer k.
Output:	The set of all weak $\mathcal{C}$ backdoor sets of ϕ of size at most k.

Similarly, define ENUM-STRONG-BACKDOORSET($\mathcal{C}$) for the set of all strong $\mathcal{C}$ backdoor sets of ϕ of size at most k. Observe that the backdoor set problems can be seen as modification problems where solutions are sequences of variable assignments. The target property then simply is the class of CNF formulas $\mathcal{C}$.

Notice that Creignou et al. [13] (Theorem 4), have studied enumeration for exact strong HORNbackdoor sets and provided an algorithm running in DelayFPT, where HORN denotes the set of all Horn formulas, that is, CNF formulas whose clauses contain at most one positive literal.

Definition 17 (Base Class [44]). *The class $\mathcal{C}$ is a base class if it can be recognised in P (that is, $\mathcal{C} \in $ P), the satisfiability of its formulas is in P and the class is closed under isomorphisms w.r.t. variable names. We say that $\mathcal{C}$ is clause defined if for every CNF formula ϕ we have: $\phi \in \mathcal{C}$ if and only if $\{C\} \in \mathcal{C}$ for all clauses C from ϕ.*

Proposition 3 ([44] (Proposition 2)). *For every clause-defined base class $\mathcal{C}$, the detection of weak $\mathcal{C}$ backdoor sets is in FPT for input formulas in 3-CNF.*

In their proof, Gaspers and Szeider [44] describe how utilising a bounded search tree allows one to solve the detection of weak $\mathcal{C}$ backdoors in FPT time. Interesting to note, this technique results in obtaining all minimal solutions in FPT time. This observation results in the following theorem.

Theorem 4. *For every clause-defined base class $\mathcal{C}$ and input formula in 3-CNF*

- ENUM-WEAK-$\mathcal{C}$-BACKDOORS^{LEX} $\in$ DelayFPT *with polynomial space and*
- ENUM-WEAK-$\mathcal{C}$-BACKDOORS^{SIZE} $\in$ DelayFPT *with exponential space.*

Proof. The set of operations Ω contains functions that replace a specific variable $i \in \mathbb{N}$ by a truth value $t \in \{0,1\}$. A solution encodes the chosen backdoor set together with the required assignment. Proposition 3 yields ENUM-MIN-WEAK-$\mathcal{C}$BACKDOORS^{LEX}, resp., ENUM-MIN-WEAK-$\mathcal{C}$-BACKDOORS^{SIZE} being FPT enumerable. As the consistency check for solutions is in polynomial time, applying Corollary 7 completes the proof. □

In the following result, we will examine the parametrised enumeration complexity of the task to enumerate all strong $\mathcal{C}$-backdoor sets of a given 3-CNF formula for some clause-defined base class $\mathcal{C}$. Crucially, every strong backdoor set has to contain at least one variable from a clause that is not in $\mathcal{C}$ which relates to "hitting all bad clauses" like in the definition of deletion backdoors (see Definition 15).

Theorem 5. *For every clause-defined base class $\mathcal{C}$ and input formula in 3-CNF:*

- ENUM-STRONG-$\mathcal{C}$-BACKDOORS$^{\text{LEX}} \in$ DelayFPT *with polynomial space and*
- ENUM-STRONG-$\mathcal{C}$-BACKDOORS$^{\text{SIZE}} \in$ DelayFPT *with exponential space.*

Proof. We show that for every clause-defined base class $\mathcal{C}$ and input formula in 3-CNF, the problem MIN-STRONG-$\mathcal{C}$-BACKDOORS is FPT enumerable. Indeed, we only need to branch on the variables from a clause $C \notin \mathcal{C}$ and remove the corresponding literals over the considered variable from ϕ. The size of the branching tree is at most 3^k. As for base classes the satisfiability test is in P, this yields an FPT algorithm. The neighbourhood function $\mathcal{N}(x, S)$ for $x = (\phi, k)$ is defined to be the set of the pairwise unions of all minimal strong $\mathcal{C}$ backdoors of $(\phi[(S \cup \{x_i\})], k - |S| - 1)$ together with $S \cup \{x_i\}$ for all variables $x_i \notin S$. If $\text{Vars}(\phi) = \{x_1, \ldots, x_n\}$, then the operations are $\omega_i \colon \phi \mapsto \phi(0/x_i) \wedge \phi(1/x_i)$. As application of the functions ω_i happens only with respect to the backdoor size k, which is the parameter, the formula size increases by an exponential factor in the parameter only. This yields the preconditions for Corollary 7. □

4.3. Weighted Satisfiability Problems

Finally, we consider satisfiability questions for formulas in the Schaefer framework [8]. A constraint language Γ is a finite set of relations. A Γ formula ϕ, is a conjunction of constraints using only relations from Γ and, consequently, is a quantifier-free first order formula.

As opposed to the approach of Creignou et al. [13], who examined maximum satisfiability, we now focus on the problem MINONES-SAT(Γ) defined below.

If $\theta \colon X \to \{0, 1\}, \theta' \colon Y \to \{0, 1\}$ are two (partial) assignments over some set of variables X and Y, then $\theta \subset \theta'$ is true, if $\theta(x) = \theta'(x)$ for all $x \in X$ and $X \subset Y$.

Definition 18 (Minimality)**.** *Given a propositional formula ϕ and an assignment θ over the variables in ϕ with $\theta \models \phi$, we say that θ is minimal if there does not exist an assignment $\theta' \subset \theta$ and $\theta' \models \phi$. The size $|\theta|$ of θ is the number of variables it sets to true.*

Formally, the problem of interest is defined with respect to any fixed constraint language Γ:

Problem:	MIN-MINONES-SAT$^{\text{SIZE}}(\Gamma)$
Input:	(ϕ, k), a Γ formula ϕ, $k \in \mathbb{N}$.
Parameter:	The integer k.
Output:	Generate all inclusion-minimal satisfying assignments θ of ϕ with $|\theta| \leq k$ by non-decreasing size.

Similarly, the problem ENUM-MINONES-SAT(Γ) asks for all satisfying assignments θ of ϕ with $|\theta| \leq k$. In this context, the operations in Ω are functions that replace the variable with index $i \in \mathbb{N}$ by true.

Theorem 6. *For all constraint languages Γ, we have:* MIN-MINONES-SAT$^{\text{SIZE}}(\Gamma)$ *is FPT enumerable and* ENUM-MINONES-SAT$^{\text{SIZE}}(\Gamma) \in$ DelayFPT *with exponential space.*

Proof. For the first claim we can simply compute the minimal assignments by a straightforward branching algorithm: Initially, begin with the all 0-assignment, then consider all unsatisfied clauses in turn and flip one of the occurring variables to true. The second claim follows by a direct application of Corollary 7. □

5. Conclusions

We presented FPT delay ordered enumeration algorithms for a large variety of problems, such as cluster editing, chordal completion, closest-string and weak and strong backdoors. An important point of our paper is that we propose a general strategy for efficient enumeration. This is rather rare in the literature, where algorithms are usually devised individually for specific problems. In particular, our scheme yields DelayFPT algorithms for all graph modification problems that are characterised by a finite set of forbidden patterns.

Initially, we focussed on graph-theoretic problems. Afterwards, the generic approach we presented covered problems which are not only of a graph-theoretic nature. Here, we defined general modification problems detached from graphs and constructed generic enumeration algorithms for arising problems in the world of strings, numbers, formulas, constraints, etc.

As an observation, we would like to mention that the DelayFPT algorithms presented in this paper require exponential space due to the inherent use of the priority queues to achieve ordered enumeration. An interesting question, continuing the research of Meier [45], is whether there is a method which requires less space but uses a comparable delay between the output of solutions and still obeys the underlying order on solutions.

Author Contributions: Conceptualization, N.C., R.K., A.M., J.-S.M., F.O. and H.V.; Funding acquisition, N.C., A.M. and H.V.; Methodology, N.C., R.K., A.M., J.-S.M., F.O. and H.V.; Supervision, N.C., A.M. and H.V.; Writing—original draft, N.C., R.K., A.M., F.O. and H.V.; Writing—review and editing, N.C., R.K., A.M., F.O. and H.V.

Funding: This research was funded by Deutsche Forschungsgemeinschaft (ME 4279/1-2) and the French Agence Nationale de la Recherche (AGGREG project reference ANR-14-CE25-0017).

Acknowledgments: We thank the anonymous reviewers for their valuable feedback.

Conflicts of Interest: The authors declare no conflict of interest.

References

1. Durand, A.; Schweikardt, N.; Segoufin, L. Enumerating answers to first-order queries over databases of low degree. In Proceedings of the 33rd ACM SIGMOD-SIGACT-SIGART Symposium on Principles of Database Systems, PODS'14, Snowbird, UT, USA, 22–27 June 2014; Hull, R., Grohe, M., Eds.; pp. 121–131. [CrossRef]
2. Fogaras, D.; Rácz, B. A Scalable Randomized Method to Compute Link-Based Similarity Rank on the Web Graph. In Proceedings of the Current Trends in Database Technology—EDBT 2004 Workshops, EDBT 2004 Workshops PhD, DataX, PIM, P2P&DB, and ClustWeb, Heraklion, Crete, Greece, 14–18 March 2004; Revised Selected Papers; Lindner, W., Mesiti, M., Türker, C., Tzitzikas, Y., Vakali, A., Eds.; Lecture Notes in Computer Science; Springer: Berlin, Germany, 2004; Volume 3268, pp. 557–567._55. [CrossRef]
3. Acuña, V.; Milreu, P.V.; Cottret, L.; Marchetti-Spaccamela, A.; Stougie, L.; Sagot, M. Algorithms and complexity of enumerating minimal precursor sets in genome-wide metabolic networks. *Bioinformatics* **2012**, *28*, 2474–2483. [CrossRef] [PubMed]
4. Dill, K.A.; Lucas, A.; Hockenmaier, J.; Huang, L.; Chiang, D.; Joshi, A.K. Computational linguistics: A new tool for exploring biopolymer structures and statistical mechanics. *Polymer* **2007**, *48*, 4289–4300. [CrossRef]
5. Johnson, D.S.; Papadimitriou, C.H.; Yannakakis, M. On Generating All Maximal Independent Sets. *Inf. Process. Lett.* **1988**, *27*, 119–123. [CrossRef]
6. Sörensen, K.; Janssens, G.K. An algorithm to generate all spanning trees of a graph in order of increasing cost. *Pesqui. Oper.* **2005**, *25*, 219–229. [CrossRef]
7. Creignou, N.; Hébrard, J.J. On generating all solutions of generalized satisfiability problems. *Theor. Inform. Appl.* **1997**, *31*, 499–511. [CrossRef]
8. Schaefer, T.J. The Complexity of Satisfiability Problems. In Proceedings of the 10th Annual ACM Symposium on Theory of Computing, San Diego, CA, USA, 1–3 May 1978; Lipton, R.J., Burkhard, W.A., Savitch, W.J., Friedman, E.P., Aho, A.V., Eds.; pp. 216–226. [CrossRef]

9. Creignou, N.; Olive, F.; Schmidt, J. Enumerating All Solutions of a Boolean CSP by Non-decreasing Weight. In Proceedings of the 14th International Conference on Theory and Applications of Satisfiability Testing, SAT 2011, Ann Arbor, MI, USA, 19–22 June 2011; Lecture Notes in Computer Science; Springer: Berlin, Germany, 2011; Volume 6695, pp. 120–133.
10. Fernau, H. On Parameterized Enumeration. In *Computing and Combinatorics*; Springer: Berlin, Germany, 2002.
11. Damaschke, P. Parameterized enumeration, transversals, and imperfect phylogeny reconstruction. *Theor. Comput. Sci.* **2006**, *351*, 337–350. [CrossRef]
12. Fomin, F.V.; Saurabh, S.; Villanger, Y. A Polynomial Kernel for Proper Interval Vertex Deletion. *SIAM J. Discret. Math.* **2013**, *27*, 1964–1976. [CrossRef]
13. Creignou, N.; Meier, A.; Müller, J.; Schmidt, J.; Vollmer, H. Paradigms for Parameterized Enumeration. *Theory Comput. Syst.* **2017**, *60*, 737–758. [CrossRef]
14. Shamir, R.; Sharan, R.; Tsur, D. Cluster graph modification problems. *Discret. Appl. Math.* **2004**, *114*, 173–182. [CrossRef]
15. Yannakakis, M. Computing the minimum fill-in is NP complete. *SIAM J. Algebr. Discret. Methods* **1981**, *2*, 77–79. [CrossRef]
16. Yannakakis, M. Node- and edge-deletion NP complete problems. In Proceedings of the 10th Annual ACM Symposium on Theory of Computing, San Diego, CA, USA, 1–3 May 1978; pp. 253–264.
17. Gramm, J.; Niedermeier, R.; Rossmanith, P. Fixed-Parameter Algorithms for CLOSEST STRING and Related Problems. *Algorithmica* **2003**, *37*, 25–42. [CrossRef]
18. Williams, R.; Gomes, C.P.; Selman, B. Backdoors To Typical Case Complexity. In Proceedings of the IJCAI-03, Eighteenth International Joint Conference on Artificial Intelligence, Acapulco, Mexico, 9–15 August 2003; pp. 1173–1178.
19. Kröll, M.; Pichler, R.; Skritek, S. On the Complexity of Enumerating the Answers to Well-designed Pattern Trees. In Proceedings of the 19th International Conference on Database Theory (ICDT 2016), Bordeaux, France, 15–18 March 2016; Martens, W., Zeume, T., Eds.; Leibniz International Proceedings in Informatics (LIPIcs); Schloss Dagstuhl–Leibniz-Zentrum fuer Informatik: Wadern, Germany, 2016; Volume 48, pp. 22:1–22:18. [CrossRef]
20. Bentert, M.; Fluschnik, T.; Nichterlein, A.; Niedermeier, R. Parameterized aspects of triangle enumeration. *J. Comput. Syst. Sci.* **2019**, *103*, 61–77. [CrossRef]
21. Bökler, F.; Ehrgott, M.; Morris, C.; Mutzel, P. Output-sensitive complexity of multiobjective combinatorial optimization. *J. Multi-Criteria Decis. Anal.* **2017**, *24*, 25–36. [CrossRef]
22. Kröll, M.; Pichler, R.; Woltran, S. On the Complexity of Enumerating the Extensions of Abstract Argumentation Frameworks. In Proceedings of the 26th International Joint Conference on Artificial Intelligence, IJCAI 17, Melbourne, Australia, 19–25 August 2017; AAAI Press: Palo Alto, CA, USA, 2017; pp. 1145–1152.
23. Carbonnel, C.; Hebrard, E. On the Kernelization of Global Constraints. In Proceedings of the 26th International Joint Conference on Artificial Intelligence, IJCAI 2017, Melbourne, Australia, 19–25 August 2017; AAAI Press: Palo Alto, CA, USA, 2017; pp. 578–584. [CrossRef]
24. Schmidt, J. Enumeration: Algorithms and Complexity. Master's Thesis, Leibniz Universität Hannover, Hanover, Germany, 2009.
25. Hirai, Y.; Yamamoto, K. Balancing weight-balanced trees. *J. Funct. Program.* **2011**, *21*, 287–307. [CrossRef]
26. Bodlaender, H.L.; Heggernes, P.; Lokshtanov, D. Graph Modification Problems (Dagstuhl Seminar 14071). *Dagstuhl Rep.* **2014**, *4*, 38–59. [CrossRef]
27. Garey, M.R.; Johnson, D.S. *Computers and Intractability; A Guide to the Theory of NP-Completeness*; W. H. Freeman & Co.: New York, NY, USA, 1990.
28. Cai, L. Fixed-parameter tractability of graph modification problems for hereditary properties. *Inf. Process. Lett.* **1996**, *58*, 171–176. [CrossRef]
29. Kaplan, H.; Shamir, R.; Tarjan, R.E. Tractability of parameterized completion problems on chordal, strongly chordal, and proper interval graphs. *SIAM J. Comput.* **1999**, *28*, 1906–1922. [CrossRef]
30. Brandtstädt, A.; Le, V.B.; Spinrad, J.P. *Graph Classes: A Survey*; Monographs on Discrete Applied Mathematics, SIAM: Philadelphia, PA, USA, 1988.
31. Avis, D.; Fukuda, K. Reverse search for enumeration. *Discret. Appl. Math.* **1996**, *65*, 21–46. [CrossRef]
32. Frances, M.; Litman, A. On covering problems of codes. *Theory Comput. Syst.* **1997**, *30*, 113–119. [CrossRef]

33. Pfandler, A.; Rümmele, S.; Szeider, S. Backdoors to Abduction. In Proceedings of the 23rd International Joint Conference on Artificial Intelligence, IJCAI 13, Beijing, China, 3–9 August 2013; Rossi, F., Ed.; pp. 1046–1052.
34. Fichte, J.K.; Szeider, S. Backdoors to tractable answer set programming. *Artif. Intell.* **2015**, *220*, 64–103. [CrossRef]
35. Fichte, J.K.; Szeider, S. Backdoors to Normality for Disjunctive Logic Programs. *ACM Trans. Comput. Log.* **2015**, *17*, 7:1–7:23. [CrossRef]
36. Dvořák, W.; Ordyniak, S.; Szeider, S. Augmenting tractable fragments of abstract argumentation. *Artif. Intell.* **2012**, *186*, 157–173. [CrossRef]
37. Fichte, J.K.; Meier, A.; Schindler, I. Strong Backdoors for Default Logic. In Proceedings of the 19th International Conference on Theory and Applications of Satisfiability Testing—SAT 2016, Bordeaux, France, 5–8 July 2016; pp. 45–59._4. [CrossRef]
38. Meier, A.; Ordyniak, S.; Ramanujan, M.S.; Schindler, I. Backdoors for Linear Temporal Logic. *Algorithmica* **2019**, *81*, 476–496. [CrossRef] [PubMed]
39. Kronegger, M.; Ordyniak, S.; Pfandler, A. Backdoors to planning. *Artif. Intell.* **2019**, *269*, 49–75. [CrossRef]
40. Ganian, R.; Ramanujan, M.S.; Szeider, S. Combining Treewidth and Backdoors for CSP. In Proceedings of the 34th Symposium on Theoretical Aspects of Computer Science, STACS 2017, Hannover, Germany, 8–11 March 2017; Leibniz International Proceedings in Informatics (LIPIcs); Vollmer, H., Vallée, B., Eds.; Schloss Dagstuhl—Leibniz-Zentrum fuer Informatik: Wadern, Germany, 2017; Volume 66, pp. 36:1–36:17. [CrossRef]
41. Gaspers, S.; Misra, N.; Ordyniak, S.; Szeider, S.; Zivny, S. Backdoors into heterogeneous classes of SAT and CSP. *J. Comput. Syst. Sci.* **2017**, *85*, 38–56. [CrossRef]
42. Nishimura, N.; Ragde, P.; Szeider, S. Solving #SAT using vertex covers. *Acta Inform.* **2007**, *44*, 509–523. [CrossRef]
43. Szeider, S. Matched Formulas and Backdoor Sets. *J. Satisf. Boolean Model. Comput.* **2009**, *6*, 1–12.
44. Gaspers, S.; Szeider, S. Backdoors to Satisfaction. In *The Multivariate Algorithmic Revolution and Beyond*; Lecture Notes in Computer Science; Springer: Berlin/Heidelberg, Germany, 2012; Volume 7370; pp. 287–317. [CrossRef]
45. Meier, A. Enumeration in Incremental FPT-Time. *arXiv* **2018**, arXiv:1804.07799.

© 2019 by the authors. Licensee MDPI, Basel, Switzerland. This article is an open access article distributed under the terms and conditions of the Creative Commons Attribution (CC BY) license (http://creativecommons.org/licenses/by/4.0/).

Article

A Machine Learning Approach to Algorithm Selection for Exact Computation of Treewidth

Borislav Slavchev, Evelina Masliankova and Steven Kelk *

Department of Data Science and Knowledge Engineering, Maastricht University, 6211 LK Maastricht, The Netherlands; b.slavchev@student.maastrichtuniversity.nl (B.S.); e.masliankova@student.maastrichtuniversity.nl (E.M.)
* Correspondence: steven.kelk@maastrichtuniversity.nl

Received: 25 July 2019; Accepted: 16 September 2019; Published: 20 September 2019

Abstract: We present an algorithm selection framework based on machine learning for the exact computation of *treewidth*, an intensively studied graph parameter that is NP-hard to compute. Specifically, we analyse the comparative performance of three state-of-the-art exact treewidth algorithms on a wide array of graphs and use this information to predict which of the algorithms, on a graph by graph basis, will compute the treewidth the quickest. Experimental results show that the proposed meta-algorithm outperforms existing methods on benchmark instances on all three performance metrics we use: in a nutshell, it computes treewidth faster than any single algorithm in isolation. We analyse our results to derive insights about graph feature importance and the strengths and weaknesses of the algorithms we used. Our results are further evidence of the advantages to be gained by strategically blending machine learning and combinatorial optimisation approaches within a hybrid algorithmic framework. The machine learning model we use is intentionally simple to emphasise that speedup can already be obtained without having to engage in the full complexities of machine learning engineering. We reflect on how future work could extend this simple but effective, proof-of-concept by deploying more sophisticated machine learning models.

Keywords: treewidth; tree decomposition; algorithm selection; machine learning; combinatorial optimisation

---

1. Introduction

1.1. The Importance of Treewidth

The *treewidth* of an undirected graph $G = (V, E)$, denoted $tw(G)$, is a measure of how "treelike" it is [1]. Trees have treewidth 1 and the treewidth of a graph can be as large as $|V| - 1$. The concept of treewidth is closely linked to that of a *tree decomposition* of G, which is the arrangement of a set of subsets of V into a tree backbone, such that certain formal criteria are satisfied; these are described in the preliminaries. The width of the tree decomposition is equal to the size of its largest subset, minus 1 and $tw(G)$ is the minimum width, ranging over all possible tree decompositions of G [2]. In recent decades treewidth has attracted an immense amount of attention, due to the observation that many NP-hard problems on graphs become polynomial-time solveable on graphs of bounded treewidth. This is because, when a tree decomposition of width t is available, it is usually possible to organise dynamic programming algorithms in a hierarchical way such that the intermediate lookup-tables consulted by the dynamic programming, have size that is bounded by a function of t [3]. In fact, because of this many NP-hard problems on graphs can actually be solved in time $f(tw(G)) \cdot \text{poly}(n)$, where $n = |V(G)|$ and f is a computable (typically exponential) function that depends only on $tw(G)$.

We say then that the problem is *fixed parameter tractable* (FPT) with respect to $tw(G)$. For this reason, treewidth has become a cornerstone of the parameterized complexity literature, which advocates a more fine-grained approach to running time complexity than the traditional approach of expressing worst-case running times purely as a function of the size of the input [4].

Given its relevance as a 'gateway to tractability', treewidth has attracted enormous interest not just from the combinatorial optimisation and theoretical computer science communities but also from many other domains such as artificial intelligence (particularly Bayesian inference and constraint satisfaction: see Reference [5] for a highly comprehensive list), computational biology [6] and operations research [7]. Unfortunately, computing the treewidth (equivalently, a minimum-width tree decomposition) is NP-hard [8]. This does not nullify its usefulness, however. Alongside practical heuristics (see e.g., References [9,10]) there is an extensive literature on computing the treewidth of a graph exactly, in spite of its NP-hardness. Some of these results have a purely theoretical flavour, most famously in Reference [11] but there has also been a consistent stream of research on computing treewidth exactly *in practice*. Such practical algorithms have an interesting history. Until a few years ago, progress had been somewhat incremental; see References [12,13] for representative articles from this period. However, in 2016 the Parameterized Algorithms and Computational Experiments (PACE) treewidth challenge, an initiative of the parameterized complexity community to promote the development of practically efficient algorithms, stimulated a major breakthrough. The *positive-instance driven* dynamic programming approach pioneered by Tamaki [14] allowed the treewidth to be computed exactly for many graphs that had hitherto been considered out of range [14,15]. This 2017 edition of PACE subsequently yielded an additional hundred-fold speed-up over PACE 2016 results [16].

Following the advances triggered by PACE 2016 and Tamaki's approach, there are now a number of highly advanced exact treewidth solvers, several of which we will encounter later in this article. When studying these solvers we observed that, once the solving time rose above a few seconds, no single solver universally dominated over (i.e., ran more quickly than) all the others. This led us to the following insight: if we could decide, *on a graph-by-graph basis*, which solver will compute the treewidth of the graph most quickly, we could contribute to a further reduction in the time required to compute treewidth by selecting this fastest algorithm. Traditionally, such an approach would involve a blend of comparing worst-case time complexities, analysing the 'inner workings' of the algorithms and an ad-hoc exploration of which types of graphs do and do not solve quickly on certain algorithms. Such a process, however, is time-consuming, requires algorithm-specific knowledge and may have limited efficacy due to the opaqueness of the algorithmic approaches used and/or highly complex correlations between input characteristics and algorithmic performance. Hence the question: can we, as far as possible, *automatically* learn the relative strengths of a set of algorithms? This brings us to the domain of *machine learning* [17] and machine learning-driven *algorithm selection* in particular. We note that, although interest has peaked in recent years due to the emergence of data science, there is already a quite extensive literature on the use of machine learning in combinatorial optimisation and integer linear programming in particular (see e.g, References [18–20]). In addition to algorithm selection, which we will elaborate upon below, this literature focuses on topics such as classification of instances by difficulty [21], running time regression [22,23] and learning to branch [24–26]. We refer to References [22,27] for excellent surveys of the area and References [28–30] for some recent case-specific applications in operations research.

1.2. Algorithm Selection

The algorithm selection problem was initially formulated by Rice [31] to deal with the following question: given a problem instance and a set of algorithms, which algorithm should be selected to solve the instance? The simplest answer to this question is dubbed the 'winner takes all' approach—all candidate algorithms' performance is measured on a set of problem instances and only the algorithm with the best average performance is used. However, there is always a risk that the algorithm that is best on average will be a sub-optimal choice for many specific instances.

Conversely, the ideal solution to the problem would be an oracle predictor that always knows which algorithm is best for any instance. In practice, it is rarely possible to create such a predictor. Instead, researchers strive to heuristically approximate an oracle. In 2003 Leyton-Brown et al. provided the first proof-of-concept of per-instance selection [32]. Since then, notable applications and successes include SAT solvers [33,34], classification [35], probabilistic inference [36], graph coloring [37], integer linear programming [38] and many others. For a thorough history of the development and applications of algorithm selection, the reader is referred to Kerschke et al. [39].

1.3. Our Contribution

Although machine learning has been applied to learn desirable characteristics of a given set of tree decompositions [40] (with a view to speeding up dynamic programming algorithms that then run over the tree decompositions), algorithm selection has not been applied to the computation of treewidth itself. This paper sets out to fill that gap by demonstrating a working algorithm selection framework for the problem. Specifically, we use standard, "out of the box" techniques from machine learning to train a classifier which subsequently decides, on a graph by graph basis, which out of a set of state-of-the-art exact treewidth solvers will terminate fastest. The exact treewidth algorithms were all submissions to Track A of the aforementioned PACE 2017 tournament. The resulting hybrid algorithm significantly outperforms any single one of the algorithms in isolation. We also investigate the *features* (i.e., measurable characteristics) of problem instances and solvers that make some (graph, algorithm) pairings better than others. The data used to train the classifier has been made publicly available at github.com/bslavchev/as-treewidth-exact. We feel that the combination of treewidth and the machine learning approach to algorithm selection is particularly timely, given that they both fit into the wider movement towards fine-grained, multivariate analysis of running-time complexity. In order to keep the exposition compact and accessible for a wider audience we have deliberately chosen for a stripped-down machine learning framework in which we only use a generic, easy-to-compute set of features and a small number of pre-selected classifiers. As we explain in the section on future work, more features, classifiers and analytical techniques can certainly be explored in the future. However, the take-away message for this article is that machine learning-driven algorithm selection can already be effective without having to resort to highly sophisticated machine learning frameworks.

The structure of this paper is as follows: Section 2 will formally define treewidth and the algorithm selection problem. Section 3 will describe the graph features, treewidth solvers and machine learning models that we used. Section 4 will present the datasets used, the experiments conducted and their results. Section 5 will further analyse the experiment results, while Section 6 contains a discussion of the insights that the research has yielded. Finally, Section 7 reflects on possible future work and the trend towards the incorporation of machine learning techniques within combinatorial optimisation.

2. Preliminaries

2.1. Tree Decompositions and Treewidth

Formally, given an undirected graph $G = (V, E)$, a tree decomposition of G is a pair (X, T) where X is a set whose individual elements X_i are subsets of V called *bags* and T is a tree whose nodes are the subsets X_i. Additionally, the following three rules hold:

1. The union of all bags X_i equals V, i.e., every vertex of the original graph G is in at least one bag in the tree decomposition.
2. For every edge (u, v) in G, there exists a bag X that contains both u and v, i.e., both endpoints of every edge in the original graph G can be found together in at least one bag in the tree decomposition.
3. If two bags X_i and X_j both contain a vertex v, then every bag on the path between X_i and X_j also contains v.

The treewidth of a tree decomposition (X, T) is the size of the largest bag in the decomposition, minus one and the treewidth of a graph G is the minimum width ranging over all tree decompositions of G [2].

2.2. Algorithm Selection

The Algorithm Selection problem, as formulated by Rice [31], deals with the following question: Given a problem instance and a set of algorithms, which algorithm should be selected to solve the instance? Rice identified four defining characteristics of the algorithm selection problem:

1. the *set of algorithms A*,
2. the instances of the problem, also known as the *problem space P*,
3. measurable characteristics (features) of a problem instance, known as the *feature space F*,
4. the *performance space Y*.

The selection procedure S is what connects these four components together and produces the final answer—given an instance p_i, S selects from A an algorithm a_i based on the features f_i that p_i has, in order to maximize the performance y_i.

3. Methodology

3.1. Features

Algorithm selection is normally based on a set of features that are extracted from problem instances under the assumption that those features contain signal for which algorithm to select. The choice of features is, of course, subjective. To enhance generality and avoid circularity, we deliberately avoided selecting features based on specific knowledge of how the treewidth algorithms operate. To that end, we extracted thirteen features from each of the graphs, listed below. These are very standard summary statistics of graphs, which are a subset of the features used by Reference [37] where algorithm selection was used to tackle the NP-hard graph colouring problem. More specifically, we selected these features because they are trivial to compute, ensuring that the feature extraction phase of our hybrid algorithm requires only a negligible amount of time to execute. The remaining features used by Reference [37] are more complex and time-consuming to compute, requiring non-trivial polynomial-time algorithms to approximate a number of other graph parameters, several of which are themselves NP-hard to compute. Compared to many machine learning frameworks, our selected subset is a fairly small set of features: it is more common to start with a very large set of features and then to progressively eliminate features that do not seem to contain signal. Nevertheless, as we shall see in due course, our parsimonious choice of features seems well-suited to the problem at hand.

1: **number of nodes:** v
2: **number of edges:** e
3,4: **ratio:** $\frac{v}{e}, \frac{e}{v}$
5: **density:** $\frac{2e}{v(v-1)}$
6–13: **degree statistics:** min, max, mean, median, $Q_{0.25}$, $Q_{0.75}$, variation coefficient, entropy.

For brevity, the features $Q_{0.25}$, $Q_{0.75}$ and variation coefficient will be referred to as **q1** (first quartile), **q3** (third quartile) and **variation** in the rest of this paper. We note that there are some mathematical correlations in the above features, particularly between v/e and e/v. We keep both to align with Reference [37], where both features were also used but also because we wish to defer the question of feature correlations and significance to our post-facto feature analysis later in the article. There we observe that leaving out exactly one of e/v and v/e in any case does not *improve* performance. Other correlations in the features are more indirect and occur via more complex mathematical transformations. There is no guarantee that the classifiers we use can easily 'learn' these complex correlations, which justifies inclusion. It is also useful to keep all 13 features to illustrate that careful pre-processing of features, which can be quite a complex process, is not strictly necessary to obtain a powerful hybrid algorithm.

3.2. Treewidth Algorithms

The treewidth implementations and algorithms used were submissions to Track A of the 2017 Parameterized Algorithms and Computational Experiments (PACE) competition [16]. All solvers used are listed below with a brief description and a link to where the implementation can be accessed. We emphasise that, where solvers have multiple functionalities, we used used them in *exact* mode. That is, we ask the solver to compute the true treewidth of the graph, not a heuristic approximation of it.

- **tdlib**, by **Lukas Larisch and Felix Salfelder**, referred to as **tdlib**. This is an implementation of the algorithm that Tamaki proposed [14] for the 2016 iteration of the PACE challenge [15], which itself builds on the algorithm by Arnborg et al. [8]. Implementation available at github.com/freetdi/p17
- **Exact Treewidth**, by **Hisao Tamaki and Hiromu Ohtsuka**, referred to as **tamaki**. Also an implementation of Tamaki's algorithm [14]. Implementation available at github.com/TCS-Meiji/PACE2017-TrackA
- **Jdrasil**, by **Max Bannach, Sebastian Berndt and Thorsten Ehlers**, referred to as **Jdrasil** [41]. Implementation available at github.com/maxbannach/Jdrasil

Tamaki's original 2016 implementation [14] was initially considered but later excluded due to being dominated by other algorithms, that is, there was no problem instance where it was the fastest solver.

3.3. Machine Learning Algorithms

We start by introducing two standard techniques from machine learning. Both these techniques generate, after analysing training data, a *classifier*: in our case, an algorithm to which we input the 13 features of the input graph and which then outputs which of the three treewidth solvers to run on that graph.

Decision Trees are a widely-used model for solving classification problems [42]. The model works by partitioning the training dataset according to the features of each instance, with each partition being assigned one of the target labels. The partitioning is done in steps—first, the entire dataset is partitioned and then each of the resulting partitions may be further partitioned recursively, until a certain condition is met. The resulting model can be expressed hierarchically as a rooted tree, where every node represents a partition and every edge represents a partitioning rule based on instance features. Then, classifying an instance is as simple as starting at the root node and following the path that the instance takes according to its features and the partitioning rules it encounters. The leaf node in which the instance ends up determines the label that it should be assigned. A comprehensive introduction to decision trees is due to Kotsiantis [42]. Decisions trees are popular due to their interpretability. Some of their disadvantages include a propensity for overfitting through building too-complex trees, as well as the possibility for small changes in the training dataset to produce radical differences in the resulting tree [43].

Random Forest is also a widely-used model for solving classification problems, based on decision trees [44]. Essentially, a Random Forest is an ensemble model that uses a majority-voting system between multiple decision trees to classify instances. The trees themselves are also built according to special rules—a detailed explanation of those and Random Forests in general is due to Liaw et al. [44]. Random Forests exhibit strong predictive performance and resistance to over-fitting.

Due to their strong predictive performance, our primary results use Random Forests to learn the mapping from graphs to algorithms. However, compared to a single decision tree, Random Forest classifiers can be difficult to interpret. For this reason we also consider classification based on a single decision tree, which in this article turns out to have marginally lower predictive power but allows us to analyse and interpret the sequence of decisions taken to map graphs to algorithms. Once built, the computational time required to execute our classifiers is negligible.

3.4. Reflections on the Choice of Machine Learning Model

First, we note that in addition to Random Forests and decision trees the machine learning literature encompasses a wide range of other classifiers which could be used to select the best algorithm. It is not the purpose of this article to undertake comparative analysis of all these different classifiers in order to identify the *best* classifier, although this is something that future work could certainly explore. Rather, we wish to demonstrate that, armed with a sensible choice of classifier, algorithm selection can already outperform the original algorithms in isolation. Random Forests are renowned for their predictive power (see e.g., the discussions in References [37,45]), motivating our choice in this case. For completeness we did also try a Support Vector Machine (SVM) model [46], which attempts to classify instances through the use of separating hyperplanes but due to very weak performance in preliminary experiments we did not explore it further.

Second, we remark that our current model—selecting the fastest algorithm—is only one of many possibilities for estimating performance. For example, one could try a *regression*-based approach whereby a mathematical model is trained that estimates the actual running time of an algorithm on a given input graph. Then, the (estimated) running times of the three algorithms could be computed and compared, allowing us a more detailed insight into the relative performance of the algorithms. Such an approach certainly has its merits and would also obviate the need for *moderate difficulty* filtering, which we describe in the next section. However, we defer this to future work, since our simple classification model already proves to be highly effective, even with the aforementioned filtering technicality. Also, there are quite some complexities and choices involved in modelling and learning the running times of sophisticated algorithms, which at this stage would over-complicate the exposition and introduce an extra layer of parameter estimation.

We return to both these points in the future work section.

4. Experiments and Results

4.1. Datasets

The dataset we used is composed of a multitude of publicly available graph datasets—this section will provide a list of all datasets used and where they can be found. Some needed to be converted to the PACE treewidth format from their original format—those are marked with an asterisk (*) in the list below. Additionally, the list will also provide a shorthand name for every dataset. Datasets used are:

- PACE 2017 treewidth exact competition instances, referred to as **ex**. Available at github.com/PACE-challenge/Treewidth-PACE-2017-instances
- PACE 2017 bonus instances, referred to as **bonus**. Available at github.com/PACE-challenge/Treewidth-PACE-2017-bonus-instances
- Named graphs, referred to as **named**. (These are graphs with special names, originally extracted from the SAGE graphs database). Available at github.com/freetdi/named-graphs
- Control flow graphs, referred to as **cfg**. Available at github.com/freetdi/CFGs
- PACE 2017 treewidth heuristic competition instances, referred to as **he**. Available at github.com/PACE-challenge/Treewidth-PACE-2017-instances
- UAI 2014 Probabilistic Inference Competition instances, referred to as **uai**. Available at github.com/PACE-challenge/UAI-2014-competition-graphs
- SAT competition graphs, referred to as **sat_sr15**. Available at people.mmci.uni-saarland.de/~hdell/pace17/SAT-competition-gaifman.tar
- Transit graphs, referred to as **transit**. Available at github.com/daajoe/transit_graphs
- TreeDecomposition.com database [10]*, referred to as **toto**. Available at treedecompositions.com/

- PACE 2016 treewidth instances [15], referred to as **pace2016**. Available at bit.ly/pace16-tw-instances-20160307
- Asgeirsson and Stein [47]*, referred to as **vc**. Available at ru.is/kennarar/eyjo/vertexcover.html
- PACE 2019 Vertex Cover challenge instances*, referred to as **vcPACE**. Available at pacechallenge.org/files/pace2019-vc-exact-public-v2.tar.bz2
- DIMACS Maximum Clique benchmark instances*, referred to as **dimacsMC**. Available at iridia.ulb.ac.be/~fmascia/maximum_clique/DIMACS-benchmark

In total, 30,340 graphs from these datasets were used.

4.2. Experimental Setup

The features described in Section 3.1 were extracted from all graphs. All algorithms described in Section 3.2 were run on all graphs from Section 4.1.

The three exact algorithms tested in this research output a solution as soon as they find one and then terminate. However, a limit of 30 min run time was imposed to keep tests within an acceptable time frame; this is the same as the time limit used in the PACE 2017 competition. If an algorithm was terminated due to going above the time limit, it was presumed to have failed to find a solution. For each run, we recorded whether the algorithm terminated within the time limit, the running time (if it did) and the treewidth of the solution. Recording the treewidth is not strictly necessary, since we are primarily interested in which algorithm terminates most quickly but it is useful auxiliary information and we have made it publicly available. All experiments were conducted on Google Cloud Compute virtual machines with 50 vCPUs and 325GB memory. Since all algorithm implementations used are single-threaded, fifty experiments were run in parallel. Where necessary (e.g., Java-based solvers), each experiment was allocated 6.5GB of heap space.

The extracted features and experimental results were combined into a single dataset, in order to train the machine learning models. There were 162 graphs on which attempting to run a solver or the feature extractor would cause either the process or the machine where the process was running to crash. Most of the time, these graphs came from datasets that were not meant for exact treewidth computation, such as **sat_sr15, he** and **dimacsMC** (Section 4.1) and were therefore too big. For instance, **all** graphs of the sat_sr15 dataset were too hard for any solver to terminate. For an even more extreme example, graphs 195 through 200 of the PACE 2017 *Heuristic* Competition dataset were all too big for our feature extractor to load them into memory, with graph 195 being the smallest of them at 1.3 million vertices and graph 200 being the biggest at 15.5 million. These graphs are three orders of magnitude larger than the graphs the implementations were built to work on—the biggest graph from the PACE 2017 Exact Competition dataset is graph number 198, which has 3104 vertices. Other errors included unusual yet trivial cases like the graph *collect-neighbor_collect_neighbor_init.gr* from the dataset **cfg**, which only contained one vertex and which broke assumptions of both our feature extractor and the solvers we used. All such entries were discarded from the dataset. An additional 680 entries were discarded because no solver managed to obtain a solution on them in the given time limit—presumably those problem instances were too hard. The resulting dataset contained a total of 29,498 instances.

In the course of preliminary experiments, it was discovered that the dataset we assembled required some further pre-processing. When each graph in the dataset was assigned a label corresponding to the exact algorithm that found a solution the fastest, a very large class imbalance was detected—**tdlib** was labeled the best algorithm for about 99% (29,234) of instances; **Jdrasil**—for under 0.2% (55); and **tamaki**—for about 0.7% (209). Under these circumstances, an algorithm selection approach could trivially achieve 99% accuracy by simply always selecting **tdlib**. To create genuine added value above this trivial baseline, we imposed additional rules in order to re-balance the dataset towards graphs that are neither too easy, nor too hard; we call these graphs of *moderate* difficulty. Graphs were considered **not** of *moderate* difficulty if **either** all algorithms found a solution quicker than some lower bound (i.e., the graph is too easy), **or** all algorithms failed to find a solution within the allotted time (i.e., the graph is too hard). The reasoning behind this approach is that if algorithms' run times lie outside of the defined *moderate* area, there is little gains to be made with algorithm selection anyway. If a graph is too easy, a comparatively weak algorithm can still solve it quickly; if a graph is too hard, there simply is no "correct" algorithm to select. Formally, if lb is the lower bound, ub is the allotted time (upper bound) and $rt(A, G)$ is the run time of algorithm A on graph G, then if $(\forall A : lb \geq rt(A, G)) \vee (\forall A : ub \leq rt(A, G))$ holds, G will be excluded.

Three different lower bounds were used—1, 10 and 30 s—and graphs that passed the conditions were stored in datasets A, B and C, respectively. (For the upper bound, we used the time-out limit of 30 min). This necessitates us making a distinction between *source* datasets and *filtered* datasets—the source datasets are those publicly available datasets that we started with; the filtered datasets are the sum of all graphs from all source datasets which passed through the respective filter. Table 1 breaks down how many graphs from each source dataset remained in each filtered dataset.

Table 1. The total number of graphs per source dataset and the number of those graphs in filtered datasets A, B and C (see Sections 4.1 and 4.2 for clarification).

Datasets	Unfiltered	A	B	C
bonus	100	36	36	35
cfg	1797	43	1	0
ex	200	200	172	137
he	200	26	17	17
named	150	47	14	11
pace2016	145	95	16	13
toto	27,123	594	92	53
transit	19	10	4	4
uai	133	27	14	43
vc	178	58	42	38
vcPACE	100	6	6	6
dimacsMC	80	20	13	10
sat_sr15	115	0	0	0
Total	30,340	1162	427	337

The same label-assigning procedure described above was repeated. The resulting class distributions are shown in Table 2. Afterwards, Random Forest models were trained using the Python machine-learning package *scikit-learn* [48], using all default settings except for allowing for up to 1000 trees in the forest. Both Leave-One-Out (LOO) cross-validation and 5-fold cross-validation were used in order to decrease the importance of randomness in the train-test split. 5-fold cross validation was repeated 100 times with different random seeds for the same purpose; at the end, the results from all 100 runs were averaged. These two cross-validation methods produced virtually identical results, hence we only present the results for LOO cross-validation. The results presented are for the entire dataset. Our algorithm selector is evaluated based on its predictions for each graph when it was in the test set—that is, a model is never evaluated on a graph that it has been trained on. We deemed the usage of a hold-out set unnecessary, as no hyper-parameter optimisation took place. (For readers not familiar with these technical machine learning terms, we refer to survey literature such as Tsamardinos et al. [49]).

Table 2. Number of graphs, within each dataset, for which the given algorithm was fastest (see Sections 4.1 and 4.2).

	Unfiltered	Dataset A	Dataset B	Dataset C
Jdrasil	55	55	49	42
tamaki	209	209	186	168
tdlib	29,234	898	192	127

An additional model was trained on Dataset B, using a single CART decision tree as a classifier and 50% of the data as a training set. The purpose of this model was not to optimise predictive performance but interpretability, as a Decision Tree model is significantly easier to interpret than a Random Forest model. We hoped that this interpretation would provide insight into what makes solvers good on some instances and bad on others. Dataset B was chosen due to having the best class distribution, with no clearly best algorithm.

An experiment using Principal Component Analysis (PCA) for dimensionality reduction was also undertaken on Dataset B. The purpose was to evaluate to what extent the feature set can be shrunk without significant loss of performance and potentially to provide insight into feature importance through examination of the principal components.

Reflections on filtering. Before turning to the results, we wish to reflect further on *moderate difficulty* filtering, described above. Recall: graphs which are *not* of moderate difficulty, are those where all the solvers terminated extremely quickly (i.e., more quickly than the lower bound) or all the solvers exceeded the upper bound (here: exceeded 30 min of run time, that is, timed out). As mentioned earlier, we originally introduced this filtering after observing heavy skew in the initial dataset. We did not know *a priori* what the 'correct' lower bound should be, which is why we tried three different values (1, 10, 30 s) (Note that, once the bound is chosen, the Random Forest classifier can be trained on our data in less than a minute, although cross-validation takes longer). As our results show, this filtering is sufficient to obtain a hybrid algorithm that clearly outperforms the individual algorithms. Interestingly, the filtering also mirrors our practical experience of solving NP-hard problems exactly. That is, many different algorithms can solve *easy* instances of NP-hard problems within a few seconds but beyond this running times of algorithms tend to increase dramatically and relative efficiencies of different algorithms become more pronounced: one algorithm might take seconds, while another takes hours or days. Once inputs are eliminated that are far too *hard* for any existing exact algorithm to solve, we are left with inputs that are harder than a certain triviality threshold but not impossibly hard—and it is particularly useful to be able to distinguish between algorithmic performance in this zone. A second issue concerns the choice of lower bound. If we wish to use the hybrid algorithm in practice, by training it on newly gathered data, what should the lower bound be? If we do not have any information at all concerning the underlying distribution of running times—which will always be a challenge for any machine learning model—we propose training using a lower bound such as 1 or 10 s. Unseen graphs which are moderately difficult (subject to the chosen bound) will utilise the classifier in the region it was trained for. Others will be solved very quickly by all solvers—a running time of at most 10 s is, for many NP-hard problems, very fast—or all solvers will time out, and then it does not, in practice, matter which algorithm the model chose. In the future work section we consider alternatives to moderate difficulty filtering.

4.3. Experimental Results

This section is divided into five subsections where each of the first three subsections covers the results of experimenting on one of the three datasets generated by setting a different lower bound for algorithms' run time, as described in Section 4.2. The hybrid algorithm—essentially, the mapping from graph to algorithm that is prescribed by the trained classifier—will be compared against the solvers and against an 'oracle' algorithm, which is a hypothetical hybrid algorithm that always selects the best solver. Three performance metrics will be used:

- Victories. A 'victory' is defined as being (or *selecting*, in the case of the hybrid algorithm or the oracle algorithm) the fastest algorithm for a certain graph.
- Total run time on the entire dataset.
- Terminations. A 'termination' is defined as successfully solving the given problem instance within the given time. No regard is given to the run-time, the only thing that matters is whether the algorithm managed to find a solution at all.

Please refer to Figure 1 for detailed results from these three experiments. We note that the running times for the hybrid algorithm do not include the time to execute the Random Forest classifier. This is acceptable because in the context of the experiments its contribution to the running time is negligible: approximately a hundredth of a second for a single graph.

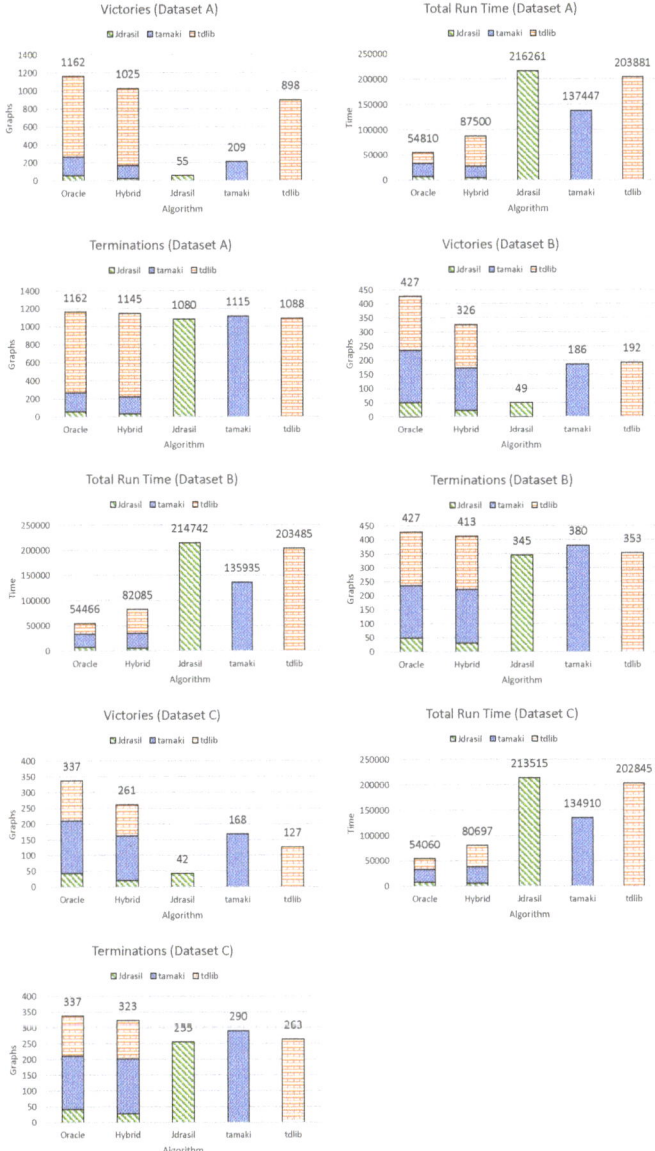

Figure 1. Experimental results using Leave One Out cross validation. Each chart presents the results from the experiment on one of the three datasets (A, B and C) based on one of the three performance metrics (victories, total run time and terminations). For the Oracle and Hybrid algorithms, results are not presented as an aggregate of all solver choices—instead, each solver's contribution to each performance metric is presented separately (see Sections 4.2 and 4.3). Total running times are measured in seconds.

The fourth subsection covers the experiment conducted on Dataset B with a decision tree classifier, while the fifth subsection covers the Principal Component Analysis experiment.

4.3.1. Dataset A

(The information in this subsection can also be found in Figure 1; similarly for Datasets B and C). The hybrid algorithm selected the fastest algorithm on 1025 out of 1162 graphs, whereas the best individual solver (**tdlib**) was fastest for 898. The hybrid algorithm's run time was 87,500 s, while the overall fastest solver (**tamaki**) required 137,447 s and the perfect algorithm—54,810 s). The hybrid algorithm terminated on 1145 out of the 1162 graphs, whereas the best solver (**tamaki**) terminated on 1115.

4.3.2. Dataset B

The hybrid algorithm selected the fastest algorithm on 326 out of 427 graphs, whereas the best individual solver (**tdlib**) was fastest for 192. The hybrid algorithm's run time was 82,085 s, while the overall fastest solver (**tamaki**) required 135,935 s and the perfect algorithm—54,466 s. The hybrid algorithm terminated on 413 out of the 427 graphs, whereas the best solver (**tamaki**) terminated on 380.

4.3.3. Dataset C

The hybrid algorithm selected the fastest algorithm on 261 out of 337 graphs, whereas the best individual solver (**tamaki**) was fastest for 168. The hybrid algorithm's run time was 80,697 s, while the overall fastest solver (**tamaki**) required 134,910 s and the perfect algorithm—54,060 s. The hybrid algorithm terminated on 323 out of the 337 graphs, whereas the best solver (**tamaki**) terminated on 290.

4.3.4. Dataset B—Decision Tree

A decision tree was trained with the default *scikit-learn* settings but it was too large to interpret, having more than 40 leaf nodes. We optimised the model's hyper-parameters until we obtained a model that was small enough that it could be easily interpreted, without sacrificing too much accuracy. The final model was built with the following restrictions: any leaf node must contain at least 2 samples; the maximum depth of the tree is 3; a node is not allowed to split if its impurity is lower than 0.25.

The resulting hybrid algorithm's performance was deemed satisfactorily close to the Random Forest selector we trained on the same dataset. The decision tree selected the fastest algorithm on 151 out of 214 graphs, whereas the Random Forest model selected the fastest for 153. The decision tree selector's run time was 54,017 s, while the Random Forest selector required 55,696 s. The decision tree selector terminated on 202 out of the 214 graphs, whereas the Random Forest selector terminated on 201.

4.3.5. Dataset B—Principal Component Analysis

Principal Component Analysis with three components was applied to the dataset and used to train a Random Forest classifier, which was trained and evaluated in the same way as the other Random Forest models. The three components cumulatively explained about 90% of the variance in the data; for a detailed breakdown, please refer to Table 3. The trained model had an accuracy of about **73%** compared to the baseline model's accuracy of **76.5%**. While the loss of accuracy is small, our drop-column feature analysis in Section 5 shows many sets of three features that can be used to build a model of similar or higher accuracy.

Table 3. The individual components in the Principal Component Analysis. Each row refers to one of the three components. The leftmost column shows how much of the total variance of the data is explained by that component; all other columns show how the respective feature is factored into the component.

VE	v	e	v/e	e/v	Density	q1	Median	q3	Minimum	Mean	Maximum	Variation	Entropy
0.58	−0.01	0.33	−0.17	0.36	0.30	0.36	0.36	0.36	0.32	0.36	0.00	−0.02	0.12
0.25	0.55	0.09	0.15	0.00	−0.07	0.00	0.00	0.00	0.00	0.00	0.55	0.55	0.24
0.08	−0.14	0.21	0.63	0.05	−0.44	0.07	0.06	0.04	0.02	0.05	−0.18	−0.17	0.52

5. Analysis

In this section, the experimental results and the machine learning models behind them are analysed with the intention of deriving insights into why certain algorithm-graph pairings are stronger than others and to determine which features of a graph are the most predictive.

We begin by analysing the importance of features for a Random Forest model trained on 50% of Dataset B. We chose to train a separate model, instead of using one from our cross-validation, because those models are all trained on all but one samples. We chose that dataset because of its class distribution—the two solvers that are strong on average (**tdlib** and **tamaki**) have nearly equal results and the third solver (**Jdrasil**) is still best for a significant number of problem instances, unlike in Dataset A. This guarantees that the hybrid algorithm's task is the hardest, as the trivial 'winner-takes-all' approach would be the least effective.

We used the feature importance functionality that is built into the *scikit-learn* package. The results are presented in Figure 2. The results indicate that almost all features are important for the classification and their importance varies within relatively tight bounds.

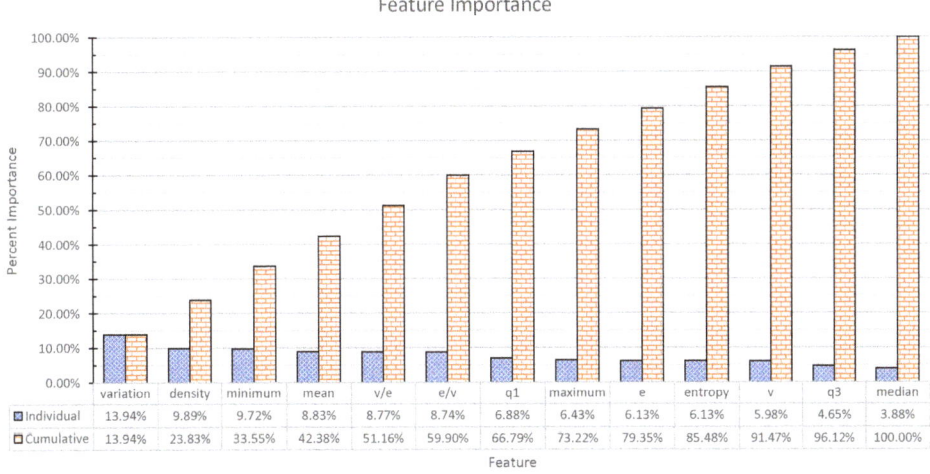

Figure 2. Feature importance for the Random Forest model trained on Dataset B (Section 4.2). We refer the reader to Section 3.1 for the list of features used. Each bar indicates the fraction of impurity in the dataset that is removed by splits on the relevant feature in all trees of the model.

In order to gain further insight, we decided to exclude certain features from the dataset, retraining the model on the reduced dataset and measuring its accuracy against the baseline of the original full model, which is **76.5%**. In order to obtain more statistically robust results, we made multiple runs of 10-fold cross-validation and kept their scores. Afterwards, the scores were compared to the baseline of using all features using the Wilcoxon Signed-Rank Test with an alpha of 0.05.

At first, we attempted excluding single features. We did 10-fold cross-validation 50 times for every excluded feature. No result was statistically significantly different from the baseline.

Next, we attempted excluding two features at a time. Again, we did 10-fold cross-validation 50 times for every excluded set of features. Two pairs of features obtained a statistically significant worse accuracy score—**variation** and **minimum degree**, as well as **variation** and **entropy**.

Finally, we attempted excluding three features at a time. This time, we did 10-fold cross-validation only 5 times, as here the computational cost of doing otherwise was prohibitive. While there were 14 sets of features that led to significantly worse scores, the magnitude of the change was rather small—the worst performance was **74.5%** and was achieved by removing **minimum degree, maximum degree** and **variation**. Of note is also the fact that 13 of 14 sets contained at least two of these same features and one set contained only one of them.

Since removing features seemed to provide us with little insight, we attempted another approach—removing all features except for a small number of designated features. Then we repeated the same procedure—we retrained the model on the reduced dataset and measured its accuracy against the original model. Again, we did 10-fold cross-validation a multitude of times. We began by only selecting one feature to retain, repeating the training 50 times. The feature **minimum degree** emerged as a clear winner, having **72%** accuracy. We highlight the fact that by removing three features at a time, we only managed to lower the accuracy to about **74.5%**, while a model with only one feature successfully reached **72%**.

Next, we selected pairs of features to keep and did 10-fold cross-validation 50 times. 7 sets of features performed at least as well as the original model. All of them contained **variation** and/or **maximum degree**.

Finally, we selected sets of 3 features to keep and did 10-fold cross-validation 5 times. More than 50 sets of features performed at least as well as the original model. Notably, despite the previously demonstrated importance of **variation, minimum degree** and **maximum degree**, adding all three of them produced results that were around the middle of the pack at **74.5%**. However, a large majority of the best results contained at least one or more often two of those features.

All experiments also showed models that surpassed the performance of the benchmark model, reaching **78.5%** by only adding **mean, variation** and **maximum degree**—compared to **76.5%** for the original model. Naturally, we view such results with caution. They could well be the result of chance, especially seeing as how we use no validation set—however, another possibility is that the classifiers can be more efficiently trained on smaller subsets of features.

The frequency with which the features **variation, minimum degree** and **maximum degree** appear in our analysis indicates that they carry some critically important signal that the classifier needs in order to be accurate. However, it appears that one or two of the features are sufficient to reproduce the signal and adding the third one does not help much.

We also attempted to determine feature importance by examining the Decision Tree model that was built for Dataset B (Figure 3). Our analysis ignores nodes that offer under 50% accuracy and nodes that contain very few (less than four) samples, as those are deemed not to bring a significant amount of insight to the analysis.

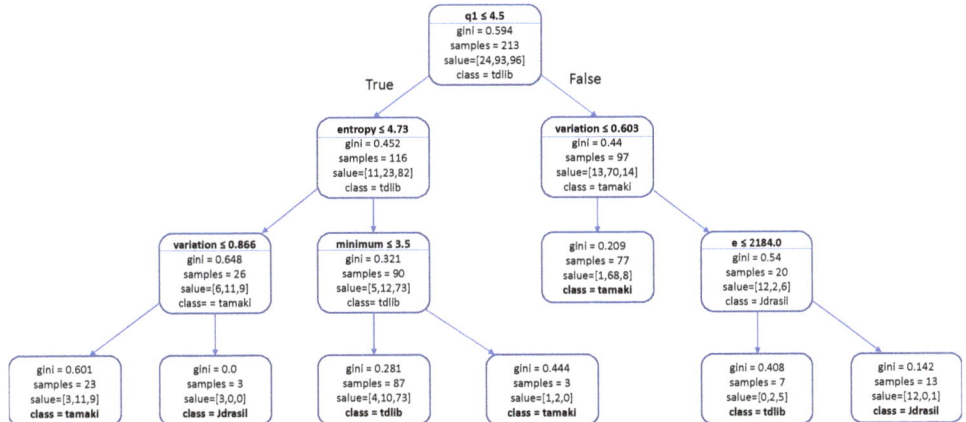

Figure 3. The CART Decision Tree trained on Dataset B. Each rectangle represents a node in the tree. The first line in each internal node indicates the condition according to which that node splits instances. In leaf nodes, this line is omitted, as no split is executed there. Each internal node's left child pertains to the subset of instances that met the condition, whereas the right child pertains to those that did not meet the condition. The next line describes the node's Gini impurity, which is the probability that a randomly chosen element from the set would be incorrectly labeled if its label was randomly drawn from the distribution of labels in the subset. The next line shows how many graphs are in the subset that the node represents, while the following line shows how many instances of each class are in that subset, with the three values corresponding to **Jdrasil**, **tamaki** and **tdlib**, respectively. The last line shows the class which is the most represented in the subset, which is also the class that the Decision Tree would assign to all instances in that subset.

The very first split in the model already provides a dramatic improvement in accuracy. Its left branch is heavily dominated by **tdlib**—after that split alone, selecting **tdlib** would be the correct choice about 70% of the time. On the right side of the split, a similar situation is observed with **tamaki** being the correct choice about 72% of the time. This shows that the first split, which sends to the left graphs where the first quartile of the degree distribution is smaller than or equal to 4.5, is very important for solvers' performance. The first quartile being low is an indicator for a small or sparse graph and according to the model, **tdlib**'s performance on those is better, while **tamaki** seems to cope better with larger or denser graphs.

There are two paths that lead to **Jdrasil** being the correct label, both of which require that the variation coefficient feature is higher than a certain threshold, which is quite high (0.603 and 0.866). This leads us to believe that **Jdrasil** performs well on graphs where there is significant variation in the degree of all vertices. One of these paths also requires the graph to have more than 2184 edges or otherwise **tdlib** is selected instead. This again indicates that **tdlib** is better at solving smaller or sparser graphs, while **Jdrasil** can deal with variability in larger graphs too.

One thing worth highlighting is that **tdlib** also seems to excel on graphs where the degree distribution has a low first quartile and the minimum degree is low. This also confirms our belief that **tdlib** excels on smaller and sparser graphs.

6. Discussion

In this section, the experimental results and analysis are discussed and some general insights are derived. One such general insight is the relative unavailability of graph sets of moderate difficulty. Depending on the definition of 'moderate', only between 1 and 3 percent of the graphs we tested could be considered as such. A likely explanation for this is that most datasets were assembled before the 2016 and 2017 editions of the PACE challenge, which introduced implementations that were multiple orders of magnitude faster than what was previously available, which in turn rendered many instances too easy.

Moving to specific points, we start by noting the promising performance of our hybrid algorithm compared to individual solvers. Our experiments clearly demonstrate the strength of the algorithm selection approach, as it outperformed all solvers on all datasets and all performance metrics, even though our underlying machine learning model is (deliberately) simple. However, the comparison with an omniscient algorithm selector also demonstrates that there is room for improvement in our framework. Section 7 lays out some suggestions for how our work can be improved upon.

Next, we reflect on the question: which features of a problem instance are the most predictive? We utilised three different approaches to answering this question—measuring how much each feature reduces impurity in our Random Forest model on Dataset B, measuring the performance of models that were only trained on a subset of all features and analysing the Decision Tree model we trained on Dataset B. Overall, our three approaches provided different and sometimes conflicting insights but some insights were confirmed by multiple approaches. One such insight is that there do not seem to be critically important individual features, as it appears that many different features can carry the same information, partly or in whole, which makes determining their individual importance difficult.

Measuring the impurity reduction of all features indicated that all features have a positive contribution to prediction. The most important feature, **degree variation coefficient**, was only about 3.5 times more important than the least important feature, **median degree**. The overall distribution of feature importances is such that the most important five features together account for about 50% of the importance, while the remaining eight account for the rest. Our interpretation of these results is that they show all features being significant contributors and while there are features that are more important than others, there are no clearly dominant features that eliminate the need for others.

The feature removal analysis indicated that three features—**variation**, **minimum degree** and **maximum degree**—all seem to be related in that removing all of them significantly reduces performance and performance increases as more of them are added, until all three are added, which does not provide a significant improvement in performance. Our interpretation is that there is a predictive signal that is present only in those three features but any two of them are sufficient to reproduce it. Besides this insight, feature removal provided little that we could interpret and that was not in direct conflict with other parts of the same analysis.

The analysis of the Decision Tree model indicated that size, density and variability are the most important characteristics of a graph; however, those could be expressed through different numerical features. For instance, the first quartile of the degree distribution (**q1**), which could be an indicator for graph size or density, was by far the most predictive feature in our Decision Tree model. To make discussing this easier, we will separate the concepts *characteristic* and *feature*. Characteristics are a more general characteristic that can be represented by many different *features*; *features* are specifically the numbers we described in Section 3.1.

Combining the results from all three approaches is difficult, as they often conflict. However, the insight from the Decision Tree analysis that size, density and variability are the most important characteristics of a graph, no matter what specific numerical proxy they are represented by, is consistent with results from other analysis approaches. The feature removal analysis indicated that **variation**, **minimum degree** and **maximum degree** together carry an important signal—these features can be considered a proxy for size, density and variability. The impurity reduction results are also consistent with this, as they showed **variation**, **density** and **minimum degree** being the three most important features. The relative lack of importance stemming from which proxy is used for these characteristics of a graph is also demonstrated by **q1**—while that feature is by far the most important in the Decision Tree analysis, the other two analytical approaches did not show it being particularly important, as it was only seventh out of thirteen in impurity reduction and did not make even one appearance in the sets of important features that the feature removal analysis yielded.

Our experiments also yield some insights into the strengths and weaknesses of the solvers. One insight that becomes clear from the class distribution in both our unfiltered dataset and the three filtered datasets (as per Section 4.2) is that the solver **tdlib** is dominant on 'easy' graphs. In the unfiltered dataset, **tdlib** was the best algorithm for 97% of graphs which became progressively less with a higher lower bound being imposed on difficulty. At the lower bound of 30 s (Dataset A), **tdlib** was the best algorithm for 77% of graphs; at 10 s (Dataset B) that number went further down to 45%; and at 30 s (Dataset C) it was only 38%. Notably, **tdlib** kept the "most victories" title in the unfiltered dataset, Dataset A and Dataset B; however, in Dataset C, **tamaki** dethroned it with 49% versus 38%. Undeniably, going from a 97% dominance to no longer being the best algorithm as difficulty increases tells us something about the strengths and weaknesses of the solver. This is also confirmed by our analysis of the Decision Tree model, which clearly showed **tdlib** had an aptitude for smaller and sparser graphs.

Another insight is **tamaki**'s robustness. It is the best solver in terms of terminations **and** run time on all three datasets, despite not being the best in terms of victories on datasets A and B. Most interesting is **tamaki**'s time performance on Dataset A—while **tdlib** has more than four times as many victories on that dataset as **tamaki** does, **tamaki**'s run time is still about 30% better than **tdlib**'s. Our analysis of the Decision Tree model showed **tamaki** having an affinity for larger or denser graphs, complementing **tdlib**'s strength on smaller or sparser graphs. The weakness of **tamaki** on sparse graphs that we discover is consistent with the findings of the solver's creator [14].

Finally, **Jdrasil** seemed to have a tighter niche than the other solvers—specifically, larger graphs with a lot of variability in their vertices' degree. However, **Jdrasil** clearly struggles on most graphs, as evidenced by its always coming in last in our experiments on all datasets and all performance metrics.

Summarizing, our analysis suggests that the most important characteristics of a graph are size, density and variability and that when focussing on these characteristics the three algorithms have the following strengths: **tdlib**—low density and size; **tamaki**—high density and size, low variability; **Jdrasil**—high density and size, high variability.

7. Conclusions and Future Work

7.1. Conclusions

In this article, we presented a novel approach based on machine learning for algorithm selection to compute the treewidth of a graph. Given a set of algorithms and a set of features of a specific problem instance, our system predicts which algorithm is most likely to have the best performance, according to a previously learned classification model. For this purpose, 13 features of graphs were identified and used. To evaluate our approach, we compared three state-of-the-art solvers, our hybrid algorithm and a hypothetical, perfect 'oracle' algorithm selection system, on thirteen publicly-available graph datasets and three performance measures. Our intentionally stripped-down machine learning approach outperformed all individual solvers on all datasets and on all performance metrics, which clearly demonstrates its utility. However, the hypothetical perfect algorithm selector still significantly outperformed our approach, which indicates that there are further improvements to be made. Based on our results, we argued that size, density and degree variability are the most important characteristics of a graph that determine which algorithm would be fastest on it. We also identified specific solvers' strengths and weaknesses based on these same characteristics.

7.2. Future Work

One weakness of our work is that algorithms were only run once on each graph, so any randomness in algorithms' run time is completely unaccounted for—future work may try to alleviate that by running algorithms several times and obtaining a confidence interval on the run time. Another sensitive part of our experimental framework concerns the selection of the three running time lower bounds (1, 10 and 30 s) that defined our three *moderately difficult* datasets A, B and C. These choices echo the practical experience of many algorithm designers that, if an exponential-time algorithm takes more than a few seconds to terminate, the running time quickly explodes and the comparative efficiencies of the algorithms become more important. It would be interesting to try to place the selection of such lower bounds on a more formal footing; in some sense they capture an interesting 'triviality threshold' beyond which the performance of existing treewidth solvers starts to deviate in a practically significant way. Future work could also explore whether the moderate difficulty filtering can be removed altogether, obviating the need to select the lower bound during training. One logical step in this direction would be to switch from our 'who wins?' classification model, to a regression-based model which is capable of predicting the running time of individual algorithms (or some kind of weighted classification model which takes the margin of victory into account). The regression model(s) could be used to estimate the running time of the different treewidth solvers on a given input graph, the fastest solver would then be chosen and auxiliary sensitivity analysis such as the margin of victory would then also be immediately available. There has been some interesting work on running-time regression of this kind (see e.g., References [22,23,50]). Such an approach could be effective but it is non-trivial to learn the running times of sophisticated algorithms, particularly if—as in this case—it is unclear what type of mathematical function (modelling the running time) we are trying to learn or whether the same function is applicable to all the different solvers. Indeed, more experimental parameters will need to be estimated and tuned and when modelling running times some kind of context-sensitive choices seem unavoidable to ensure that the regression model focuses on running times of the appropriate order of magnitude. For example, in the present context it would be most important that the regression model predicts well when running times are seconds or minutes, not microseconds or years. In other words, it is possible that the parameter estimation inherent in our moderate difficulty filtering will re-emerge somewhere else in the regression model. Nevertheless we regard this as a very interesting route for future research.

Introducing more solvers could offer further improvement, as additional solvers might exhibit a different set of strengths and weaknesses which the algorithm selection approach can then exploit. Also, as already discussed in Section 6, using a larger and more diverse set of graphs to experiment on, with a focus on graphs that are neither too easy nor too hard for algorithm selection to be useful, could strengthen the research: given the advances of PACE 2016, there is an urgent need for new, more difficult (but not *too* difficult!) datasets. Following standard machine learning methodology, we could consider incorporating different/more features, experimenting with different classifiers and carefully optimising hyper-parameters. In order to make such research easier, we have provided full access to the dataset we generated for this project. The dataset can be found on github.com/bslavchev/as-treewidth-exact.

More generally, we suggest a more thorough and rigorous analysis of feature importance (and their interactions). It would be fascinating to delve into the underlying mechanics of the three treewidth solvers (**tdlib, tamaki, Jdrasil**) to try to shed more light on *why* certain (graph, algorithm) pairings are superior to others and to analyse how different features of the graph contribute to the running times of the algorithms. It is highly speculative but perhaps such an analysis could help to identify new parameters which can then be analysed using the formal machinery of parameterized complexity. In this article we chose emphatically *not* to "open the black box" in this way, preferring instead to see how far a simple machine learning framework could succeed using generic graph features and without in-depth, algorithm-specific knowledge. Nevertheless, such an analysis would be a valuable contribution to the algorithm selection literature. (Beyond treewidth, it would be interesting to explore whether our simple machine learning framework can be effective, without extensive modification, in the computation of other NP-hard graph parameters.)

Finally we suggest the development of an easy-to-use software package that utilises our work and the state of the art in treewidth solvers to provide the best possible performance out of the box and to aid researchers and practitioners alike. Rather than re-inventing the wheel anew, it is probably wisest to integrate an algorithm selection framework into an existing general treewidth framework (such as the wider **tdlib** ecosystem, which is available at github.com/freetdi/tdlib), as it would provide numerous opportunities and benefits—for example, selecting not just solvers but also pre-processors and kernelisation algorithms for treewidth (see Reference [51] and the references therein) and using the results from pre-processors as features for the solver selection, among others. Such a framework could also incorporate advances in computing treewidth using parallel processing [52].

7.3. The Best of Both Worlds

As a final note, we view our work as part of the accelerating trend towards the use of machine learning to derive data-driven insights, which are then used to enhance the in-practice performance of combinatorial optimisation algorithms. The deepening synthesis between predictive and prescriptive analytics in operations research is also part of this trend. Combinatorial optimisation traditionally focusses on aggressively optimised algorithms, which are then analysed using worst-case, univariate complexity analysis. Apart from the simplest algorithms, such one-dimensional analyses cannot adequately capture the role of many implicit parameters in determining the running time of algorithms (and indeed, the emergence of parameterized/multivariate complexity is a formal reaction to this). Machine learning, although not the appropriate instrument for combinatorial optimisation *per se*, can fulfill a powerful role in automatically fathoming such implicit parameters.

Author Contributions: Conceptualization, B.S., E.M., S.K.; methodology, B.S., E.M.; software, B.S.; validation, B.S., E.M.; formal analysis, B.S., E.M., S.K.; investigation, B.S., E.M.; resources, B.S.; data curation, B.S., E.M.; Writing—Original draft preparation, B.S.; Writing—Review and editing, B.S., E.M. S.K.; visualization, B.S.; supervision, S.K.; project administration, S.K.

Funding: This research received no external funding.

Acknowledgments: The authors thank the two anonymous reviewers for constructive and insightful comments, and Bas Willemse for moral and practical support during the writing of the article.

Conflicts of Interest: The authors declare no conflict of interest.

References

1. Diestel, R. *Graph Theory (Graduate Texts in Mathematics)*; Springer: New York, NY, USA, 2005.
2. Bodlaender, H.L. A tourist guide through treewidth. *Acta Cybern.* **1994**, *11*, 1.
3. Bodlaender, H.L.; Koster, A.M. Combinatorial optimization on graphs of bounded treewidth. *Comput. J.* **2008**, *51*, 255–269. [CrossRef]
4. Cygan, M.; Fomin, F.V.; Kowalik, Ł.; Lokshtanov, D.; Marx, D.; Pilipczuk, M.; Pilipczuk, M.; Saurabh, S. *Parameterized Algorithms*; Springer: Cham, Switzerland, 2015; Volume 4.
5. Bannach, M.; Berndt, S. Positive-Instance Driven Dynamic Programming for Graph Searching. *arXiv* **2019**, arXiv:1905.01134.
6. Hammer, S.; Wang, W.; Will, S.; Ponty, Y. Fixed-parameter tractable sampling for RNA design with multiple target structures. *BMC Bioinform.* **2019**, *20*, 209. [CrossRef] [PubMed]
7. Bienstock, D.; Ozbay, N. Tree-width and the Sherali–Adams operator. *Discret. Optim.* **2004**, *1*, 13–21. [CrossRef]
8. Arnborg, S.; Corneil, D.G.; Proskurowski, A. Complexity of finding embeddings in a k-tree. *SIAM J. Algeb. Discret. Methods* **1987**, *8*, 277–284. [CrossRef]
9. Strasser, B. Computing Tree Decompositions with FlowCutter: PACE 2017 Submission. *arXiv* **2017**, arXiv:1709.08949.
10. van Wersch, R.; Kelk, S. ToTo: An open database for computation, storage and retrieval of tree decompositions. *Discret. Appl. Math.* **2017**, *217*, 389–393. [CrossRef]
11. Bodlaender, H. A Linear-Time Algorithm for Finding Tree-Decompositions of Small Treewidth. *SIAM J. Comput.* **1996**, *25*, 1305–1317. [CrossRef]
12. Bodlaender, H.L.; Fomin, F.V.; Koster, A.M.; Kratsch, D.; Thilikos, D.M. On exact algorithms for treewidth. *ACM Trans. Algorithms (TALG)* **2012**, *9*, 12. [CrossRef]

13. Gogate, V.; Dechter, R. A complete anytime algorithm for treewidth. In Proceedings of the 20th conference on Uncertainty in artificial intelligence, UAI 2004, Banff, AB, Canada, 7–11 July 2004; AUAI Press: Arlington, VA, USA, 2004; pp. 201–208.
14. Tamaki, H. Positive-instance driven dynamic programming for treewidth. *J. Comb. Optim.* **2019**, *37*, 1283–1311. [CrossRef]
15. Dell, H.; Husfeldt, T.; Jansen, B.M.; Kaski, P.; Komusiewicz, C.; Rosamond, F.A. The first parameterized algorithms and computational experiments challenge. In Proceedings of the 11th International Symposium on Parameterized and Exact Computation (IPEC 2016), Aarhus, Denmark, 24–26 August 2016; Schloss Dagstuhl-Leibniz-Zentrum für Informatik: Wadern, Germany, 2017.
16. Dell, H.; Komusiewicz, C.; Talmon, N.; Weller, M. The PACE 2017 Parameterized Algorithms and Computational Experiments Challenge: The Second Iteration. In Proceedings of the 12th International Symposium on Parameterized and Exact Computation (IPEC 2017), Leibniz International Proceedings in Informatics (LIPIcs), Vienna, Austria, 6–8 September 2017; Lokshtanov, D., Nishimura, N., Eds.; Schloss Dagstuhl–Leibniz-Zentrum fuer Informatik: Dagstuhl, Germany, 2018; Volume 89, pp. 1–12, doi:10.4230/LIPIcs.IPEC.2017.30. [CrossRef]
17. Jordan, M.I.; Mitchell, T.M. Machine learning: Trends, perspectives, and prospects. *Science* **2015**, *349*, 255–260. [CrossRef]
18. Hutter, F.; Hoos, H.H.; Leyton-Brown, K. Automated configuration of mixed integer programming solvers. In Proceedings of the International Conference on Integration of Artificial Intelligence (AI) and Operations Research (OR) Techniques in Constraint Programming, Thessaloniki, Greece, 4–7 June 2019; Springer: Cham, Switzerland, 2010; pp. 186–202.
19. Kruber, M.; Lübbecke, M.E.; Parmentier, A. Learning when to use a decomposition. In Proceedings of the International Conference on AI and OR Techniques in Constraint Programming for Combinatorial Optimization Problems, Padova, Italy, 5–8 June 2017; Springer: Cham, Switzerland, 2017; pp. 202–210.
20. Tang, Y.; Agrawal, S.; Faenza, Y. Reinforcement Learning for Integer Programming: Learning to Cut. *arXiv* **2019**, arXiv:1906.04859.
21. Smith-Miles, K.; Lopes, L. Measuring instance difficulty for combinatorial optimization problems. *Comput. Oper. Res.* **2012**, *39*, 875–889. [CrossRef]
22. Hutter, F.; Xu, L.; Hoos, H.H.; Leyton-Brown, K. Algorithm runtime prediction: Methods & evaluation. *Artif. Intell.* **2014**, *206*, 79–111.
23. Leyton-Brown, K.; Hoos, H.H.; Hutter, F.; Xu, L. Understanding the empirical hardness of NP-complete problems. *Commun. ACM* **2014**, *57*, 98–107. [CrossRef]
24. Lodi, A.; Zarpellon, G. On learning and branching: A survey. *Top* **2017**, *25*, 207–236. [CrossRef]
25. Alvarez, A.M.; Louveaux, Q.; Wehenkel, L. A machine learning-based approximation of strong branching. *INFORMS J. Comput.* **2017**, *29*, 185–195. [CrossRef]
26. Balcan, M.F.; Dick, T.; Sandholm, T.; Vitercik, E. Learning to branch. *arXiv* **2018**, arXiv:1803.10150.
27. Bengio, Y.; Lodi, A.; Prouvost, A. Machine Learning for Combinatorial Optimization: A Methodological Tour d'Horizon. *arXiv* **2018**, arXiv:1811.06128.
28. Fischetti, M.; Fraccaro, M. Machine learning meets mathematical optimization to predict the optimal production of offshore wind parks. *Comput. Oper. Res.* **2019**, *106*, 289–297. [CrossRef]
29. Sarkar, S.; Vinay, S.; Raj, R.; Maiti, J.; Mitra, P. Application of optimized machine learning techniques for prediction of occupational accidents. *Comput. Oper. Res.* **2019**, *106*, 210–224. [CrossRef]
30. Nalepa, J.; Blocho, M. Adaptive guided ejection search for pickup and delivery with time windows. *J. Intell. Fuzzy Syst.* **2017**, *32*, 1547–1559. [CrossRef]
31. Rice, J.R. The algorithm selection problem. In *Advances in Computers*; Elsevier: Amsterdam, The Netherlands, 1976; Volume 15, pp. 65–118.
32. Leyton-Brown, K.; Nudelman, E.; Andrew, G.; McFadden, J.; Shoham, Y. A portfolio approach to algorithm selection. In Proceedings of the IJCAI, Acapulco, Mexico, 9–15 August 2003; Volume 3, pp. 1542–1543.
33. Nudelman, E.; Leyton-Brown, K.; Devkar, A.; Shoham, Y.; Hoos, H. Satzilla: An algorithm portfolio for SAT. Available online: http://www.cs.ubc.ca/~kevinlb/pub.php?u=SATzilla04.pdf (accessed on 12 July 2019).
34. Xu, L.; Hutter, F.; Hoos, H.H.; Leyton-Brown, K. SATzilla: Portfolio-based algorithm selection for SAT. *J. Artif. Intell. Res.* **2008**, *32*, 565–606. [CrossRef]

35. Ali, S.; Smith, K.A. On learning algorithm selection for classification. *Appl. Soft Comput.* **2006**, *6*, 119–138. [CrossRef]
36. Guo, H.; Hsu, W.H. A machine learning approach to algorithm selection for NP-hard optimization problems: A case study on the MPE problem. *Ann. Oper. Res.* **2007**, *156*, 61–82. [CrossRef]
37. Musliu, N.; Schwengerer, M. Algorithm selection for the graph coloring problem. In Proceedings of the International Conference on Learning and Intelligent Optimization 2013 (LION 2013), Catania, Italy, 7–11 January 2013; pp. 389–403.
38. Xu, L.; Hutter, F.; Hoos, H.H.; Leyton-Brown, K. Hydra-MIP: Automated algorithm configuration and selection for mixed integer programming. In Proceedings of the RCRA Workshop on Experimental Evaluation of Algorithms for Solving Problems with Combinatorial Explosion at the International Joint Conference on Artificial Intelligence (IJCAI), Paris, France, 16–20 January 2011; pp. 16–30.
39. Kerschke, P.; Hoos, H.H.; Neumann, F.; Trautmann, H. Automated algorithm selection: Survey and perspectives. *Evol. Comput.* **2019**, *27*, 3–45. [CrossRef]
40. Abseher, M.; Musliu, N.; Woltran, S. Improving the efficiency of dynamic programming on tree decompositions via machine learning. *J. Artif. Intell. Res.* **2017**, *58*, 829–858. [CrossRef]
41. Bannach, M.; Berndt, S.; Ehlers, T. Jdrasil: A modular library for computing tree decompositions. In Proceedings of the 16th International Symposium on Experimental Algorithms (SEA 2017), London, UK, 21–23 June 2017; Schloss Dagstuhl–Leibniz-Zentrum fuer Informatik: Wadern, Germany, 2017.
42. Kotsiantis, S.B. Decision trees: A recent overview. *Artif. Intell. Rev.* **2013**, *39*, 261–283. [CrossRef]
43. Li, R.H.; Belford, G.G. Instability of decision tree classification algorithms. In Proceedings of the Eighth ACM SIGKDD International Conference on Knowledge Discovery and Data Mining, Edmonton, AB, Canada, 23–26 July 2002; ACM: New York, NY, USA, 2002; pp. 570–575.
44. Liaw, A.; Wiener, M. Classification and regression by randomForest. *R News* **2002**, *2*, 18–22.
45. Bertsimas, D.; Dunn, J. Optimal classification trees. *Mach. Learn.* **2017**, *106*, 1039–1082. [CrossRef]
46. Cristianini, N.; Shawe-Taylor, J. *An Introduction to Support Vector Machines and Other Kernel-Based Learning Methods*; Cambridge University Press: Cambridge, UK, 2000.
47. Ásgeirsson, E.I.; Stein, C. Divide-and-conquer approximation algorithm for vertex cover. *SIAM J. Discret. Math.* **2009**, *23*, 1261–1280. [CrossRef]
48. Pedregosa, F.; Varoquaux, G.; Gramfort, A.; Michel, V.; Thirion, B.; Grisel, O.; Blondel, M.; Prettenhofer, P.; Weiss, R.; Dubourg, V.; et al. Scikit-learn: Machine learning in Python. *J. Mach. Learn. Res.* **2011**, *12*, 2825–2830.
49. Tsamardinos, I.; Rakhshani, A.; Lagani, V. Performance-estimation properties of cross-validation-based protocols with simultaneous hyper-parameter optimization. *Int. J. Artif. Intell. Tools* **2015**, *24*, 1540023. [CrossRef]
50. Smith-Miles, K.A. Cross-disciplinary perspectives on meta-learning for algorithm selection. *ACM Comput. Surv. (CSUR)* **2009**, *41*, 6. [CrossRef]
51. Bodlaender, H.L.; Jansen, B.M.; Kratsch, S. Preprocessing for treewidth: A combinatorial analysis through kernelization. *SIAM J. Discret. Math.* **2013**, *27*, 2108–2142. [CrossRef]
52. Van Der Zanden, T.C.; Bodlaender, H.L. Computing Treewidth on the GPU. *arXiv* **2017**, arXiv:1709.09990.

© 2019 by the authors. Licensee MDPI, Basel, Switzerland. This article is an open access article distributed under the terms and conditions of the Creative Commons Attribution (CC BY) license (http://creativecommons.org/licenses/by/4.0/).

Article

The Inapproximability of k-DOMINATINGSET for Parameterized AC^0 Circuits [†]

Wenxing Lai

Shanghai Key Laboratory of Intelligent Information Processing, School of Computer Science, Fudan University, Shanghai 201203, China; wenxing.lai@outlook.com or wenxinglai.fudan@gmail.com

† This paper is an extended version of our paper published in the Thirteenth International Frontiers of Algorithmics Workshop (FAW 2019), Sanya, China, 29 April–3 May 2019.

Received: 18 July 2019; Accepted: 11 October 2019; Published: 4 November 2019

Abstract: Chen and Flum showed that any FPT-approximation of the k-CLIQUE problem is not in para-AC^0 and the k-DOMINATINGSET (k-DOMSET) problem could not be computed by para-AC^0 circuits. It is natural to ask whether the $f(k)$-approximation of the k-DOMSET problem is in para-AC^0 for some computable function f. Very recently it was proved that assuming $W[1] \neq FPT$, the k-DOMSET problem cannot be $f(k)$-approximated by FPT algorithms for any computable function f by S., Laekhanukit and Manurangsi and Lin, seperately. We observe that the constructions used in Lin's work can be carried out using constant-depth circuits, and thus we prove that para-AC^0 circuits could not approximate this problem with ratio $f(k)$ for any computable function f. Moreover, under the hypothesis that the 3-CNF-SAT problem cannot be computed by constant-depth circuits of size $2^{\varepsilon n}$ for some $\varepsilon > 0$, we show that constant-depth circuits of size $n^{o(k)}$ cannot distinguish graphs whose dominating numbers are either $\leq k$ or $> \left(\frac{\log n}{3 \log \log n}\right)^{1/k}$. However, we find that the hypothesis may be hard to settle by showing that it implies $NP \not\subseteq NC^1$.

Keywords: parameterized AC^0; dominating set; inapproximability

1. Introduction

The dominating set problem is often regarded as one of the most important NP-complete problems in computational complexity. A dominating set in a graph is a set of vertices such that every vertex in the graph is either in the set or adjacent to a vertex in it. The dominating set problem is, given a graph $G = (V, E)$ and a number $k \in \mathbb{N}$, to decide the minimum dominating set of G has a size of at most k. This problem is tightly connected to the set cover problem, which was firstly shown to be NP-complete in Karp's famous NP-completeness paper [1]. Unless P = NP, we do not expect to solve this problem and its optimization variant in polynomial time. Furthermore, the set cover conjecture asserts that for every fixed $\varepsilon > 0$, no algorithm can solve the set cover problem in time $2^{(1-\varepsilon)n}\text{poly}(m)$, even if set sizes are bounded by $\Delta = \Delta(\varepsilon)$ [2,3]. One way to handle NP-hard problems is to use approximation algorithms. One key measurement of an approximation algorithm for the dominating set problem is its approximation ratio, i.e., the ratio between the size of the solution output by the algorithm and the size of the minimum dominating set. It is known that greedy algorithms can achieve an approximation ratio of $\approx \ln n$ [4–8]. Though this problem has a PTAS (polynomial-time approximation scheme, an algorithm which takes an instance of an optimization problem and a parameter $\varepsilon > 0$ and, in polynomial time, approximate the problem with ratio $1 + \varepsilon$) applied to apex-minor-free graphs for contraction-bidimensional parameters [9], after a long line of works [10–14], the approximation ratio of this problem was matched by the lower bound by Dinur and Steurer [15], who followed the construction presented in Feige's work [12], showing that for every $\varepsilon > 0$, we could not obtain a $(1-\varepsilon)\ln n$-approximation for this problem unless P = NP. Besides approximation, another widely-considered technique to circumvent the intractability of NP-hard problems is parameterization. If we take the minimum solution size k as a parameter, then the brute-force algorithm can solve the k-DOMINATINGSET (k-DOMSET) problem in $O(n^{k+1})$ time. However, it is recently proved that, assuming FPT $\ne$ W[1], for any computable function f, there is no $f(k)$-FPT-approximation algorithm, that is, there is no approximation algorithm running in FPT-time and with a ratio of $f(k)$ [16–18].

Circuit complexity was thought to be a promising direction to solve P vs. NP. Though it has been long known that some problems, like the parity problem, are not in AC^0 [19–21], proving that non-uniform lower bounds for functions in nondeterministic complexity classes such as NP, NQP = NTIME$[n^{\log^{O(1)} n}]$, or NEXP is a well-known challenge. After Williams' proving that NEXP does not have $n^{\log^{O(1)} n}$-size ACC ∘ THR circuits (ACC composed with a layer of linear threshold gates at the bottom) [22,23], Murray and Williams showed that for every k,d, and m there is an e and a problem in NTIME$[n^{\log^e n}]$ which does not have depth-d $n^{\log^k n}$-size AC$[m]$ circuits with linear threshold gates at the bottom layer [24].

Rossman showed that the k-CLIQUE problem has no bounded-depth and unbounded fan-in circuits of size $O(n^{k/4})$ [25], which may be viewed as an AC^0 version of FPT $\ne$ W[1]. Chen and Flum [26] showed that any FPT-approximation of the k-CLIQUE problem is not in para-AC^0. The parameterized circuit complexity class para-AC^0 introduced by Elberfeld, Stockhusen, and Tantau [27] as the AC^0 analog of the class FPT, is the class of parameterized problems computed by constant-depth circuits of size $f(k)\text{poly}(n)$ for some computable function f. In the same paper, based on Rossman's result, they also showed that the k-DOMSET problem could not be computed in para-AC^0. This brings us to the main question addressed in our work: *Is there a computable function f such that the $f(k)$-approximation of k-DOMSET is in* para-AC^0? Furthermore, since we could enumerate every k tuple of vertices by depth-3 circuits of size $O(n^{k+1})$ using brute force, we might wonder whether it is possible to have a computable function f such that the $f(k)$-approximation of k-DOMSET could be computed by constant-depth circuits of size $n^{o(k)}$.

Our Work

In this paper, we show that for any computable function f, the $f(k)$-FPT-approximation of the k-DOMSET problem is not in para-AC^0. Furthermore, under the hypothesis that constant-depth circuits of size $2^{o(n)}$ could not compute 3-CNF-SAT (we call it AC^0-ETH, the constant-depth version of ETH—the exponential time hypothesis), there is no computable function f such that the $f(k)$-approximation of k-DOMSET could be computed by constant-depth circuits of size $n^{o(k)}$. Theorems 1 and 2 are direct consequences of Theorems 3 and 4, respectively.

Theorem 1. *Given a graph G with n vertices, there is no constant-depth circuits of size $f(k)n^{o(\sqrt{k})}$ for any computable function f which distinguish between:*

- *The size of the minimum dominating set is at most k,*
- *The size of the minimum dominating set is greater than $\left(\frac{\log n}{\log\log n}\right)^{1/\binom{k}{2}}$.*

Note that this theorem implies the nonexistence of para-AC^0 circuits which $f(k)$-approximates the k-DOMSET problem for any computable function f. This is because if there is an $f(k)$-approximation para-AC^0 circuit $C_{n,k}$ whose size is $g(k)\text{poly}(n)$, we can construct a constant-depth para-AC^0 circuit $C'_{n,k}$ to distinguish the size of the minimum dominating set is at most k or greater than $\left(\frac{\log n}{\log\log n}\right)^{1/\binom{k}{2}}$ as follows. Compare $f(k)$ and $\left(\frac{\log n}{\log\log n}\right)^{1/\binom{k}{2}}$ — if $f(k)$ is smaller, we let $C'_{n,k}$ be $C_{n,k}$; otherwise, since $f(k) \geq \left(\frac{\log n}{\log\log n}\right)^{1/\binom{k}{2}}$, we let $C'_{n,k}$ be the circuit which, using brute force, computes the size of a minimum dominating set with the depth-3 circuit of size $O(n^{k+1})$. Since $f(k) \geq \left(\frac{\log n}{\log\log n}\right)^{1/\binom{k}{2}}$, we have $f(k)^{k^3} \geq \left(\frac{\log n}{\log\log n}\right)^{2k}$; by simple calculations we know that $k \cdot \log^{2k-1} n \geq (k+1)(\log\log n)^{2k}$ for $k \geq 2, n \geq 2$, which implies $k \cdot \left(\frac{\log n}{\log\log n}\right)^{2k} \geq (k+1) \cdot \log n$, that is, $2^{k \cdot \left(\frac{\log n}{\log\log n}\right)^{2k}} \geq n^{k+1}$. Thus, we know $2^{kf(k)^{k^3}} \geq n^{k+1}$, which means the circuit is still a para-AC^0 circuit.

Hypothesis 1 (AC^0-ETH, the constant-depth version of ETH). *There exists $\delta > 0$ such that no constant-depth circuits of size $2^{\delta n}$ can decide whether the 3-CNF-SAT instance φ is satisfiable, where n is the number of variables of φ.*

Theorem 2. *Assuming AC^0-ETH, given a graph G with n vertices, there is no constant-depth circuits of size $f(k)n^{o(k)}$ for any computable function f which distinguish between:*

- *The size of the minimum dominating set of G is at most k,*
- *The size of the minimum dominating set of G is greater than $\left(\frac{\log n}{3\log\log n}\right)^{1/k}$.*

Though AC^0-ETH seems much weaker than ETH (ETH implies the nonexistence of uniform circuits of size $2^{\delta n}$ and any depth which could compute the 3-CNF-SAT problem), we show that the hypothesis is hard to settle by proving it implies $NP \not\subseteq NC^1$, which, believed to be true, remains open for decades. Moreover, it is still unknown whether the weaker version, $NP \not\subseteq ACC \circ THR$, holds or not.

Since our hard set cover instances can be easily reduced to the instances of the total dominating set problem, the connected dominating set problem and the independent dominating set problem, we can apply our inapproximability results to these variants of the dominating set problem. More discussion of the variants can be found in the work of Downey and Fellows [28] and the work of Chlebík and Chlebíková [29].

Compared with the conference version [30] of this article, the proofs of Lemmas 1–7 are firstly given here; some results are slightly improved by more careful analyses.

2. Preliminaries

We denote by $\mathbb{N}$ the set of nonnegative integers. For each $n, m \in \mathbb{N}$, we define $[n] := \{1,\ldots,n\}$ and $[n,m] := [m] \setminus [n-1]$ for $m > n > 0$. For any set A and $k \in \mathbb{N}$, we let $\binom{A}{k} := \{B \subseteq A \mid |B| = k\}$ be the set of subsets with exactly k elements of A. For a sequence of bits b, we let $b[l]$ be the l-th bit of b.

For a graph G, the set of vertices of G is denoted by V_G and the set of edges is denoted by E_G; for a vertex $v \in V_G$, we let $N_G(v) := \{u \in V_G \mid \{u,v\} \in E_G\}$ be the neighbors of v. Since a graph G is represented using a binary string, we express the bit of the edge $\{u,v\}$ by $\text{bit}_G\{u,v\}$. For a bipartite graph $G = (L, R, E)$, we often tacitly represent G only using $O(|L| \cdot |R|)$ bits.

In this article, logarithms have base 2, and fractions and irrational numbers are rounded up if necessary.

2.1. Problem Definitions

The decision problems studied in this paper are listed below:

- In the k-DOMINATINGSET (k-DOMSET) problem, our goal is to decide if there is a dominating set of size k in the given graph G.
- In the k-SETCOVER problem, we are given a bipartite graph $I = (S, U, E)$ and the goal is to decide whether there is a subset X of S with cardinality k such that for each vertex v in U, there exists a vertex u in X that covers v, i.e., $\{u, v\} \in E$.
- In the k-CLIQUE problem, our goal is to determine if there is a clique of size k in the given graph G.
- In the 3-CNF-SAT problem, we are given a propositional formula φ in which every clause contains at most 3 literals and the goal is to decide whether φ is satisfiable.

We say a set cover instance $I = (S, U, E)$ has set cover number m if the size of a minimum set $X \subseteq S$ such that X could cover U is m. Similarly, we say a graph G has dominating number m if the size of a minimum dominating set of G is m.

As we mentioned, the dominating set problem is tightly connected to the set cover problem. Given a k-DOMSET instance $G = (V, E)$, we can construct a k-SETCOVER instance $I = (S, U, E')$ with $S = V$, $U = V$ and $E' = \bigcup_{\{u,v\} \in E} \{\{u_S, u_U\}, \{v_S, v_U\}, \{u_S, v_V\}, \{v_S, u_U\}\}$; here, for each vertex $v \in V$, we denote the corresponding vertices in S, U by v_S and v_U, respectively. It is quite clear that G has dominating number k if and only if I has set cover number k. Also, given a k-SETCOVER instance $I = (S, U, E)$, we can construct a k-DOMSET instance $G = (D \cup U_1 \cup U_2, C \cup E_1 \cup E_2)$ by letting $D = S$, $U_1 = U_2 = U$, $C = \{\{u, v\} \mid u, v \in D\}$, $E_1 = \{\{s, u\} \mid s \in D, u \in U_1\}$ and $E_2 = \{\{s, u\} \mid s \in D, u \in U_2\}$. It is trivial that I has set cover number k if and only if G has dominating number k. The reductions can also be found in the work of Chlebík and Chlebíková [29].

It is notable that each hard instance with gap reduced from a CLIQUE or 3-CNF-SAT instance satisfies that the size of the sets M is at most $\text{poly}(N)$ where N is the size of the universe. Hence, it is safe to tacitly apply the inapproximability of k-SETCOVER to the k-DOMSET problem.

2.2. Circuit Complexity

For $n, m \in \mathbb{N}$, an n-input, m-output Boolean circuit C is a directed acyclic graph with n vertices with no incoming edges and m vertices with no outgoing edges. All nonsource vertices are called gates and are labeled with one of either $\vee, \wedge,$ or $\neg$. The size of C, denoted by $|C|$, is the number of vertices in it. The depth of C is the length of the longest directed path from an input node to the output node. We often tacitly identify C with the function $C: \{0,1\}^n \to \{0,1\}^m$ it computes.

All the circuits considered in this paper are non-uniform and with unbounded fan-in $\wedge$ and $\vee$ gates unless otherwise stated.

The classes of AC^0, para-AC^0, and NC^1 are defined as follows:

- AC^0 is the class of problems which can be computed by constant-depth circuit families $(C_n)_{n \in \mathbb{N}}$ where every C_n has size $\text{poly}(n)$, and whose gates have unbounded fan-in.
- Para-AC^0 is the class of parameterized problems which can be computed by a circuit family $(C_{n,k})_{n,k \in \mathbb{N}}$ satisfying that there exist $d \in \mathbb{N}$ and a computed function f such that for every $n \in \mathbb{N}, k \in \mathbb{N}$, $C_{n,k}$ has depth d and size $f(k)\text{poly}(n)$, and whose gates have unbounded fan-in.
- NC^1 is the class of problems which can be computed by a circuit family $(C_n)_{n \in \mathbb{N}}$ where C_n has depth $O(\log n)$ and size $\text{poly}(n)$, and whose gates have a fan-in of 2.

2.3. Covering Arrays

A covering array $CA(N; t, p, v)$ is an $N \times p$ array A whose cells take values from a set V of size v and the set of rows of every $N \times t$ subarray of A is the whole set V^t. The smallest number N such that $CA(N; t, p, v)$ exists is denoted by $CAN(t, p, v)$. Covering arrays are discussed extensively since the

1990s, as they play an important role in the interaction testing of complex engineered systems. The recent discussion about the upper bounds of the size of covering arrays can be found as presented by Sarkar and Colbourn [31].

In this article, we always assume $V = \{0,1\}$. It is noted that in Lin's work [18], a covering array $CA(N;k,n,2)$ is also called an (n,k)-universal set.

3. Introducing Gap to the k-SETCOVER Problem

Theorem 1 and Theorem 2 show that the para-AC^0 circuits cannot approximate the k-DOMSET problem with ratio $f(k)$ for any computable function f. To achieve this, we need to introduce gaps for k-DOMSET instances. In this section, we present the lemmas which allow us to introduce gaps to the k-SETCOVER problem, using *gap-gadgets* as presented in Lin's work [18]. Lemma 1 gives an upper bound for CAN$(k,n,2)$. The next two lemmas also follow the idea from Lin's work [18]. Lemma 2 allows us to construct gap gadgets with $h \leq \frac{\log n}{\log \log n}$ and $k \log \log n \leq \log n$. In Lemma 3, we present the construction which introduces gaps to set cover instances.

Definition 1. *A (k,n,m,ℓ,h)-Gap-Gadget is a bipartite graph $T = (A, B, E)$ satisfying the following conditions.*

(G1) *A is partitioned into $(A_1, \ldots, A_m)$ where $|A_i| = \ell$ for every $i \in [m]$.*
(G2) *B is partitioned into $(B_1, \ldots, B_k)$ where $|B_j| = n$ for every $j \in [k]$.*
(G3) *For each $b_1 \in B_1, \ldots, b_k \in B_k$, there exists $a_1 \in A_1, \ldots, a_m \in A_m$ such that a_i is adjacent to b_j for $i \in [m], j \in [k]$.*
(G4) *For any $X \subseteq B$ and $a_1 \in A_1, \ldots, a_m \in A_m$, if a_i has $k+1$ neighbors in X for $i \in [m]$, then $|X| > h$.*

Lemma 1. $CAN(k,n,2) \leq k2^k \log n$ for $n \geq 5$.

Proof. We let $M = 2^k \binom{n}{k}(1 - \frac{1}{2^k})^{k2^k \log n}$. Kleitman and Spencer showed that if $2^k \binom{n}{k}(1 - \frac{1}{2^k})^r < 1$, $CAN(k,n,2) \leq r$ [32]. Thus, we only need to show that $M < 1$. Since $2^x \log(\frac{2^x}{2^x-1}) > \frac{1}{\ln 2}$ and $n > 5$, we have

$$\log M < k + k \log n + k2^k \log n \log(1 - \frac{1}{2^k})$$
$$< k + k \log n - \frac{1}{\ln 2} k \log n$$
$$< 0$$

This implies $M < 1$. □

Lemma 2. *There is a constant-depth circuit family $(C_{k,n,h})_{k,n,h \in \mathbb{N}}$ which, for sufficiently large n and $k, h \in \mathbb{N}$ with $h \leq \frac{\log n}{\log \log n}$ and $k \log \log n \leq \log n$, given $S = S_1 \cup \cdots \cup S_k$ with $|S_i| = n$ for $i \in [k]$, outputs a $(k, n, n \log h, h^k, h)$-Gap-Gadget $T = (A, B, E)$ with $|A| = h^k n \log h$, $B = S$. Furthermore, $C_{k,n,h}$ has size at most $kh^k n^2 \log h$ and could output whether $\mathbf{a}$ and b are adjacent using $O(1)$ gates, for every $\mathbf{a} \in A, b \in B$.*

Proof. Let $m = n \log h$. Note that $\log((h \log h)2^{h \log h} \log m) \leq (h+2) \log h + \log \log n$, that is, $(h \log h)2^{h \log h} \log m \leq n \leq n \log h$; by Lemma 1, we know that there exists a covering array $CA(n \log h; k, n, 2)$, denoted by $\mathcal{S}$.

We partition every row of $\mathcal{S}$ into $n = \frac{m}{\log h}$ blocks so that each block has length $\log h$, interpreted as an integer in $[h]$. From the $m \times n$ numbers of $\mathcal{S}$, we could obtain an $m \times n$ matrix M by setting $M_{r,c}$ to be the c-th integer of the r-th row.

Claim. *For any $C \subseteq [n]$ with $|C| \leq h$, there exists $r \in [m]$ such that $|\{M_{r,c} \mid c \in C\}| = |C|$.*

This claim says that for any $C = \{i_1, \ldots, i_j\} \subseteq [n]$ for $j \leq h$, there is a row r such that the i_1-th, ..., i_j-th numbers of r are distinct. This is because we could choose the corresponding bits $C' = \cup_{c \in C}[(c-1)\log h + 1, c\log h]$ (since for each $c \in [n]$, the c-th number of a row is from the $((c-1)\log h + 1)$-th bit to the $(c \log n)$-th bit) of the row, with $|C'| \leq h \log h$; by the property of a CA$(n \log h; k, n, 2)$ covering array, there must be a row r such that for each $i_{j'} \in C$, $M_{r,i_{j'}} = j'$.

Now we construct a bipartite graph $T = (A, B, E)$ as follows.

- $A = \cup_{i \in [m]} A_i$ with each $A_i = \{\boldsymbol{a} \mid \boldsymbol{a} = (a_1, \ldots, a_k), a_j \in [h] \text{ for } j \in [k]\}$;
- $B = \cup_{i \in [k]} B_i$ with $B_i = S_i$ for $i \in [k]$;
- $E = \{\{\boldsymbol{a}, b\} \mid \boldsymbol{a} \in A_i, b \in B_j \text{ and } M_{i,b} = \boldsymbol{a}[j], \text{ for } i \in [m], j \in [k]\}$, that is, for every $i \in [m], j \in [k]$ and every $\boldsymbol{a} \in A_i, b \in B_j$, if $M_{i,b} = \boldsymbol{a}[j]$ then we add an edge between $\boldsymbol{a}$ and b.

We prove that T is a $(k, n, n \log h, h^k, h)$-Gap-Gadget. It is clear that (G1) and (G2) hold for T. For (G3), given any $b_1 \in B_1, \ldots, b_k \in B_k$, we know that for each $i \in [m]$, $(M_{i,b_1}, \ldots, M_{i,b_k}) \in A_i$, which is adjacent to $b_1, \ldots, b_k$.

If T does not satisfy (G4), then there exists $X \subseteq B$ with $|X| \leq h$ such that there is $\boldsymbol{a}_1 \in A_1, \ldots, \boldsymbol{a}_m \in A_m$ and $\boldsymbol{a}_i$ has at least $k+1$ neighbors in X for each $i \in [m]$. Since $|X| \leq h$, we know that there is a row $r \in [m]$ such that $|\{M_{r,c} \mid c \in X\}| = X$. For this r, there exist some $j \in [k]$ such that $\boldsymbol{a}_r$ has at least 2 neighbors $b_1 \neq b_2$ in B_j. However, $\{\boldsymbol{a}_r, b_1\}$ and $\{\boldsymbol{a}_r, b_2\} \in E$ means that $b_1 = b_2 = M_{i,b_1}$. This implies $|\{M_{r,c} \mid c \in X\}| < X$, which is a contradiction.

The $\mathsf{C}_{k,n,h}$ outputs T with $kh^k n^2 \log h$ bits where whether $\boldsymbol{a}$ and b are connected is determined by

$$\text{bit}_T\{\boldsymbol{a}, b\} = \begin{cases} 1 & \text{if } M_{i,b} = \boldsymbol{a}[j] \\ 0 & \text{otherwise} \end{cases}$$

for every $\boldsymbol{a} \in A_i, b \in B_j$. □

Given a set cover instance $I = (S, U, E)$, we construct the gap gadget $T = (A, B, E_T)$ with $B = S$. To use the gap gadget, we construct a new set cover instance $I' = (S', U', E')$ with $S' = S$ such that for every $X \subseteq S'$ which covers U', there exists $\boldsymbol{a}_1 \in A_1, \ldots, \boldsymbol{a}_m \in A_m$ witnessing that there is an $X' \subseteq X$ which covers U and each vertex of which is adjacent to $\boldsymbol{a}_i$ for some $i \in [m]$.

In the following lemma, we use the hypercube set system, which is firstly presented in Feige's work [12] and is also used in [16–18]. The set $X^Y = \{f : Y \to X\}$ is considered to be all the functions from Y to X with $|X^Y| = |X|^{|Y|}$.

Lemma 3. *There is a constant-depth circuit family $(\mathsf{C}_{n,k})_{n,k \in \mathbb{N}}$ which, for each $k \in \mathbb{N}$, given a set cover instance $I = (S, U, E)$ where $S = S_1 \cup \cdots \cup S_k$ and $|S_i| = n$ for $i \in [k]$ and a (k, n, m, ℓ, h)-Gap-Gadget constructed with S as Lemma 2 describes, outputs a set cover instance $I' = (S', U', E)$ with $S' = S$ and $|U'| = m|U|^\ell$ such that*

- *If there exists $s_1 \in S_1, \ldots, s_k \in S_k$ which could cover U, then the set cover number of I' is at most k;*
- *If the set cover number of I is larger than k, then the set cover number of I' is greater than h.*

Furthermore, the circuit $\mathsf{C}_{n,k}$ has size at most $knm\ell|U|^\ell$ and could output whether s and f are connected with at most $(\ell + 1)$ gates.

Proof. Let $T = (A = \cup_{i \in [m]} A_i, B, E_T)$ be the (k, n, m, ℓ, h)-Gap-Gadget with $B_i = S_i$ for $i \in [k]$. $I' = (S', U', E')$ is defined as follows.

- $S' = S$;
- $U' = \cup_{i \in [m]} U^{A_i}$;
- For every $s \in S'$ and $f \in U^{A_i}$ for each $i \in [m]$, $\{s, f\} \in E'$ if there is an $a \in A_i$ such that $\{s, f(a)\} \in E$ and $\{a, s\} \in E_T$.

If there exist $s_1 \in S_1, \ldots, s_k \in S_k$ that can cover U, then we show that for each $f \in U'$, it is covered by some vertex in $C = \{s_1, \ldots, s_k\}$. Suppose $f \in U^{A_i}$. By (G3) we know that, for $s_1 \in S_1, \ldots, s_k \in S_k$, there exists $a_1 \in A_1, \ldots, a_m \in A_m$ such that a_p is adjacent to s_q for $p \in [m], q \in [k]$. Since C covers U, there must be $s_j \in C$ for some $j \in [k]$ covers $f(a_i)$. That is, we have $\{f(a_i), s_j\} \in E$ and $\{a_i, s_j\} \in E_T$, which means s_j covers f.

If the set cover number of I is greater than k, we show that for every $X \subseteq S'$ that covers U', we must have $|X| > h$.

Claim. For any $X \subseteq S'$ that covers U', there exist $a_1 \in A_1, \ldots, a_m \in A_m$ that $|N_T(a_i) \cap X| \geq k + 1$ for every $i \in [m]$.

Otherwise, there is some $i \in [m]$ such that for any $a \in A_i$, we have $|N_T(a) \cap X| \leq k$, which means there is some $u \in U$ not covered by $N_T(a) \cap X$ since the covering number of I is greater than k. For $f \in U^{A_i}$ such that $f(a') = u$ for any $a' \in A_i$, it is covered by S only if it is covered by some $s \in S \setminus N_T(a)$ since u can only be covered by $S \setminus N_T(a)$. However, for any $s \in S \setminus N_T(a)$, s is not a neighbor of a. That is, f is not adjacent to $S \setminus N_T(a)$, either. Hence, f is not covered by X, which is a contradiction.

With the claim, we know that for any $X \subseteq S'$ that covers U', there exist $a_1 \in A_1, \ldots, a_m \in A_m$ that a_i has $k + 1$ neighbors in X for every $i \in [m]$. With (G4), we must have $|X| \geq h$.

The $C_{n,k}$ outputs I' with $knm\ell|U|^{\ell}$ bits where whether there is an edge between s and f is determined by

$$\text{bit}_{I'}\{s, f\} = \bigvee_{a \in A_i} \text{bit}_T\{s, a\} \wedge \text{bit}_I\{s, f(a)\},$$

for every $s \in S', f \in U^{A_i}$ using at most $(\ell + 1)$ gates. □

4. Inapproximability of k-DOMINATINGSET

In this section, we show the inapproximability of the dominating set problem by proving Theorems 3 and 4. To show Theorem 3, we have Lemmas 4 and 5. Lemma 4 follows the idea in recent papers [17,18], presenting the circuits that output a $\binom{k}{2}$-SETCOVER instance given a k-CLIQUE instance as input. With Lemma 4, Lemma 5 introduces circuits reducing k-CLIQUE instances to set cover instances with gaps. To prove Theorem 4, Lemma 6 (firstly shown by Pătraşcu and Williams [33]) is used to prove the inapproximability of set cover problem using constant-depth $n^{o(k)}$ circuits, assuming Hypothesis 1. At the end of this section, we show that Hypothesis 1 may be hard to settle by showing that it implies NP $\not\subseteq$ NC^1, which has remained open for decades.

4.1. The Unconditional Inapproximability of k-DOMINATINGSET

Now we give the circuits which reduce k-CLIQUE instances to $\binom{k}{2}$-SETCOVER instances, and introduce gaps to them. Finally we use Rossman's result [25], i.e., the unconditional lower bounds of the size of constant-depth circuits determining the k-CLIQUE problem, to show the inapproximability of the k-DOMSET problem.

Lemma 4. *There is a $(C_{n,k})_{n,k\in\mathbb{N}}$ circuit family which, given a k-CLIQUE instance G with $|V_G| = n$, outputs a set cover instance $I = (S, U, E)$ with $|U| \leq k^3 \log n$ and $S \leq \binom{k}{2}\binom{n}{2}$ such that G contains a k-clique if and only if the set cover number of I is at most $\binom{k}{2}$. Furthermore, $C_{n,k}$ has constant depth and size at most $k^5 n^2 \log n$.*

Proof. Firstly, we construct $G' = (V_1 \cup \cdots \cup V_k, E')$, a k-colored version of G as follows. Let each $V^{(i)}$ be a copy of V and for every $v \in V_G$, we call the corresponding vertex in $V^{(i)}$ by $v^{(i)}$; let $E' = \cup_{1 \leq i < j \leq k} E_{i,j}$ with $E_{i,j} = \{\{u^{(i)}, v^{(j)}\} \mid \{u, v\} \in E_G\}$. Note that each $V^{(i)}$ is an independent set for $i \in [k]$ and G' contains a k-clique if and only if G contains a k-clique.

Now we construct the the set cover instance $I = (S, U, E)$ according to G' in the following way. Given $v \in G'$, we denote by $b(v)$ the bit representation of v. Note that when i is fixed, every vertex in V_i could be determined using $\log n$ bits.

- $S = E' = \cup_{1 \leq i < j \leq k} E_{i,j}$;
- $U = \cup_{i \in [k]} U_i$ with $U_i = \{(f^{(i)}, l) \mid f^{(i)} : \{0, 1\} \rightarrow [k] \setminus \{i\}, l \in [\log n]\}$;
- For every $i \in [k]$, we connect every $(f^{(i)}, l) \in U_i$ to each $\{v_i, v_j\} \in S$, with $v_i \in V_i, v_j \in V_j$, such that $f^{(i)}(b(v_i)[l]) = j$.

Suppose that there is a k-clique in G' with vertices $u_1 \in V_1, \ldots, u_k \in V_k$. We claim that $\{\{u_i, u_j\} \mid 1 \leq i < j \leq k\}$ covers U. This is because for any $(f^{(i)}, l) \in U_i$, we have $f^{(i)}(b(u_i)[l]) \in [k] \setminus \{i\}$ and thus, it is covered by $\{u_i, u_j\}$.

If there is $X \subseteq S$ covers U with cardinality at most $\binom{k}{2}$, then we show that there is a k-clique in G'. Firstly, $|X \cap E_{i,j}| = 1$ for $1 \leq i < j \leq k$. Otherwise, let $f^{(i)}(0) = f^{(i)}(1) = j$ and for any $l \in [\log n]$, $(f^{(i)}, l) \in U_i$ is not covered by X.

Now we let X be the vertices $e_{i,j} \in E_{i,j}$ for every $1 \leq i < j \leq k$. For each $i \in [k]$ and distinct $j, j' \in [k] \setminus \{i\}$, we let $e_{i,j} = \{v, u^{(j)}\}$ and $e_{i,j'} = \{v', u^{(j')}\}$. We claim that $v = v'$. Otherwise, there must be a bit $l \in [\log n]$ such that $b(v)[l] \neq b(v')[l]$. Without loss of generality, we assume $b(v)[l] = 0, b(v')[l] = 1$. Now we take $f^{(i)}$ such that $f^{(i)}(0) = j'$ and $f^{(i)}(1) = j$. Then $(f^{(i)}, l)$ is not covered by X, which is a contradiction.

Hence, for every $i \in [k]$, we could safely take the vertex $v^{(i)} \in V_i$ such that $v^{(i)}$ is in the edge $e_{i,j} \in X \cap E_{i,j}$ for arbitrary j as the i-th vertex of the k-clique.

The $C_{n,k}$ outputs I' with at most $k^5 n^2 \log n$ bits where whether $e_{u^{(i)}, v^{(j)}}$ and $(f^{(i)}, l)$ is connected is determined by

$$\text{bit}_I\{e_{u^{(i)}, v^{(j)}}, (f^{(i)}, l)\} = \begin{cases} 1, & \text{if } f^{(i)}(b(u^{(i)})[l]) = j \\ 0, & \text{otherwise} \end{cases}$$

for every $1 \leq i < j \leq k$, $e_{u^{(i)}, v^{(j)}} \in E_{i,j}$ and every $(f^{(i)}, l) \in U_i$. Hence, $C_{n,k}$ is with each output gate of depth at most 3 and of size at most $k^5 n^2 \log n$. □

Lemma 5. *There is a $(C_{n,k})_{n,k\in\mathbb{N}}$ circuit family which, given a k-CLIQUE instance G with $|V_G| = n$, could output a set cover instance $I = (S, U, E)$ with $|U| \leq n^5$ and $S \leq \binom{k}{2}\binom{n}{2}$ such that*

- *If G contains a k-clique, then the set cover number of I is at most $\binom{k}{2}$;*
- *If G contains no k-clique, then the set cover number of I is greater than $\left(\frac{\log n}{\log \log n}\right)^{1/\binom{k}{2}}$.*

Furthermore, $C_{n,k}$ has constant depth and size at most $n^7 \log n$.

Proof. By Lemma 4, we can construct a $\binom{k}{2}$-SETCOVER instance $I = (S, U, E)$ with $|S| \leq \binom{k}{2}\binom{n}{2}$ and $|U| \leq k^3 \log n$, using a constant-depth circuit of size at most $k^5 n^2 \log n$. Let $m = \binom{n}{2}$. By Lemma 2, we can construct a $(\binom{k}{2}, m, m \log h, h^{\binom{k}{2}}, h)$-Gap-Gadget T with $h = (\frac{\log m}{\log \log m})^{1/\binom{k}{2}}$ given S, using a constant-depth circuit of size at most $O(n^2 \log^2 n)$. By Lemma 3, we could have a constant-depth circuit of size at most $n^7 \log n$ which computes a set cover instance $I' = (S', U', E')$ with $S' = S$, $|U'| \leq m \log h (k^3 \log n)^{h^{\binom{k}{2}}}$ such that

- If G contains a k-clique, then the set cover number of I is at most $\binom{k}{2}$;
- If G contains no k-clique, then the set cover number of I is greater than $(\frac{\log m}{\log \log m})^{1/\binom{k}{2}} \geq (\frac{\log n}{\log \log n})^{1/\binom{k}{2}}$.

Here, since $(k^3 \log n)^{\frac{\log m}{\log \log m}} \leq (k^3 \log n)^{\frac{2 \log n}{\log \log n}} \leq (k^{\frac{\log n}{\log \log n}})^6 \cdot n^2 \leq n^{\frac{6k}{\log \log n}} \cdot n^2 \leq n^{2+o(1)}$, we can conclude that $|U'| \leq m \log h (k^3 \log n)^{h^{\binom{k}{2}}} \leq n^2 \log n \cdot n^{2+o(1)} \leq n^{4+o(1)} < n^5$. □

Theorem 3. *Given a set cover instance $I = (S, U, E)$ with $n = |S| + |U|$, for $k > 28$, any constant-depth circuit of size $O(n^{\frac{\sqrt{k}}{20}})$ cannot distinguish between*

- *The set cover number of I is at most k, or*
- *The set cover number of I is greater than $\left(\frac{\log n}{\log \log n}\right)^{1/\binom{k}{2}}$.*

Proof. Rossman showed that for every $k \in \mathbb{N}$, the k-CLIQUE problem on n-vertex graphs requires constant-depth circuits of size $\omega(n^{\frac{k}{4}})$ [25]. Now if there is a constant-depth circuit $C_{n,k}$ of size $O(n^{\frac{\sqrt{k}}{20}})$ that could distinguish between the set cover number of I where $|V_I| = n$ is at most k or greater than $\left(\frac{\log n}{\log \log n}\right)^{1/\binom{k}{2}}$, then by Lemma 5, given $k \in \mathbb{N}$ and a graph G with $|V_G| = n$, we can construct a set cover instance I with vertex number at most $2n^5$ satisfying that if G has a k-clique then the set cover number of I is at most $\binom{k}{2}$ and otherwise it is greater than $\left(\frac{\log n}{\log \log n}\right)^{1/\binom{k}{2}}$—we could use $C_{2n^5, \binom{k}{2}}$ to decide whether the set cover number of I is either $\leq \binom{k}{2}$ or $> \left(\frac{\log n}{\log \log n}\right)^{1/\binom{k}{2}}$. The circuits are of size $O((n^5)^{\sqrt{\binom{k}{2}}/20}) + O(n^7 \log n) = O(n^{\frac{k}{4}})$ when $k > 28$, which contradicts the result shown by Rossman [25]. □

Note that Theorem 3 implies Theorem 1 since for every set cover instance $I = (S, U, E)$ we can construct a dominating set instance $I' = (S \cup U, E \cup \{\{u, v\} \mid u, v \in S\})$ simply by adding edges to S so that it becomes a clique. Then the dominating number of I' is the same as the set cover number of I.

4.2. The Inapproximability of k-DOMINATINGSET Assuming AC^0-ETH

Next we show the $f(k)$-inapproximability of the set cover problem for constant-depth circuits of size $n^{o(k)}$ for any computable function f, assuming AC^0-ETH. To achieve this, we use Lemma 6 to reduce 3-CNF-SAT formulas to set cover instances with gaps.

Lemma 6. *There is a circuit family $(C_{n,k})_{n,k \in \mathbb{N}}$ which for every $k \in \mathbb{N}$, given a 3-CNF-SAT instance φ with n variables where n is much larger than k, outputs $N \leq 2^{\frac{11n}{2k}}$ and a set cover instance $I = (S, U, E)$ satisfying*

- $|S| + |U| \leq N$;
- If φ is satisfiable, then the set cover number of I is at most k;
- If φ is not satisfiable, then the set cover number of I is greater than $\left(\frac{\log N}{3 \log \log N}\right)^{1/k}$;

Furthermore, $C_{n,k}$ has constant depth and size at most $2^{\frac{11n}{2k}}$.

Proof. Firstly, we construct a set cover instance $I' = (S', U', E')$ whose set cover number is k if and only if φ is satisfiable, as follows.

Partition n variables into k parts and each part has n/k variables. We let $S' = S_1 \cup \cdots \cup S_k$ and for $i \in [k]$, each S_i be the set of all the assignments of the variables from the i-th part. Thus, $|S_i| = 2^{n/k}$ for each $i \in [k]$. Let $U' := \{C \mid C \in \varphi\} \cup \{x_i \mid i \in [k]\}$ be the clauses of φ and vertices $x_1, \ldots, x_k$. We define $E' := \cup_{i \in [k]} (\{\{s, C\} \mid s \text{ satisfies } C \text{ for } s \in S_i, C \in U'\} \cup \{\{s, x_i\} \mid s \in S_i\})$ the edges connecting each $s \in S_i$ with x_i and every $C \in U'$ such that s satisfies C, for each $i \in [k]$.

It is clear that if φ is satisfied by assignment σ, then we know that there are $s_1 \in S_1, \ldots, s_k \in S_k$ such that we can combine $s_1, \ldots, s_k$ to get the σ, satisfying φ. Now suppose $s_1 \in S_1, \ldots, s_k \in S_k$ can cover U'. Note that the set cover number of I' cannot be less than k because of the existence of $x_1, \ldots, x_k$. Since the different sets $S_1, \ldots, S_k$ of variables are pairwise disjoint, we could simply combine the assignments $s_1, \ldots, s_k$, which together satisfy all the clauses, to have the assignment satisfying φ.

We have I' with $|S'| = k2^{n/k}$ and $|U'| \leq \binom{n}{3} + k \leq 2n^3$ and let $m = 2^{n/k}$. By Lemma 2, there is a constant-depth circuit which can compute a $(k, m, m \log h, h^k, h)$-Gap-Gadget T with $h = \sqrt[k]{\frac{\log m}{\log \log m}} \geq \left(\frac{\log N}{3 \log \log N}\right)^{1/k}$ which has size $O(k^2 m^2 h^{2k+1}) = O(k^2 m^2 \log^3 m)$. By Lemma 3, there is a constant-depth circuit C that can construct a set cover instance $I = (S, U, E)$ which, given I' and T such that

- If φ is satisfiable, then the set cover number of I is at most k;
- If φ is unsatisfiable, then the set cover number of I is greater than h;
- $S = S'$, $|U| = m \log h |U'|^{h^k} \leq \frac{1}{k} m (\log \log m - \log \log \log m)(2n^3)^{\frac{\log m}{\log \log m}} \leq m \cdot \log \log m \cdot (2^{\frac{\log m}{\log \log m}} + n^{\frac{3n}{k(\log n - \log k)}}) \leq m \cdot \log \log m \cdot (2^{\frac{\log m}{\log \log m}} + n^{\frac{3n}{k}}) = m^{4+o(1)}$.

Thus, $|S| + |U| = km + m^{4+o(1)} \leq 2^{\frac{11n}{2k}} = N$. Furthermore, C has size at most $(h^k + 1)(km \cdot m^{4+o(1)}) = O(m^{5+o(1)}) = 2^{\frac{11n}{2k}}$. □

Theorem 4. *Assuming AC^0-ETH, there is $\varepsilon > 0$ such that, given a set cover instance $I = (S, U, E)$ with $n = |S| + |U|$, any constant-depth Boolean circuit of size $n^{\varepsilon k}$ cannot distinguish between*

- *The set cover number of I is at most k, or*
- *The set cover number of I is greater than $\left(\frac{\log n}{3 \log \log n}\right)^{1/k}$.*

Proof. By AC^0-ETH, there exists $\delta > 0$ such that no constant-depth circuits of size $2^{\delta n}$ can decide whether the 3-CNF-SAT instance φ is satisfiable where n is the number of variables of φ. For every 3-CNF-SAT formula φ, there is a constant-depth circuit $C_{n,k}$ of size $2^{\frac{11n}{2k}}$ which, given φ, computes a set cover instance $I = (S, U, E)$ with $|S| + |U| \leq N$ for $N \leq 2^{\frac{11n}{2k}}$ whose set cover number is either at most k or greater than $\left(\frac{\log N}{3 \log \log N}\right)^{1/k}$ by Lemma 6. Now take $\varepsilon = \delta/12$.

If a constant-depth circuit C_n of size $n^{\varepsilon k}$ could distinguish between the set cover number of I being at most k or greater than $\left(\frac{\log n}{3 \log \log n}\right)^{1/k}$ where n is the vertex number of the given set cover instance I, then we could use $C_{\binom{N}{2}}$ to determine the set cover number of I is whether at most k, i.e., to decide if φ is satisfiable. The used circuits have size at most $2^{\frac{11n}{2k}} + (N^2)^{\epsilon k} \leq 2^{12\epsilon n} = 2^{\delta n}$, which contradicts AC^0-ETH. □

Using the same trick for Theorem 1, we know that Theorem 4 implies Theorem 2.

4.3. The Difficulty of Proving AC^0-ETH

Though AC^0-ETH seems much weaker than ETH, we find that it is still very hard to settle by showing Theorem 5, i.e., AC^0-ETH implies NP $\not\subseteq NC^1$. Firstly, we show the trade off between depth compression and size expansion when simulating NC^1 circuits using constant-depth ones. Then we prove the Theorem 5 by showing that AC^0-ETH implies that 3-CNF-SAT $\notin NC^1$.

Lemma 7. *For every $L \in NC^1$, i.e., there exists $c \in \mathbb{N}$ such that L could be computed by a family of circuits $(C_n)_{n \in \mathbb{N}}$ such that C_n has size at most n^c and depth at most $c \log n$, there exists $d \in \mathbb{N}$ such that there is a family of circuits $(C'_n)_{n \in \mathbb{N}}$ which satisfies*

- *$s \in L$ if and only if $C'_{|s|}$ outputs 1;*
- *C'_n has depth d and size at most $n^{3c/2}(2^{n^{2c/d}+1}+1)$.*

Proof. We show that for every $n \in \mathbb{N}$, C_n could be simulated by a circuit C_n' that has depth d and size $n^{3c/2}(2^{n^{2c/d}+1}+1)$. Suppose C_n has size n^c and depth $c \log n$ (otherwise, we could add dummy gates to C_n). For every gate σ of depth $\frac{c \log n}{\frac{1}{2}d}$, let f_σ be the Boolean function computed by σ. Note that f_σ has at most $n^{2c/d}$ input bits, denoted by $b_1, \ldots, b_{n^{2c/d}}$ since C_n is of fan-in 2. Now we could replace σ using brute force by

$$\bigvee_{f_\sigma(\ell)=1, \ell \in \{0,1\}^{n^{2c/d}}} \bigwedge_{i \in [n^{2c/d}]} \beta_i$$

where $\beta_i = b_i$ if $\ell[i] = 1$ and $\beta_i = \neg b_i$ if $\ell[i] = 0$. That is, σ could be simulated by a 2-depth circuit which has size at most $2^{n^{2c/d}+1}+1$.

Assume for every $l \in [d/2-1]$, every gate σ of depth $\frac{l \cdot 2c \log n}{d}$ could be replaced by a $2l$-depth circuit $C^{(\sigma)}$ which has size $n^{\frac{(l-1)\cdot 3c}{d}}(2^{n^{2c/d}+1}+1)$. Now we could simulate each gate γ of depth $\frac{(l+1)\cdot 2c \log n}{d}$, whose output is determined by the gates $\sigma_1, \ldots, \sigma_{n^{2c/d}}$ from the $\frac{l \cdot 2c \log n}{d}$-th layer, in the similar way. That is, σ is replaced by

$$\bigvee_{f_\sigma(\ell)=1, \ell \in \{0,1\}^{n^{2c/d}}} \bigwedge_{i \in [n^{2c/d}]} C^{(i)}$$

where $C^{(i)} = C^{(\sigma_i)}$ if $\ell[i] = 1$ and $C^{(i)} = \neg C^{(\sigma_i)}$ if $\ell[i] = 0$. Now, $C^{(\gamma)}$ has depth $2(l+1)$ and size at most $(2^{n^{2c/d}+1}+1) + n^{2c/d} \cdot n^{\frac{(l-1)\cdot 3c}{d}}(2^{n^{2c/d}+1}+1) \leq n^{\frac{l \cdot 3c}{d}}(2^{n^{2c/d}+1}+1)$.

Thus, the output gate of C_n could be simulated by a depth d circuit whose size is at most $n^{3c/2}(2^{n^{2c/d}+1}+1)$. □

Theorem 5. *AC^0-ETH implies NP $\not\subseteq NC^1$.*

Proof. We show that AC^0-ETH implies 3-CNF-SAT $\notin NC^1$. If there exists $c \in \mathbb{N}$ such that 3-CNF-SAT could be computed by a family of circuits $(C_n)_{n \in \mathbb{N}}$ satisfying C_n has size at most n^c and depth at most

$c \log n$. By Lemma 7, 3-CNF-SAT could be computed by $5c$-depth, size $n^{3c/2}(2^{n^{2/5}+1} + 1) = O(2^{\sqrt{n}})$ circuits for sufficiently large n, which contradicts AC^0-ETH. □

5. Conclusions and Open Questions

We have presented that para-AC^0 circuits could not approximate the k-DOMSET problem with ratio $f(k)$ for any computable function f. With the hypothesis that the 3-CNF-SAT problem cannot be computed by constant-depth circuits of size $2^{\delta n}$ for some $\delta > 0$, we could show that constant-depth circuits of size $n^{o(k)}$ cannot distinguish graphs whose dominating numbers are either $\leq k$ or $> \left(\frac{\log n}{3 \log \log n}\right)^{1/k}$.

A natural question is to settle the hypothesis, which may be hard since we show that it implies NP $\not\subseteq$ NC^1. Another question is to ask: Are constant-depth circuits of size $n^{o(k)}$ unable to approximate dominating number with ratio $f(k)$ for any computable function f *without assuming* AC^0-*ETH*? Sparsification is one of the key techniques when ETH is involved. Could sparsification be implemented using constant-depth circuits with size $2^{\varepsilon n}$ for any $\varepsilon > 0$? Moreover, we could have more results assuming the set cover conjecture [3]. Can we prove the inapproximability of the set cover problem based on this conjecture?

Funding: This research was funded by National Natural Science Foundation of China under grant 61872092.

Acknowledgments: I thank Yijia Chen for his valuable advice and help when preparing this work. I am grateful to Bundit Laekhanukit and Chao Liao for many helpful discussions and comments. I also thank the anonymous referees for their detailed comments.

Conflicts of Interest: The author declares no conflicts of interest.

References

1. Karp, R.M. Reducibility among Combinatorial Problems. In *Complexity of Computer Computations*; Miller, R.E.; Thatcher, J.W.; Bohlinger, J.D., Eds.; The IBM Research Symposia Series; Springer: Boston, MA, USA, 1972; pp. 85–103.
2. Cygan, M.; Dell, H.; Lokshtanov, D.; Marx, D.; Nederlof, J.; Okamoto, Y.; Paturi, R.; Saurabh, S.; Wahlström, M. On Problems As Hard As CNF-SAT. *ACM Trans. Algorithms* **2016**, *12*, 41:1–41:24. [CrossRef]
3. Krauthgamer, R.; Trabelsi, O. The Set Cover Conjecture and Subgraph Isomorphism with a Tree Pattern. In Proceedings of the 36th International Symposium on Theoretical Aspects of Computer Science (STACS 2019), Berlin, Germany, 13–16 March 2019; pp. 45:1–45:15.
4. Johnson, D.S. Approximation Algorithms for Combinatorial Problems. *J. Comput. Syst. Sci.* **1974**, *9*, 256–278. [CrossRef]
5. Stein, S.K. Two Combinatorial Covering Theorems. *J. Comb. Theory Ser. A* **1974**, *16*, 391–397. [CrossRef]
6. Lovász, L. On the Ratio of Optimal Integral and Fractional Covers. *Discret. Math.* **1975**, *13*, 383–390. [CrossRef]
7. Chvatal, V. A Greedy Heuristic for the Set-Covering Problem. *Math. Oper. Res.* **1979**, *4*, 233–235. [CrossRef]
8. Slavík, P. A Tight Analysis of the Greedy Algorithm for Set Cover. *J. Algorithms* **1997**, *25*, 237–254. [CrossRef]
9. Demaine, E.D.; Hajiaghayi, M. The Bidimensionality Theory and Its Algorithmic Applications 1. *Comput. J.* **2008**, *51*, 292–302. [CrossRef]
10. Lund, C.; Yannakakis, M.; Yannakakis, M. On the Hardness of Approximating Minimization Problems. *J. ACM* **1994**, *41*, 960–981. [CrossRef]
11. Raz, R.; Safra, S. A Sub-Constant Error-Probability Low-Degree Test, and a Sub-Constant Error-Probability PCP Characterization of NP. In Proceedings of the Twenty-Ninth Annual ACM Symposium on Theory of Computing (STOC '97), El Paso, TX, USA, 4–6 May 1997; ACM: New York, NY, USA, 1997; pp. 475–484.
12. Feige, U. A Threshold of $\ln n$ for Approximating Set Cover. *J. ACM* **1998**, *45*, 634–652. [CrossRef]
13. Alon, N.; Moshkovitz, D.; Safra, S. Algorithmic Construction of Sets for k-restrictions. *ACM Trans. Algorithms* **2006**, *2*, 153–177. [CrossRef]

14. Moshkovitz, D. The Projection Games Conjecture and the NP-Hardness of ln n-Approximating Set-Cover. In *Approximation, Randomization, and Combinatorial Optimization. Algorithms and Techniques*; Hutchison, D., Kanade, T., Kittler, J., Kleinberg, J.M., Mattern, F., Mitchell, J.C., Naor, M., Nierstrasz, O., Pandu Rangan, C., Steffen, B., et al., Eds.; Springer: Berlin/Heidelberg, Germany, 2012; Volume 7408, pp. 276–287.
15. Dinur, I.; Steurer, D. Analytical Approach to Parallel Repetition. In Proceedings of the Forty-Sixth Annual ACM Symposium on Theory of Computing (STOC '14), New York, NY, USA, 31 May–3 June 2014; ACM: New York, NY, USA, 2014; pp. 624–633.
16. Chalermsook, P.; Cygan, M.; Kortsarz, G.; Laekhanukit, B.; Manurangsi, P.; Nanongkai, D.; Trevisan, L. From Gap-ETH to FPT-Inapproximability: Clique, Dominating Set, and More. In Proceedings of the 2017 IEEE 58th Annual Symposium on Foundations of Computer Science (FOCS), Berkeley, CA, USA, 15–17 October 2017; pp. 743–754.
17. Laekhanukit, B.; Manurangsi, P. On the Parameterized Complexity of Approximating Dominating Set. In Proceedings of the 50th Annual ACM SIGACT Symposium on Theory of Computing (STOC 2018), Los Angeles, CA, USA, 25–29 June 2018; ACM Press: Los Angeles, CA, USA, 2018; pp. 1283–1296.
18. Lin, B. A Simple Gap-Producing Reduction for the Parameterized Set Cover Problem. In Proceedings of the 46th International Colloquium on Automata, Languages, and Programming (ICALP 2019), Patras, Greece, 8–12 July 2019; pp. 81:1–81:15.
19. Furst, M.; Saxe, J.B.; Sipser, M. Parity, Circuits, and the Polynomial-Time Hierarchy. *Math. Syst. Theory* **1984**, *17*, 13–27. [CrossRef]
20. Ajtai, M. Σ_1^1-Formulae on finite structures. *Ann. Pure Appl. Log.* **1983**, *24*, 1–48. [CrossRef]
21. Håstad, J. Almost Optimal Lower Bounds for Small Depth Circuits. In *Leibniz International Proceedings in Informatics (LIPIcs), Proceedings of the 41st International Symposium on Mathematical Foundations of Computer Science (MFCS 2016)*; Faliszewski, P., Muscholl, A., Niedermeier, R., Eds.; Schloss Dagstuhl–Leibniz-Zentrum fuer Informatik: Dagstuhl, Germany, 2016; Volume 58, pp. 1–14.
22. Williams, R. New Algorithms and Lower Bounds for Circuits with Linear Threshold Gates. In Proceedings of the Forty-Sixth Annual ACM Symposium on Theory of Computing (STOC '14), New York, NY, USA, 31 May–3 June 2014; ACM: New York, NY, USA, 2014; pp. 194–202.
23. Williams, R. Nonuniform ACC Circuit Lower Bounds. *J. ACM* **2014**, *61*, 2:1–2:32. [CrossRef]
24. Murray, C.; Williams, R. Circuit Lower Bounds for Nondeterministic Quasi-Polytime: An Easy Witness Lemma for NP and NQP. In Proceedings of the 50th Annual ACM SIGACT Symposium on Theory of Computing (STOC '18), Los Angeles, CA, USA, 25–29 June 2018; ACM: New York, NY, USA, 2018; pp. 890–901.
25. Rossman, B. On the Constant-depth Complexity of k-clique. In Proceedings of the Fortieth Annual ACM Symposium on Theory of Computing (STOC '08), Victoria, BC, Canada, 17–20 May 2008; ACM: New York, NY, USA, 2008; pp. 721–730.
26. Chen, Y.; Flum, J. Some Lower Bounds in Parameterized AC^0. In *Leibniz International Proceedings in Informatics (LIPIcs), Proceedings of the 41st International Symposium on Mathematical Foundations of Computer Science (MFCS 2016)*; Faliszewski, P., Muscholl, A., Niedermeier, R., Eds.; Schloss Dagstuhl–Leibniz-Zentrum fuer Informatik: Dagstuhl, Germany, 2016; Volume 58, pp. 1–14.
27. Elberfeld, M.; Stockhusen, C.; Tantau, T. On the Space and Circuit Complexity of Parameterized Problems: Classes and Completeness. *Algorithmica* **2015**, *71*, 661–701. [CrossRef]
28. Downey, R.G.; Fellows, M.R. *Parameterized Complexity*; Monographs in Computer Science; Springer: New York, NY, USA, 1999.
29. Chlebík, M.; Chlebíková, J. Approximation Hardness of Dominating Set Problems in Bounded Degree Graphs. *Inf. Comput.* **2008**, *206*, 1264–1275. [CrossRef]
30. Lai, W. The Inapproximability of k-DominatingSet for Parameterized AC^0 Circuits. In Proceedings of the Thirteenth International Frontiers of Algorithmics Workshop (FAW 2019), Sanya, China, 29 April–3 May 2019; pp. 133–143.
31. Sarkar, K.; Colbourn, C.J. Upper Bounds on the Size of Covering Arrays. *SIAM J. Discret. Math.* **2017**, *31*, 1277–1293. [CrossRef]

32. Kleitman, D.J.; Spencer, J. Families of *k*-independent sets. *Discret. Math.* **1973**, *6*, 255–262. [CrossRef]
33. Pătraşcu, M.; Williams, R. On the Possibility of Faster SAT Algorithms. In *Proceedings of the Twenty-First Annual ACM-SIAM Symposium on Discrete Algorithms*; Charikar, M., Ed.; Society for Industrial and Applied Mathematics: Philadelphia, PA, USA, 2010; pp. 1065–1075.

 © 2019 by the author. Licensee MDPI, Basel, Switzerland. This article is an open access article distributed under the terms and conditions of the Creative Commons Attribution (CC BY) license (http://creativecommons.org/licenses/by/4.0/).

Article

Solving Integer Linear Programs by Exploiting Variable-Constraint Interactions: A Survey

Robert Ganian [1,*] and Sebastian Ordyniak [2,*]

[1] Institute of Logic and Computation, Vienna University of Technology, 1040 Vienna, Austria
[2] Department of Computer Science, University of Sheffield, Western Bank, Sheffield S10 2TN, UK
* Correspondence: rganian@gmail.com or rganian@ac.tuwien.ac.at (R.G.); sordyniak@gmail.com or s.ordyniak@sheffield.ac.uk (S.O.)

Received: 29 September 2019; Accepted: 20 November 2019; Published: 22 November 2019

Abstract: Integer Linear Programming (ILP) is among the most successful and general paradigms for solving computationally intractable optimization problems in computer science. ILP is NP-complete, and until recently we have lacked a systematic study of the complexity of ILP through the lens of variable-constraint interactions. This changed drastically in recent years thanks to a series of results that together lay out a detailed complexity landscape for the problem centered around the structure of graphical representations of instances. The aim of this survey is to summarize these recent developments, put them into context and a unified format, and make them more approachable for experts from many diverse backgrounds.

Keywords: integer linear programming; parameterized complexity; treewidth; treedepth

1. Introduction

Integer Linear Programming (ILP) is among the most successful and general paradigms for solving computationally intractable optimization problems in computer science. In particular, a wide variety of problems in areas such as process scheduling [1], planning [2,3], vehicle routing [4], packing [5], and network hub location [6], to name a few, are efficiently solved in practice via a translation into an Integer Linear Program.

ILP is **NP**-complete, and a significant amount of research has been carried out on tractable fragments of ILP defined in terms of the algebraic properties of the instance (see, e.g., the work of Papadimitriou and Steiglitz exploiting total unimodularity (Section 13.2, [7])) or in terms of restrictions on the number of constraints or variables (see Lenstra's algorithm [8] together with its subsequent improvements by Kannan [9], Frank and Tardos [10]).

On the other hand, until recently we have lacked a systematic study of the complexity of ILP through the lens of variable-constraint interactions. This represented a stark contrast to our understanding of other fundamental problems such as BOOLEAN SATISFIABILITY and CONSTRAINT SATISFACTION, where we have classical results that explore and showcase how interactions between variables and constraints (formalized via graphical representations) can be used to define natural tractable fragments of the problem—consider, e.g., the early work of Freuder [11], Dechter and Pearl [12]. The situation for ILP changed drastically in recent years thanks to a flurry of results that together lay out a detailed complexity landscape for the problem centered around variable-constraint interactions, captured in terms of graphical representations of instances. The aim of this survey is to summarize these recent developments, put them into context and a unified format, and make them more approachable for experts from many diverse backgrounds. We will also call attention to prominent open problems in the area.

Survey Organization. After introducing some basic preliminaries for ILPs, graphical representations and structural parameters in Section 2, Section 3 proceeds to a brief overview of classical algorithms for ILP that rely on explicit restrictions such as bounds on the number of variables and/or constraints. In Section 4 we focus on algorithms and lower bounds for ILP that target instances whose variable-constraint interactions give rise to graphical representations of bounded treewidth and/or treedepth. Section 5 then covers results utilizing other structural parameters related to variable-constraint interactions, and the final Section 6 provides an outlook to future work in the area.

2. Preliminaries

For a positive integer n, we use $[n]$ to denote the set $\{1,\ldots,n\}$. We use bold face letters for vectors and normal font when referring to their components, that is, $\mathbf{x}$ is a vector and x_3 is its third component.

2.1. Graphs

We use standard graph terminology, see for instance Diestel's handbook [13]. A graph G is a tuple (V, E), where V or $V(G)$ is the vertex set and E or $E(G)$ is the edge set. A graph H is a subgraph of a graph G, denoted $H \subseteq G$, if H can be obtained by deleting vertices and edges from G. All our graphs are simple and loopless.

A *path* from vertex v to vertex w in G is a sequence of pairwise distinct vertices $v_1, \ldots, v_j$ of G such that $v = v_1$ and $w = v_j$ and $\{v_i, v_{i+1}\} \in E(G)$ for every i with $1 \leq i < j$; and we define the *length* of a path to be equal the number of vertices it contains (i.e., j). A *tree* is a graph in which, for any two vertices $v, w \in G$, there is precisely one unique path from v to w; a tree is *rooted* if it contains a specially designated vertex r, the *root*. Given a vertex v in a tree G with root r, the *parent* of v is the unique vertex w with the property that $\{v, w\}$ is the first edge on the path from v to r.

2.2. Integer Linear Programming

We consider instances of Integer Linear Programming (ILP) in the following two normal forms. In the first case, which we call the equality normal form, instances consist of a matrix $A \in \mathbb{Z}^{m \times n}$ with m rows (constraints) and n columns (variables) and vectors $\mathbf{c}, \mathbf{b} \in \mathbb{Z}^m, \mathbf{l}, \mathbf{u} \in \mathbb{Z}^n \cup \{\infty, -\infty\}$. The set of solutions for the equality normal form is given by:

$$\{\, \mathbf{y} \mid A\mathbf{y} = \mathbf{b}, \mathbf{l} \leq \mathbf{y} \leq \mathbf{u} \,\} \tag{EQ}$$

In the second case, which we call the inequality normal form, instances consist of a matrix $A \in \mathbb{Z}^{m \times n}$ with m rows (constraints) and n columns (variables) and vectors $\mathbf{c}, \mathbf{b} \in \mathbb{Z}^m$. Here, the set of solutions is given by:

$$\{\, \mathbf{y} \mid A\mathbf{y} \leq \mathbf{b} \,\} \tag{INEQ}$$

We denote by ILPF$_=$ and ILPF$_\leq$ the feasibility problem for ILPs, whose sets of solutions are given in the equality respectively the inequality form, e.g., ILPF$_=$ is the problem of deciding whether $\{\, \mathbf{y} \mid A\mathbf{y} = \mathbf{b}, \mathbf{l} \leq \mathbf{y} \leq \mathbf{u} \,\}$ is non-empty and if so to output a vector $\mathbf{y}$ in $\{\, \mathbf{y} \mid A\mathbf{y} = \mathbf{b}, \mathbf{l} \leq \mathbf{y} \leq \mathbf{u} \,\}$. Moreover, ILP$_=$ and ILP$_\leq$ denote the corresponding minimization versions, e.g., ILP$_=$ is the problem of deciding whether $\mathbf{y} \in \{\, \mathbf{y} \mid A\mathbf{y} = \mathbf{b}, \mathbf{l} \leq \mathbf{y} \leq \mathbf{u} \,\}$ contains a vector $\mathbf{y}$ that minimizes $\mathbf{cy}$ and if so outputs such a vector.

For the matrix A of an instance $\mathcal{I}$, we let $\mathbf{x} = (x_1, \ldots, x_n)$ be a vector representing the columns in A and let the variable set, denoted var$(\mathcal{I})$, be the set of such columns. If $\mathcal{I}$ is given in equality normal form, then its constraint set denoted by con$(\mathcal{I})$ contains one equation for every equation in the system $A\mathbf{x} = \mathbf{b}$ respectively one inequality for every inequality in the system $A\mathbf{x} \leq \mathbf{b}$, if $\mathcal{I}$ is given in inequality normal form.

We will also make use of the following notions which describe specific properties of ILP instances. We denote by $\|A\|_\infty, \|\mathbf{b}\|, $ and $\|\mathbf{c}\|$ the maximum absolute value of any coefficient (entry) in the matrix A, in the vector $\mathbf{b}$, and in the vector $\mathbf{c}$, respectively.

For an ILP instance $\mathcal{I}$ in inequality form, we say that the domain of the variable x_i is *bounded* if there are constraints p and r of the form $x_i \leq b_p$ and $-x_i \leq b_r$, otherwise we say that its domain is *unbounded*. Moreover, we denote by $\|x_i\|$ the maximum *domain span* of the i-th variable, i.e., given by $\|b_p - b_r\|$ if x_i has bounded domain and ∞ otherwise. We denote by $\|\mathbf{x}\|_\infty$ the maximum domain span of any variable, i.e., $\max_i \|x_i\|$. On the other hand, if $\mathcal{I}$ is in equality form, we say that the domain of the i-th variable x_i is *bounded* if $l_i, u_i \notin \{-\infty, \infty\}$, where l_i, u_i are the i-th entries of $\mathbf{l}, \mathbf{u}$, respectively; otherwise, we say that the domain of $\mathbf{x}$ is *unbounded*. Moreover, we denote by $\|x_i\|$ the maximum *domain span* of the i-th variable, i.e., $\|u_i - l_i\|$ any variable, and by $\|\mathbf{x}\|_\infty$ the maximum domain span of any variable, i.e., $\max_i \|x_i\|$.

In either case, we say that an ILP instance has *bounded domain* if all variables have bounded domain, and we say that the instance has *unary bounded domain* if the coefficients bounding the domain of variables are encoded in unary.

Finally, we call an ILP instance *unary* if all coefficients in A, $\mathbf{b}$, $\mathbf{c}$, as well as $\mathbf{l}$, $\mathbf{u}$ (if they are part of the input) are given in unary. We say that an ILP instance is *fully unary* if it is unary and all variables have (unary) bounded domain.

2.3. Parameterized Complexity

In parameterized algorithmics [14–17] the runtime of an algorithm is studied with respect to a parameter $k \in \mathbb{N}$ and input size n. The basic idea is to find a parameter that describes the structure of the instance such that the combinatorial explosion can be confined to this parameter. In this respect, the most favorable complexity class is **FPT** (*fixed-parameter tractable*) which contains all problems that can be decided by an algorithm running in time $f(k) \cdot n^{\mathcal{O}(1)}$, where f is a computable function. Algorithms with this running time are called *fpt-algorithms*.

There is a variety of classes capturing *parameterized intractability*. Here we require only the class **paraNP**, which is defined as the class of problems that are solvable by a nondeterministic Turing-machine in fpt-time. We will make use of the characterization of **paraNP**-hardness given by Flum and Grohe (Theorem 2.14, [15]): any parameterized (decision) problem that remains **NP**-hard when the parameter is set to some constant is **paraNP**-hard. Showing **paraNP**-hardness for a problem rules out the existence of an fpt-algorithm under the assumption that $\mathbf{P} \neq \mathbf{NP}$. In fact, it even allows us to rule out algorithms running in time $n^{f(k)}$ for any function f (these are called *XP* algorithms).

2.4. Graph Parameters

Treewidth. Treewidth is the most prominent structural parameter and has been extensively studied in a number of fields. In order to define treewidth, we begin with the definition of its associated decomposition. A *tree-decomposition* $\mathcal{T}$ of a graph $G = (V, E)$ is a pair (T, χ), where T is a tree and χ is a function that assigns each tree node t a set $\chi(t) \subseteq V$ of vertices such that the following conditions hold:

(T1) For every edge $\{u, v\} \in E(G)$ there is a tree node t such that $u, v \in \chi(t)$.
(T2) For every vertex $v \in V(G)$, the set of tree nodes t with $v \in \chi(t)$ forms a non-empty subtree of T.

The sets $\chi(t)$ are called *bags* of the decomposition $\mathcal{T}$ and $\chi(t)$ is the bag associated with the tree node t. The *width* of a tree-decomposition (T, χ) is the size of a largest bag minus 1. A tree-decomposition of minimum width is called *optimal*. The *treewidth* of a graph G, denoted by $\mathrm{tw}(G)$, is the width of an optimal tree decomposition of G.

Proposition 1 ([18–20]). *It is possible to compute an optimal tree-decomposition of an n-vertex graph G with treewidth k in time $k^{\mathcal{O}(k^3)}n$, and to compute a 5-approximate one in time $2^{\mathcal{O}(k)}n$. Moreover, the number of nodes in the obtained tree decompositions is at most $\mathcal{O}(n)$.*

Treedepth. Another important notion that we make use of extensively is that of treedepth. Treedepth is a structural parameter closely related to treewidth, and the structure of graphs of bounded treedepth is well understood [21]. A useful way of thinking about graphs of bounded treedepth is that they are (sparse) graphs with no long paths.

We formalize a few notions needed to define treedepth. A *rooted forest* is a disjoint union of rooted trees. For a vertex x in a tree T of a rooted forest, the *height* (or *depth*) of x in the forest is the number of vertices in the path from the root of T to x. The *height of a rooted forest* is the maximum height of a vertex of the forest.

Definition 1 (Treedepth). *Let the* closure *of a rooted forest $\mathcal{F}$ be the graph $\mathrm{clos}(\mathcal{F}) = (V_c, E_c)$ with the vertex set $V_c = \bigcup_{T \in \mathcal{F}} V(T)$ and the edge set $E_c = \{xy \colon x \text{ is an ancestor of } y \text{ in some } T \in \mathcal{F}\}$. A treedepth decomposition of a graph G is a rooted forest $\mathcal{F}$ such that $G \subseteq \mathrm{clos}(\mathcal{F})$. The* treedepth $\mathrm{td}(G)$ *of a graph G is the minimum height of any treedepth decomposition of G.*

We will later use T_x to denote the vertex set of the subtree of T rooted at a vertex x of T. Similarly to treewidth, it is possible to determine the treedepth of a graph in FPT time.

Proposition 2 ([21]). *Given a graph G with n nodes and a constant w, it is possible to decide whether G has treedepth at most w, and if so, to compute an optimal treedepth decomposition of G in time $\mathcal{O}n$.*

The following alternative (equivalent) characterization of treedepth will be useful later.

Proposition 3 ([21]). *Let G_i be the connected components of G. Then*

$$\mathrm{td}(G) = \begin{cases} 1, & \text{if } |V(G)| = 1; \\ 1 + \min_{v \in V(G)} \mathrm{td}(G - v), & \text{if } G \text{ is connected and } |V(G)| > 1; \\ \max_i \mathrm{td}(G_i), & \text{otherwise.} \end{cases}$$

We conclude with a few useful facts about treedepth.

Proposition 4 ([21]).

1. *If a graph G has no path of length d, then $\mathrm{td}(G) \leq d$.*
2. *If $\mathrm{td}(G) \leq d$, then G has no path of length 2^d.*
3. $\mathrm{tw}(G) \leq \mathrm{td}(G)$.
4. *If $\mathrm{td}(G) \leq d$, then $\mathrm{td}(G') \leq d+1$ for any graph G' obtained by adding one vertex into G.*

(Signed) Clique-width. Let k be a positive integer. A *k-graph* is a graph whose vertices are labeled by $[k]$; formally, the graph is equipped with a labeling function $\gamma \colon V(G) \to [k]$, and we also use $\gamma^{-1}(i)$ to denote the set of vertices labeled i for $i \in [k]$.

We consider an arbitrary graph as a k-graph with all vertices labeled by 1. We call the k-graph consisting of exactly one vertex v (say, labeled by i) an initial k-graph and denote it by $i(v)$. The *clique-width* of a graph G is the smallest integer k such that G can be constructed from initial k-graphs by means of repeated application of the following three operations:

1. Disjoint union (denoted by $\oplus$);
2. Relabeling: changing all labels i to j (denoted by $\rho_{i \to j}$);
3. Edge insertion: adding an edge between each vertex labeled by i and each vertex labeled by j, where $i \neq j$ (denoted by $\eta_{i,j}$ or $\eta_{j,i}$).

A construction of a k-graph G using the above operations can be represented by an algebraic term composed of $\oplus$, $\rho_{i \to j}$ and $\eta_{i,j}$ (where $i \neq j$ and $i, j \in [k]$). Such a term is called a *k-expression defining G*, and the *clique-width* of G is the smallest integer k such that G can be defined by a k-expression.

A *k-expression tree* (also called parse trees in the literature [22]) is a rooted tree representation of a *k*-expression; specifically, the *k*-expression tree can be built from a *k*-expression in a leaves-to-root fashion by using a leaf to represent each $i(v)$, each $\oplus$ operator is represented by an $\oplus$ node with two children, and each $\rho_{i \to j}$ and $\eta_{j,i}$ operator is represented by a corresponding node with a single child.

There are many graph classes which are known to have bounded clique-width. Examples of such graph classes include every graph class of bounded treewidth [23], co-graphs [23], complete (bipartite) graphs and distance hereditary graphs [24].

If the edges of *G* have signs, then one can define two different variants of clique-width for *G*. The *unsigned clique-width* of *G* is simply the clique-width of the graph G' obtained by removing all signs on the edges of *G*. On the other hand, the *signed clique-width* of *G* is the minimum *k* such that *G* can be defined by a *signed k-expression*, which is analogous to a *k*-expression with the sole distinction that the operation $\eta_{i,j}$ is replaced by $\eta_{i,j}^{\ell}$ which adds an edge with sign ℓ between all vertices labeled *i* and *j*. An example is provided in Figure 1.

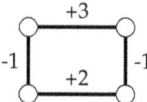

Figure 1. An example of a graph with clique-width 2 and signed clique-width 4.

We list a few known facts and observations about clique-width below:

- The difference between the signed clique-width (scw) and unsigned clique-width (cw) of a signed graph *G* can be arbitrarily large; more precisely, for every gap *g* there exists a signed graph *G* such that $\text{scw}(G) \geq \text{cw}(G) + g$ [25].
- There exists a function *f* and a polynomial-time algorithm which takes as input a (signed) graph *G* and either outputs a (signed) $f(k)$-expression or correctly determines that the (signed) clique-width of *G* is greater than *k* [26,27].
- Every signed graph of (signed) clique-width *k* can be defined by a (signed) *k*-expression which does not use the $\eta_{i,j}$ operator to create edges between vertices that are already adjacent (i.e., each edge is created only once).
- A signed *k*-expression of a bipartite signed graph *G* with bi-partition V_1, V_2 can be converted to a signed $(k+1)$-expression of *G* such that the labels used for V_1 are completely disjoint from those used for V_2 (this is because any label that was originally used for V_1 and V_2 cannot be used to create new edges).

2.5. Graphical Representations

Here, we overview some natural graphical representations which have been used to capture the variable-constraint interactions of ILP instances. We remark that such representations are not unique to the ILP setting: indeed, they have been used and studied extensively also in settings such as, e.g., constraint programming [28,29] and Boolean satisfiability [11,30].

Let *A* be an $m \times n$ integer matrix that is provided as part of an ILP instance $\mathcal{I}$. The *signed incidence graph* of *A* (or, equivalently, of $\mathcal{I}$) is the edge-labeled bipartite graph $G_{SI}(\mathcal{I}) = (R \cup C, E, \lambda)$, where $R = \{r_1, \ldots, r_m\}$ contains one vertex for each row of *A* and $C = \{c_1, \ldots, c_n\}$ contains one vertex for each column of *A*. There is an edge $\{r, c\}$ with label $\lambda(\{r, c\}) = A_{r,c}$ between the vertex $r \in R$ and $c \in C$ if $A_{r,c} \neq 0$, that is, if row *r* contains a nonzero coefficient in column *c*. In other words, the vertex set of G_{SI} is $\text{con}(\mathcal{I}) \cup \text{var}(\mathcal{I})$, a variable is adjacent to a constraint if and only if it occurs in that constraint with a non-zero coefficient, and the labels on edges encode this coefficient.

The *incidence graph* of *A* (or $\mathcal{I}$), denoted $G_I(\mathcal{I})$, is equal to the signed incidence graph without the edge-labels. The *primal graph* of *A* (or $\mathcal{I}$) is the graph $G_P(\mathcal{I}) = (C, E)$, where *C* is the set of columns of *A* and $\{c, c'\} \in E$ whenever there exists a row of *A* with a nonzero coefficient in both columns *c* and c'. This graph is also sometimes called the *Gaifman graph* in the literature. The *dual graph* of A (or $\mathcal{I}$) is the graph $G_D(\mathcal{I}) = (R, E)$, where R is the set of rows of A and $\{r, r'\} \in E$ whenever there exists a column of A with a nonzero coefficient in both rows r and r'. In other words, the vertex sets of these graphs are $\text{var}(\mathcal{I})$ and $\text{con}(\mathcal{I})$, respectively; an edge then signifies that two variables directly interact via a constraint or that two constraints directly interact via a variable, respectively. For all graph representations introduced above, we drop the $\mathcal{I}$ in the parenthesis when the instance is clear from context. Figure 2 illustrates the four graphical representations of a constraint matrix.

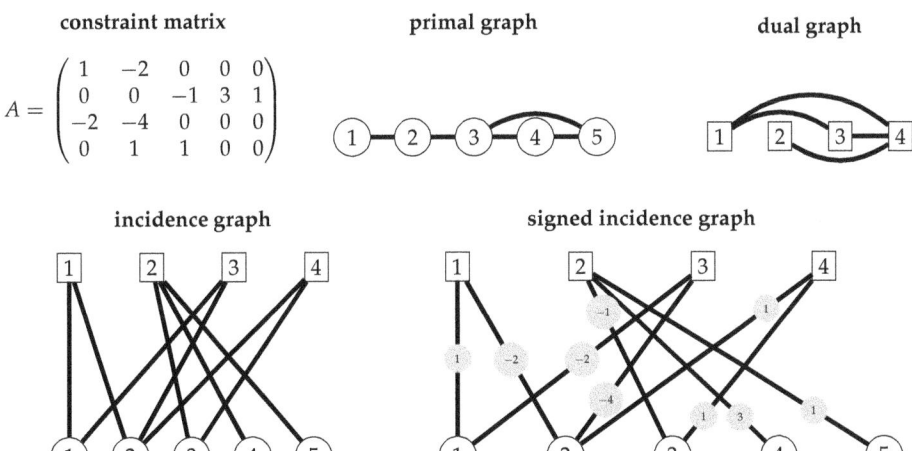

Figure 2. The primal, dual, incidence, and signed incidence graph of the constraint matrix shown in the top left corner. Vertices corresponding to variables (constraints) of the matrix are indicated by circles (rectangles). The label of a vertex corresponds to its row/column-index in the constraint matrix.

For a decompositional width measure $\text{dw} \in \{\text{td}, \text{tw}, \text{cw}, \text{scw}\}$, we denote by $\text{dw}_{SI}(\mathcal{I})$, $\text{dw}_I(\mathcal{I})$, $\text{dw}_P(\mathcal{I})$, $\text{dw}_D(\mathcal{I})$, the width of the signed incidence graph, the incidence graph, the primal graph, and the dual graph of $\mathcal{I}$, respectively.

2.6. Representation Stability

Changing between the equality and inequality representations for ILP does not have a significant effect on most of the structural parameters considered in this paper. In particular, it is easy to show that the parameters td, tw, cw, scw as well as the parameters fracture number (frac) and torso-width (defined in Section 5) differ at most by a factor of two when switching between the two representations. To see this it suffices to consider the standard transformations between $\text{ILP}_=$ and $\text{ILP}_\leq$, which are given as follows.

Given an instance $\mathcal{I}_=$ of $\text{ILP}_=$, we obtain an equivalent instance of $\text{ILP}_\leq$ by replacing every equality constraint of $\mathcal{I}_=$ with two inequality constraints and by adding the lower and upper bounds for the variables to the constraint matrix. It is easy to see that this transformation increases the above mentioned parameters for the primal, dual, and incidence graph by at most a factor of two.

Similarly, given an instance $\mathcal{I}_\leq$ of $\text{ILP}_\leq$, we obtain an equivalent instance of $\text{ILP}_=$ by introducing (i.e., adding) one novel "slack" variable to every constraint with a lower bound of 0. It is, similarly to the previous case, straightforward to show that this does not increase any of the considered parameters by more than a factor of two. As a consequence, for the statement of most of our complexity results we will simply consider instances of ILP and/or ILPF (which may be given in equality as well as in inequality form).

3. Solving ILPs with Explicit Restrictions

Initial work on mapping the complexity of integer linear programming predominantly focused on identifying tractable classes by placing restrictions on explicit properties of instances, such as the number of variables or of constraints. Lenstra [8] showed that ILP can be solved by an algorithm which has an exponential dependency on the number of variables, but only a linear dependency on the size of the instance. His running time was subsequently improved by Kannan [9] and Frank and Tardos [10].

Theorem 1 ([8–10]). *An ILP instance $\mathcal{I}$ with n variables can be solved in time $\mathcal{O}(n^{2.5n+o(n)} \cdot |\mathcal{I}|)$; in other words, ILP is fixed-parameter tractable parameterized by n.*

Papadimitriou showed that ILP is fixed-parameter tractable parameterized by $m + \|A\|_\infty + \|\mathbf{b}\|$ [31]. His result was recently improved by Eisenbrand and Weismantel [32], and then further improved by Jansen and Rohwedder [33]. Even more recently, Knop, Pilipczuk and Wrochna showed that the running time of this result cannot be substantially improved [34].

Theorem 2 ([32–34]). *An $\text{ILP}_=$ instance $\mathcal{I}$ with m constraints can be solved in time $\mathcal{O}((m \cdot \|A\|_\infty)^m \cdot \log \|\mathbf{b}\|)$; in other words, ILP is fixed-parameter tractable parameterized by $m + \|A\|_\infty$.*

On the other hand, ILP remains intractable when all other obvious numerical measures are bounded. In particular, ILPF remains **NP**-complete even when $\|A\|_\infty = \|\mathbf{b}\| = \|\mathbf{u}\| = 1$ and $\|\mathbf{l}\| = 0$, as witnessed by the folklore encoding of the VERTEX COVER problem into ILP, i.e., given a graph G and an integer k, the ILP instance has one binary variable for every vertex of G (representing of whether or not the vertex is chosen to be in a vertex cover) and for every edge between u and v a constraint ensuring that the sum of the variables for u and v is at least one (ensuring that the vertex cover contains at least one vertex from every edge).

While this is not the focus of this survey, we also mention that there is a significant body of work on exploiting algebraic properties to solve ILP. Perhaps the most prominent example of a complexity result obtained in this vein is the well-known fact that instances $\mathcal{I}$ whose matrix A is *totally unimodular* (i.e., each of its square submatrices has a determinant in $\{-1, 0, 1\}$) can be solved in polynomial time [35]).

Theorem 3 ([35,36]). *An ILP instance $\mathcal{I}$ having a totally unimodular constraint matrix can be solved in polynomial-time.*

We say that an ILP instance is non-negative if all entries of A and b are non-negative. Cunningham and Geelen [37] showed that non-negative $\text{ILP}_=$ is fixed-parameter tractable parameterized by $\|b\|$ and ω, where ω is the branchwidth of the column-matroid of A, i.e., the matroid whose elements are the column vectors of A and whose independent sets are the set of all linearly independent column vectors.

Theorem 4 ([37]). *A non-negative $\text{ILP}_=$ instance $\mathcal{I}$ with m constraints and n variables can be solved in time $\mathcal{O}((\|b\|_\infty + 1)^{2\omega} \omega mn + m^2 n)$, where ω is equal to the branchwidth of the column-matroid of A.*

4. Parameters for Sparse Variable-Constraint Interactions: Treewidth and Treedepth

We note that due to the discussion at the end of Section 2.5 all the results presented in this section hold regardless of whether our instance is provided in inequality or equality form. In 2015, Jansen and Kratsch [36] showed that the treewidth of the primal graph can be used to efficiently solve ILP when the variable domains are bounded by the parameter. More precisely:

Theorem 5 ([36]). *Let c be a constant. Given an ILPF instance $\mathcal{I}$ with unary bounded domain satisfying the property that all but at most c variables have domain span at most d, and let G_P be the primal graph of $\mathcal{I}$. Then $\mathcal{I}$ admits a fixed-parameter algorithm when parameterized by $d + \text{tw}_P(\mathcal{I})$.*

The result follows from standard leaves-to-root dynamic programming along a tree decomposition of $G_P(\mathcal{I})$, and can be straightforwardly adapted to also solve ILP. This result provides a useful tool for dealing with instances where all variables have bounded domain. One year after Jansen and Kratsch's result, Ganian and Ordyniak used a reduction from SUBSET SUM to rule out the application of treewidth In the setting of unbounded domain—even for extremely restricted instances of ILPF.

Theorem 6 ([38]). *ILPF is* **NP**-*complete even when restricted to instances $\mathcal{I}$ such that $\mathrm{tw}_P(\mathcal{I}) \leq 2$ and $\|\mathbf{b}\| = \|A\|_\infty = 1$.*

In the same paper, Ganian and Ordyniak complemented this result with a fixed-parameter algorithm for ILP parameterized by $\mathrm{td}_P(\mathcal{I}) + \|A\|_\infty + \|\mathbf{b}\|$; the proof uses a pruning technique which transforms the instance into an equivalent one of size bounded by the parameter (a "kernel"). Their result was later superseded by Koutecký, Levin and Onn [39], who used Graver-best oracles to show:

Theorem 7 ([39]). *ILP is* **FPT** *when parameterized by $\mathrm{td}_P(\mathcal{I}) + \|A\|_\infty$.*

We note that both parameters $\mathrm{td}_P(\mathcal{I})$ and $\|A\|_\infty$ are required to achieve even XP algorithms: it is well known that ILP is **NP**-hard when restricted to instances with $\|A\|_\infty = 1$, and Ganian and Ordyniak [38] showed that it is also **NP**-hard when restricted to instances with $\mathrm{td}_P(\mathcal{I}) \leq 4$. It is worth noting that both Theorem 7 and its predecessor have a non-elementary dependency on the parameter.

In their paper, Koutecký, Levin and Onn also used the same techniques to obtain a fixed-parameter algorithm that uses the treedepth of the dual graph (as opposed to the primal one):

Theorem 8 ([39]). *ILP is* **FPT** *when parameterized by $\mathrm{td}_D(\mathcal{I}) + \|A\|_\infty$.*

They also established an analogue to Theorem 6 for dual graphs, showing that restricting the dual or primal graphs leads to a similar complexity behavior for ILP:

Theorem 9 ([39]). *ILPF is* **NP**-*complete even when restricted to instances $\mathcal{I}$ such that $\mathrm{tw}_D(\mathcal{I}) \leq 3$ and with $\|A\|_\infty = 2$.*

Since the classical encoding of SUBSET SUM into an instance $\mathcal{I}$ of ILPF only uses a single constraint (i.e., $|\mathrm{con}(\mathcal{I})| = 1$), it is immediate that ILPF is also **NP**-complete when $\mathrm{td}_D(\mathcal{I}) = 1$; in other words, it is not possible to strengthen Theorem 8 by dropping any of the parameters. By the standard reduction from SUBSET SUM we mean the reduction to the ILP instance that has one binary variable for every integer in the SUBSET SUM instance (representing whether or not the integer is in a solution) and one constraint over all variables ensuring that the sum of all chosen integers equals the target value of the SUBSET SUM instance.

The third fundamental graph representation that has been considered for restricting the variable-constraint interactions of an ILP instance $\mathcal{I}$ is the incidence graph. It is worth noting that a trivial transformation of the respective decompositions yields $\mathrm{td}_I(\mathcal{I}) \leq \max(\mathrm{td}_P(\mathcal{I}), \mathrm{td}_D(\mathcal{I}))$ and similarly $\mathrm{tw}_I(\mathcal{I}) \leq \max(\mathrm{tw}_P(\mathcal{I}), \mathrm{tw}_D(\mathcal{I}))$; on the other hand, there are instances where both $\mathrm{td}_I(\mathcal{I})$ and $\mathrm{tw}_I(\mathcal{I})$ are bounded but the dual and primal graphs exhibit neither bounded treewidth nor treedepth. Hence, tractability results using the treewidth and treedepth of the incidence graph have the potential to supersede similar results for both previously considered graph representations, while any obtained hardness results carry over from primal and dual graphs to incidence graphs.

Ganian, Ordyniak and Ramanujan [40] identified conditions under which $\mathrm{tw}_I(\mathcal{I})$ can be used to obtain algorithms for ILP. Notably, after factoring in Proposition 1 their result states:

Theorem 10. ILP *can be solved in time* $\Gamma^{\mathcal{O}(\operatorname{tw}_I(\mathcal{I}))} \cdot |\mathcal{I}|$, *where* Γ *is the maximum absolute value of any partial evaluation of a constraint by any feasible assignment of* $\mathcal{I}$; *here, a partial evaluation of a constraint/row* $\mathbf{r}$ *of* A *with a feasible assignment* $\mathbf{x}$ *is equal to* $\mathbf{rx}'$, *where* $\mathbf{x}'$ *is any vector obtained from* $\mathbf{x}$ *after setting some of its entries to* 0.

Note that $\Gamma \leq n \|A\|_\infty d$, where d is the maximum domain span of every variable in $\mathcal{I}$. Hence, Theorem 10, e.g., implies that fully unary ILP can be solved in polynomial-time if $\operatorname{tw}_I(\mathcal{I})$ is bounded by a constant.

On the other hand, ILPF remains **NP**-complete even when restricted to instances with strong restrictions on the treewidth and coefficients—to some extent justifying the dependency of the above algorithm on Γ. Indeed, the first part of the following theorem follows from the classical encoding of SUBSET SUM into ILPF, while the second part was shown by Ganian et al. [40].

Theorem 11. ILPF *remains* **NP**-*complete even on instances* $\mathcal{I}$ *with* (1) $\operatorname{tw}_I(\mathcal{I}) = 1$ *and Boolean domains for all variables, as well as with* (2) $\operatorname{tw}_I(\mathcal{I}) \leq 3$ *and* $\max(\|A\|_\infty, \|\mathbf{b}\|) = 2$.

A natural question is whether one can use $\operatorname{td}_I(\mathcal{I})$ instead of $\operatorname{tw}_I(\mathcal{I})$ in order to obtain tractability for ILP under a weaker restriction than by bounding Γ—notably, can one lift Theorems 7 and 8 to the incidence treedepth setting? Very recently, Eiben et al. [41] answered the question in the negative by showing:

Theorem 12 ([41]). ILP *remains* **NP**-*complete even when restricted to instances* $\mathcal{I}$ *such that* $\max(\|A\|_\infty, \|\mathbf{b}\|) = 1$ *and* $\operatorname{td}_I(\mathcal{I}) \leq 5$.

In the full version of that paper, they also showed that restricting the structure by the size of a minimum vertex cover of $G_I(\mathcal{I})$—a significantly stronger restriction than treedepth—leads to tractability.

Theorem 13. ILP *is* **FPT** *parameterized by* $\|A\|_\infty$ *and the vertex cover number of* $G_I(\mathcal{I})$.

Note that even though the vertex cover number is sensitive to changes between the equality and inequality form of ILP in general, the above theorem still holds for both forms. This is because the proof of Theorem 13 works by first observing that the number of (linearly independent) equalities is bounded by a function of the vertex cover number and $\|A\|_\infty$ and then uses Theorem 2 to show tractability. Almost the same approach can be used for inequalities, i.e., one can again observe that the number of inequalities is bounded in terms of the parameters (otherwise there are redundant inequalities) and then use the standard reduction from $\operatorname{ILP}_\leq$ to $\operatorname{ILP}_=$; since the reduction does not increase the number of constraints, one can again apply Theorem 2.

We conclude this section by touching on the complexity of integer linear programs whose graph representations have an extremely simple structure—notably, have treewidth 1 (i.e., are acyclic). This setting was investigated by Eiben et al. [42], who showed that $\operatorname{ILP}_\leq$ restricted to unary instances whose graph representations are acyclic exhibit a different complexity behavior than ILP restricted to instances of bounded treewidth. We summarize their results in Table 1.

Table 1. The complexity map for ILP$_\leq$ for unary instances (first row), instances whose coefficients are encoded in unary (second row), and instances with unary bounded domain (third row). Without these restrictions, the problem becomes intractable due to the simple acyclic structure of the classical encoding of SUBSET SUM into ILP.

	Acyclic Primal Graph	Acyclic Incidence Graph	
		ILPF	ILP
UNARY	P	P	P
UNARY-C	NP-hard	P	NP-hard
UNARY-D	P	P	NP-hard

5. Other Parameters Exploiting Variable-Constraint Interactions

Of course, ILP has also been studied through the lens of structural parameters that are different than treewidth. The first example of such a parameter is the *fracture number* of Dvořák et al. [43], which captures the "distance" of an ILP instance from being fractured into small independent components.

Three variants of the fracture number will be of interest for the purposes of this survey: the *constraint fracture number* of an ILP $\mathcal{I}$ ($\text{frac}_C(\mathcal{I})$) is the minimum number ℓ of constraints that need to be deleted from $\mathcal{I}$ so that the resulting instance $\mathcal{I}'$ satisfies the following: each connected component of $G_I(\mathcal{I}')$ contains at most ℓ vertices. The *variable fracture number* ($\text{frac}_V(\mathcal{I})$) and *mixed fracture number* ($\text{frac}(\mathcal{I})$) are then defined analogously, with the distinction that we may only delete variables or are allowed to delete both variables and constraints, respectively.

The constraint fracture number is bounded whenever the dual graph has bounded treedepth, and the mixed fracture number is bounded whenever the incidence graph has bounded treedepth (and similarly for the variables fracture number and the primal graph); however, the converse of these statements is not true. Intuitively, this means that the fracture number can be viewed as a stronger restriction than treedepth. Dvořák et al. [43] showed that the fracture number can be used to obtain XP-algorithms for ILP in settings which would remain **NP**-hard if treedepth were used instead (see Theorem 12). See Figure 3 for an illustration of the relationships between the different variants of fracture number as well as their relation to treewidth and treedepth.

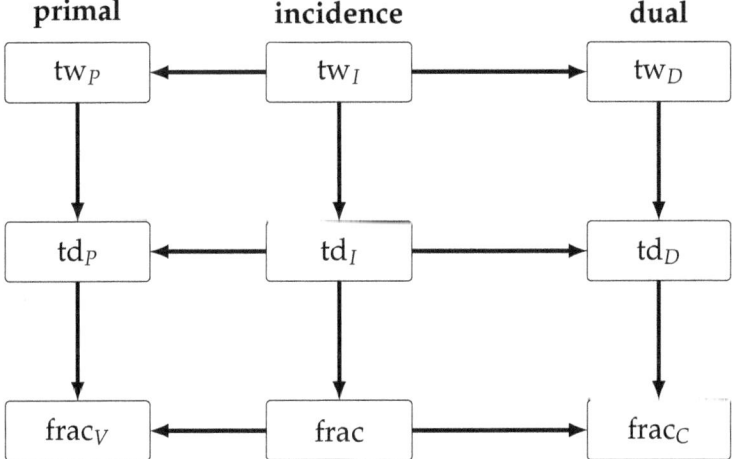

Figure 3. The relationship between the structural parameters treewidth, treedepth, and fracture number for the primal, dual, and incidence graph. An arc from one parameter to another indicates that the former is a more general parameter, i.e., whenever the later is bounded so is the former. The variable, constraint, and mixed fracture number are defined in Section 5.

Theorem 14 (Corollary 8, [43]). *ILP admits an XP-algorithm parameterized by* $\|A\|_\infty + \mathrm{frac}(\mathcal{I})$.

Theorem 15 (Theorem 12, [43]). *ILP restricted to unary instances admits an XP-algorithm parameterized by* $\mathrm{frac}_C(\mathcal{I})$.

Note that theorem 15 cannot be improved to an FPT-algorithm due to (Theorem 14, [43]). Moreover, an analogous result does not hold for the mixed respectively variable fracture number as already unary ILP is **NP**-hard if the variable fracture number is bounded by a constant (Theorem 13, [43]), which also excludes the use of a less restrictive parameter than constraint fracture number (such as treedepth) in Theorem 15.

Another structural parameter that can be used to solve ILP is the *torso-width*. The base idea behind torso-width is to decompose instances into (possibly many) separate parts with only limited interaction between them, and to solve some parts with Lenstra's algorithm (Theorem 1) and others with dynamic programming along a tree decomposition as per Theorem 5.

To define torso-width, we will need the operation of *collapsing*: for a graph G and a vertex set X, the operation of collapsing X deletes X from the graph and adds an edge between each pair of neighbors of X. We denote the resulting graph $G \circ X$. Now, let q be a fixed constant. A graph G is a q-torso of $\mathcal{I}$ iff there exists a set P of variables, each with domain span at most q, such that $G = G_P(\mathcal{I}) \circ P$. The q-torso-width of $\mathcal{I}$, denoted $\mathrm{tor}_q(\mathcal{I})$, is then the minimum integer k such that $\mathcal{I}$ has a q-torso G such that:

- G has treewidth at most k, and
- the largest connected component of the subgraph of $G_P(\mathcal{I})$ induced on $V(G_P(\mathcal{I})) \setminus V(G)$ contains at most k vertices.

Ganian, Ordyniak and Ramanujan [40] showed that the q-torso-width of $\mathcal{I}$ can be approximated by a fixed-parameter algorithm, and that this parameter can also be used to solve ILP. This result can be seen as a generalization of Theorem 1 as well as Theorem 5.

Theorem 16 (Theorem 5, [40]). *For every fixed integer q, ILP is* **FPT** *parameterized by q-torso-width*.

Eiben, Ganian, Knop and Ordyniak [42] also investigated the complexity of ILP with respect to the parameter clique-width. They showed (and it is also not difficult to observe) that ILPF remains **NP**-complete even when restricted to extremely simple instances whose incidence, primal, and/or dual graphs have bounded clique-width. However, ILP becomes polynomially tractable when restricted to unary instances of bounded signed clique-width (of their signed incidence graph), under the assumption that a suitable k-expression is provided in the input.

Theorem 17. *There exists an algorithm which takes as input a unary instance $\mathcal{I}$ of ILP and a signed k-expression tree T of $G_{SI}(\mathcal{I})$, runs in time $\mathcal{O}(|\mathcal{I}|^{4k} \cdot |T|)$, and solves $\mathcal{I}$.*

6. Summary and Future Work

This survey provides an overview of recently obtained (as well as previously known) (in-)tractability results for ILP with a focus on structural restrictions of the primal, dual, and incidence graph. The classes based on fracture number and treedepth can alternatively be defined in terms of block matrices and are also known as n-fold, tree-fold, 2-stage stochastic, and multi-stage stochastic integer linear programs; a recent and comprehensive overview for these classes, their exact relation to the classes considered in this survey, as well as the current best algorithmic approaches and techniques employed for these classes is given in [44].

Even though the complexity of ILP w.r.t. decompositional parameters such as treedepth, treewidth, and clique-width is by now quite well understood, we believe that the study of parameterized complexity of ILP is still in its infancy. Apart from studying more restrictive settings such as tree-like instances (in combination with, e.g., $\|A\|_\infty$) as well as related parameters such as feedback edge set, feedback vertex set, and bandwidth, we see at least two very promising directions for developing novel and even more general structural parameterizations: backdoor sets and hybrid parameters. Both of these approaches have already been successfully applied in settings such as BOOLEAN SATISFIABILITY and CONSTRAINT SATISFACTION [29,45,46]. Informally, a backdoor set captures the situation when an instance is "close" to being tractable, and it looks promising to develop backdoor sets into one of the newly defined tractable classes. For instance, can we solve instances of ILP that differ from a known tractable class only by a small set of variables or constraints? Concerning the hybrid approach, where the aim is to solve instances consisting of many parts each of them tractable for a different reason, the number of possible directions seems even greater—torso-width is thus far the only explored hybrid parameter and many more tractable classes of ILP have been discovered since its introduction. Finally, it is important to explore if, how, and how far the known tractable fragments for ILP can be employed for well-known generalizations of ILP such as mixed or quadratic integer programs.

Author Contributions: Conceptualization, R.G. and S.O.; Writing—original draft, R.G. and S.O.

Funding: Robert Ganian acknowledges support from the Austrian Science Fund (FWF, Project P31336).

Acknowledgments: Open Access Funding by the Austrian Science Fund (FWF).

Conflicts of Interest: The authors declare no conflict of interest.

References

1. Floudas, C.; Lin, X. Mixed Integer Linear Programming in Process Scheduling: Modeling, Algorithms, and Applications. *Ann. Oper. Res.* **2005**, *139*, 131–162. [CrossRef]
2. van den Briel, M.; Vossen, T.; Kambhampati, S. Reviving Integer Programming Approaches for AI Planning: A Branch-and-Cut Framework. In Proceedings of the Fifteenth International Conference on Automated Planning and Scheduling (ICAPS 2005), Monterey, CA, USA, 5–10 June 2005; Biundo, S., Myers, K.L., Rajan, K., Eds.; AAAI: Menlo Park, CA, USA, 2005; pp. 310–319.
3. Vossen, T.; Ball, M.O.; Lotem, A.; Nau, D.S. On the Use of Integer Programming Models in AI Planning. In Proceedings of the Sixteenth International Joint Conference on Artificial Intelligence (IJCAI 99), Stockholm, Sweden, 31 July–6 August 1999; Dean, T., Ed.; Morgan Kaufmann: San Francisco, CA, USA, 1999; Volume 2, pp. 304–309, 1450p.
4. Toth, P.; Vigo, D. (Eds.) *The Vehicle Routing Problem*; Society for Industrial and Applied Mathematics: Philadelphia, PA, USA, 2001.
5. Lodi, A.; Martello, S.; Monaci, M. Two-dimensional packing problems: A survey. *Eur. J. Oper. Res.* **2002**, *141*, 241–252. [CrossRef]
6. Alumur, S.A.; Kara, B.Y. Network hub location problems: The state of the art. *Eur. J. Oper. Res.* **2008**, *190*, 1–21. [CrossRef]
7. Papadimitriou, C.H.; Steiglitz, K. *Combinatorial Optimization: Algorithms and Complexity*; Prentice-Hall: Upper Saddle River, NJ, USA, 1982.
8. Lenstra, H.W., Jr. Integer programming with a fixed number of variables. *Math. Oper. Res.* **1983**, *8*, 538–548. [CrossRef]
9. Kannan, R. Minkowski's Convex Body Theorem and Integer Programming. *Math. Oper. Res.* **1987**, *12*, 415–440. [CrossRef]
10. Frank, A.; Tardos, É. An application of simultaneous Diophantine approximation in combinatorial optimization. *Combinatorica* **1987**, *7*, 49–65. [CrossRef]
11. Freuder, E.C. A Sufficient Condition for Backtrack-Bounded Search. *J. ACM* **1985**, *32*, 755–761. [CrossRef]
12. Dechter, R.; Pearl, J. Tree Clustering for Constraint Networks. *Artif. Intell.* **1989**, *38*, 353–366. [CrossRef]
13. Diestel, R. *Graph Theory*, 4th ed.; Graduate Texts in Mathematics; Springer: Berlin/Heidelberg, Germany, 2012; Volume 173.
14. Cygan, M.; Fomin, F.V.; Kowalik, L.; Lokshtanov, D.; Marx, D.; Pilipczuk, M.; Pilipczuk, M.; Saurabh, S. *Parameterized Algorithms*; Springer: Berlin/Heidelberg, Germany, 2015.
15. Flum, J.; Grohe, M. *Parameterized Complexity Theory*; Texts in Theoretical Computer Science. An EATCS Series; Springer: Berlin/Heidelberg, 2006; Volume XIV.
16. Niedermeier, R. *Invitation to Fixed-Parameter Algorithms*; Oxford University Press: Oxford, UK, 2006.

17. Downey, R.G.; Fellows, M.R. *Fundamentals of Parameterized Complexity*; Texts in Computer Science; Springer: London, UK, 2013; pp. 3–707.
18. Kloks, T. *Treewidth: Computations and Approximations*; Springer: Berlin, Germany, 1994.
19. Bodlaender, H.L. A linear-time algorithm for finding tree-decompositions of small treewidth. *SIAM J. Comput.* **1996**, *25*, 1305–1317. [CrossRef]
20. Bodlaender, H.L.; Drange, P.G.; Dregi, M.S.; Fomin, F.V.; Lokshtanov, D.; Pilipczuk, M. A c^k n 5-Approximation Algorithm for Treewidth. *SIAM J. Comput.* **2016**, *45*, 317–378. [CrossRef]
21. Nešetřil, J.; Ossona de Mendez, P. *Sparsity: Graphs, Structures, and Algorithms*; Algorithms and Combinatorics; Springer: Berlin/Heidelberg, Germany, 2012; Volume 28.
22. Courcelle, B.; Makowsky, J.A.; Rotics, U. Linear time solvable optimization problems on graphs of bounded clique-width. *Theory Comput. Syst.* **2000**, *33*, 125–150. [CrossRef]
23. Courcelle, B.; Olariu, S. Upper bounds to the clique width of graphs. *Discret. Appl. Math.* **2000**, *101*, 77–114. [CrossRef]
24. Golumbic, M.C.; Rotics, U. On the clique-width of some perfect graph classes. *Int. J. Found. Comput. Sci.* **2000**, *11*, 423–443. [CrossRef]
25. Bliem, B.; Ordyniak, S.; Woltran, S. Clique-Width and Directed Width Measures for Answer-Set Programming. In *Proceedings ECAI 2016—22nd European Conference on Artificial Intelligence*; Kaminka, G.A., Fox, M., Bouquet, P., Hüllermeier, E., Dignum, V., Dignum, F., van Harmelen, F., Eds.; IOS Press: Amsterdam, The Netherlands, 2016; Volume 285, pp. 1105–1113.
26. Oum, S.; Seymour, P.D. Approximating clique-width and branch-width. *J. Comb. Theory Ser. B* **2006**, *96*, 514–528. [CrossRef]
27. Kanté, M.M.; Rao, M. The Rank-Width of Edge-Coloured Graphs. *Theory Comput. Syst.* **2013**, *52*, 599–644. [CrossRef]
28. Samer, M.; Szeider, S. Constraint satisfaction with bounded treewidth revisited. *J. Comput. Syst. Sci.* **2010**, *76*, 103–114. [CrossRef]
29. Ganian, R.; Ramanujan, M.S.; Szeider, S. Discovering Archipelagos of Tractability for Constraint Satisfaction and Counting. *ACM Trans. Algorithms* **2017**, *13*, 29:1–29:32. [CrossRef]
30. Samer, M.; Szeider, S. Fixed-Parameter Tractability. In *Handbook of Satisfiability*; Biere, A., Heule, M., van Maaren, H., Walsh, T., Eds.; IOS Press: Amsterdam, The Netherlands, 2009; Chapter 13, pp. 425–454.
31. Papadimitriou, C.H. On the Complexity of Integer Programming. *J. ACM* **1981**, *28*, 765–768. [CrossRef]
32. Eisenbrand, F.; Weismantel, R. Proximity results and faster algorithms for Integer Programming using the Steinitz Lemma. In Proceedings of the Twenty-Ninth Annual ACM-SIAM Symposium on Discrete Algorithms (SODA 2018), New Orleans, LA, USA, 7–10 January 2018, 2018; pp. 808–816.
33. Jansen, K.; Rohwedder, L. On integer programming and convolution. In Proceedings of the 10th Innovations in Theoretical Computer Science Conference (ITCS 2019), San Diego, CA, USA, 10–12 January 2019; pp. 43:1–43:17.
34. Knop, D.; Pilipczuk, M.; Wrochna, M. Tight complexity lower bounds for integer linear programming with few constraints. In Proceedings of the 36th International Symposium on Theoretical Aspects of Computer Science (STACS 2019), Berlin, Germany, 13–16 March 2019; pp. 44:1–44:15.
35. Schrijver, A. *Combinatorial Optimization—Polyhedra and Efficiency*; Springer: Berlin, Germany, 2003.
36. Jansen, B.M.P.; Kratsch, S. A structural approach to kernels for ILPs: Treewidth and total unimodularity. In Proceedings of the Algorithms—ESA 2015—23rd Annual European Symposium, Patras, Greece, 14–16 September 2015; Bansal, N., Finocchi, I., Eds.; Lecture Notes in Computer Science; Springer: Cham, Switzerland, 2015; Volume 9294, pp. 779–791.
37. Cunningham, W.H.; Geelen, J. On integer programming and the branch-width of the constraint matrix. In Proceedings of the 12th International IPCO Conference on Integer Programming and Combinatorial Optimization, Ithaca, NY, USA, 25–27 June 2007; Fischetti, M., Williamson, D.P., Eds.; Lecture Notes in Computer Science; Springer: Berlin/Heidelberg, Germany, 2007; Volume 4513, pp. 158–166.
38. Ganian, R.; Ordyniak, S. The complexity landscape of decompositional parameters for ILP. *Artif. Intell.* **2018**, *257*, 61–71. [CrossRef]
39. Koutecký, M.; Levin, A.; Onn, S. A parameterized strongly polynomial algorithm for block structured integer programs. In Proceedings of the 45th International Colloquium on Automata, Languages, and Programming, Prague, Czech Republic, 9–13 July 2018; pp. 85:1–85:14.

40. Ganian, R.; Ordyniak, S.; Ramanujan, M.S. Going beyond primal treewidth for (M)ILP. In Proceedings of the Thirty-First AAAI Conference on Artificial Intelligence, San Francisco, CA, USA, 4–9 February 2017; Singh, S.P., Markovitch, S., Eds.; AAAI Press: Menlo Park, CA, USA, 2017; pp. 815–821.
41. Eiben, E.; Ganian, R.; Knop, D.; Ordyniak, S.; Pilipczuk, M.; Wrochna, M. Integer Programming and Incidence Treedepth. In Proceedings of the Integer Programming and Combinatorial Optimization—20th International Conference (IPCO 2019), Ann Arbor, MI, USA, 22–24 May 2019; pp. 194–204.
42. Eiben, E.; Ganian, R.; Knop, D.; Ordyniak, S. Unary Integer Linear Programming with Structural Restrictions. In Proceedings of the Twenty-Seventh International Joint Conference on Artificial Intelligence (IJCAI 2018), Stockholm, Sweden, 13–19 July 2018; pp. 1284–1290.
43. Dvořák, P.; Eiben, E.; Ganian, R.; Knop, D.; Ordyniak, S. Solving Integer Linear Programs with a Small Number of Global Variables and Constraints. In Proceedings of the Twenty-Sixth International Joint Conference on Artificial Intelligence, Melbourne, Australia, 19–25 August 2017; Sierra, C., Ed.; IJCAI: Freiburg, Germany, 2017; pp. 607–613.
44. Eisenbrand, F.; Hunkenschröder, C.; Klein, K.; Koutecký, M.; Levin, A.; Onn, S. An algorithmic theory of integer programming. *Mathematics* **2019**, arXiv:1904.01361.
45. Gaspers, S.; Szeider, S. Backdoors to satisfaction. In *The Multivariate Algorithmic Revolution and Beyond—Essays Dedicated to Michael R. Fellows on the Occasion of His 60th Birthday*; Bodlaender, H.L., Downey, R., Fomin, F.V., Marx, D., Eds.; Lecture Notes in Computer Science; Springer: Berlin/Heidelberg, Germany, 2012; Volume 7370, pp. 287–317.
46. Gaspers, S.; Ordyniak, S.; Szeider, S. Backdoor Sets for CSP. In *The Constraint Satisfaction Problem: Complexity and Approximability*; Schloss Dagstuhl—Leibniz-Zentrum fuer Informatik: Dagstuhl, Germany, 2017; pp. 137–157.

© 2019 by the authors. Licensee MDPI, Basel, Switzerland. This article is an open access article distributed under the terms and conditions of the Creative Commons Attribution (CC BY) license (http://creativecommons.org/licenses/by/4.0/).

Article

FPT Algorithms for Diverse Collections of Hitting Sets

Julien Baste [1],*, Lars Jaffke [2],*, Tomáš Masařík [3,4],*, Geevarghese Philip [5,6],* and Günter Rote [7],*

[1] Institute of Optimization and Operations Research, Ulm University, 89081 Ulm, Germany
[2] Department of Informatics, University of Bergen, 5008 Bergen, Norway
[3] Department of Applied Mathematics of the Faculty of Mathematics and Physics, Charles University, 11800 Prague, Czech Republic
[4] Faculty of Mathematics, Informatics and Mechanics, University of Warsaw, 02-097 Warszawa, Poland
[5] Chennai Mathematical Institute, Chennai 603103, India
[6] International Joint Unit Research Lab in Computer Science (UMI ReLaX), Chennai 603103, India
[7] Fachbereich Mathematik und Informatik, Freie Universität Berlin, D-14195 Berlin, Germany
* Correspondence: julien.baste@uni-ulm.de (J.B.); lars.jaffke@uib.no (L.J.); masarik@kam.mff.cuni.cz (T.M.); gphilip@cmi.ac.in (G.P.); rote@inf.fu-berlin.de (G.R.)

Received: 25 October 2019; Accepted: 23 November 2019; Published: 27 November 2019

Abstract: In this work, we study the d-HITTING SET and FEEDBACK VERTEX SET problems through the paradigm of finding diverse collections of r solutions of size at most k each, which has recently been introduced to the field of parameterized complexity. This paradigm is aimed at addressing the loss of important *side information* which typically occurs during the abstraction process that models real-world problems as computational problems. We use two measures for the diversity of such a collection: the sum of all pairwise Hamming distances, and the minimum pairwise Hamming distance. We show that both problems are fixed-parameter tractable in $k + r$ for both diversity measures. A key ingredient in our algorithms is a (problem independent) network flow formulation that, given a set of 'base' solutions, computes a maximally diverse collection of solutions. We believe that this could be of independent interest.

Keywords: solution diversity; fixed-parameter tractability; hitting sets; vertex cover; feedback vertex set; Hamming distance

1. Introduction

The typical approach in modeling a real-world problem as a computational problem has, broadly speaking, two steps: (i) abstracting the problem into a mathematical formulation that captures the crux of the real-world problem, and (ii) asking for a best solution to the mathematical problem.

Consider the following scenario: Dr. $\mathcal{O}$ organizes a panel discussion and has a shortlist of candidates to invite. From that shortlist, Dr. $\mathcal{O}$ wants to invite as many candidates as possible, such that each of them will bring an individual contribution to the panel. Given two candidates, A and B, it may not be beneficial to invite both A and B, for various reasons: their areas of expertise or opinions may be too similar for both to make a distinguishable contribution, or it may be preferable not to invite more than one person from each institution. It may even be the case that A and B do not see eye-to-eye on some issues which could come up at the discussion, and Dr. $\mathcal{O}$ wishes to avoid a confrontation.

A natural mathematical model to resolve Dr. $\mathcal{O}$'s dilemma is as an instance of the VERTEX COVER problem: each candidate on the shortlist corresponds to a vertex, and for each pair of candidates A and B, we add the edge between A and B if it is *not* beneficial to invite both of them. Removing a smallest *vertex cover* in the resulting graph results in a largest possible set of candidates such that each of them may be expected to individually contribute to the appeal of the event.

Formally, a *vertex cover* of an undirected graph G is any subset $S \subseteq V(G)$ of the vertex set of G such that every edge in G has at least one end-point in G. The VERTEX COVER problem asks for a vertex cover of the smallest size:

VERTEX COVER	
Input:	Graph G.
Solution:	A vertex cover S of G of the smallest size.

While the above model does provide Dr. $\mathcal{O}$ with a set of candidates to invite that is *valid* in the sense that each invited candidate can be expected to make a unique contribution to the panel, a vast amount of *side information* about the candidates is lost in the modeling process. This side information could have helped Dr. $\mathcal{O}$ to get more out of the panel discussion. For instance, Dr. $\mathcal{O}$ may have preferred to invite more well-known or established people over 'newcomers', if they wanted the panel to be highly visible and prestigious; or they may have preferred to have more 'newcomers' in the panel, if they wanted the panel to have more outreach. Other preferences that Dr. $\mathcal{O}$ may have had include: to have people from many different cultural backgrounds, to have equal representation of genders, or preferential representation for affirmative action; to have a variety in the levels of seniority among the attendants, possibly skewed in one way or the other. Other factors, such as the total carbon footprint caused by the participants' travels, may also be of interest to Dr. $\mathcal{O}$. This list could go on and on.

Now, it is possible to plug in some of these factors into the mathematical model, for instance by including weights or labels. Thus, a vertex weight could indicate 'how well-established' a candidate is. However, the complexity of the model grows fast with each additional criterion. The classic field of multicriteria optimization [1] addresses the issue of bundling multiple factors into the objective function, but it is seldom possible to arrive at a balance in the various criteria in a way which captures more than a small fraction of all the relevant side information. Moreover, several side criteria may be conflicting or incomparable (or both); consider in Dr. $\mathcal{O}$'s case 'maximizing the number of different cultural backgrounds' vs. 'minimizing total carbon footprint'.

While Dr. $\mathcal{O}$'s story is admittedly a made-up one, the VERTEX COVER problem is in fact used to model *conflict resolution* in far more realistic settings. In each case, there is a *conflict graph G* whose vertices correspond to entities between which one wishes to avoid a conflict of some kind. There is an edge between two vertices in G if and only if they could be in conflict, and finding and deleting a smallest vertex cover of G yields a largest conflict-free subset of entities. We describe three examples to illustrate the versatility of this model. In each case, it is intuitively clear, just like in Dr. $\mathcal{O}$'s problem, that formulating the problem as VERTEX COVER results in a lot of significant side information being thrown away, and that while finding a smallest vertex cover in the conflict graph will give a *valid* solution, it may not really help in finding a *best* solution, *or even a reasonably good solution*. We list some side information that is lost in the modeling process; the reader should find it easy to come up with any amount of other side information that would be of interest, in each case.

- **Air traffic control.** Conflict graphs are used in the design of decision support tools for aiding Air Traffic Controllers (ATCs) in preventing untoward incidents involving aircraft [2,3]. Each node in the graph G in this instance is an aircraft, and there is an edge between two nodes if the corresponding aircraft are at risk of interfering with each other. A vertex cover of G corresponds to a set of aircraft that can be issued *resolution commands* which ask them to change course, such that afterwards there is no risk of interference.

  In a situation involving a large number of aircraft, it is unlikely that *every* choice of ten aircraft to redirect is *equally* desirable. For instance, in general, it is likely that (i) it is better to ask smaller aircraft to change course in preference to larger craft, and (ii) it is better to ask aircraft which are cruising to change course, in preference to those which are taking off or landing.

- **Wireless spectrum allocation.** Conflict graphs are a standard tool in figuring out how to distribute wireless frequency spectrum among a large set of wireless devices so that no two devices whose usage could potentially interfere with each other are allotted the same frequencies [4,5]. Each node in G is a user, and there is an edge between two nodes if (i) the users request the same frequency, and (ii) their usage of the same frequency has the potential to cause interference. A vertex cover of G corresponds to a set of users whose requests can be denied, such that afterwards there is no risk of interference.

    When there is large collection of devices vying for spectrum, it is unlikely that *every* choice of ten devices to deny the spectrum is *equally* desirable. For instance, it is likely that denying the spectrum to a remote-controlled toy car on the ground is preferable to denying the spectrum to a drone in flight.

- **Managing inconsistencies in database integration.** A database constructed by integrating data from different data sources may end up being inconsistent (that is, violating specified integrity constraints) even if the constituent databases are individually consistent. Handling these inconsistencies is a major challenge in database integration, and conflict graphs are central to various approaches for restoring consistency [6–9]. Each node in G is a database item, and there is an edge between two nodes if the two items together form an inconsistency. A vertex cover of G corresponds to a set of database items in whose *absence* the database achieves consistency.

    In a database of large size, it is unlikely that all data are created equal; some database items are likely to be of better relevance or usefulness than others, and so it is unlikely that *every* choice of ten items to delete is *equally* desirable.

Getting back to our first example, it seems difficult to help Dr. $\mathcal{O}$ with their decision by employing the 'traditional' way of modeling computational problems, where one looks for one best solution. If, on the other hand, Dr. $\mathcal{O}$ was presented with a *small set of good solutions* that, in some sense, are *far apart*, then they might hand-pick the list of candidates that they consider the best choice for the panel and make a more informed decision. Moreover, several forms of side-information may *only become apparent once Dr. $\mathcal{O}$ is presented some concrete alternatives*, and are more likely to be retrieved from alternatives that look very different. That is, a bunch of good quality, dissimilar solutions may end up capturing a lot of the "lost" side information. In addition, this applies to each of the other three examples as well. In each case, finding one best solution could be of little utility in solving the original problem, whereas finding a *small set of solutions, each of good quality, which are not too similar to one another* may offer much more help.

To summarize, real-world problems typically have complicated side constraints, and the optimality criterion may not be clear. Therefore, the abstraction to a mathematical formulation is almost always a simplification, omitting important side information. There are at least two obstacles to simply adapting the model by incorporating these secondary criteria into the objective function or taking into account the side constraints: (i) they make the model complicated and unmanagable, and, (ii) more importantly, these criteria and constraints are often not precisely formulated, potentially even unknown a priori. There may even be no sharp distinction between optimality criteria and constraints (the so-called "soft constraints").

One way of dealing with this issue is to present a small number r of *good* solutions and let the *user* choose between them, based on all the experience and additional information that the user has and that is ignored in the mathematical model. Such an approach is useful even when the objective can be formulated precisely, but is difficult to optimize: After generating r solutions, each of which is *good enough* according to some quality criterion, they can be compared and screened in a second phase, evaluating their exact objective function or checking additional side constraints. In this context, it makes little sense to generate solutions that are very similar to each other and differ only in a few features. It is desirable to present a *diverse* variety of solutions.

It should be clear that the issue is scarcely specific to VERTEX COVER. Essentially *any* computational problem motivated by practical applications likely has the same issue: the modeling process throws out so much relevant side information that any algorithm that finds just one optimal solution to an input instance may not be of much use in solving the original problem in practice. One scenario where the traditional approach to modeling computational problems fails completely is when computational problems may combined with a human sense of aesthetics or intuition to solve a task, or even to stimulate inspiration. Some early relevant work is on the problem of designing a tool which helps an architect in creating a floor plan which satisfies a specified set of constraints. In general, the number of feasible floor plans—those which satisfy constraints imposed by the plot on which the building has to be erected, various regulations which the building should adhere to, and so on—would be too many for the architect to look at each of them one by one. Furthermore, many of these plans would be very similar to one another, so that it would be pointless for the architect to look at more than one of these for inspiration. As an alternative to optimization for such problems, Galle proposed a "Branch & Sample" algorithm for generating a "limited, representative sample of solutions, uniformly scattered over the entire solution space" [10].

The Diverse X Paradigm. Mike Fellows has proposed *the Diverse X Paradigm* as a solution for these issues and others [11]. In this paradigm, "X" is a placeholder for an optimization problem, and we study the complexity—specifically, the fixed-parameter tractability—of the problem of finding a few different good quality solutions for X. Contrast this with the traditional approach of looking for just one good quality solution. Let X denote an optimization problem where one looks for a minimum-size subset of some set; VERTEX COVER is an example of such a problem. The generic form of X is then:

X
Input: An instance I of X.
Solution: A solution S of I of the smallest size.

Here, the form that a "solution S of I" takes is dictated by the problem X; compare this with the earlier definition of VERTEX COVER.

The *diverse* variant of problem X, as proposed by Fellows, has the form:

DIVERSE X
Input: An instance I of X, and positive integers k, r, t.
Parameter: (k, r)
Solution: A set $\mathcal{S}$ of r solutions of I, each of size at most k, such that a *diversity measure* of $\mathcal{S}$ is at least t.

Note that one can construct diverse variants of other kinds of problems as well, following this model: it doesn't have to be a minimization problem, nor does the solution have to be a subset of some kind. Indeed, the example about floor plans described above has neither of these properties. What is relevant is that one should have (i) some notion of "good quality" solutions (for X, this equates to a small size) and (ii) some notion of a set of solutions being "diverse".

Diversity measures. The concept of diversity appears also in other fields, and there are many different ways to measure the diversity of a collection. For example, in ecology, the diversity of a set of species ("biodiversity") is a topic that has become increasingly important in recent times—see, for example, Solow and Polasky [12].

Another possible viewpoint, in the context of multicriteria optimization, is to require that the sample of solutions should try to represent the *whole solution space*. This concept can be quantified for example by the geometric *volume* of the represented space [13,14], or by the *discrepancy* [15]. See ([16], Section 3) for an overview of diversity measures in multicriteria optimization.

In this paper, we follow the simple possibility of looking for a collection of good solutions that have large *distances* from each other, in a sense that will be made precise below, see Equations (1)–(2). Direction (2), i.e., taking the pairwise sum of all Hamming distances, has been taken by many practical papers in the area of genetic algorithms—see, e.g., [17,18]. This now classical approach can be traced as far back as 1992 [19]. In [20], it has been boldly stated that this measure (and its variations) is one of the most broadly used measures in describing population diversity within genetic algorithms. One of its advantages is that it can be computed very easily and efficiently unlike many other measures, e.g., some geometry or discrepancy based measures.

Our Problems and Results

In this work, we focus on diverse versions of two minimization problems, d-HITTING SET and FEEDBACK VERTEX SET, whose solutions are subsets of a finite set. d-HITTING SET is in fact a *class* of such problems which includes VERTEX COVER, as we describe below. We will consider two natural diversity measures for these problems: the minimum Hamming distance between any two solutions, and the sum of pairwise Hamming distances of all the solutions.

The *Hamming distance* between two sets S and S', or *the size of their symmetric difference*, is

$$d_H(S, S') := |(S \setminus S') \cup (S' \setminus S)|.$$

We use

$$\text{div}_{\min}(S_1, \ldots, S_r) := \min_{1 \le i < j \le r} d_H(S_i, S_j) \quad (1)$$

to denote the minimum Hamming distance between any pair of sets in a collection of finite sets, and

$$\text{div}_{\text{total}}(S_1, \ldots, S_r) := \sum_{1 \le i < j \le r} d_H(S_i, S_j) \quad (2)$$

to denote the sum of all pairwise Hamming distances. (In Section 5, we will discuss some issues with the latter formulation.)

A *feedback vertex set* of a graph G is any subset $S \subseteq V(G)$ of the vertex set of G such that the graph $G - S$ obtained by deleting the vertices in S is a *forest*; that is, it contains no cycle.

FEEDBACK VERTEX SET
Input: A graph G.
Solution: A feedback vertex set of G of the smallest size.

More generally, a *hitting set* of a collection $\mathcal{F}$ of subsets of a universe U is any subset $S \subseteq U$ such that every set in the family $\mathcal{F}$ has a non-empty intersection with S. For a fixed positive integer d, the d-HITTING SET problem asks for a hitting set of the smallest size of a family $\mathcal{F}$ of d-sized subsets of a finite universe U:

---
d-HITTING SET
Input: A finite universe U and a family $\mathcal{F}$ of subsets of U, each of size at most d.
Solution: A hitting set S of $\mathcal{F}$ of the smallest size.

Observe that both VERTEX COVER and FEEDBACK VERTEX SET are special cases of finding a smallest hitting set for a family of subsets. VERTEX COVER is also an instance of d-HITTING SET, with $d = 2$: the universe U is the set of vertices of the input graph and the family $\mathcal{F}$ consists of all sets $\{v, w\}$, where vw is an edge in G. There is no obvious way to model FEEDBACK VERTEX SET as a d-HITTING SET instance, however, because the cycles in the input graph are not necessarily of the same size.

In this work, we consider the following problems in the DIVERSE X paradigm. Using $\text{div}_{\text{total}}$ as the diversity measure, we consider DIVERSE d-HITTING SET and DIVERSE FEEDBACK VERTEX SET, where X is d-HITTING SET and FEEDBACK VERTEX SET, respectively. Using $\text{div}_{\min}$ as the diversity measure, we consider MIN-DIVERSE d-HITTING SET and MIN-DIVERSE FEEDBACK VERTEX SET, where X is d-HITTING SET and FEEDBACK VERTEX SET, respectively.

In each case, we show that the problem is fixed-parameter tractable (FPT), with the following running times:

Theorem 1. DIVERSE d-HITTING SET *can be solved in time* $r^2 d^{kr} \cdot |U|^{O(1)}$.

Theorem 2. DIVERSE FEEDBACK VERTEX SET *can be solved in time* $2^{7kr} \cdot n^{O(1)}$.

Theorem 3. MIN-DIVERSE d-HITTING SET *can be solved in time*

- $2^{kr^2} \cdot (kr)^{O(1)}$ *if* $|U| < kr$ *and*
- $d^{kr} \cdot |U|^{O(1)}$ *otherwise.*

Theorem 4. MIN-DIVERSE FEEDBACK VERTEX SET *can be solved in time* $2^{kr \cdot \max(r, 7 + \log_2(kr))} \cdot (nr)^{O(1)}$.

Defining the diverse versions DIVERSE VERTEX COVER and MIN-DIVERSE VERTEX COVER of VERTEX COVER in a similar manner as above, we get

Corollary 1. DIVERSE VERTEX COVER *can be solved in time* $2^{kr} \cdot n^{O(1)}$. MIN-DIVERSE VERTEX COVER *can be solved in time*

- $2^{kr^2} \cdot (kr)^{O(1)}$ *if* $n < kr$ *and*
- $2^{kr} \cdot n^{O(1)}$ *otherwise.*

Related Work. The parameterized complexity of finding a diverse collection of good-quality solutions to algorithmic problems seems to be largely unexplored. To the best of our knowledge, the only existing work in this area consists of: (i) a privately circulated manuscript by Fellows [11] which introduces the Diverse X Paradigm and makes a forceful case for its relevance, and (ii) a manuscript by Baste et al. [21] which applies the Diverse X Paradigm to *vertex-problems* with the *treewidth* of the input graph as an extra parameter. In this context, a *vertex-problem* is any problem in which the input contains a graph G and the solution is some subset of the vertex set of G that satisfies some problem-specific properties. Both VERTEX COVER and FEEDBACK VERTEX SET are vertex-problems in this sense, as are many other graph problems. The *treewidth* of a graph is, informally put, a measure of how tree-like the graph is. See, e.g., ([22], Chapter 7) for an introduction of the use of the treewidth of a graph as a parameter in designing FPT algorithms. The work by Baste et al. [21] shows how to convert essentially any treewidth-based dynamic programming algorithm for solving a vertex-problem, into an algorithm for computing a diverse set of r solutions for the problem, with the diversity measure being the sum $\text{div}_{\text{total}}$ of Hamming distances of the solutions. This latter algorithm is FPT in the combined parameter (r, w), where w is the treewidth of the input graph. As a special case, they obtain a running time of $\mathcal{O}((2^{k+2}(k+1))^r kr^2 n)$ for DIVERSE VERTEX COVER. Furthermore, they show that the r-DIVERSE versions (i.e., where the diversity measure is $\text{div}_{\text{total}}$) of a handful of problems have polynomial kernels. In particular, they show that DIVERSE VERTEX COVER has a kernel with $\mathcal{O}(k(k+r))$ vertices, and that DIVERSE d-HITTING SET has a kernel with a universe size of $\mathcal{O}(k^d + kr)$.

Organization of the rest of the paper. In Section 2, we list some definitions which we use in the rest of the paper. In Section 3, we describe a generic framework which can be used for computing solution families of maximum diversity for a variety of problems whose solutions form subsets of some finite set. We prove Theorem 1 in Section 3.3 and Theorem 2 in Section 4. In Section 5, we discuss some potential pitfalls in using div$_{\text{total}}$ as a measure of diversity. In Section 6, we prove Theorems 3 and 4. We conclude in Section 7.

2. Preliminaries

Given two integers p and q, we denote by $[p, q]$ the set of all integers r such that $p \leq r \leq q$ holds. Given a graph G, we denote by $V(G)$ (resp. $E(G)$) the set of *vertices* (resp. *edges*) of G. For a subset $S \subset V(G)$, we use $G[S]$ to denote the subgraph of G induced by S, and $G \setminus S$ for the graph $G[V(G) \setminus S]$. A set $S \subseteq V(G)$ is a vertex cover (resp. a feedback vertex set) if $G \setminus S$ has no edge (resp. no cycle). Given a graph G and a vertex v such that v has exactly two neighbors, say w and w', *contracting* v consists of removing the edges $\{v, w\}$ and $\{v, w'\}$, removing v, and adding the edge $\{w, w'\}$. Given a graph G and a vertex $v \in V(G)$, we denote by $\delta_G(v)$ the *degree* of v in G. For two vertices u, v in a connected graph G, we use $\text{dist}_T(u, v)$ to denote the *distance* between u and v in G, which is the length of a shortest path in G between u and v.

A *deepest leaf* in a tree T is a vertex $v \in V(T)$ such that there exists a root $r \in V(T)$ satisfying $\text{dist}_T(r, v) = \max_{u \in V(T)} \text{dist}_T(r, u)$. A *deepest leaf* in a forest F is a deepest leaf in some connected component of F. A deepest leaf v has the property that there is another leaf in the tree at distance at most 2 from v unless v is an isolated vertex or v's neighbor has degree 2.

The objective function div$_{\text{total}}$ in (2) has an alternative representation in terms of frequencies of occurrence [21]: If y_v is the number of sets of $\{S_1, \ldots, S_r\}$ in which v appears, then

$$\text{div}_{\text{total}}(S_1, \ldots, S_r) = \sum_{v \in U} y_v (r - y_v). \tag{3}$$

Auxiliary problems. We define two auxiliary problems that we will use in some of the algorithms presented in Section 3. In the MAXIMUM COST FLOW problem, we are given a directed graph G, a *target* $d \in \mathbb{R}^+$, a *source vertex* $s \in V(G)$, a *sink vertex* $t \in V(G)$, and for each edge $(u, v) \in E(G)$, a *capacity* $c(u, v) > 0$, and a *cost* $a(u, v)$. A (s, t)-*flow*, or simply *flow* in G is a function $f \colon E(G) \to \mathbb{R}$, such that for each $(u, v) \in E(G)$, $f(u, v) \leq c(u, v)$, and for each vertex $v \in V(G) \setminus \{s, t\}$, $\sum_{(u,v) \in E(G)} f(u, v) = \sum_{(v,u) \in E(G)} f(v, u)$. The *value* of the flow f is $\sum_{(s,u) \in E(G)} f(s, u)$ and the *cost* of f is $\sum_{(u,v) \in E(G)} f(u, v) \cdot a(u, v)$. The objective of the MAXIMUM COST FLOW problem is to find the maximum cost (s, t)-flow of value d.

The second problem is the MAXIMUM WEIGHT b-MATCHING problem. Here, we are given an undirected edge-weighted graph G, and for each vertex $v \in V(G)$, a *supply* $b(v)$. The goal is to find a set of edges $M \subseteq E(G)$ of maximum total weight such that each vertex $v \in V(G)$ is incident with at most $b(v)$ edges in M.

3. A Framework for Maximally Diverse Solutions

In this section, we describe a framework for computing solution families of maximum diversity for a variety of hitting set problems. This framework requires that the solutions form a family of subsets of a ground set U that is upward closed: any superset $T \supseteq S$ of a solution S is also a solution.

The approach is as follows: In a first phase, we enumerate the class $\mathcal{S}$ of all *minimal solutions* of size at most k. (A larger class $\mathcal{S}$ is also fine as long as it is guaranteed to contain all minimal solutions of size at most k). Then, we form all r-tuples $(S_1, \ldots, S_r) \in \mathcal{S}^k$. For each such family $(S_1, \ldots, S_r)$, we try to *augment* it to a family $(T_1, \ldots, T_r)$ under the constraints $T_i \supseteq S_i$ and $|T_i| \leq k$, for each $i \in [1, r]$, in such a way that div$_{\text{total}}(T_1, \ldots, T_r)$ is maximized.

For this augmentation problem, we propose a network flow model that computes an optimal augmentation in polynomial time, see Section 3.1. This has to be repeated for each family, $O(|\mathcal{S}|^r)$ times. The first step, the generation of $\mathcal{S}$, is problem-specific. Section 3.3 shows how to solve it for d-HITTING SET. In Section 4, we will adapt our approach to deal with a FEEDBACK VERTEX SET.

3.1. Optimal Augmentation

Given a universe U and a set $\mathcal{S}$ of subsets of U, the problem $\texttt{diverse}_{r,k}(\mathcal{S})$ consists of finding an r-tuple $(S_1, \ldots, S_r)$ that maximizes $\text{div}_{\text{total}}(S_1, \ldots, S_r)$, over all r-tuples $(S_1, \ldots, S_r)$ such that, for each $i \in [1, r]$, $|S_i| \leq k$, and there exists $S \in \mathcal{S}$ such that $S \subseteq S_i \subseteq U$.

Theorem 5. *Let U be a finite universe, r and k be two integers, and $\mathcal{S}$ be a set of s subsets of U. $\texttt{diverse}_{r,k}(\mathcal{S})$ can be solved in time $r^2 s^r \cdot |U|^{O(1)}$.*

Proof. The algorithm that proves Theorem 5 starts by enumerating all r-tuples $(S_1, S_2, \ldots, S_r) \in \mathcal{S}^r$ of elements from $\mathcal{S}$. For each of these s^r r-tuples, we try to augment each S_i, using elements of U, in such a way that the diversity d of the resulting tuple $(T_1, \ldots, T_r)$ is maximized and such that, for each $i \in [1, r]$, $S_i \subseteq T_i \subseteq U$ and $|T_i| \leq k$. It is clear that this algorithm will find the solution to $\texttt{diverse}_{r,k}(\mathcal{S})$.

We show how to model this problem as a maximum-cost network flow problem with piecewise linear concave costs. This problem can be solved in polynomial time. (See, for example, [23] for basic notions about network flows).

Without loss of generality, let $U = \{1, 2, \ldots, n\}$. We use a variable $0 \leq x_{ij} \leq 1$ to decide whether element j of U should belong to set T_i. In an optimal flow, these values are integral. Some of these variables are already fixed because T_i must contain S_i:

$$x_{ij} = 1 \text{ for } j \in S_i. \tag{4}$$

The size of T_i must not exceed k:

$$\sum_{j=1}^{n} x_{ij} \leq k, \text{ for } i = 1, \ldots, r. \tag{5}$$

Finally, we can express the number y_j of sets T_i in which an element j occurs:

$$y_j = \sum_{i=1}^{r} x_{ij}, \text{ for } j = 1, \ldots, n. \tag{6}$$

These variables y_j are the variables in terms of which the objective function (3) is expressed:

$$\text{maximize } \sum_{j=1}^{n} y_j (r - y_j). \tag{7}$$

These constraints can be modeled by a network as shown in Figure 1. There are nodes T_i representing the sets T_i and a node V_j for each element $j \in U$. In addition, there is a source s and a sink t. The arcs emanating from s have capacity k. Together with the flow conservation equations at the nodes T_i, this models the constraints (5). Flow conservation at the nodes V_j gives rise to the flow variables y_j in the arcs leading to t according to (6). The arcs with fixed flow (4) could be eliminated from the network, but, for ease of notation, we leave them in the model. The only arcs that carry a cost are the arcs leading to t, and the costs are given by the concave function (7).

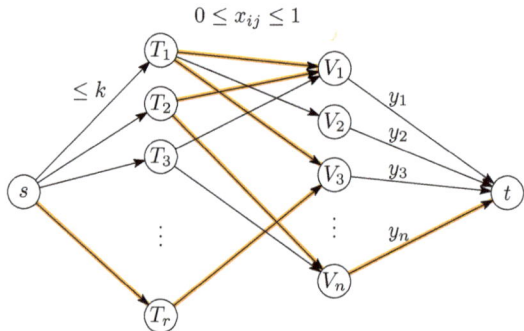

Figure 1. The network. The middle layer between the vertices T_i and V_j is a complete bipartite graph, but only a few selected arcs are shown. A potential augmenting path is highlighted.

There is now a one-to-one correspondence between integral flows from s to t in the network and solutions $(T_1, \ldots, T_r)$, and the cost of the flow is equal to the diversity (2) or (3). We are thus looking for a flow of maximum cost. The *value* of the flow (to total flow out of s) can be arbitrary. (It is equal to the sum of the sizes of the sets T_i.)

The concave arc costs (7) on the arcs leading to t can be modeled in a standard way by multiple arcs. Denote the concave cost function by $f_y := y(r-y)$, for $y = 0, 1, \ldots, r$. Then, each arc (V_i, t) in the last layer is replaced by r parallel arcs of capacity 1 with costs $f_1 - f_0, f_2 - f_1, \ldots, f_r - f_{r-1}$. This sequence of values $f_y - f_{y-1} = r - 2y + 1$ is decreasing, starting out with positive values and ending with negative values. If the total flow along such a bundle is y, the maximum-cost way to distribute this flow is to fill the first y arcs to capacity, for a total cost of $(f_1 - f_0) + (f_2 - f_1) + \cdots + (f_y - f_{y-1}) = f_y - f_0 = f_y$, as desired.

An easy way to compute a maximum-cost flow is the longest augmenting path method. (Commonly, it is presented as the *shortest* augmenting path method for the *minimum*-cost flow). This holds for the classical flow model where the cost on each arc is a linear function of the flow. An augmenting path is a path in the residual network with respect to the current flow, and the cost coefficient of an arc in such a path must be taken with opposite sign if it is traversed in the direction opposite to the original graph.

Proposition 1 (The shortest augmenting path algorithm, cf. [23] (Theorem 8.12)). *Suppose a maximum-cost flow among all flows of value v from s to t is given. Let P be a maximum-cost augmenting path from s to t. If we augment the flow along this path, this results in a new flow, of some value v'. Then, the new flow is a maximum-cost flow among all flows of value v' from s to t.*

Let us apply this algorithm to our network. We initialize the constrained flow variables x_{ij} according to Equation (4) to 1 and all other variables x_{ij} to 0. This corresponds to the original solution $(S_1, S_2, \ldots, S_r)$, and it is clearly the optimal flow of value $\sum_{i=1}^r |S_i|$ because it is the only feasible flow of this value.

We can now start to find augmenting paths. Our graph is bipartite, and augmenting paths have a very simple structure: They start in s, alternate back and forth between the T-nodes and the V-nodes, and finally make a step to t. Moreover, in our network, all costs are zero except in the last layer, and an augmenting path contains precisely one arc from this layer. Therefore, *the cost of an augmenting path is simply the cost of the final arc.*

The flow variables in the final layer are never decreased. The resulting algorithm has therefore a simple greedy-like structure. Starting from the initial flow, we first try to saturate as many of the arcs of cost $f_1 - f_0$ as possible. Next, we try to saturate as many of the arcs of cost $f_2 - f_1$ as possible, and so on. Once the incremental cost $f_{y+1} - f_y$ becomes negative, we stop.

Trying to find an augmenting path whose last arc is one of the arcs of cost $f_{y+1} - f_y$, for fixed y, is a reachability problem in the residual graph, and it can be solved by graph search in $O(nr)$ time because the network has $O(nr)$ vertices. Every augmentation increases the flow value by 1 unit. Thus, there are at most kr augmentations, for a total runtime of $O(kr^2n)$. □

3.2. Faster Augmentation

We can obtain faster algorithms by using more advanced network algorithms from the literature. We will derive one such algorithm here. The best choice depends on the relation between n, k, and r. We will apply the following result about *b-matchings*, which are generalizations of matchings: Each node v has a given *supply* $b(v)$, specifying that v should be incident to at most v edges.

Proposition 2 ([24]). *A maximum-weight b-matching in a bipartite graph with $N_1 + N_2$ nodes on the two sides of the bipartition and M edges that have integer weights between 0 and W can be found in time $O(N_1 M \log(2 + \frac{N_1^2}{M} \log(N_1 W)))$.*

We will describe below how the network flow problem from above can be converted into a b-matching problem with $N_1 = r+1$ plus $N_2 = n$ nodes and $M = 2rn$ edges of weight at most $W = 2r$. Plugging these values into Proposition 2 gives a running time of $O(r^2 n \log(2 + \frac{r}{n} \log(r^2))) = O(r^2 n \max\{1, \log \frac{r \log r}{n}\})$ for finding an optimal augmentation. This improves over the run time $O(r^2 nk)$ from the previous section unless r is extremely large (at least 2^k).

From the network of Figure 1, we keep the two layers of nodes T_i and V_j. Each vertex T_i gets a supply of $b(T_i) := k$, and each vertex V_j gets a supply of $b(V_j) := r$. To mimic the piecewise linear costs on the arcs (V_j, t) in the original network, we introduce r parallel *slack edges* from a new source vertex s' to each vertex V_j. The costs are as follows. Let $g_1 > g_2 > \cdots > g_r$ with $g_y = f_y - f_{y-1}$ denote the costs in the last layer of the original network, and let $\hat{g} := r$. Since $g_1 = r - 1$, this is larger than all costs. Then, every edge (T_i, V_j) from the original network gets a weight of $\hat{g}$, and the r new slack edges entering each V_j get positive weights $\hat{g} - g_1, \hat{g} - g_2, \ldots, \hat{g} - g_r$. We set the supply of the extra source node to $b(s') := rn$, which imposes no constraint on the number of incident edges.

Now, suppose that we have a solution for the original network in which the total flow into vertex V_j is y. In the corresponding b-matching, we can then use $b(V_j) - y = r - y$ of the slack edges incident to V_j. The $r - y$ maximum-weight slack edges have weights $\hat{g} - g_r, \hat{g} - g_{r-1}, \ldots, \hat{g} - g_{y+1}$. The total weight of the edges incident to V_j is therefore

$$r\hat{g} - g_r - g_{r-1} - \cdots - g_{y+1} = r\hat{g} + (g_1 + g_2 + \cdots + g_y),$$

using the equation $g_1 + g_2 + \cdots + g_r = f_r - f_0 = 0$. Thus, up to an addition of the constant $nr\hat{g}$, the maximum weight of a b-matching agrees with the maximum cost of a flow in the original network.

3.3. Diverse Hitting Set

In this section, we show how to use the optimal augmentation technique developed in Section 3 to solve the DIVERSE d-HITTING SET. For this, we use the following folklore lemma about minimal hitting sets.

Lemma 1. *Let $(U, \mathcal{F})$ be an instance of d-HITTING SET, and let k be an integer. There are at most d^k inclusion-minimal hitting sets of $\mathcal{F}$ of size at most k, and they can all be enumerated in time $d^k |U|^2$.*

Combining Lemma 1 and Theorem 5, we obtain the following result.

Theorem 1. DIVERSE d-HITTING SET *can be solved in time $r^2 d^{kr} \cdot |U|^{O(1)}$.*

Proof. Using Lemma 1, we can construct the set $\mathcal{S}$ of all inclusion-minimal hitting sets of $\mathcal{F}$, each of size at most k. Note that the size of $\mathcal{S}$ is bounded by d^k. As every superset of an element of $\mathcal{S}$ is also a hitting set, the theorem follows directly from Theorem 5. □

4. Diverse Feedback Vertex Set

A *feedback vertex set* (FVS) (also called a *cycle cutset*) of a graph G is any subset $S \subseteq V(G)$ of vertices of G such that every cycle in G contains at least one vertex from S. The graph $G - S$ obtained by deleting S from G is thus an acyclic graph. Finding an FVS of small size is an NP-hard problem [25] with a number of applications in Artificial Intelligence, many of which stem from the fact that many hard problems become easy to solve in acyclic graphs. An example for this is the Propositional Model Counting (or #SAT) problem that asks for the number of satisfying assignments for a given CNF formula, and has a number of applications, for instance in planning [26,27] and in probabilistic inference problems such as Bayesian reasoning [28–31]. A popular approach to solving #SAT consists of first finding a small FVS S of the CNF formula. Assigning values to all the variables in S results in an acyclic instance of CNF. The algorithm assigns all possible sets of values to the variables in S, computes the number of satisfying assignments of the resulting acyclic instances, and returns the sum of these counts [32].

In this section, we focus on the DIVERSE FEEDBACK VERTEX SET problem and prove the following theorem.

Theorem 2. DIVERSE FEEDBACK VERTEX SET *can be solved in time* $2^{7kr} \cdot n^{O(1)}$.

In order to solve r-DIVERSE k-FEEDBACK VERTEX SET, one natural way would be to generate every feedback vertex set of size at most k and then check which set of k solutions provide the required sum of Hamming distances. Unfortunately, the number of feedback vertex sets is not FPT parameterized by k. Indeed, one can consider a graph containing k cycle of size $\frac{n}{k}$, leading to $\left(\frac{n}{k}\right)^k$ different feedback vertex sets of size k.

We avoid this problem by generating all such small feedback vertex sets up to some equivalence of degree two vertices. We obtain an exact and efficient description of all feedback vertex sets of size at most k, which is formally captured by Lemma 2. A *class of solutions* of a graph G, is a pair (S, ℓ) such that $S \subseteq V(G)$ and $\ell : S \to 2^{V(G)}$ is a function such that for each $u \in S$, $u \in \ell(u)$, and for each $u, v \in S$, $u \neq v$, $\ell(u) \cap \ell(v) = \emptyset$. Given a class of solutions (S, ℓ), we define $\text{sol}(S, \ell) = \{S' \subseteq V(G) : |S'| = |S|$ and $\forall v \in S, |S' \cap \ell(v)| = 1\}$. A *class of FVS solutions* is a class of solutions (S, ℓ) such that each $S' \in \text{sol}(S, \ell)$ is a feedback vertex set of G. Moreover, if $S' \in \text{sol}(S, \ell)$ and $S' \subseteq S'' \subseteq V(G)$, we say that S'' is *described* by (S, ℓ). Note that S'' is also a feedback vertex set. In a class of FVS solutions (S, ℓ), the meaning of the function ℓ is that, for each cycle C in G, there exists $v \in S$ such that each element of $\ell(v)$ hits C. This allows us to group related solutions into only one set $\text{sol}(S, \ell)$.

Lemma 2. *Let G be a n-vertex graph. There exists a set $\mathcal{S}$ of classes of FVS solutions of G of size at most 2^{7k} such that each feedback vertex set of size at most k is described by an element of $\mathcal{S}$. Moreover, $\mathcal{S}$ can be constructed in time $2^{7k} \cdot n^{O(1)}$.*

Proof. Let G be a n-vertex graph. We start by generating a feedback vertex set $F \subseteq V$ of size at most k. The current best deterministic algorithm for this by Kociumaka and Pilipczuk [33] finds such a set in time $3.62^k \cdot n^{O(1)}$. In the following, we use the ideas used for the iterative compression approach [34].

For each subset $F' \subseteq F$, we initiate a branching process by setting $A := F'$, $B := F - F'$, and $G' := G$. Observe that, initially, as $B \subseteq F$ and $|F| \leq k$, the graph $G[B]$ has at most k components. In the branching process, we will add more vertices to A and B, and we will remove vertices and edges from G', but we will maintain the property that $A \subseteq V(G')$ and $B \subseteq V(G')$. The set C will always denote the vertex set $V(G') \setminus (A \cup B)$. Note that $G'[C]$ is initially a forest; we ensure that it always remains a forest.

We also initialize a function $\ell \colon V(G) \to 2^{V(G)}$ by setting $\ell(v) = \{v\}$ for each $v \in V(G)$. This function will keep information about vertices that are deleted from G. While searching for a feedback vertex set, we consider only feedback vertex sets that contain all vertices of A but no vertex of B. Vertices in C are still undecided. The function ℓ will maintain the invariant that, for each $v \in V(G')$, $\ell(v) \cap V(G') = \{v\}$, and, for each $v \in C$, all vertices of $\ell(v)$ intersect at exactly the same cycles in $G \setminus A$. Moreover, for each $v \in A$, the value $\ell(v)$ is fixed and will not be modified anymore in the branching process. During the branching process, we will progressively increase the size of A, B, and the sets $\ell(v)$, $v \in V(G)$.

By *reducing* (G', A, B, ℓ), we mean that we apply the following rules exhaustively.

- If there is a $v \in C$ such that $\delta_{G'[B \cup C]}(v) \leq 1$, we delete v from G'.
- If there is an edge $\{u,v\} \in E(G'[C])$ such that $\delta_{G'[B \cup C]}(u) = \delta_{G'[B \cup C]}(v) = 2$, we contract u in G' and set $\ell(v) := \ell(v) \cup \ell(u)$.

These are classical preprocessing rules for the FEEDBACK VERTEX SET problem; see, for instance, ([22], Section 9.1). Indeed, vertices of degree one cannot appear in a cycle, and consecutive vertices of degree 2 hit exactly the same cycles. After this preprocessing, there are no adjacent degree-two vertices and no degree-one vertices in C. (Degrees are measured in $G'[B \cup C]$).

We start to describe the branching procedure. We work on the tuple (G', A, B, ℓ). After each step, the value $|A| - cc(B)$ will increase, where $cc(B)$ denotes the number of connected components of $G'[B]$.

At each step of the branching, we do the following. If $|A| > k$ or if $G'[B]$ contains a cycle, we immediately stop this branch as there is no solution to be found in it. If A is a feedback vertex set of size at most k, then $(A, \ell|_A)$ is a class of FVS solutions, we add it to $\mathcal{S}$ and stop working on this branch. Otherwise, we reduce (G', A, B, ℓ). We pick a deepest leaf v in $G'[C]$ and apply one of the two following cases, depending on the vertex v:

- **Case 1:** The vertex v has at least two neighbors in B (in the graph G').

  If there is a path in B between two neighbors of v, then we have to put v in A, as otherwise this path together with v will induce a cycle. If there is no such path, we branch on both possibilities, inserting v either into A or into B.

- **Case 2:** The vertex v has at most one neighbor in B.

  Since v is a leaf in $G'[C]$, it has at most one neighbor also in C. On the other hand, we know that v has degree at least 2 in $G'[B \cup C]$. Thus, v has exactly one neighbor in B and one neighbor in C, for a degree of 2 in $G'[B \cup C]$. Let p be the neighbor in C. Again, as we have reduced (G', A, B, ℓ), the degree of p in $G'[B \cup C]$ is at least 3. Thus, either it has a neighbor in B, or, as v is a deepest leaf, it has another child, say w that is also a leaf in $G'[C]$, and w has therefore a neighbor in B. We branch on the at most $2^3 = 8$ possibilities to allocate v, p, and w if considered, between A and B, taking care not to produce a cycle in B.

In both cases, either we put at least one vertex in A, and so $|A|$ increases by one, or all considered vertices are added to B. In the latter case, the considered vertices are connected, at least two of them have a neighbor in B, and no cycles were created; therefore, the number of components in B drops by one. Thus, $|A| - cc(B)$ increases by at least one. As $-k \leq |A| - cc(B) \leq k$, there can be at most $2k$ branching steps.

Since we branch at most $2k$ times and at each branch we have at most 2^3 possibilities, the branching tree has at most 2^{6k} leaves. Thus, for each of the at most 2^k subsets F' of F, we add at most 2^{6k} elements to $\mathcal{S}$.

It is clear that we have obtained all solutions of FVS and they are described by the classes of FVS solutions in $\mathcal{S}$, which is of size 2^{7k}. □

Proof of Theorem 2. We generate all 2^{7kr} r-tuples of the classes of solutions given by Lemma 2, with repetition allowed.

We now consider each r-tuple $((S_1, \ell_1), (S_2, \ell_2), \ldots, (S_r, \ell_r)) \in \mathcal{S}^r$ and try to pick an appropriate solution T_i from each class of solutions (S_i, ℓ_i), $i \in [1, k]$, in such a way that the diversity of the resulting tuple of feedback vertex sets $(T_1, \ldots, T_r)$ is maximized. The network of Section 3.1 must be adapted to model the constraints resulting from solution classes. Let (S, ℓ) be a solution class, with $|S| = b$. For our construction, we just need to know the family $\{\ell(v) \mid v \in S\} = \{L_1, L_2, \ldots, L_b\}$ of disjoint nonempty vertex sets. The solutions that are described by this class are all sets that can be obtained by picking at least one vertex from each set L_q. Figure 2 shows the necessary adaptations for *one* solution $T = T_i$. In addition to a single node T that is either directly of indirectly connected to all nodes $V_1, \ldots, V_n$, like in Figure 1, we have additional nodes representing the sets L_q. For each vertex j that appears in one of the sets L_q, there is an additional node U_j in an intermediate layer of the network. The flow from s to L_q is forced to be equal to 1, and this ensures that at least one element of the set L_q is chosen in the solution. Here, it is important that the sets L_q are disjoint.

A similar structure must be built for each set $T_1, \ldots, T_r$, and all these structures share the vertices s and $V_1, \ldots, V_n$. The rightmost layer of the network is the same as in Figure 1.

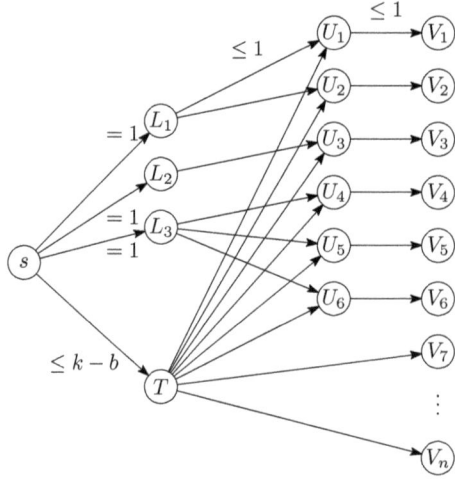

Figure 2. Part of the modified network for a solution T which is specified by $b = 3$ sets $L_1 = \{1, 2\}$, $L_2 = \{3\}$, and $L_3 = \{4, 5, 6\}$.

The initial flow is not so straightforward as in Section 3.1 but is still easy to find. We simply saturate the arc from s to each of the nodes L_q in turn by a shortest augmenting path. Such a path can be found by a simple reachability search in the residual network, in $O(rn)$ time. The total running time $O(kr^2n)$ from Section 3.1 remains unchanged. □

5. Modeling Aspects: Discussion of the Objective Function

In Sections 3 and 4, we have used the sum of the Hamming distances, $\text{div}_{\text{total}}$, as the measure of diversity. While this metric is of natural interest, it appears that, in some specific cases, it may not be a useful choice. We present a simple example where the *most diverse* solution according to $\text{div}_{\text{total}}$ is not what one might expect.

Let r be an even number. We consider the path with $2r - 2$ vertices, and we are looking for r vertex covers of size at most $r - 1$, of maximum diversity.

Figure 3 shows an example with $r = 6$. The smallest size of a vertex cover is indeed $r - 1$, and there are r different solutions. One would hope that the "maximally diverse" selection of r solutions would pick all these different solutions. However, no, the selection that maximizes $\text{div}_{\text{total}}$ consists of $r/2$ copies of just *two* solutions, the "odd" vertices and the "even" vertices (the first and last solution in Figure 3).

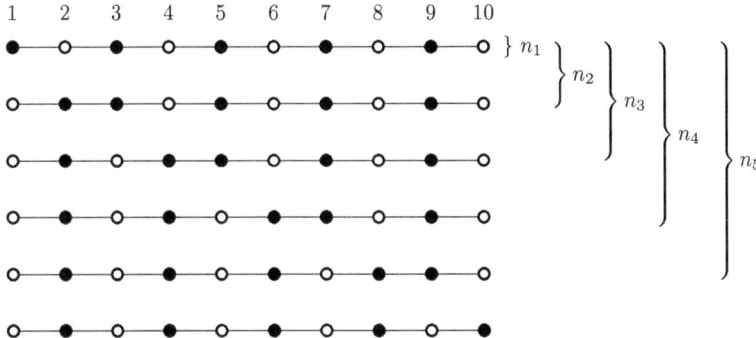

Figure 3. The $r = 6$ different vertex covers of size $r - 1 = 5$ in a path with $2(r - 1) = 10$ vertices.

This can be seen as follows. If the selected set contains in total n_i copies of the first i solutions in the order of Figure 3, then the objective can be written as

$$2n_1(r - n_1) + 2n_2(r - n_2) + \cdots + 2n_{r-1}(r - n_{r-1}).$$

Here, each term $2n_i(r - n_i)$ accounts for two consecutive vertices $2i - 1, 2i$ of the path in the formulation (3). The unique way of maximizing each term individually is to set $n_i = r/2$ for all i. This corresponds to the selection of $r/2$ copies of the first solution and $r/2$ copies of the last solution, as claimed.

In a different setting, namely the distribution of r points inside a square, an analogous phenomenon has been observed ([16], Figure 1): Maximizing the sum of pairwise Euclidean distances places all points at the corners of the square. In fact, it is easy to see that, in this geometric setting, any locally optimal solution must place all points on the boundary of the feasible region. By contrast, for our combinatorial problem, we don't know whether this pathological behavior is typical or rare in instances that are not specially constructed. Further research is needed. A notion of diversity which is more robust in this respect is the *smallest* difference between two solutions, which we consider in Section 6.

6. Maximizing the Smallest Hamming Distance

The undesired behavior highlighted in Section 5 is the fact that the collection that maximizes the sum of the Hamming distances uses several copies of the same set. In this section, we explore how to handle this unexpected behavior by changing the distance to the minimal Hamming distance between two sets of the collection. This modification naturally removes the possibility of selecting the same solution twice. We show how to solve MIN-DIVERSE d-HITTING SET and r-MIN-DIVERSE k-FEEDBACK VERTEX SET for this metric.

Theorem 3. MIN-DIVERSE d-HITTING SET *can be solved in time*

- $2^{kr^2} \cdot (kr)^{O(1)}$ *if* $|U| < kr$ *and*
- $d^{kr} \cdot |U|^{O(1)}$ *otherwise.*

Proof. Let $(U, \mathcal{F}, k, r, t)$ be an instance of MIN-DIVERSE d-HITTING SET where $|U| = n$. If $n < kr$, we solve the problem by complete enumeration: There are trivially at most 2^n hitting sets of size at most k. We form all r-tuples $(T_1, \ldots, T_r)$ of them and select the one that maximizes $\text{div}_{\min}(T_1, \ldots, T_r)$. The running time is at most $O((2^n)^r r^2 n) = O(2^{kr^2} kr^3)$.

We now assume that $n \geq kr$. We use the same strategy as in Section 3: We generate all r-tuples $(S_1, \ldots, S_r)$ of *minimal solutions* and try to augment each one to a r-tuple $(T_1, \ldots, T_r)$ such that, for each $i \in [1, r]$, $|T_i| \leq k$ and $S_i \subseteq T_i \subseteq V(G)$ hold. The difference is that we try to maximize $\text{div}_{\min}(T_1, \ldots, T_r)$ instead of $\text{div}_{\text{total}}(T_1, \ldots, T_r)$ in the augmentation. Given that we have a large supply of $n \geq kr$ elements in U, this is easy. To each set S_i, we add $k - |S_i|$ new elements, taking care that we pick different elements for each S_i that are not in any of the other sets S_j. The Hamming distance between two resulting sets is then $d_H(T_i, T_j) = d_H(S_i, S_j) + (k - |S_i|) + (k - |S_i|)$, and it is clear that this is the largest possibly distance that two sets $T_i' \supseteq S_i$ and $T_j' \supseteq S_j$ with $|T_i'|, |T_j'| \leq k$ can achieve. Thus, since our choice of augmentation individually maximizes each pairwise Hamming distance, it also maximizes the smallest Hamming distance. This procedure can be carried out in $O(kr + n) = O(n)$ time. In addition, we need $O(kr^2) = O(n^2)$ time to compute the smallest distance.

Using Lemma 1, we construct the set $\mathcal{S}$ of all minimal solutions of the d-Hitting Set instance $(U, \mathcal{F})$, each of size at most k. We then go through every r-tuple $(S_1, \ldots, S_r) \in \mathcal{S}^r$ and augment it optimally, as just described. The running time is $d^{kr} \cdot O(n^2)$. □

Theorem 4. Min-Diverse Feedback Vertex Set *can be solved in time* $2^{kr \cdot \max(r, 7 + \log_2(kr))} \cdot (nr)^{O(1)}$.

Proof. Let G be a n-vertex graph. If $n < kr$, we again solve the problem by complete enumeration: There are trivially at most 2^n feedback vertex sets of size at most k. We form all r-tuples $(T_1, \ldots, T_r)$ of them and select the one that maximizes $\text{div}_{\min}(T_1, \ldots, T_r)$. The running time is at most $O((2^n)^r r^2 n) = O(2^{kr^2} r^2 n)$.

We assume now that $n \geq kr$. As in Section 4, we construct a set $\mathcal{S}$ of at most 2^{7k} classes of FVS solutions of G, using Lemma 2. Then, we go through all $(2^{7k})^r$ r-tuples of classes $S = ((S_1, \ell_1), \ldots, (S_r, \ell_r)) \in \mathcal{S}^r$. For each such r-tuple, we look for the r-tuple $(T_1, \ldots, T_r)$ of feedback vertex sets such that each T_i is described by (S_i, ℓ_i), and the objective value $\text{div}_{\min}(T_1, \ldots, T_r)$ is maximized. Thus far, the procedure is completely analogous to the algorithm of Theorem 2 in Section 4 for maximizing $\text{div}_{\text{total}}(T_1, \ldots, T_r)$.

Now, in going from a class (S_i, ℓ_i) to T_i, we have to select a vertex from every set $\ell_i(v)$, for $v \in S_i$, and we may add an arbitrary number of additional vertices, up to size k. We make this selection as follows: Whenever $|\ell_i(v)| < kr$, we simply try all possibilities of choosing an element of $\ell_i(v)$ and putting it into T_i. If $|\ell_i(v)| \geq kr$, we defer the choice for later. In this way, we have created at most $(kr)^{kr}$ "partial" feedback vertex sets $(T_1^0, \ldots, T_r^0)$.

For each such $(T_1^0, \ldots, T_r^0)$, we now add the remaining elements. In each list $\ell_i(v)$ which has been deferred, we greedily pick an element that is distinct from all other chosen elements. This is always possible since the list is large enough. Finally, we fill up the sets to size k, again choosing fresh elements each time. Each such choice is an optimal choice because it increases the Hamming distance between the concerned set T_i and *every* other set T_j by 1, which is the best that one can hope for. As we proceed to this operation for each $S \in \mathcal{S}^r$, where $|\mathcal{S}| \leq 2^{7k}$, and that, for each such S, we create at most $(kr)^{kr}$ r-tuples, and we obtain an algorithm running in time $2^{7kr} \cdot (kr)^{kr} \cdot n^{O(1)}$. The theorem follows. □

7. Conclusions and Open Problems

In this work, we have considered the paradigm of finding small diverse collections of reasonably good solutions to combinatorial problems, which has recently been introduced to the field of fixed-parameter tractability theory [21].

We have shown that finding diverse collections of d-hitting sets and feedback vertex sets can be done in FPT time. While these problems can be classified as FPT via the kernels and a treewidth-based meta-theorem proved in [21], the methods proposed here are of independent interest. We introduced a method of generating a maximally diverse set of solutions from a set that either contains all minimal solutions of bounded size (d-HITTING SET) or from a collection of structures that in some way *describes* all solutions of bounded size (FEEDBACK VERTEX SET). In both cases, the maximally diverse collection of solutions is obtained via a network flow model, which does not rely on any specific properties of the studied problems. It would be interesting to see if this strategy can be applied to give FPT-algorithms for diverse problems that are not covered by the meta-theorem or the kernels presented in [21].

While the problems in [21] as well as the ones in Sections 3 and 4 seek to maximize the *sum* of all pairwise Hamming distances, we also studied the variant that asks to maximize the *minimum* Hamming distance, taken over each pair of solutions. This was motivated by an example where the former measure does not perform as intended (Section 5). We showed that also, under this objective, the diverse variants of d-HITTING SET and FEEDBACK VERTEX SET are FPT. It would be interesting to see whether this objective also allows for a (possibly treewidth-based) meta-theorem.

In [21], the authors ask whether there is a problem that is in FPT parameterized by a solution size whose r-diverse variant becomes W[1]-hard upon adding r as another component of the parameter. We restate this question here.

Question 1 (Open Question [21]). *Is there a problem Π with solution size k, such that Π is FPT parameterized by k, while DIVERSE Π, asking for r solutions, is W[1]-hard parameterized by $k + r$?*

To the best of our knowledge, this problem is still wide open. We believe that the div_{min} measure is more promising to obtain such a result rather than the div_{total} measure. A possible way to tackle both measures at once might be a parameterized (and strenghtened) analogue of the following approach that is well-studied in classical complexity. Yato and Seta propose a framework [35] to prove NP-completeness of finding a *second* solution to an NP-complete problem. In other words, there are some problems where given one solution it is still NP-hard to determine whether the problem has a different solution.

From a different perspective, one might want to identify problems where obtaining one solution is polynomial-time solvable, but finding a diverse collection of r solutions becomes NP-hard. The targeted running time should be FPT parameterized by r (and maybe t, the diversity target) only. We conjecture that this is most probably NP- or W[−] hard in general. However, we believe it is interesting to search for well-known problems where it is not the case.

Author Contributions: Conceptualization, J.B., L.J., T.M., G.P. and G.R.; Methodology, J.B., L.J., T.M., G.P. and G.R.; Investigation, J.B., L.J., T.M., G.P. and G.R.; Writing–original draft preparation, J.B., L.J., T.M., G.P. and G.R.; Writing–review and editing, J.B., L.J., T.M., G.P. and G.R.

Funding: Tomáš Masařík received funding from the European Research Council (ERC) under the European Union's Horizon 2020 research and innovation programme Grant Agreement No. 714704, and from Charles University student Grant No. SVV-2017-260452. Lars Jaffke is supported by the Bergen Research Foundation (BFS). Geevarghese Philip received funding from the following sources: the European Research Council (ERC) under the European Union's Horizon 2020 research and innovation programme (Grant No. 819416), the Norwegian Research Council via grants MULTIVAL and CLASSIS, BFS (Bergens Forsknings Stiftelse) "Putting Algorithms Into Practice" Grant No. 810564 and NFR (Norwegian Research Foundation) Grant No. 274526d "Parameterized Complexity for Practical Computing".

Acknowledgments: The first, second, third and fourth authors would like to thank Mike Fellows for introducing them to the notion of diverse FPT algorithms and sharing the manuscript "The Diverse X Paradigm" [11].

Conflicts of Interest: The authors declare no conflict of interest.

References

1. Ehrgott, M. *Multicriteria Optimization*; Springer: Berlin/Heidelberg, Germany, 2005; Volume 491.
2. Vela, A.E. Understanding Conflict-Resolution Taskload: Implementing Advisory Conflict-Detection and Resolution Algorithms in an Airspace. Ph.D. Thesis, Georgia Institute of Technology, Atlanta, GA, USA, 2011.
3. Idan, M.; Iosilevskii, G.; Ben-Yishay, L. Efficient air traffic conflict resolution by minimizing the number of affected aircraft. *Int. J. Adapt. Control Signal Process.* **2010**, *24*, 867–881. [CrossRef]

4. Gandhi, S.; Buragohain, C.; Cao, L.; Zheng, H.; Suri, S. A general framework for wireless spectrum auctions. In Proceedings of the 2nd IEEE International Symposium on New Frontiers in Dynamic Spectrum Access Networks, Dublin, Ireland, 17–20 April 2007; pp. 22–33.
5. Hoefer, M.; Kesselheim, T.; Vöcking, B. Approximation algorithms for secondary spectrum auctions. *ACM Trans. Internet Technol. (TOIT)* **2014**, *14*, 16:1–16:24. [CrossRef]
6. Chomicki, J.; Marcinkowski, J. Minimal-change integrity maintenance using tuple deletions. *Inf. Comput.* **2005**, *197*, 90–121. [CrossRef]
7. Arenas, M.; Bertossi, L.; Chomicki, J.; He, X.; Raghavan, V.; Spinrad, J. Scalar aggregation in inconsistent databases. *Theor. Comput. Sci.* **2003**, *296*, 405–434. [CrossRef]
8. Pema, E.; Kolaitis, P.G.; Tan, W.C. On the tractability and intractability of consistent conjunctive query answering. In Proceedings of the 2011 Joint EDBT/ICDT Ph. D. Workshop, Uppsala, Sweden, 25 March 2011; pp. 38–44. [CrossRef]
9. Ioannou, E.; Staworko, S. Management of inconsistencies in data integration. In *Data Exchange, Integration, and Streams*; Schloss Dagstuhl-Leibniz-Zentrum für Informatik: Dagstuhl, Germany, 2013; Volume 5, pp. 217–225. [CrossRef]
10. Galle, P. Branch & sample: A simple strategy for constraint satisfaction. *BIT Numer. Math.* **1989**, *29*, 395–408. [CrossRef]
11. Fellows, M.R. University of Bergen, Bergen, Norway. Unpublished work, 2018.
12. Solow, A.R.; Polasky, S. Measuring biological diversity. *Environ. Ecol. Stat.* **1994**, *1*, 95–103. [CrossRef]
13. Bringmann, K.; Cabello, S.; Emmerich, M.T.M. Maximum Volume Subset Selection for Anchored Boxes. In Proceedings of the 33rd International Symposium on Computational Geometry (SoCG 2017), Brisbane, Australia, 4–7 July 2017; Aronov, B., Katz, M.J., Eds.; Schloss Dagstuhl–Leibniz-Zentrum für Informatik: Dagstuhl, Germany, 2017; Volume 77, pp. 22:1–22:15. [CrossRef]
14. Kuhn, T.; Fonseca, C.M.; Paquete, L.; Ruzika, S.; Duarte, M.M.; Figueira, J.R. Hypervolume Subset Selection in Two Dimensions: Formulations and Algorithms. *Evol. Comput.* **2016**, *24*, 411–425._a_00157. [CrossRef] [PubMed]
15. Neumann, A.; Gao, W.; Doerr, C.; Neumann, F.; Wagner, M. Discrepancy-based Evolutionary Diversity Optimization. In Proceedings of the Genetic and Evolutionary Computation Conference, Kyoto, Japan, 15–19 July 2018; ACM: New York, NY, USA, 2018; pp. 991–998. [CrossRef]
16. Ulrich, T.; Bader, J.; Thiele, L. Defining and Optimizing Indicator-Based Diversity Measures in Multiobjective Search. In *Parallel Problem Solving from Nature, PPSN XI*; Schaefer, R., Cotta, C., Kołodziej, J., Rudolph, G., Eds.; Springer: Berlin/Heidelberg, Germany, 2010; pp. 707–717._71. [CrossRef]
17. Gabor, T.; Belzner, L.; Phan, T.; Schmid, K. Preparing for the Unexpected: Diversity Improves Planning Resilience in Evolutionary Algorithms. In Proceedings of the 2018 IEEE International Conference on Autonomic Computing, ICAC 2018, Trento, Italy, 3–7 September 2018; pp. 131–140. [CrossRef]
18. Morrison, R.W.; Jong, K.A.D. Measurement of Population Diversity. In Proceedings of the 5th International Conference, Evolution Artificielle, EA 2001, Le Creusot, France, 29–31 October 2001; pp. 31–41._3. [CrossRef]
19. Louis, S.J.; Rawlins, G.J.E. Syntactic Analysis of Convergence in Genetic Algorithms. In Proceedings of the Second Workshop on Foundations of Genetic Algorithms, Vail, CO, USA, 26–29 July 1992; pp. 141–151. [CrossRef]
20. Wineberg, M.; Oppacher, F. The Underlying Similarity of Diversity Measures Used in Evolutionary Computation. In Proceedings of the Genetic and Evolutionary Computation-GECCO 2003, Genetic and Evolutionary Computation Conference, Chicago, IL, USA, 12–16 July 2003; pp. 1493–1504._21. [CrossRef]
21. Baste, J.; Fellows, M.; Jaffke, L.; Masařík, T.; de Oliveira Oliveira, M.; Philip, G.; Rosamond, F. Diversity in Combinatorial Optimization. *arXiv* **2019**, arXiv:1903.07410.
22. Cygan, M.; Fomin, F.; Kowalik, L.; Lokshtanov, D.; Marx, D.; Pilipczuk, M.; Pilipczuk, M.; Saurabh, S. *Parameterized Algorithms*; Springer: Berlin/Heidelberg, Germany, 2015; doi:10.1007/978-3-319-21275-3. [CrossRef]
23. Tarjan, R.E. *Data Structures and Network Algorithms*; SIAM: Philadelpia, PA, USA, 1983; doi:10.1137/1.9781611970265. [CrossRef]
24. Ahuja, R.K.; Orlin, J.B.; Stein, C.; Tarjan, R.E. Improved algorithms for bipartite network flow. *SIAM J. Comput.* **1994**, *23*, 906–933. [CrossRef]

25. Karp, R.M. Reducibility among combinatorial problems. In *Complexity of Computer Computations*; Springer: Berlin/Heidelberg, Germany, 1972; pp. 85–103.
26. Domshlak, C.; Hoffmann, J. Fast probabilistic planning through weighted model counting. In Proceedings of the Sixteenth International Conference on Automated Planning and Scheduling, Cumbria, UK, 6–10 June 2006; pp. 243–252.
27. Palacios, H.; Bonet, B.; Darwiche, A.; Geffner, H. Pruning conformant plans by counting models on compiled d-DNNF representations. In Proceedings of the Fifteenth International Conference on Automated Planning and Scheduling, Menlo Park, CA, USA, 5–10 June 2005; pp. 141–150.
28. Bacchus, F.; Dalmao, S.; Pitassi, T. Algorithms and complexity results for #SAT and Bayesian inference. In Proceedings of the 44th Annual IEEE Symposium on Foundations of Computer Science, Cambridge, MA, USA, 11–14 October 2003; pp. 340–351. [CrossRef]
29. Littman, M.L.; Majercik, S.M.; Pitassi, T. Stochastic Boolean satisfiability. *J. Autom. Reason.* **2001**, *27*, 251–296.:1017584715408. [CrossRef]
30. Sang, T.; Bearne, P.; Kautz, H. Performing Bayesian inference by weighted model counting. In Proceedings of the 20th National Conference on Artificial Intelligence, Pittsburgh, PA, USA, 9–13 July 2005; pp. 475–481.
31. Apsel, U.; Brafman, R.I. Lifted MEU by weighted model counting. In Proceedings of the Twenty-Sixth AAAI Conference on Artificial Intelligence, Toronto, ON, Canada, 22–26 July 2012; pp. 1861–1867.
32. Dechter, R.; Cohen, D. *Constraint Processing*; Morgan Kaufmann: Burlington, MA, USA, 2003.
33. Kociumaka, T.; Pilipczuk, M. Faster deterministic Feedback Vertex Set. *Inf. Process. Lett.* **2014**, *114*, 556–560. [CrossRef]
34. Reed, B.; Smith, K.; Vetta, A. Finding odd cycle transversals. *Oper. Res. Lett.* **2004**, *32*, 299–301. [CrossRef]
35. Yato, T.; Seta, T. Complexity and completeness of finding another solution and its application to puzzles. *IEICE Trans. Fundam. Electron. Commun. Comput. Sci.* **2003**, *86*, 1052–1060.

© 2019 by the authors. Licensee MDPI, Basel, Switzerland. This article is an open access article distributed under the terms and conditions of the Creative Commons Attribution (CC BY) license (http://creativecommons.org/licenses/by/4.0/).

Article

Parameterized Algorithms in Bioinformatics: An Overview

Laurent Bulteau and Mathias Weller *

Centre national de la recherche scientifique (CNRS), Université Paris-Est Marne-la-Vallée, 77454 Marne-la-Vallée, France; laurent.bulteau@u-pem.fr
* Correspondence: mathias.weller@u-pem.fr

Received: 30 September 2019; Accepted: 19 November 2019; Published: 1 December 2019

Abstract: Bioinformatics regularly poses new challenges to algorithm engineers and theoretical computer scientists. This work surveys recent developments of parameterized algorithms and complexity for important NP-hard problems in bioinformatics. We cover sequence assembly and analysis, genome comparison and completion, and haplotyping and phylogenetics. Aside from reporting the state of the art, we give challenges and open problems for each topic.

Keywords: fixed-parameter tractability; genome assembly; sequence analysis; comparative genomics; haplotyping; phylogenetics; agreement forests

1. Introduction

With the dawn of molecular biology came the need for biologists to process and analyze data way over human capacity, implying the need for computer programs to assemble sequences, measure genomic distances, spot (exceptions to) patterns, and build and compare phylogenetic relationships, to name only few examples. While many of these problems have efficient solutions, others are inherently hard. In this work, we survey selected NP-hard problems in genome comparison and completion (Section 2), sequence assembly and analysis (Section 3), haplotyping (Section 4), and phylogenetics (Section 5), along with important results regarding parameterized algorithms and hardness. While trying to be accessible to a large audience, we assume that the reader is somewhat familiar with common notions of algorithms and fixed-parameter tractability, such as the $O^*(.)$-notation, giving only exponential parts of a running time, ignoring polynomial factors (for details about parameterized complexity, we refer to the recent monographs [1,2]). In a nutshell, parameterized complexity exploits the fact that the really hard instances are pathological and real-world processes generate data which, although enormous in size, are very structured and governed by simple processes. It seems straightforward that biological datasets abide by this "law of low real-world complexity".

While excellent surveys of fixed-parameter tractable (FPT) problems in bioinformatics exist [3–6], this work tries to present an update, summarizing the state of the FPT-art as well as reporting on some additional "hot topics". We further aim at: 1. iterating that parameterized algorithms are a viable and widespread tool to attack NP-hard problems in the context of biological data; and 2. supplying engineers and problem-solvers with computational problems that are actually being solved in practice, inspiring research on concrete open questions. Herein, we focus on problems that are particular for bioinformatics, avoiding more universal and general problems such as the various versions of CLUSTERING which, while important for bioinformatics, have applications in many areas.

Generally, each topic has a problem description, a results section, some open problems and, possibly, an assortment of notes, giving additional information about less popular variants, often only mentioning results or problems related to the main topic without going into details. Finally, we want to apologize for the limited coverage of relevant work but, as our colleague Jesper Jansson wrote,

"Writing a survey about bioinformatics-related FPT results sounds like a monumental task!"

2. Genome Comparison and Completion

This section considers genetic data at its largest scale, i.e. as a sequence of genes, independent of their underlying DNA sequences. At this scale, long-range evolutionary events occur, cutting and pasting whole segments of chromosomes ("chromosomal rearrangement") [7]. The number of such rearrangements between the genomes of two species can be used to estimate their "genetic distance" which, in turn, allows for phylogenetic reconstruction. Further, we consider the problems of identifying genes with a common ancestor after duplications, and filling the gaps in an incomplete genome.

In the most general setting, a genome G may be represented as a collection of strings and circular strings, where each string represents a chromosome and each character represents a gene or gene family. However, for ease of representation, most models use a single-sequence representation of the genome (which generalizes easily to multi chromosomal inputs in most cases). There is however an important distinction between the *string model*, allowing gene repetitions (also called duplicates, or paralogs), and the *permutation model* enforcing that each gene appears exactly once. Many problems in the following sections can be formulated in both models. The string model is obviously the most general in practice, but often comes with a prohibitive algorithmic cost. Parameterized algorithms can offer a good solution, using the fact that most genes only have a small number of occurrences. Thus, the maximum number of occurrences of any character in an input string, denoted occ in this paper, is a ubiquitous parameter. In both models, a genome is *signed* if each gene is given a sign (+ or -) representing its orientation along the DNA strand.

In the following, a string s' is called *substring* of a string s if $s' = s$ or s' is the result of removing the first or last character of some substring of s. Further, s' is called *subsequence* of s if $s' = s$ or s' is the result of removing any character from a subsequence of s. We say that two strings are *balanced* if each character appears with the same number of occurrences in both.

2.1. Genomic Distances

The problems in this section aim at computing a distance between genomes, reflecting the "amount of evolution" that occurred since the last common ancestor. Among the most basic distances for the permutation model is the *breakpoint distance*, counting the number of pairs of genes that occur consecutively in one genome but not in the other (a pair ab is considered to be identical to ba in the unsigned model, however +a+b is identical only to +a+b and -b-a in the signed model). Its complement is a measure of similarity, the *number of common adjacencies* (the number of pairs of consecutive genes that occur in both genomes). Common adjacencies can easily be generalized in the string model: adjacencies of each genome are seen as a multi-set, and the number of common adjacencies is the size of the intersection of these multi-sets.

More advanced genomic distances are rearrangement distances. A rearrangement acts on the genome by reordering some genes in a certain way, mimicking a large-scale evolutionary event. The core question in rearrangement distances is to compute the minimum number of rearrangements transforming a genome into another. These distances are more precise—in the sense that they directly describe parsimonious evolutionary scenarios—but in most cases significantly more difficult to compute than the breakpoint distance. Note that *balanced* genomes are usually considered, i.e., any character appears with the same number of occurrences in every genome. See the work of Fertin et al. [8] for an extensive survey on rearrangement distances.

2.1.1. Double-Cut and Join Distance

Problem Description. The double-cut and join distance needs the most general genome model as intermediary steps, that is, a multi-set of strings and circular strings. A *double-cut and join* operation splits the genome in two positions and joins the four created endpoints in any way (see Figure 1). For signed genomes, this operation is required to maintain consistent orientation of each gene. We focus on the case where the source and target genomes have a single chromosome (i.e. strings or permutations).

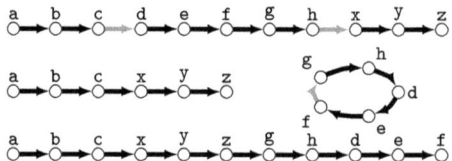

Figure 1. Two DCJ operations turn abcdefghxyz into abcxyzghdef (breakpoints are gray arcs): 1. split cd and hx and join cx and dh, and 2. split fg and join zg.

DOUBLE CUT AND JOIN DISTANCE (DCJ)

Input: genomes G_1, G_2 (balanced strings or permutations), and some $k \in \mathbb{N}$
Question: Can G_1 be transformed into G_2 with a series of $\leq k$ double-cut and join operations?

Results. The DCJ distance can be computed in linear time in the signed permutations model [9,10]. However, it is NP-hard in the unsigned permutation or string models [11]. On the positive side, it admits an $O(2^{2k}n)$-time algorithm for unsigned permutations [12].

An issue with the DCJ distance is that the solution space might be very large, as many different scenarios may yield the same distance. Thus, more precise models have been designed to "focus" the solution towards the most realistic scenarios. Fertin et al. [13] introduced the *wDCJ* distance, where intergene distances are added to the genome model and must be accounted for in the DCJ scenario (when breaking between consecutive genes, the number of intergenomic bases must be shared among both sides). This variant makes the problem NP-hard for signed permutations, but fixed-parameter tractable for the parameter k. Another constraint focuses on common intervals, which are segments of the genomes that have exactly the same gene content, but in different orders. A DCJ scenario is *perfect* if it never breaks a common interval. Bérard et al. [14] showed that finding a perfect scenario with a minimum number of operations is FPT for a parameter given by the common interval structure.

2.1.2. Reversal Distance

Problem Description. A *reversal* [15] is a rearrangement reversing the order of the characters in any substring of the genome. When considering signed genomes, a reversal additionally switches all the signs in the reversed substring (see Figure 2).

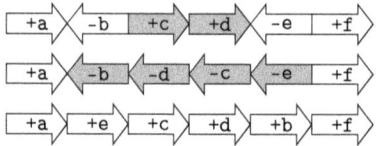

Figure 2. A signed genome +a-b+c+d-e+f turned into +a+e+c+d+b+f by a signed reversals on +c+d followed by one on -b-d-c-e.

SORTING BY REVERSALS (SBR)

Input: genomes G_1, G_2 (balanced strings or permutations), and some $k \in \mathbb{N}$
Question: Can G_1 be transformed into G_2 with a series of at most k reversals?

Results. SORTING BY REVERSALS is polynomial-time solvable in the signed-permutations model [16]. However, it is NP-hard for unsigned permutations [17] and for strings, even when occ = 2 [18] and for binary signed strings [19]. For unsigned permutations, SORTING BY REVERSALS is easily FPT for k, using the fact that reversals never need to cut at positions that are not breakpoints [20]. For balanced strings, a possible parameter, b_{max}, is the number of blocks of the input strings, where a *block* is a maximal factor of the form a^n for some character a. SORTING BY REVERSALS can be solved in $O^*\left(b_{max}^{O(b_{max})}\right)$ time in both the signed and unsigned variants. [21]. As for DCJ, the variant restricting scenarios to reversals preserving common intervals is also NP-hard, and FPT for a parameter given by the common interval structure [22].

Notes. Other rearrangements can be considered, such as *transpositions* (two consecutive factors are swapped), and the *prefix* and/or *suffix* variants of reversals and transpositions (that is, reversals and transpositions affecting the first or last character of the sequence, respectively) (see, e.g., [23]).

No polynomial-time algorithm is known for computing these rearrangement distances, although NP-completeness is still open, notably on permutations for signed prefix reversals and prefix transpositions. Fixed-parameter tractability results may be achieved in the permutation model using the relationship between these rearrangements and breakpoints [24], and in the string model using the number of blocks b_{max} as a parameter [21,25].

2.2. Common Partitions

Two strings S_1 and S_2 have a *common string partition* S if S is a (multi-)set of strings called *blocks* such that both S_1 and S_2 can be seen as the concatenation of the blocks of S. Furthermore, S is a *common strip partition* if each of its blocks has length at least two. Note that any two balanced strings have a common string partition, but may not have a common strip partition. Common strings or common strips may be used to identify syntenic regions between two genomes. Intuitively, most genes from the genomes of two close species should have a simple common string partition.

2.2.1. Minimum Common String Partition

Problem Description. The most natural partition problem is MINIMUM COMMON STRING PARTITION [18] (see Figure 3), which admits slightly different formulations [26,27]. It can be seen as a generalization of the breakpoint distance on balanced strings. Indeed, a common string partition may be seen as a mapping between the characters of both strings, i.e. a permutation, whose breakpoints correspond to the limits between consecutive blocks. Rather than a genomic distance, it can also be seen as a method for assigning orthologs across two genomes, assuming that all duplication events happened before their last common ancestor. For example, depending on how orthologs are matched, genomes abacd and acdab may be seen as permutations 12345 and 34512 (i.e., matching two blocks ab and acd), or 12345 and 14532 (matching four blocks a, b, a, and cd). Gene sequences alone may not be sufficient to distinguish both cases: in such a situation, the first option presumably reflects better the evolutionary history of these species, since it involves fewer large-scale events in the genome.

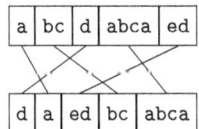

Figure 3. A minimum common string partition for the strings abcdabcaed and daedbcabca.

MINIMUM COMMON STRING PARTITION (MCSP)
Input: genomes G_1, G_2 (balanced strings), and some $k \in \mathbb{N}$
Question: Do G_1 and G_2 admit a common string partition of size at most k?

Results. MCSP is FPT for several combinations of parameters, depending on the relative size of the blocks, the size k of the partition and/or the maximum number occ of occurrences of a character [28,29]. It can also be solved in $k^{O(k^2)}$ time [30], although this is too slow to be of practical interest. Finally, MCSP admits a more practical algorithm for the parameter $k + $ occ, running in $O(occ^{2k}kn)$ time [31].

This last algorithm also applies to unbalanced strings where the problem is extended as follows: excessive characters may be removed from their respective input strings, but only between consecutive blocks.

Another possible parameter is the number of *preserved duos*, $k' = |G_1| - k$. Note that MCSP is denoted *Maximum-Duo Preserving String Mapping* in the context of approximations and parameterized algorithms maximizing k'. This version is FPT as well, and admits a kernel of size $O(k'^8)$ [32].

2.2.2. Maximal Strip Recovery

Problem Description. In the MAXIMAL STRIP RECOVERY problem [33] (see Figure 4), the goal is again to identify common blocks between two genomes. But here, we may remove characters (genes) from the input strings so that the resulting subsequences can be partitioned into common strips. Only the number of deleted genes is counted in the objective function, rather than the number of strips.

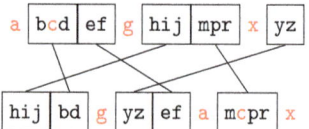

Figure 4. Illustration of the CMSR problem, recovering the strips bd, ef, hij, mpr, and yz by deleting $k = 4$ characters in the strings abcdefghijmprxyz and hijbdgyzefamcprx.

─ COMPLEMENTARY MAXIMAL STRIP RECOVERY (CMSR) ─
Input: genomes G_1, G_2 (balanced strings or permutations), and some $k \in \mathbb{N}$
Question: Are there subsequences G'_1 of G_1 and G'_2 of G_2 that admit a common strip partition with $|G_1| - |G'_1| \leq k$?

Results. In the permutation model, this problem admits a linear kernel [34] and an $O^*(2.36^k)$ FPT algorithm [35]. The picture is less simple for parameter $|G'_1|$ (i.e., the MAXIMAL STRIP RECOVERY problem) in the permutation model. The generalization to four strings instead of two is W[1]-hard [36]. On the other hand, it is FPT for parameter $|G'_1| + \delta$, where δ is the maximum number of consecutive characters deleted within any strip [35]. The main case (MSR with two strings and parameter $|G'_1|$ only) is still open.

Open Problems. Very little is known about MSR and CMSR in the string model. Both are NP-hard, and MSR is polynomially solvable for constant parameter values (by enumerating all possible subsequences of a given size and their strip partitions), but no other parameterized result is known. In particular, is CMSR NP-hard on strings for $k = 0$? In other words, is there a polynomial-time algorithm finding *any* common strip partition of two balanced strings?

2.3. Genome Completion

In both problems in this section, the goal is no longer to compare two genomes (either by building a rearrangement scenario or identifying conserved regions), but to generate a complete genome, based on incomplete data and/or genomic sequences of close species.

2.3.1. Scaffold Filling

Problem Description. A scaffold is a partial representation of the genome, obtained after creating and ordering contigs from reads. Errors during DNA sequencing, low coverage, or sequence repetitions yield unavoidable gaps in the genome assembly (see Section 3). One solution to obtain an approximation of the original genome is to use a better-known genome to infer the missing genes. A *completion* of genome G_1 with respect to G_2 is a supersequence G'_1 of G_1 such that G'_1 and G_2 are balanced. The quality of the completion is evaluated by a genomic distance (DCJ, breakpoints, etc.) denoted by d in the following.

---
d-SCAFFOLD FILLING (d-SF)

Input: genomes G_1, G_2 (strings or permutations), and some $k \in \mathbb{N}$
Question: Is there a completion G_1' of G_1 wrt. G_2, such that $d(G_1', G_2) \leq k$?

Results. In the signed permutations model, d-SF can be solved in polynomial time when the considered distance d is the DCJ distance [37]. On the other hand, for the string model, only weaker distances have been considered: the problem turns out to be NP-hard even for the common adjacencies measure, but is FPT for this measure as parameter [38]. The string algorithms can be generalized to the variant where one seeks extensions of both input genomes at the same time (the "two-sided" case).

Open Problems. What is the parameterized complexity of MCSP-SF on strings? As MCSP is NP-hard, this problem must be too, but it may admit efficient algorithms parameterized by the MCSP distance and/or the parameter occ. Another open question is the DCJ algorithm can be fully generalized to the two-sided case.

2.3.2. Breakpoint Median

Problem Description. In the BREAKPOINT MEDIAN problem, the goal is to build a most-likely common ancestor from three genomes, which can in turn be used to recreate all ancestral genomes in a phylogenetic tree [39].

BREAKPOINT MEDIAN (BM)

Input: genomes G_1, G_2, G_3 (signed permutations), some $d \in \mathbb{N}$
Question: Is there a genome G with $\sum_{i=1}^{3} d_{\text{bp}}(G, G_i) \leq d$?

Results. The problem can be seen as a special case of TRAVELING SALESMAN [39] and can be solved in $O(2.15^d n)$ time.

3. Genome Assembly and Sequence Analysis

Producing the genetic code (sequence of bases A, C, G, and T) present in a sample lies at the core of almost all research in molecular biology. This, however, requires solving an intricate puzzle with millions of pieces ("reads"), involving many computational problems on sequences, such as the SCAFFOLDING problem [40].

Moreover, to draw conclusions from sequences, they have to be analyzed and compared. Such comparisons are made difficult, at the most basic level, by the amount of genetic variation and errors that can appear all along the sequences. A central task is to detect character insertions, deletions, and substitutions within a set of strings representing "similar" DNA or peptide sequences, thereby identifying a common structure [41].

3.1. Multiple Sequence Alignment

Problem Description. MULTIPLE SEQUENCE ALIGNMENT is the central alignment problem in bioinformatics, used to align DNA sequences along one another, taking insertions, deletions, and substitutions into account (see Figure 5). An *extension* of a string s is a string obtained from s by inserting any number of *gap* characters, denoted -. An *alignment* of strings $s_1, \ldots, s_k$ is obtained by creating an extension $s_1', \ldots, s_k'$ of each input string, all of the same size. *Column i* of an alignment is the tuple $(s_1'[i], \ldots, s_k'[i])$. A *cost function* is any function assigning a positive cost to a column. The *unit cost function* counts the total number of mismatches in a column, i.e. the number of pairs (j, j') such that $s_j[i] \neq s_{j'}[i]$.

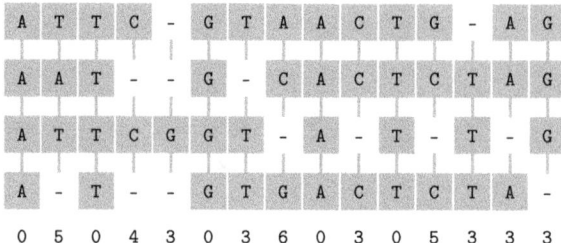

Figure 5. A multiple sequence alignment for $k = 4$ strings with unit costs (shown below each column).

Multiple Sequence Alignment with cost function ϕ (MSA)

Input: Strings $s_1, \ldots, s_k$, and some $m \in \mathbb{N}$
Question: $\exists$ an alignment $s'_1, \ldots, s'_k$ such that the sum of the cost of each column is $\leq m$?

Results. MULTIPLE SEQUENCE ALIGNMENT can be solved by a straightforward dynamic programming in time $\ell^{O(k)}$ for any cost function (where ℓ is an upper-bound on the input string length). It is NP-hard for a wide range of cost functions, including all metric cost functions (even on binary strings), and in particular the unit cost function [42–44]. Due to its central position in DNA analysis, this problem has been the subject of over 100 heuristics over the last decades (this was already said in 2009 [45]), but there does not seem to be any success so far in terms of exact parameterized algorithms.

Open Problems. Does MULTIPLE SEQUENCE ALIGNMENT admit an FPT algorithm for parameter k? If not, can $\ell^{o(k)}$ be achieved?

3.2. Identifying Common Patterns

3.2.1. Closest String (and variants)

Problem Description. In CLOSEST STRING, the goal is to find, given a set of strings, a center string that is close enough to all others (counting the number of mismatches, i.e. the Hamming distance, denoted by Ham). This problem and many variants have been widely studied under the point of view of parameterized algorithms: we only highlight some prominent results, and refer the reader to the work of Bulteau et al. [46] for a more exhaustive review.

Closest String (CS)

Input: strings $s_1, \ldots, s_k$, all of length ℓ, and some $d_r \in \mathbb{N}$
Question: Is there a string s^* such that $\forall_i \; \text{Ham}(s^*, s_i) \leq d_r$?

Results. CLOSEST STRING is NP-hard, even for binary alphabets [47]. On the other hand, it is FPT for *any* of the following parameters: d_r, k, and ℓ [48]. However, the algorithm for k makes use of an integer linear program with at most 2^k variables, implying a mostly impractical combinatorial explosion. An easier (actually trivial) variant, denoted CONSENSUS STRING, aims at minimizing the sum (or average) of the hamming distances to the center, rather than the worst-case distance.

Closest Substring. A common generalization of CS called CLOSEST SUBSTRING asks for a common pattern in all input strings, i.e. it allows trimming input strings until they reach a desired given length m. In this case, the *consensus* variant, optimizing the sum of distances, is no longer trivial and is denoted CONSENSUS PATTERN. Although these problems become much harder than CLOSEST STRING and are W[1]-hard for any single parameter among ℓ, k, d, and m, they still admit FPT algorithms for several combinations of these parameters [49–52].

Closest String with Outliers. Another noteworthy variant allows ignoring a small number t of *outliers* from the input set of strings [53]. Interestingly, all parameterized algorithms for CLOSEST STRING extend to this variant when t is considered as an additional parameter.

Radius or sum. A generalization for CLOSEST STRING (that also applies to the variants above) considers both a constraint on the maximum hamming distance (the *radius*, d_r) and on the sum of hamming distances (denoted d_s). Indeed, the radius constraint only focuses on worst-case strings, and the sum constraint tends to overfit large sets of clustered strings, so taking both constraints can help reduce those problems. All algorithms mentioned for the radius-only version can be extended, in one way or another, to take both constraints into account without additional parameters [54].

Another point of view consists in seeing the radius and sum measures as the L_1 and L_∞ norms, respectively, of the vector $(d(s^*, s_1), \ldots, d(s^*, s_k))$. Thus, a possible compromise consists in optimizing the L_p norm of this vector for any rational p, $1 < p < \infty$. Chen et al. [55] studied CLOSEST STRING under these norms on binary alphabets, proved its NP-hardness, and gave an FPT algorithm for parameter k.

3.2.2. Longest Common Subsequence

Problem Description. The LONGEST COMMON SUBSESQUENCE problem in its simplest version asks for a string of maximal length that is a subsequence of two input strings (see Figure 6(left)). It admits a classical dynamic programming algorithm, with complexity $O(n^2)$. However, the problem becomes NP-hard when we increase the number of strings, as follows.

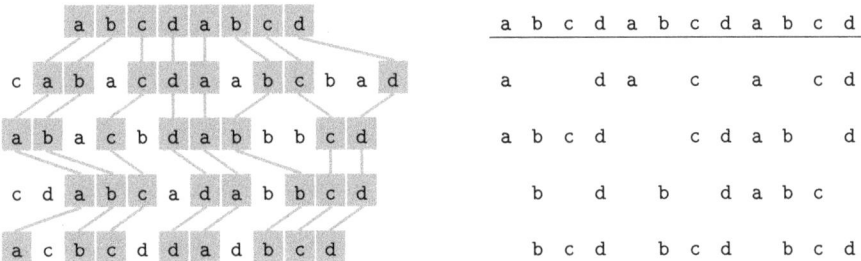

Figure 6. Illustration of a longest common subsequence of length $m = 8$ (**left**) and of a shortest common supersequence of length $m = 12$ (**right**), each on a set of $k = 4$ strings.

───────── LONGEST COMMON SUBSESQUENCE (LCS) ─────────
Input: Strings $s_1, \ldots, s_k$, all of length at most ℓ, and some $m \in \mathbb{N}$
Question: $\exists$ string s^* of length at least m such that for all i, s^* is a subsequence of s_i?

Results. The foremost parameters for LONGEST COMMON SUBSESQUENCE are k, ℓ, m, and the alphabet size denoted $|\Sigma|$. However, it is W[1]-hard (or worse) for all combinations of these parameters that do not yield a simple brute force enumeration algorithm, i.e. $k + m$ with any $|\Sigma|$, and k with $|\Sigma| = 2$ [56,57].

Omitted letters. Another possible parameter is the number of *omitted* letters, $\ell - m$: LONGEST COMMON SUBSESQUENCE is FPT for parameters $\ell - m$ and k [58], and the complexity is open for parameter $\ell - m$ only.
Sub-quadratic time? ($k = 2$) For two strings, there is no sub-quadratic algorithm unless the strong exponential time hypothesis fails [59]. However, parameterized algorithms can help subdue this lower bound to some extent: Bringmann and Künnemann [60] proposed an extensive study of the possible combinations of parameters yielding "FPT in P" algorithms (also sometimes referred to as "fully polynomial FPT", see [61–63]), with matching lower-bounds in each case.
Constrained LCS. ($k = 2$) In this variant, the solution must contain each of a given set of f restriction strings as subsequences. The problem is in P for $f = 1$ [64], NP-hard in general, and W[1]-hard for parameter f [65]. Finally, it is FPT for the total size of the restriction strings [66].

Restricted LCS. Here, the restriction strings are *forbidden* subsequences (rather than mandatory). The problem is FPT for the parameter $\ell + k + f$ [67], as well as for the total size of restriction strings (with $k = 2$) [66].

Open Problems. What is the complexity of LONGEST COMMON SUBSESQUENCE parameterized by $\ell - m$, i.e., the maximal number of letters deleted in any string?

3.2.3. Shortest Common Supersequence

Problem Description. In SHORTEST COMMON SUPERSEQUENCE, we now want to find a larger string of which every input string is a subsequence (see Figure 6(right)).

— SHORTEST COMMON SUPERSEQUENCE (SCS) —
Input: Strings $s_1, \ldots, s_k$, all of length at most ℓ, and some $m \in \mathbb{N}$
Question: $\exists$ string s^* of length at most m such that each s_i is a subsequence of s^*?

Results. As for LCS, SHORTEST COMMON SUPERSEQUENCE is W[1]-hard for parameter k even on binary alphabets [56], and does not admit an $\ell^{o(k)}$ algorithm [68]. It does admit an FPT algorithm for the number of *repetitions*, $m - |\Sigma|$ when occ $= 1$ (each character appears at most once in any string) [69].

Open Problems. Can the FPT algorithm for occ $= 1$ be extended to more general strings?

3.2.4. Center and Median Strings

Problem Description. In the problems described so far in this section, the goal is to find a string minimizing the distance to all input strings, where the distance counts the number of substitutions (CLOSEST STRING and CONSENSUS STRING), deletions (LONGEST COMMON SUBSESQUENCE) or insertions (SHORTEST COMMON SUPERSEQUENCE). Each problem has its own applications, depending on the input data and allowed operations. A more generic variant allows for all three operations, i.e., uses the edit distance (also referred to as the Levenshtein distance, denoted Lev) to measure the similarity to the solution string.

— CENTER STRING and MEDIAN STRING —
Input: Strings $s_1, \ldots, s_k$, and some $d \in \mathbb{N}$
Question: $\exists$ string s^* with the following property?
 (CENTER STRING) $\forall_i \text{Lev}(s^*, s_i) \leq d$
 (MEDIAN STRING) $\sum_{i=1}^{k} \text{Lev}(s^*, s_i) \leq d$

Results. Both problems are NP-hard and W[1]-hard for k, even on binary strings [70]. On the other hand, CENTER STRING is FPT for parameter $d + k$ [71].

Open Problems. Are these problems FPT when parameterized by d only, or by the string length?

3.3. Scaffolding

Problem Description. As mentioned previously, overlapping reads usually does not lead to a fully reconstructed (that is, one sequence per chromosome) genome. Instead, assembly algorithms usually produce fairly long and correct substrings called *contigs* that cannot be extended by overlapping reads. However, current sequencing techniques ("Next-Generation Sequencing" (NGS)) yield additional information about reads. Indeed, NGS-produced reads come in pairs ("mate pairs") and a rough estimate about the number of base-pairs between the two reads of a pair is known. Thus, it is possible that a read r of a pair (r, r') aligns well with parts of some contig c while r' aligns well with parts of contig c'. This presents evidence that, in the target genome, contig c is followed by contig c' and gives us an approximation of the distance between them (see Hunt et al. [72] for more details). To represent this information, we can construct a *scaffold graph* by representing contigs as pairs of vertices connected by a perfect matching, and connecting contig extremities u and v with an edge weighted by the number of mate-pairs indicating that u is followed by v in the chromosome. In this graph, we are looking for a cover of all matching edges using a small number of paths and cycles (corresponding to linear and circular chromosomes in the genome).

SCAFFOLDING (SCA)

Input: graph G with edge-weights ω, perfect matching $\mathcal{M}$ in G, some $\sigma_p, \sigma_c, k \in \mathbb{N}$
Question: Is there a collection $\mathcal{C}$ of $\leq \sigma_p$ paths and $\leq \sigma_c$ cycles covering $\mathcal{M}$ in G of total weight $\geq k$?

Huson et al. [73] considered G as multigraph in which each non-contig edge has an approximate length. This makes for more realistic modeling since two hypotheses involving different gap-sizes between contigs should be incompatible. The authors showed NP-completeness for this problem.

To better support repetitions, (Chateau et al. [74] unpublished) added multiplicities to the contig edges, which can be roughly estimated from coverage numbers. Then, a solution is a set of $\leq \sigma_p$ open and $\leq \sigma_c$ closed walks in G. They presented an ILP solving a slight generalization of this problem, but no parameterized algorithms are known in this case. However, the multiplicity-supporting version of SCAFFOLDING gives rise to another problem: optimal solutions are much less likely to be unique and each individual optimal solution is thus likely to correspond to chimeric sequences (a genomic sequence is called *chimeric* if it contains material from different chromosomes or from unrelated parts of one chromosome). This gives rise to the problem of removing inter-contig edges such as to extract the sub-walks that are shared by all decompositions of a solution into walks [75].

SCAFFOLD LINEARIZATION (LIN)

Input: graph G with edge-weights ω and multiplicities m, perfect matching $\mathcal{M}$ in G, $k \in \mathbb{N}$
Question: Is there a set $X \subseteq E(G) \setminus \mathcal{M}$ of total cost $\leq k$ s.t. $G - E$ can be uniquely decomposed into walks?

Results. Gao et al. [76] presented an $O(|V(G)|^w|E(G)|)$-time algorithm for SCA where w is the maximum number of contigs spanned by any inter-contig edge in the solution ordering. SCA is known to be W[1]-hard for k [75] but solvable in $O^*((tw!)^2)$ time [77], where tw denotes the treewidth of G. A simple reduction from HAMILTONIAN PATH proves SCA NP-hard even on bipartite planar graphs with constant edge weights and $\sigma_p + \sigma_c = 1$ [75]. For a variation of this problem, they proved a kernel for a parameter involving the feedback edge set number of G. Finally, SCAFFOLD LINEARIZATION has been shown to be NP-hard under multiple cost-measures for the solution X [78], but can be solved in $O^*(c^{tw})$ time for different $c \leq 5$, depending on the cost-measure [79].

Notes. Donmez and Brudno [80] approached scaffolding by first computing an orientation of the contigs, formulated as ODD CYCLE TRANSVERSAL (the problem of removing few vertices from a graph such that the result is bipartite), and then ordering them in the scaffold graph using a formulation as FEEDBACK ARC SET. Note further that advances in engineering resulted in so-called "third-generation sequencing" techniques that allow researchers to read large chunks of DNA from a single chromosome [81]. Since these "long reads" have an elevated error rate, focus has recently shifted to detecting and correcting these errors with short reads instead of scaffolding paired reads.

Open Questions. Dallard et al. [82] observed that a simple, fast greedy heuristic often produces good scaffolds. It would be interesting to know if those scaffolds are generally similar to optimal scaffolds and, if so, consider parameters measuring this distance.

4. Haplotyping

Cells of *diploid* (or, more generally, *polyploid*) species have two (multiple) copies of each chromosome. These copies are imperfect in that they can differ in a tiny percentage of their positions ("sites") and such positions are called *Single-nucleotide polymorphisms* (SNPs). When sequencing a diploid genome (see Section 3 for more on sequencing), each read comes from one of the copies of some chromosome but, a priori, there is no way of knowing whether two reads came from the same copy or not. Thus, an assembled diploid genome of an individual contains SNPs where both chromosomes agree ("homozygous sites") and SNPs where they disagree ("heterozygous sites"). Since the vast majority of SNPs exist in only two states ("biallelic SNPs") [83,84], the SNPs of a chromosome can be represented as a string over the alphabet $\{0, 1\}$ called a *haplotype* and the SNPs of a diploid individual can be represented as a string over the alphabet $\{0, 1, 2\}$ called the *genotype*, where 0 and 1 mean that the two chromosomes agree on one of the biallelic states, whereas 2 means that the chromosomes disagree. Now, experimentally determining the genotype of an individual is (relatively) cheap and easy while experimentally determining the haplotypes of the chromosomes is expensive [85,86]. For many applications, however, knowledge of the haplotypes is required and, thus, computational methods have been developed to infer haplotype data ("haplotyping" or "phasing"). In Section 4.1, we consider the problem of inferring the haplotypes by overlapping reads from a diploid genome, similar to the reconstruction of DNA sequences (see Figure 7). In Section 4.2, we consider the problem of inferring the haplotypes of a population, given its genotypes (see Figure 8). This requires assuming that the distribution of haplotypes follows a parsimony criterion. Although most parsimonious solutions often do not reflect the truth [87], they can form the basis for more precise algorithms in the future.

For surveys on both variants of haplotyping, refer to the works in [88–95].

4.1. Haplotype Assembly

Problem Description. A matrix M with m length-n rows ("fragments") over $\{0, 1, \text{-}\}$ is called *in conflict* if there are rows r and r' and a position $i < n$ such that $r[i]$, $r'[i]$ and - are pairwise distinct (that is, - $\neq r[i] \neq r'[i] \neq$ -) and *conflict-free* (or *1-conflict-free*), otherwise. For all $p \geq 1$, M is called *p-conflict-free* if its rows can be partitioned into p conflict-free sets of rows. While we prefer stating the general ("polyploid") version of this problem, the special case that $p = 2$ ("diploid" genome) is by far the most commonly considered.

```
01101---00101--
010001--1101001
-11010---0101--
--001----001---
--101000--111--
----111-110----
-----1111111001
```

Figure 7. A matrix of reads from two chromosomes, condensed to known SNPs of the species. Flipping the underlined (red) entries allows a bipartition into two conflict-free sets (gray and black). Note that not all SNPs exhibit both possible states.

p-(SINGLE) INDIVIDUAL HAPLOTYPING (also referred to as HAPLOTYPE ASSEMBLY) (p-IH)
Input: matrix M, integer k
Question: can M be turned p-conflict-free with k operations?

p-IH can be interpreted as a clustering problem that finds p consensus strings such as to minimize the sum of the radii needed to capture all input strings (rows), where the distance function depends on the nature of allowed "operations". Most of the results are for diploid genomes, that is, for the 2-IH problem. In the following, a row r is called *active* in position j if there are i and ℓ such that $i \leq j < \ell$ and $r[i] \neq \text{-} \neq r[\ell]$, the set of positions in which r is active is called $\alpha(r)$ and a *gap* in a row r is any position $i \in \alpha(r)$ with $r[i] = \text{-}$. We let g denote the maximum number of gaps in any row of M, we let $\ell := \max_r |\alpha(r)|$ denote the maximum number of active positions of any row, and we let

$c := \max_j \alpha^{-1}(j)$ denote the "coverage", that is, the maximum over all positions j of the number of rows r that are active in j. Note that gapless rows can still contain chains of - at the beginning and end.

Results. The complexity of 2-IH depends on the type of operations allowed on the matrix. Mainly, three operations have been considered in the literature:

Deleting rows. The problem of turning M 2-conflict-free by row-deletions is known as MINIMUM FRAGMENT REMOVAL (MFR). It is equivalent to turning the "conflict graph" bipartite by removing vertices (this problem is called ODD CYCLE TRANSVERSAL or VERTEX BIPARTIZATION in the literature). Herein, the vertices of the conflict graph are the rows of M and rows u and v have an edge if they cannot be assigned to the same chromosome, that is, $- \neq u[i] \neq v[i] \neq -$ for some position i. MFR is NP-hard even if each row has at most one gap [96], but becomes polynomial-time solvable for gapless M [96,97]. It can be solved in $O(2^{2g}m^2n + 2^{3g}m^3)$ time [97] and in $O(nc3^c + m\log m + m\ell)$ time [98]. Of course, results for ODD CYCLE TRANSVERSAL apply to MFR: it admits $O^*(3^k)$-time [99,100] and $O(2.32^k)$-time [101] algorithms as well as a randomized kernel of size $O(k^{4.5})$ [102].

Deleting columns. The problem of turning M 2-conflict-free by column-deletions is known as MINIMUM SNP REMOVAL (MSR). It has been shown to be NP-hard even if each row has at most two gaps [96], but polynomial-time solvable for gapless M [96,103]. Bafna et al. [97] showed an $O(2^g mn^2)$-time algorithm.

Flipping values. By far the most studied approach is turning M 2-conflict-free by flipping values, that is, turning a 1 into a 0 or vice versa. This problem is known as MINIMUM ERROR CORRECTION (MEC) or MINIMUM LETTER FLIP in the literature and it is has been shown to be NP-hard, even on gapless inputs [104]. A simple reduction from EDGE BIPARTIZATION [105] further shows that MEC is still NP-hard if each row contains at most three non-(-) characters and each column contains at most two non-(-) characters, thus excluding parameterized algorithms, even for the combined parameter. A parameterized algorithm with running time $O^*(2^m)$ is trivial (assign each row to one of the two partitions/chromosomes) and an $O^*(2^n)$-time algorithm was presented by Wang et al. [106]. Further research focused on the coverage c, which is very small in practice. $O^*(2^c)$-time algorithms with various polynomial factors were presented by Xie et al. [107], Deng et al. [108], Patterson et al. [109], Pirola et al. [110], and Garg et al. [111]. MEC can be linearly reduced to ODD CYCLE TRANSVERSAL [105], implying that results for ODD CYCLE TRANSVERSAL carry over to MEC as well ($O^*(3^k)$ time [99,100], $O^*(2.32^k)$ time [101], and $O(k^{4.5})$-size randomized kernel [102]).

On gapless inputs, MEC can be solved in $O(2^\ell mn)$ time [112] (assuming that all positions in the corrected matrix are "heterozygous", that is, contain both 0 and 1) or in $O(3^\ell \ell m)$ time [105].

Notes. A variant of MFR in which rows are to be removed such as to make M 2-conflict-free *and* to maximize the number of SNPs covered by the solution, called LONGEST HAPLOTYPE RECONSTRUCTION, has been shown to be NP-hard but solvable in $O(n^2(n+m))$ time [104] for gapless M. A more practical variant of MEC, incorporating the aspect of sequencing that each position in a read is given as four probabilities (probability for C, G, A, and T (or U)) instead of a single character was considered by Xie et al. [107], who included a "GenoSpectrum" in the input and give an algorithm running in $O(nc2^c + m\log m + m\ell)$ time. Another version of MEC, constraining output matrices by a "guide-genotype" was considered by Zhang et al. [113], showing that this version can be solved in $O(4^\ell m)$ time. A slight modification of the MEC problem, asking for two haplotypes h_1 and h_2 and a bipartition (R_1, R_2) of the rows such that h_i is at most k flips away from the reads in R_i for $i \in \{1,2\}$, was considered by Hermelin and Rozenberg [114]. Their algorithm solves this problem in $O(2^g k^k mn)$ time.

Since diploid species usually inherit one chromosome from each of their parents, it seems plausible that knowledge of the pedigree of a group of species helps infer their haplotypes. In this context, Garg et al. [115] considered a set I of individuals and a set T of mother–father–child relations, given in addition to the fragment-matrices of each individual, and ask for a set of flips in each fragment-matrix such as to minimize the cost of the flips plus a cost function penalizing unlikely inheritance scenarios. They solved this modified problem in $O^*(2^{4|T|+|I|+c})$ time.

Finally, regarding polyploidy (that is, $p > 2$), Li et al. [116] showed that p-ploid MFR can be solved in $O(2^{2g}nm^2 + 2^t(2m)^{p+1})$ time and in $O(nm^2 + m^{p+1})$ time if the input matrix M is gapless. Further, p-ploid MEC can be solved in $O^*(2^{p(\ell+1)})$ time and in $O^*(p^c)$ time [105].

Open Questions. Parameterizations for the IH problem focus mainly around the number g of gaps per row and the coverage c. While the coverage c was reasonably estimated with 10–30 for past sequencing projects, future technology is expected to increase this number. Even today, whole-exome-sequencing uses coverages of at least 100 [117]. Quoting Garg [118],

> "developing a parameterized algorithm for this integrative framework and deciding parameters that work well in practice is very important."

Thus, a future challenge will be to find more relevant parameters and exploit them to design parameterized algorithms. We suspect that graph-measures of the conflict graph might come to the rescue here, using the known reductions to bipartization problems [105]. Indeed, a more thorough investigation into the multivariate complexity of ODD CYCLE TRANSVERSAL would certainly yield important consequences for the presented variants of (SINGLE) INDIVIDUAL HAPLOTYPING.

A second challenge for (SINGLE) INDIVIDUAL HAPLOTYPING is the development of efficient preprocessing strategies. While some effective reduction rules for variants of IH are known [119–121], many of them are formulated and tuned for ILP formulations—kernelization results, whether positive or negative, have been amiss.

Finally, an obvious open question is to extend the results for the polyploid case, where only little is known in terms of parameterized complexity. Indeed, algorithms deciding the vertex-deletion distance into p-colorable graphs would be a good starting point to attack this problem.

4.2. Haplotype Inference

Problem Description. A *genotype* g is a string over the alphabet $\{0, 1, 2\}$ and a *haplotype* is a string over the alphabet $\{0, 1\}$. Two haplotypes h and h' *resolve* a genotype g if, for each position i, $g[i] = h[i]$ if $h[i] = h'[i]$ and $g[i] = 2$, otherwise. Then, a list of haplotypes *resolves* a list of genotypes if, for each genotype on the list, there are two haplotypes on the list that resolve it.

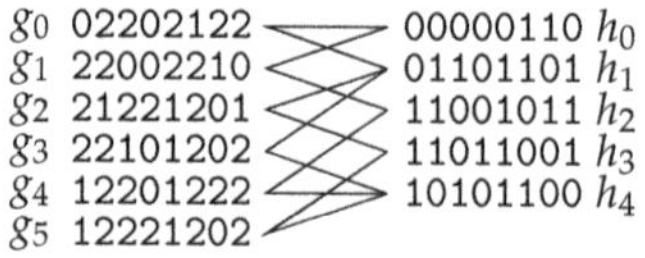

$$
\begin{array}{ll}
g_0\ 02202122 & 00000110\ h_0 \\
g_1\ 22002210 & 01101101\ h_1 \\
g_2\ 21221201 & 11001011\ h_2 \\
g_3\ 22101202 & 11011001\ h_3 \\
g_4\ 12201222 & 10101100\ h_4 \\
g_5\ 12221202 &
\end{array}
$$

Figure 8. Six genotypes g_j resolved by five haplotypes h_i.

HAPLOTYPE INFERENCE (alias HAPLOTYPE PHASING, POPULATION HAPLOTYPING) (HI)
Input: n length-m genotypes g_j, integer k
Question: find $2n$ length-m haplotypes resolving the g_j whose total cost is $\leq k$

Results. The literature considers two major variants of the HI problem.

Pure Parsimony. In the PURE PARSIMONY HAPLOTYPING problem, the cost is the number k of *different* haplotypes needed to explain the given genotypes. This problem is NP-hard [122–126], even for three letters 2 per genotype and three letters 2 per position in the genotypes, but becomes polynomial-time solvable if each genotype has at most two letters 2 [104,127] or each position has at most one letter 2 [126,128]. Further special cases were considered by Sharan et al. [125] who also presented an $O(k^{k^2+k}m)$-time algorithm for the general case [125]. A variant where haplotypes can only be picked from a prescribed pool H' was considered by Fellows et al. [129] who showed a $O(k^{O(k^2)}n^2m)$-time algorithm. Fleischer et al. [130] later presented an $O(k^{4k+3}m)$-time algorithm for the unconstrained version that can also solve the constraint version in $O(k^{4k+3}m|H'|^2)$ time (indeed, these running times can be decreased to $O(k^{4k+2}m)$ and $O(k^{4k+2}m|H'|)$ on average using perfect hashing) as well as a size-$O(2^k k^2)$ kernel. Their algorithm can also output all optimal solutions.

Perfect Phylogeny. In the HAPLOTYPING BY PERFECT PHYLOGENY problem, haplotypes are required to fit a "perfect phylogeny", that is, a tree whose leaves are labeled by the haplotypes resolving the input genotypes such that, for each position i, the subtrees induced by the leaves with 0 and 1 at position i do not intersect. Gusfield [131] introduced this problem for $k = \infty$ and showed an almost-linear-time algorithm, which was later improved to linear time [132,133]. Otherwise, the problem is NP-hard [126] but admits some polynomial-time solvable special cases based on the number of 2s in the genotypes [126].

Notes. A variant of PURE PARSIMONY HAPLOTYPING is XOR-HAPLOTYPING, where genotypes do not contain the letter 2 and two haplotypes resolve a genotype if it is the result of the bitwise XOR of the haplotypes. While not known to be NP-hard, this variant can be solved in $O(m(2^{k^2}k + n))$ time [120]. Note that the $O^*(k^{4k})$-time algorithm of Fleischer et al. [130] can be adapted for this problem.

Another problem called SHARED CENTER that aims at identifying SNPs that correlate with certain diseases was considered from the viewpoint of parameterized complexity by Chen et al. [134] who showed hardness and tractability for it.

Open Questions. A prominent open question is whether PURE PARSIMONY HAPLOTYPING can be solved in polynomial time when the number of 2s per position is bounded by two [125,135,136]. For parameterized complexity, one of the most important questions is whether PURE PARSIMONY HAPLOTYPING admits a single-exponential-time algorithm or a polynomial-size kernel when parameterized by k. Further, parameterizations for this problem have mostly focused on the "natural parameter" k but, in practice, other parameters may be more relevant and promising. Some measure of how convoluted the given genotypes are may be a promising avenue for future research. From a biological point of view, a parameter related to the number of 2s may be promising, but the known results suggest that such a parameter can only lead to FPT-results if combined with others. Finally, from a practical point of view, it may be interesting to combine the HAPLOTYPE INFERENCE and (SINGLE) INDIVIDUAL HAPLOTYPING problems. For example, it would be interesting to find a set of few haplotypes that explain the given genotypes modulo few errors or to infer few haplotypes given a set of genotypes along with a pedigree indicating their inheritance.

5. Phylogenetics

To put sequences of different species in perspective and to understand historical evolution, as well as try to predict future directions of the development of life on Earth, a "phylogeny" [137] (evolutionary tree or network) needs to be constructed. Indeed, some see the reconstruction and interpretation of a species phylogeny as the pinnacle of biological research [138]. A likely evolutionary scenario can be constructed from a multiple alignment, a character-state matrix, or a collection of sub-phylogenies, and methods for this are plentiful [139,140]. In the scope of this survey, however, we focus on recent results for NP-hard problems surrounding phylogenetic trees and networks.

5.1. Preliminaries

An evolutionary (or phylogenetic) network N is a graph whose degree-one nodes $\mathcal{L}(T)$ ("leaves") are labeled (by "taxa"). A *rooted* phylogenetic network N is a rooted acyclic directed graph whose non-root nodes either have in-degree one or out-degree one and whose out-degree-zero nodes $\mathcal{L}(T)$ ("leaves") are labeled (by "taxa"). A (rooted) phylogenetic tree is a (rooted) phylogenetic network whose underlying undirected graph is acyclic. In the context of this section, we drop the prefix "phylogenetic" for brevity and sometimes refer to networks as "phylogenies". Some works consider trees in which each leaf-label may occur more than once. These objects are called *multi-labeled* trees (or MUL trees). For a set $\mathcal{T}$ of networks, we abbreviate $\bigcup_{T \in \mathcal{T}} \mathcal{L}(T) =: \mathcal{L}(\mathcal{T})$. An important parameter of networks is their *level*, referring to the largest number of reticulations in any biconnected component of the underlying undirected graph (see [141]). The *restriction* of T to L, denoted by $T|_L$, is the result of removing all leaves not in L from T and repeatedly removing unlabeled leaves and suppressing (that is, contracting any one of its incident edges/arcs) degree-two nodes. A network N *displays* a network T if T is a topological minor of N, respecting leaf-labels, that is, N contains a subdivision of T as a subgraph. Herein, a directed edge can only be subdivided in accordance to its direction, that is, the subdivision of an arc uv creates a new node w and replaces uv by the arcs uw and wv. This notion can be generalized to the case that N is the disjoint union of some networks (see Section 5.2.3). For non-binary networks, there are two different notions of display: the "hard"-version is defined analogously to the binary case, while we say that a network N "soft"-displays a network T if any binary resolution of N displays any binary resolution of T, where a *binary resolution* of a network N is any binary network that can be turned into N by contracting edges/arcs. "Soft" and "hard" versions of display are derived from the concept of "soft" and "hard" polytomies, meaning high degree nodes that represent either a lack of knowledge of the correct evolutionary history leading to the children taxa ("soft") or a large fan-out of species due to high evolutionary pressure ("hard").

Note that many kernelization results in phylogenetics bound only on the number of labels in a reduced instance. If the input contains many trees or an intricate network, kernelization results should more fittingly be described as "partial kernel" (see [142]). We thus usually refer to such results as "kernel with ... taxa".

5.2. Combining and Comparing Phylogenies

Problem Description. An approach to reconstruct a phylogeny from the genomes of a set of species is to first reconstruct the phylogenies of the genes (using multiple alignments and after clustering them together into families) into so-called "gene trees" and then to combine these trees into a tree representing the evolutionary history of the set of species called the "species tree" (see also Section 5.3 for more on the divergence between gene trees and species trees). In general, given trees T_i each on the taxon-set X_i, we want to "amalgamate" the trees into a single tree T, which, since it agrees with and contains all the T_i, is called an *agreement supertree*. This problem is known as TREE CONSISTENCY (and, sometimes TREE COMPATIBILITY).

TREE CONSISTENCY (TCY)

Input: trees $T_1, T_2, \ldots, T_t$ on $\mathcal{X}_1, \mathcal{X}_2, \ldots$, respectively
Question: Is there a tree T^* on $\mathcal{X} := \bigcup_i \mathcal{X}_i$ with $\forall_i\ T^*|_{\mathcal{X}_i} = T_i$?

For surveys about the combination of phylogenies ("consensus methods") we refer to Bryant [143] and Degnan [144].

Results. For unrooted trees, TCY can be solved in polynomial time if all T_i share a common taxon [145] but is NP-complete in general, even if all input trees contain four taxa [145] (such a "quartet" is the smallest meaningful unrooted tree since unrooted trees with at most three taxa do not carry any phylogenetic information). This restricted problem is also known as QUARTET INCONSISTENCY. TCY can be solved in polynomial-time for two unrooted (non-binary) trees [146]. More generally, using powerful meta-theorems [147,148] (problems formulatable in "Monadic Second Order Logic" (MSOL) are FPT for the treewidth of the input structure), TCY is fixed-parameter tractable for the treewidth of the display graph [149,150] (that is, the result of identifying all leaves of the same label in the disjoint union of the input trees), which is smaller than the number t of trees. Baste et al. [151]

improved the impractical running time resulting from the application of the meta-theorems, showing an $O^*(2^{O(t^2)})$-time algorithm.

For rooted trees, TREE CONSISTENCY can be solved in polynomial time [152,153] (even for non-binary trees) but, due to noisy data and more complicated evolutionary processes, practically relevant instances are not expected to have an agreement supertree [154,155]. Thus, derivations of the problem arose, asking for a smallest amount of modification to the input such that an agreement supertree exists. The most prominent modification types are removing trees (ROOTED TRIPLET INCONSISTENCY), removing taxa (MAXIMUM AGREEMENT SUPERTREE), and removing edges (MAXIMUM AGREEMENT FOREST), which we discuss in the following.

5.2.1. Consensus by Removing Trees

When reconstructing a species tree from gene trees, we may hope that the gene trees of most of the sampled gene families actually agree with the species phylogeny and only few such families describe outliers that developed nonconformingly. In this case, we can hope to recover the true phylogeny by removing a small number of gene trees.

ROOTED TRIPLET INCONSISTENCY (RTI)
Input: triplets $T_1, T_2, \ldots$ on $\mathcal{X}_1, \mathcal{X}_2, \ldots$, respectively, and some $k \in \mathbb{N}$
Question: Is there a tree T^* on $\mathcal{X} := \bigcup_i \mathcal{X}_i$ such that $T^*|_{\mathcal{X}_i} \neq T_i$ for at most k of the T_i?

Results. ROOTED TRIPLET INCONSISTENCY is NP-hard [156–158], even on "dense" triplet sets [159] (a triplet set $\mathcal{T}$ is called *dense* or *complete* if for each leaf-triple $\{a, b, c\}$, $\mathcal{T}$ contains exactly one of $ab|c$, $ac|b$ and $bc|a$). While the general problem is W[2]-hard for k [159], the dense version admits parameterized algorithms. Indeed, Guillemot and Mnich [160] showed parameterized algorithms running in $O(4^k n^3)$ and in $O^*(2^{O(k^{1/3} \log k)})$ time, as well as an $O(n^4)$-time computable, sunflower-based kernel containing $O(k^2)$ taxa (see [161] for details on the sunflower kernelization technique). Their result has recently been improved to linear size by Paul et al. [162].

Notes. Generalizing RTI to ask for a level-ℓ *network* displaying the input trees yields a somewhat harder problem, which can be solved for dense inputs in $O(|T|^{\ell+1} n^{4\ell/3+1})$ time [163] and in $O(|T|^{\ell+1})$ time for a particular class of networks [164].

The unrooted-tree version of dense RTI (where the input consists of quartets) is known to be solvable in $O(4^k n + n^4)$ time [165].

5.2.2. Consensus by Removing Taxa

In many sciences, the most interesting knowledge can be gained by looking more closely to the non-conforming data points. In this spirit, biologists are particularly interested in taxa causing non-compatibility, that is, whose removal allows for an agreement supertree. In the spirit of parsimony, we are thus tempted to ask for a smallest number of taxa to remove from the input trees such that an agreement supertree exists.

MAXIMUM AGREEMENT SUPERTREE (MASP (also SMAST or MLI in the literature))
Input: trees $T_1, T_2, \ldots, T_t$ on $\mathcal{X}_1, \mathcal{X}_2, \ldots$, respectively, and some $k \in \mathbb{N}$
Question: Is there a tree T^* on $\mathcal{X}$ such that $|\bigcup_i \mathcal{X}_i \setminus \mathcal{X}'| \leq k$ and $\forall_i \, T^*|_{\mathcal{X}'} = T_i|_{\mathcal{X}'}$?

Results. While MASP can be solved in $O(n^{1.5})$ time for two rooted trees [166–168] (n denoting the total number of labels in the input), it is NP-hard for $t > 2$, even if all T_i are triplets [166,169], and the NP-hardness persists for fixed t [166] (but large trees). Guillemot and Berry [170] showed that, on dense, binary, rooted inputs, MASP can be solved in $O(4^k n^3)$ and $O(3.12^k + n^4)$ time by reduction to 4-HITTING SET. They further improved an $O(t(2n^2)^{3t^2})$-time algorithm of Jansson et al. [166] for binary T_i to $O((8n)^t)$ time [170], which was subsequently improved to $O((6n)^t)$ time by Hoang and Sung [171]. The latter also gave an $O((t\Delta)^{t\Delta+3}(2n)^t)$-time algorithm for general rooted inputs (Δ

denoting the maximum out-degree among the input trees). MASP is W[2]-hard for k, even if the input consists of rooted triplets [169], and W[1]-complete in the rooted case for the dual parameter $n - k$, even if we add t to it [170]. On the positive side, the problem can be solved in $O((2t)^k tn^2)$ time for binary trees [170] which has been generalized to arbitrary trees by Fernández-Baca et al. [172]. For completeness, we want to point out that many of the results for MASP also hold for MASP's sister problem MAXIMUM COMPATIBLE SUPERTREE (MCSP), in which equality with the restricted agreement supertree T^* is relaxed to being a contraction of T^* (with the notable exception that MCSP can be solved in $O(2^{2t\Delta} n^t)$ time in both the rooted and unrooted case [171]).

Notes. The special case of MASP in which $\mathcal{X}_1 = \mathcal{X}_2 = \ldots$ is called MAXIMUM AGREEMENT SUBTREE (MAST) and has been studied extensively. While still NP-hard for $t = 3$ non-binary trees [173], MAST can be solved in $O(kn^3 + n^\Delta)$ [157,174], in which time we can also compute a "kernel agreement subtree", denoting the intersection of all leaf-sets of all optimal maximum agreement subtrees [175]. MAST is fixed-parameter tractable for k with parameterized algorithms running in time $O(\min\{3^k tn, 2.18^k + tn^3\})$ [176–178] (by reduction to 3-HITTING SET).

More fine-grained versions of MAST that allow removal of different taxa from each T_i were introduced by Chauve et al. [179]. In AGREEMENT SUBTREE BY LEAF-REMOVAL (AST-LR), the objective is to minimize *the total number q* of removed leaves and, in AST-LR-d, the objective is to minimize *the maximum number d* of leaves that have to be removed from any of the trees. Both versions are NP-hard [179] but can be solved in $O((4q - 2)^q t^2 n^2)$ (AST-LR) and $O(c^d d^{3d}(n^3 + tn \log n))$ time for some constant c (AST-LR-d) [179,180].

Lafond et al. [181] considered MAST for multi-labeled trees showing that it remains NP-hard and can be solved in $O^*((2n)^t k^{kt})$ time.

Finally, Choy et al. [141] showed that a "maximum agreement sub*network*" for two binary networks of level ℓ_1 and ℓ_2, respectively, can be computed in $O^*(2^{\ell_1 + \ell_2})$ time.

5.2.3. Consensus by Removing Edges—Agreement Forests and Tree Distances

An important biological phenomenon that governs the discordance of gene trees are non-tree-like processes such as hybridization and horizontal gene transfer (HGT) (see also Section 5.3). If a branch in a gene tree corresponds to a horizontal transfer, then we expect that deleting this branch results in a forest, which is in agreement with the other gene trees. This gives rise to the idea of "agreement forests", resulting from the deletion of branches in the input phylogenies.

MAXIMUM AGREEMENT FOREST (MAF)

Input: rooted or unrooted trees $T_1, T_2, \ldots$ on $\mathcal{X}$ and some $k \in \mathbb{N}$
Question: Is there a forest F with $\mathcal{L}(F) = \bigcup_i \mathcal{L}(T_i)$, F has $\leq k$ trees and each T_i displays F?

Maximum agreement forests come in three major flavors: *unrooted* maximum agreement forests (uMAFs), *rooted* maximum agreement forests (rMAFs), and maximum *acyclic* agreement forests (MAAFs). Herein, "acyclic" makes reference to the constraint that the "inheritance graph is acyclic" (see Figure 9). Formally, the *inheritance graph* of an agreement forest F for two trees T_1 and T_2 has the trees of F as nodes and an arc uv if and only if the root of u is an ancestor of the root of v in T_1 or in T_2. Demanding acyclicity of this graph forbids, for example, that a tree u of F is "above" another tree v in T_1 but "below" v in T_2. This definition generalizes straightforwardly to more than two trees T_i. In the following, the *size* of an agreement forest F is the number of trees in F and it is equal to the number of branches to remove in each input tree to form (a subdivision of) F and F is called *maximum* if it minimizes this number. For surveys about tree distances and agreement forests, we refer to Shi et al. [182] and Whidden [183].

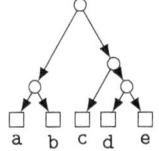 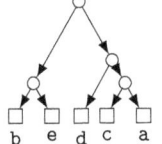 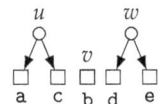 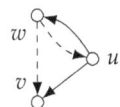 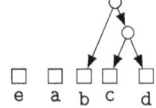

(a) Rooted trees T_1 and T_2 on the same taxon set $\{a,b,c,d,e\}$. (b) A size-3 rooted agreement forest F for T_1 and T_2. (c) The inheritance graph of F. Solid (dashed) arcs are induced by T_1 (T_2). (d) A size-3 *acyclic* agreement forest for T_1 and T_2

Figure 9. Illustration of (acyclic) agreement forests.

Results. Evidently, results heavily depend on the type of agreement forest we are looking for. Interestingly, each of the three versions corresponds to a known and well-studied distance measure between trees and we thus also include results stated for the corresponding distance-measure.

Unrooted Agreement Forest. The size of a uMAF of two binary trees T_1 and T_2 is exactly equal to the minimum number of "TBR moves" necessary to turn T_1 into T_2 [184,185] (and vice versa; indeed, this defines a metric and it is called the "TBR distance" between T_1 and T_2). Herein, a *TBR* (tree bisection and reconnection) *move* consists of removing an edge uv from a tree ("bisecting" the tree) and inserting a new edge between any two edges of the resulting subtrees ("reconnecting" the trees), that is, subdividing an edge in each of the subtrees and adding a new edge between the two new nodes.

For two trees, deciding uMAF is NP-hard [184], but fixed-parameter tractable in k. More precisely, the problem can be solved in $O^*(k^O(k))$ time [185], $O(4^k k^5 + n^{O(1)})$ time [186], and $O(4^k k + n^{O(1)})$ time [187]. These results make use of the known kernelizations with $15k$ [185,188] and $11k$ taxa [189]. For $t > 2$ binary trees, Shi et al. [190] presented an $O(4^k nt)$-time algorithm. Chen et al. [191] considered the uMAF problem on multifurcating trees, showing that it still corresponds to the TBR problem and can be solved in $O(3^k n)$ time.

Rooted Agreement Forest. The size of an rMAF of two rooted binary trees T_1 and T_2 is exactly one more than the minimum number of "rSPR moves" necessary to turn T_1 into T_2 [192,193] (and vice versa; indeed, this defines a metric and it is called the "rSPR distance" between T_1 and T_2). Herein, an *rSPR* (rooted subtree prune and regraft) *move* consists of removing ("pruning") an arc uv from a tree and "regrafting" it onto another arc xy, that is, subdividing xy with a new node z and inserting the arc zv.

The problem is known to be NP-hard and algorithms parameterized by k have been extensively studied and improved. An initial $O(4^k n^4)$-time algorithm [187,194] was improved to $O(3^k n)$ time [187], $O(2.42^k n)$ time [195], $O(2.344^k n)$ time [196], and the current best $O(2^k n)$-time algorithm by Whidden [183]. In contrast, a kernel with $28k$ taxa [193] has stood since 2005. For $t > 2$ trees, rMAF can be decided in $O^*(6^k)$ time [197] and $O(3^k nt)$ time [190].

Collins [198] showed that using "soft"-display, rMAFs still correspond to computing the rSPR distance between two *multifurcating* trees. This problem can be solved in $O^*(4^k)$ time [199] and in $O(2.42^k n)$ time [183,200] and admits a kernel with $64k$ taxa [198,201]. For $t > 2$ trees, the multifurcating rMAF problem is solvable in $O(2.74^k t^3 n^5)$ time [202]. Notably, Shi et al. [202] also considered the "hard" version of the problem and presented an $O(2.42^k t^3 n^4)$-time algorithm for it.

Acyclic Agreement Forest. The size of a MAAF of two rooted binary trees T_1 and T_2 is exactly one more than the minimum number of reticulations found in any phylogenetic network displaying both T_1 and T_2 [192] and this relation holds also if T_1 and T_2 are *non-binary* [203].

Deciding this number is known as the HYBRIDIZATION NUMBER (HN) problem and it has been shown to be NP-hard by Bordewich and Semple [204]. The problem can be solved in $O(3^n n)$ time by crawling a bounded search-tree [205]. In 2009, Whidden and Zeh claimed an $O(3^k n \log n)$-time algorithm, which they later retracted and replaced by an $O(3.18^k n)$-time algorithm [183,195]. For $t = 3$ binary trees, HN can be decided in $O^*(c^k)$ time, where c is an "astronomical constant"[206].

Concerning preprocessing, a kernel with at most $9k$ taxa is known [188,201] and this kernelization result has been generalized to the case of deciding HN for $t > 2$ binary trees (in which case HN and MAAF no longer coincide) by van Iersel and Linz [207], showing a kernel with $20k^2$ taxa for this case, which has again been generalized to $t > 2$ non-binary trees by van Iersel et al. [208], showing a kernel with at most $20k^2(\Delta - 1)$ (and at most $4k(5k)^t$) taxa [208]. For MAAF with $t = 2$ non-binary trees, Linz and Semple [203] showed a linear bikernel (that is, a kernelization into a different problem, see [209]) with $89k$ taxa, which implies a quadratic-size classical kernel. For this setting, algorithms running in $O^*(6^k k!)$ [210] and $O^*(4.83^k)$ [211] time are also known.

Notes. Any algorithm for binary uMAF, rMAF, and MAAF with running time $f(k)n^{O(1)}$ can be turned into an algorithm running in $f(\ell)n^{O(1)}$, where ℓ is the level of any binary network displaying the two input trees [212]. Furthermore, all three agreement forest variants are fixed-parameter tractable for the treewidth of the display graph of the input trees [213] (see corresponding results for unrooted TREE CONSISTENCY [149,150]). The rSPR distance has been generalized to a distance measure for phylogenetic networks called SNPR and its computation is fixed-parameter tractable [214] parameterized by the distance. Variations of the discussed distance measures include: (1) the uSPR distance, which does not have an agreement-forest formulation, is NP-hard to decide [215], admits a kernel with $76k^2$ taxa [216] (in a preprint, Whidden and Matsen [217] claimed an improvement to $28k$ taxa), and can be calculated in $O^*((28k)!!16^k)$ time [218] (using the mentioned preprint-kernel); (2) its close sibling, the replug distance, which admits a formulation as "maximum endpoint agreement forest", is conjectured to be NP-hard to decide but admits an $O^*(16^k)$-time algorithm [218]; (3) the "temporal hybridization number", denoting the smallest amount of reticulation required to explain trees with a temporal network, which was shown to be NP-hard for two trees [219] but admits an $O^*((9k)^{9k})$-time algorithm [220]; and (4) the parsimony distance, which is NP-hard [221,222] but can be solved in $O^*(1.618^{38 d_{\text{TBR}}})$ time [223], where d_{TBR} is the TBR distance between the input trees.

Open Questions. Consensus methods in phylogenetics can profit from a wide range of parameters, describing the particularities of likely set of inputs. While we would indeed expect that the consensus has a small distance to the input trees, a lot depends on how we choose to measure this distance. More general distance measures make for stronger parameters and, while the HYBRIDIZATION NUMBER problem can be solved in single-exponential time for the "standard parameter" k, it would be interesting to parameterize by a stronger parameter, such as the rSPR radius in which the input trees lie.

Inspired by the groundbreaking results of Bryant and Lagergren [149], research into the display graph and its treewidth has been conducted with some success [150,224]. However, we have yet to design concrete, practical algorithms for consensus problems parameterized by the treewidth of the display graph. As this would potentially yield very fast, practical algorithms, we suspect that this would be a fruitful topic in the coming years. Another interesting parameter is the "book thickness" of the display graph, that is, the minimum number of edge-colors needed to color the display graph such that each color class permits an outerplanar drawing. For obvious reasons, this parameter is smaller than the number t of input trees. Can the results for t be strengthened to work with the book thickness instead?

5.3. Reconciliation

Problem Description. In practice, trees depicting the evolutionary history of families of genes sampled from a set of species do not agree with the evolutionary history of the species themselves; hybridization, horizontal gene transfer, and incomplete lineage sorting being only few known causes for such discrepancies. In theory, even gene duplication and gene loss are enough to explain gene trees differing arbitrarily from the corresponding species tree. To better understand how a family of genes developed in the genome of a concurrently developing set of species, we can compute an "embedding" of the gene tree nodes to the edges of the species phylogeny called a *reconciliation* (see Figure 10). Reconciliations also allow drawing conclusions when comparing phylogenies of co-evolving species such as hosts/parasites or flowers/pollinators. More formally, a *DL-reconciliation* of a (gene-)phylogeny G with a (species-)phylogeny S is a pair (G', r) where G' is a subdivision of G and $r : V(G) \to V(S)$ is a mapping such that:

(a) for all arcs uv of G, either $r(u) = r(v)$ (in which case u is called "duplication") or $r(v)$ is a child of $r(u)$ (in which case u is called "speciation");
(b) for arcs uv and uw in G, we have $r(u) = r(v) \iff r(u) = r(w)$ (that is, no node of G can be a speciation and a duplication at the same time); and
(c) if u is a leaf in G, then $r(u)$ is a leaf labeled with the contemporary species that $r(u)$ was sampled in.

We can then define the number of losses in (G', r) as the sum over all speciations u of the outdegree of $r(u)$ in S minus the outdegree of u in G'. If horizontal transfers are allowed, Condition (a) is replaced by

(a') for all arcs uv of G, either $r(u) = r(v)$ (in which case u is called "duplication") or $r(v)$ is a child of $r(u)$ (in which case u is called "speciation"), **or $r(v)$ is incomparable to $r(u)$ (in which case u is called a "transfer" and uv is called a "transfer arc")**,

and Condition (b) is restricted to non-transfer arcs, and each transfer with out-degree one causes an additional loss. In this case, we call r a *DTL-reconciliation*. Since, in reality, transfers occur only between species existing at the same time, Condition (a') introduces further restrictions. In particular, a reconciliation r is called *time-consistent* if G can be "dated", that is, there is a mapping $t : V(G) \to \mathbb{N}$ such that, for all arcs uv of G, we have 1. $t(u) \le t(v)$; and 2. $t(u) = t(v)$ if and only if uv is a transfer arc of G under r.

A DTL-reconciliation may be time-inconsistent if, for example, there are transfer arcs uv and xy such that u and x are ancestors of y and v, respectively, in G.

Now, the parsimonious principle is used to define an optimization criterion. To this end, each evolutionary event is given a cost such as to reflect how unlikely it is to see a certain event. By the biological setup, it is usually assumed that speciations have cost zero.

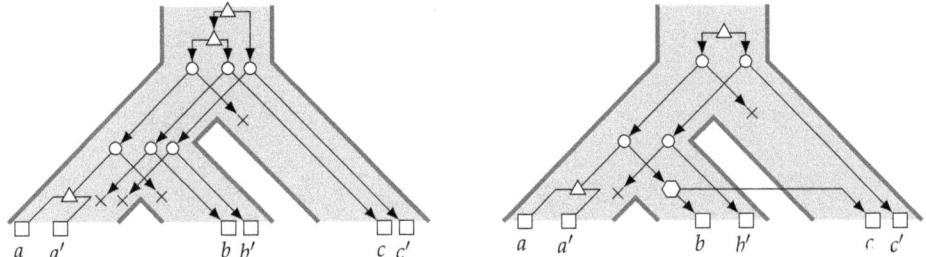

Figure 10. Two different reconciliations of the same gene tree with the same species tree. A small node u drawn in an edge xy of the species tree (large outer tree) indicates that the reconciliation maps u to y. Boxes, triangles, circles, and hexagons represent leaves, duplications, speciations, and transfers, respectively, while crosses are losses. The left reconciliation has three duplications and four losses, while the right has two duplications, two losses, and one transfer.

— RECONCILIATION (REC) ——————————————————————
Input: species tree S, gene tree G, mapping $\sigma : \mathcal{L}(G) \to \mathcal{L}(S)$, costs $\lambda, \delta, \tau \in \mathbb{N}$, some $k \in \mathbb{N}$
Question: Is there an embedding of G in S with cost at most k?

We specify the allowed type of embedding by prepending "DL", "DTL", etc. to the problem name. The study of the formal reconciliation problem was initiated by Ma et al. [225] and Bonizzoni et al. [226] and is surveyed in [227–230].

Results. Optimal binary DL-reconciliations can be computed—independently of the costs—by the *LCA-mapping*, which maps each node of G to the LCA of the nodes that its children are mapped to [231]. Thus, this problem can be solved in linear time [232,233] using $O(1)$-time LCA queries [234,235]. The non-binary variant, while not quite as straightforward, can still be solved in polynomial time [236,237].

The complexity of computing DTL-reconciliations depends heavily on their time-consistency. If we allow producing time-inconsistent reconciliations or the given species tree already comes with a dating function, then optimal DTL-reconciliations can be computed in polynomial time [238–240]. In general, however, the problem is NP-hard [238,241], but can be solved in $O(3^k|G| + |S|)$ time [238]. Hallett and Lagergren [242] showed that DTL-reconciliations with at most α speciations mapped to any one node in the species tree is FPT in $\alpha + \tau$.

Duplication, transfer, and loss are not the only evolutionary events shaping a gene tree. Hasić and Tannier [243,244] recently introduced "replacing transfers" (T_R) and "gene conversion" (C) which model important evolutionary events and showed that DLC-reconciliations can be decided in polynomial time [243] while deciding T_R-reconciliations is NP-hard and FPT for k [244]. Finally, the concept of "incomplete lineage sorting" (ILS) is an important factor influencing discrepancies between gene and species phylogenies, especially when speciation occurs in rapid succession [245]. Roughly, ILS refers to the possibility that an earlier duplication or transfer does not pervade a population at the time a speciation occurs. Thus, one branch of a speciation may carry a gene lineage while the other does not (see Figure 11 for an illustration). In DL-reconciliation, this scenario requires a loss, but this would not reflect the true evolutionary history. Unfortunately, no particular mathematical model of ILS is widely accepted, so the following results might be incomparable. In 2017, Bork et al. [246] showed that incorporating ILS into DL-reconciliation makes the problem NP-hard, even for dated species trees. Furthermore, DTL-reconciliation for dated, non-binary species trees allows ILS to be computed in $O^*(4^\Delta)$ time [247,248].

To and Scornavacca [249] started looking into the problem of reconciling rooted gene trees with a rooted species network, showing that, for the DL model, this problem is NP-hard, but solvable in $O^*(2^k)$ time.

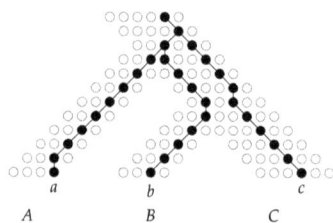

Figure 11. Illustration of incomplete lineage sorting (ILS). Each row of dots represents a generation and each dot stands for a number of individuals. The ancestral lineages of genes a, b and c, sampled in individuals of the respective extant species A, B, and C are drawn in black. When the speciation splits A from C, all three alleles a, b, and c exist in the population, but are not "fixated". This is necessary to observe ILS and gets more likely the shorter the branch to the ancestor species. The resulting gene phylogeny differs from the species phylogeny.

Open Questions. There are three major challenges for bioinformatics concerning reconciliation. The first and more obvious task is to include all known genomic players in the reconciliation game, meaning to establish a standard model incorporating duplication, transfer, loss, replacing transfers, conversion, and incomplete lineage sorting. While Hasić and Tannier [243,244] made good progress towards this goal, their model seems too clunky and lacks ILS support. The second challenge is to remove the need to provide δ, τ, etc. in the input for the RECONCILIATION problem. In practice, some biologists using implementations of algorithms for reconciliations just "play around" with these numbers until the results roughly fit their expectations, which is understandable since nobody knows the correct values. Indeed, in all likelihood, there are no "correct values" because the underlying assumption that the rates of genetic modification is constant throughout a phylogeny is invalid [250]. A more realistic approach might define expected frequencies of events for each branch and combine them with the length of this branch in order to dynamically price duplication, transfer, loss, etc. in this branch.

5.4. Miscellaneous

Given a phylogeny T, the *parsimony score* of T with respect to a labeling c of its nodes is the number of arcs in T whose extremities have a different label under c. In the SMALL PARSIMONY problem, we are given a phylogeny T and a leaf-labeling c_L and have to extend c_L to a labeling of all nodes of T such as to minimize the parsimony score. If T is a network, aside from the above definition ("hardwired"), the "softwired" version exists, asking for the minimum parsimony score of any tree T^* (on the same leaf-set as T) displayed by T. While SMALL PARSIMONY is polynomial for trees [251], the problem is NP-hard in the softwired case, even for binary T and c_L, as well as in the hardwired case, unless c_L is binary [252,253]. Fischer et al. [253] also showed that hardwired SMALL PARSIMONY is FPT for the solution parsimony score and softwired SMALL PARSIMONY is FPT for the level of the input network.

The problem of deciding whether a phylogeny T is displayed by another phylogeny N is called TREE CONTAINMENT and it is NP-hard, even if T is a tree [254]. TREE CONTAINMENT has polynomial-time algorithms for many special cases of N [254–261] and can be solved in $O^*(1.62^\ell)$ time [262] (where ℓ denotes the level of N; the authors also showed an algorithm for the related CLUSTER CONTAINMENT problem) and in $O^*(3^t)$ time [261], where t is the number of "invisible tree components" (that is, the number of tree-nodes u whose parent v is a reticulation that is not "visible" in N (that is, for each leaf a, there is a root-a-path avoiding v)). TREE CONTAINMENT stays NP-hard even if the arcs of both T and N are annotated with "branch lengths", but admits an $O^*(2^\ell)$-time algorithm in this case [263].

Recent research into the problem of rooting an unrooted network was conducted by Huber et al. [264], showing that orienting an undirected binary network as a directed network of a certain class is FPT for the level ℓ for some classes of N.

Author Contributions: Investigation, L.B. and M.W.; and Writing—original draft, L.B. and M.W.

Funding: This research received no external funding

Acknowledgments: The authors want to thank Fran Rosamond for suggesting the topic as well as everyone who helped collect interesting results for the manuscript, in particular Jesper Jansson, Steven Kelk, Fabio Pardi, Yann Ponty, Eric Rivals, Celine Scornavacca, Krister Swenson, and Norbert Zeh. Finally, we thank the anonymous reviewers for their helpful and productive comments.

Conflicts of Interest: The authors declare no conflict of interest.

References

1. Downey, R.G.; Fellows, M.R. *Fundamentals of Parameterized Complexity*; Texts in Computer Science; Springer: Berlin, Germany, 2013. [CrossRef]
2. Cygan, M.; Fomin, F.V.; Kowalik, L.; Lokshtanov, D.; Marx, D.; Pilipczuk, M.; Pilipczuk, M.; Saurabh, S. *Parameterized Algorithms*; Springer: Berlin, Germany, 2015. [CrossRef]
3. Ávila, L.F.; García, A.; Serna, M.J.; Thilikos, D.M. Parameterized Problems in Bioinformatics. unpublished manuscript.
4. Cai, L.; Huang, X.; Liu, C.; Rosamond, F.; Song, Y. Parameterized Complexity and Biopolymer Sequence Comparison. *Comput. J.* **2008**, *51*, 270–291. [CrossRef]

5. Hüffner, F.; Komusiewicz, C.; Niedermeier, R.; Wernicke, S., Parameterized Algorithmics for Finding Exact Solutions of NP-Hard Biological Problems. In *Bioinformatics: Volume II: Structure, Function, and Applications*; Springer: New York, NY, USA, 2017; pp. 363–402. [CrossRef]
6. Gramm, J.; Nickelsen, A.; Tantau, T. Fixed-Parameter Algorithms in Phylogenetics. *Comput. J.* **2007**, *51*, 79–101. [CrossRef]
7. Griffiths, A.J.; Gelbart, W.M.; Miller, J.H.; Lewontin, R.C. Chromosomal Rearrangements. In *Modern Genetic Analysis*; W.H.Freeman: New York, NY, USA, 1999; Chapter 8.
8. Fertin, G.; Labarre, A.; Rusu, I.; Tannier, E.; Vialette, S. *Combinatorics of Genome Rearrangements*; Computational Molecular Biology; MIT Press: Cambridge, MA, USA, 2009; p. 312.
9. Yancopoulos, S.; Attie, O.; Friedberg, R. Efficient sorting of genomic permutations by translocation, inversion and block interchange. *Bioinformatics* **2005**, *21*, 3340–3346. [CrossRef]
10. Bergeron, A.; Mixtacki, J.; Stoye, J. A unifying view of genome rearrangements. In *International Workshop on Algorithms in Bioinformatics*; Springer: Berlin, Germany, 2006; pp. 163–173.
11. Chauve, C.; Fertin, G.; Rizzi, R.; Vialette, S. Genomes containing duplicates are hard to compare. In *International Conference on Computational Science*; Springer: Berlin, Germany, 2006; pp. 783–790.
12. Jiang, H.; Zhu, B.; Zhu, D. Algorithms for sorting unsigned linear genomes by the DCJ operations. *Bioinformatics* **2010**, *27*, 311–316. [CrossRef]
13. Fertin, G.; Jean, G.; Tannier, E. Algorithms for computing the double cut and join distance on both gene order and intergenic sizes. *Algorithms Mol. Biol.* **2017**, *12*, 16. [CrossRef]
14. Bérard, S.; Chateau, A.; Chauve, C.; Paul, C.; Tannier, E. Perfect DCJ rearrangement. In *RECOMB International Workshop on Comparative Genomics*; Springer: Berlin, Germany, 2008; pp. 158–169.
15. Watterson, G.; Ewens, W.; Hall, T.; Morgan, A. The chromosome inversion problem. *J. Theor. Biol.* **1982**, *99*, 1–7. [CrossRef]
16. Hannenhalli, S.; Pevzner, P.A. Transforming Cabbage into Turnip: Polynomial Algorithm for Sorting Signed Permutations by Reversals. *J. ACM* **1999**, *46*, 1–27. [CrossRef]
17. Christie, D.A. Genome Rearrangement Problems. Ph.D. Thesis, University of Glasgow, Glasgow, UK, 1998.
18. Chen, X.; Zheng, J.; Fu, Z.; Nan, P.; Zhong, Y.; Lonardi, S.; Jiang, T. Assignment of Orthologous Genes via Genome Rearrangement. *IEEE/ACM Trans. Comput. Biol. Bioinform.* **2005**, *2*, 302–315. [CrossRef]
19. Radcliffe, A.; Scott, A.; Wilmer, E. Reversals and transpositions over finite alphabets. *SIAM J. Discret. Math.* **2006**, *19*, 224. [CrossRef]
20. Kececioglu, J.; Sankoff, D. Exact and approximation algorithms for sorting by reversals, with application to genome rearrangement. *Algorithmica* **1995**, *13*, 180. [CrossRef]
21. Bulteau, L.; Fertin, G.; Komusiewicz, C. Reversal distances for strings with few blocks or small alphabets. In *Symposium on Combinatorial Pattern Matching*; Springer: Berlin, Germany, 2014; pp. 50–59.
22. Bérard, S.; Chauve, C.; Paul, C. A more efficient algorithm for perfect sorting by reversals. *Inf. Process. Lett.* **2008**, *106*, 90–95. [CrossRef]
23. Dias, Z.; Meidanis, J. Sorting by prefix transpositions. In *International Symposium on String Processing and Information Retrieval*; Springer: Berlin, Germany, 2002; pp. 65–76.
24. Whidden, C. Sorting by Transpositions: Fixed-Parameter Algorithms and Structural Properties. Bachelor's Thesis, Dalhousie University, Halifax, NS, Canada, 2007.
25. Fertin, G.; Jankowiak, L.; Jean, G. Prefix and suffix reversals on strings. *Discret. Appl. Math.* **2018**, *246*, 140–153. [CrossRef]
26. Lopresti, D.; Tomkins, A. Block edit models for approximate string matching. *Theor. Comput. Sci.* **1997**, *181*, 159–179. [CrossRef]
27. Swenson, K.M.; Marron, M.; Earnest-DeYoung, J.V.; Moret, B.M. Approximating the true evolutionary distance between two genomes. *J. Exp. Algorithmics (JEA)* **2008**, *12*, 3–5. [CrossRef]
28. Damaschke, P. Minimum common string partition parameterized. In *International Workshop on Algorithms in Bioinformatics*; Springer: Berlin, Germany, 2008; pp. 87–98.
29. Jiang, H.; Zhu, B.; Zhu, D.; Zhu, H. Minimum common string partition revisited. *J. Comb. Optim.* **2012**, *23*, 519–527. [CrossRef]
30. Bulteau, L.; Komusiewicz, C. Minimum common string partition parameterized by partition size is fixed-parameter tractable. In Proceedings of the Twenty-Fifth Annual ACM-SIAM Symposium on Discrete Algorithms, SIAM, Portland, OR, USA, 5–7 January 2014; pp. 102–121.

31. Bulteau, L.; Fertin, G.; Komusiewicz, C.; Rusu, I. A fixed-parameter algorithm for minimum common string partition with few duplications. In *International Workshop on Algorithms in Bioinformatics*; Springer: Berlin, Germany, 2013; pp. 244–258.
32. Beretta, S.; Castelli, M.; Dondi, R. Parameterized tractability of the maximum-duo preservation string mapping problem. *Theor. Comput. Sci.* **2016**, *646*, 16–25. [CrossRef]
33. Zheng, C.; Zhu, Q.; Sankoff, D. Removing noise and ambiguities from comparative maps in rearrangement analysis. *IEEE/ACM Trans. Comput. Biol. Bioinform.* **2007**, *4*, 515–522. [CrossRef] [PubMed]
34. Li, W.; Liu, H.; Wang, J.; Xiang, L.; Yang, Y. An improved linear kernel for complementary maximal strip recovery: Simpler and smaller. *Theor. Comput. Sci.* **2019**, *786*, 55–66. [CrossRef]
35. Bulteau, L.; Fertin, G.; Jiang, M.; Rusu, I. Tractability and approximability of maximal strip recovery. *Theor. Comput. Sci.* **2012**, *440*, 14–28. [CrossRef]
36. Jiang, M. On the parameterized complexity of some optimization problems related to multiple-interval graphs. *Theor. Comput. Sci.* **2010**, *411*, 4253–4262. [CrossRef]
37. Muñoz, A.; Zheng, C.; Zhu, Q.; Albert, V.A.; Rounsley, S.; Sankoff, D. Scaffold filling, contig fusion and comparative gene order inference. *BMC Bioinform.* **2010**, *11*, 304. [CrossRef] [PubMed]
38. Bulteau, L.; Carrieri, A.P.; Dondi, R. Fixed-parameter algorithms for scaffold filling. *Theor. Comput. Sci.* **2015**, *568*, 72–83. [CrossRef]
39. Sankoff, D.; Blanchette, M. Multiple genome rearrangement and breakpoint phylogeny. *J. Comput. Biol.* **1998**, *5*, 555–570. [CrossRef] [PubMed]
40. Waterston, R.H.; Lander, E.S.; Sulston, J.E. On the sequencing of the human genome. *Proc. Natl. Acad. Sci. USA* **2002**, *99*, 3712–3716. [CrossRef]
41. Notredame, C. Recent Evolutions of Multiple Sequence Alignment Algorithms. *PLOS Comput. Biol.* **2007**, *3*, 1–4. [CrossRef]
42. Bonizzoni, P.; Della Vedova, G. The complexity of multiple sequence alignment with SP-score that is a metric. *Theor. Comput. Sci.* **2001**, *259*, 63–79. [CrossRef]
43. Just, W. Computational complexity of multiple sequence alignment with SP-score. *J. Comput. Biol.* **2001**, *8*, 615–623. [CrossRef]
44. Elias, I. Settling the intractability of multiple alignment. In *International Symposium on Algorithms and Computation*; Springer: Berlin, Germany, 2003; pp. 352–363.
45. Kemena, C.; Notredame, C. Upcoming challenges for multiple sequence alignment methods in the high-throughput era. *Bioinformatics* **2009**, *25*, 2455–2465. [CrossRef]
46. Bulteau, L.; Hüffner, F.; Komusiewicz, C.; Niedermeier, R. Multivariate Algorithmics for NP-Hard String Problems. *Bull. EATCS* **2014**, *114*, 1–43.
47. Frances, M.; Litman, A. On covering problems of codes. *Theory Comput. Syst.* **1997**, *30*, 113–119. [CrossRef]
48. Gramm, J.; Niedermeier, R.; Rossmanith, P. Fixed-parameter algorithms for closest string and related problems. *Algorithmica* **2003**, *37*, 25–42. [CrossRef]
49. Evans, P.A.; Smith, A.D.; Wareham, H.T. On the complexity of finding common approximate substrings. *Theor. Comput. Sci.* **2003**, *306*, 407–430. [CrossRef]
50. Fellows, M.R.; Gramm, J.; Niedermeier, R. On the parameterized intractability of motif search problems. *Combinatorica* **2006**, *26*, 141–167. [CrossRef]
51. Marx, D. Closest substring problems with small distances. *SIAM J. Comput.* **2008**, *38*, 1382–1410. [CrossRef]
52. Schmid, M.L. Finding consensus strings with small length difference between input and solution strings. *ACM Trans. Comput. Theory (TOCT)* **2017**, *9*, 13. [CrossRef]
53. Boucher, C.; Ma, B. Closest string with outliers. *BMC Bioinform.* **2011**, *12*, S55. [CrossRef]
54. Bulteau, L.; Schmid, M. Consensus Strings with Small Maximum Distance and Small Distance Sum. In Proceedings of the 43rd International Symposium on Mathematical Foundations of Computer Science (MFCS 2018), Liverpool, UK, 27–31 August 2018.

55. Chen, J.; Hermelin, D.; Sorge, M. On Computing Centroids According to the p-Norms of Hamming Distance Vectors. In Proceedings of the 27th Annual European Symposium on Algorithms, ESA 2019, Schloss Dagstuhl—Leibniz-Zentrum für Informatik, Munich/Garching, Germany, 9–11 September 2019; Volume 144; pp. 28:1–28:16. [CrossRef]
56. Pietrzak, K. On the parameterized complexity of the fixed alphabet shortest common supersequence and longest common subsequence problems. *J. Comput. Syst. Sci.* **2003**, *67*, 757–771. [CrossRef]
57. Bodlaender, H.L.; Downey, R.G.; Fellows, M.R.; Wareham, H.T. The parameterized complexity of sequence alignment and consensus. *Theor. Comput. Sci.* **1995**, *147*, 31–54. [CrossRef]
58. Irving, R.W.; Fraser, C.B. Two algorithms for the longest common subsequence of three (or more) strings. In *Annual Symposium on Combinatorial Pattern Matching*; Springer: Berlin, Germany, 1992; pp. 214–229.
59. Abboud, A.; Backurs, A.; Williams, V.V. Tight hardness results for LCS and other sequence similarity measures. In Proceedings of the 2015 IEEE 56th Annual Symposium on Foundations of Computer Science, Berkeley, CA, USA, 17–20 October 2015; pp. 59–78.
60. Bringmann, K.; Künnemann, M. Multivariate fine-grained complexity of longest common subsequence. In Proceedings of the Twenty-Ninth Annual ACM-SIAM Symposium on Discrete Algorithms, Society for Industrial and Applied Mathematics, New Orleans, LA, USA, 7–10 January 2018; pp. 1216–1235.
61. Giannopoulou, A.C.; Mertzios, G.B.; Niedermeier, R. Polynomial fixed-parameter algorithms: A case study for longest path on interval graphs. *Theor. Comput. Sci.* **2017**, *689*, 67–95. [CrossRef]
62. Mertzios, G.B.; Nichterlein, A.; Niedermeier, R. The Power of Linear-Time Data Reduction for Maximum Matching. In Proceedings of the 42nd International Symposium on Mathematical Foundations of Computer Science (MFCS 2017), Aalborg, Denmark, 21–25 August 2017; Schloss Dagstuhl–Leibniz-Zentrum fuer Informatik: Dagstuhl, Germany, 2017; Volume 83, pp. 46:1–46:14. [CrossRef]
63. Coudert, D.; Ducoffe, G.; Popa, A. Fully Polynomial FPT Algorithms for Some Classes of Bounded Clique-width Graphs. *ACM Trans. Algorithms* **2019**, *15*, 33:1–33:57. [CrossRef]
64. Tsai, Y.T. The constrained longest common subsequence problem. *Inf. Process. Lett.* **2003**, *88*, 173–176. [CrossRef]
65. Bonizzoni, P.; Della Vedova, G.; Dondi, R.; Pirola, Y. Variants of constrained longest common subsequence. *Inf. Process. Lett.* **2010**, *110*, 877–881. [CrossRef]
66. Chen, Y.C.; Chao, K.M. On the generalized constrained longest common subsequence problems. *J. Comb. Optim.* **2011**, *21*, 383–392. [CrossRef]
67. Gotthilf, Z.; Hermelin, D.; Landau, G.M.; Lewenstein, M. Restricted lcs. In *International Symposium on String Processing and Information Retrieval*; Springer: Berlin, Germany, 2010; pp. 250–257.
68. Chen, J.; Huang, X.; Kanj, I.A.; Xia, G. On the computational hardness based on linear FPT-reductions. *J. Comb. Optim.* **2006**, *11*, 231–247. [CrossRef]
69. Fellows, M.; Hallett, M.; Korostensky, C.; Stege, U. Analogs and Duals of the MAST Problem for Sequences and Trees. In *European Symposium on Algorithms*; Springer: Berlin, Germany, 1998; pp. 103–114.
70. Nicolas, F.; Rivals, E. Complexities of the centre and median string problems. In *Annual Symposium on Combinatorial Pattern Matching*; Springer: Berlin, Germany, 2003; pp. 315–327.
71. Maji, H.; Izumi, T. Listing center strings under the edit distance metric. In *Combinatorial Optimization and Applications*; Springer: Berlin, Germany, 2015; pp. 771–782.
72. Hunt, M.; Newbold, C.; Berriman, M.; Otto, T.D. A comprehensive evaluation of assembly scaffolding tools. *Genome Biol.* **2014**, *15*, R42. [CrossRef] [PubMed]
73. Huson, D.H.; Reinert, K.; Myers, E.W. The Greedy Path-merging Algorithm for Contig Scaffolding. *J. ACM* **2002**, *49*, 603–615. [CrossRef]
74. Chateau, A.; Giroudeau, R.; Poss, M.; Weller, M. Scaffolding with repeated contigs using flow formulations. unpublished manuscript.
75. Weller, M.; Chateau, A.; Dallard, C.; Giroudeau, R. Scaffolding Problems Revisited: Complexity, Approximation and Fixed Parameter Tractable Algorithms, and Some Special Cases. *Algorithmica* **2018**, *80*, 1771–1803. [CrossRef]
76. Gao, S.; Sung, W.K.; Nagarajan, N. Opera: Reconstructing Optimal Genomic Scaffolds with High-Throughput Paired-End Sequences. *J. Comput. Biol.* **2011**, *18*, 1681–1691. [CrossRef]
77. Weller, M.; Chateau, A.; Giroudeau, R. Exact approaches for scaffolding. *BMC Bioinform.* **2015**, *16*, S2. [CrossRef]

78. Weller, M.; Chateau, A.; Giroudeau, R. On the Linearization of Scaffolds Sharing Repeated Contigs. In Proceedings of the 11th International Conference on Combinatorial Optimization and Applications (COCOA'17) Part II, Shanghai, China, 16–18 December 2017; pp. 509–517. [CrossRef]
79. Davot, T.; Chateau, A.; Giroudeau, R.; Weller, M. Linearizing Genomes: Exact Methods and Local Search. In Proceedings of the SOFSEM'20, Nový Smokovec, Slovakia, 27–30 January 2019.
80. Donmez, N.; Brudno, M. SCARPA: scaffolding reads with practical algorithms. *Bioinformatics* **2012**, *29*, 428–434. [CrossRef]
81. Cao, M.D.; Nguyen, S.H.; Ganesamoorthy, D.; Elliott, A.G.; Cooper, M.A.; Coin, L.J.M. Scaffolding and completing genome assemblies in real-time with nanopore sequencing. *Nat. Commun.* **2017**, *8*, 14515. [CrossRef]
82. Dallard, C.; Weller, M.; Chateau, A.; Giroudeau, R. Instance Guaranteed Ratio on Greedy Heuristic for Genome Scaffolding. In *Combinatorial Optimization and Applications*; Springer International Publishing: Cham, Switzerland, 2016; pp. 294–308.
83. Hodgkinson, A.; Eyre-Walker, A. Human triallelic sites: Evidence for a new mutational mechanism? *Genetics* **2010**, *184*, 233–241. [CrossRef]
84. International SNP Map Working Group. A map of human genome sequence variation containing 1.42 million single nucleotide polymorphisms. *Nature* **2001**, *409*, 928. [CrossRef]
85. Andrés, A.M.; Clark, A.G.; Shimmin, L.; Boerwinkle, E.; Sing, C.F.; Hixson, J.E. Understanding the accuracy of statistical haplotype inference with sequence data of known phase. *Genet. Epidemiol.* **2007**, *31*, 659–671. [CrossRef] [PubMed]
86. Orzack, S.H.; Gusfield, D.; Olson, J.; Nesbitt, S.; Subrahmanyan, L.; Stanton, V.P. Analysis and Exploration of the Use of Rule-Based Algorithms and Consensus Methods for the Inferral of Haplotypes. *Genetics* **2003**, *165*, 915–928.
87. Climer, S.; Jäger, G.; Templeton, A.R.; Zhang, W. How frugal is mother nature with haplotypes? *Bioinformatics* **2008**, *25*, 68–74. [CrossRef]
88. Zhang, X.S.; Wang, R.S.; Wu, L.Y.; Chen, L. Models and Algorithms for Haplotyping Problem. *Curr. Bioinform.* **2006**, *1*, 105–114. [CrossRef]
89. Halldórsson, B.V.; Bafna, V.; Edwards, N.; Lippert, R.; Yooseph, S.; Istrail, S. A Survey of Computational Methods for Determining Haplotypes. In *Computational Methods for SNPs and Haplotype Inference*; Springer: Berlin/Heidelberg, Germany, 2004; pp. 26–47.
90. Lancia, G. Algorithmic approaches for the single individual haplotyping problem. *RAIRO Recherche Opérationnelle* **2016**, *50*. [CrossRef]
91. Zhao, Y.; Xu, Y.; Zhang, Q.; Chen, G. An overview of the haplotype problems and algorithms. *Front. Comput. Sci. China* **2007**, *1*, 272–282. [CrossRef]
92. Schwartz, R. Theory and Algorithms for the Haplotype Assembly Problem. *Commun. Inf. Syst.* **2010**, *10*, 23–38. [CrossRef]
93. Geraci, F. A comparison of several algorithms for the single individual SNP haplotyping reconstruction problem. *Bioinformatics* **2010**, *26*, 2217–2225. [CrossRef]
94. Xie, M.; Wang, J.; Chen, J.; Wu, J.; Liu, X. Computational Models and Algorithms for the Single Individual Haplotyping Problem. *Curr. Bioinform.* **2010**, *5*, 18–28. [CrossRef]
95. Rhee, J.K.; Li, H.; Joung, J.G.; Hwang, K.B.; Zhang, B.T.; Shin, S.Y. Survey of computational haplotype determination methods for single individual. *Genes Genom.* **2016**, *38*, 1–12. [CrossRef]
96. Lancia, G.; Bafna, V.; Istrail, S.; Lippert, R.; Schwartz, R. SNPs Problems, Complexity, and Algorithms. In *Algorithms—ESA 2001*; Springer: Berlin/Heidelberg, Germany, 2001; pp. 182–193.
97. Bafna, V.; Istrail, S.; Lancia, G.; Rizzi, R. Polynomial and APX-hard cases of the individual haplotyping problem. *Theor. Comput. Sci.* **2005**, *335*, 109–125. [CrossRef]
98. Xie, M.; Wang, J. An Improved (and Practical) Parameterized Algorithm for the Individual Haplotyping Problem MFR with Mate-Pairs. *Algorithmica* **2008**, *52*, 250–266. [CrossRef]
99. Reed, B.; Smith, K.; Vetta, A. Finding odd cycle transversals. *Oper. Res. Lett.* **2004**, *32*, 299–301. [CrossRef]
100. Hüffner, F. Algorithm Engineering for Optimal Graph Bipartization. *J. Graph Algorithms Appl.* **2009**, *13*, 77–98. [CrossRef]
101. Lokshtanov, D.; Narayanaswamy, N.S.; Raman, V.; Ramanujan, M.S.; Saurabh, S. Faster Parameterized Algorithms Using Linear Programming. *ACM Trans. Algorithms* **2014**, *11*, 15:1–15:31. [CrossRef]

102. Kratsch, S.; Wahlström, M. Compression via Matroids: A Randomized Polynomial Kernel for Odd Cycle Transversal. *ACM Trans. Algorithms* **2014**, *10*, 20:1–20:15. [CrossRef]
103. Xie, M.; Chen, J.; Wang, J. Research on parameterized algorithms of the individual haplotyping problem. *J. Bioinform. Comput. Biol.* **2007**, *05*, 795–816. [CrossRef]
104. Cilibrasi, R.; van Iersel, L.; Kelk, S.; Tromp, J. The Complexity of the Single Individual SNP Haplotyping Problem. *Algorithmica* **2007**, *49*, 13–36. [CrossRef]
105. Bonizzoni, P.; Dondi, R.; Klau, G.W.; Pirola, Y.; Pisanti, N.; Zaccaria, S. On the Minimum Error Correction Problem for Haplotype Assembly in Diploid and Polyploid Genomes. *J. Comput. Biol.* **2016**, *23*, 718–736. [CrossRef] [PubMed]
106. Wang, R.S.; Wu, L.Y.; Li, Z.P.; Zhang, X.S. Haplotype reconstruction from SNP fragments by minimum error correction. *Bioinformatics* **2005**, *21*, 2456–2462. [CrossRef] [PubMed]
107. Xie, M.; Wang, J.; Chen, J. A model of higher accuracy for the individual haplotyping problem based on weighted SNP fragments and genotype with errors. *Bioinformatics* **2008**, *24*, i105–i113. [CrossRef]
108. Deng, F.; Cui, W.; Wang, L. A highly accurate heuristic algorithm for the haplotype assembly problem. *BMC Genom.* **2013**, *14*, S2. [CrossRef] [PubMed]
109. Patterson, M.; Marschall, T.; Pisanti, N.; van Iersel, L.; Stougie, L.; Klau, G.W.; Schönhuth, A. WhatsHap: Weighted Haplotype Assembly for Future-Generation Sequencing Reads. *J. Comput. Biol.* **2015**, *22*, 498–509. [CrossRef]
110. Pirola, Y.; Zaccaria, S.; Dondi, R.; Klau, G.W.; Pisanti, N.; Bonizzoni, P. HapCol: accurate and memory-efficient haplotype assembly from long reads. *Bioinformatics* **2015**, *32*, 1610–1617. [CrossRef]
111. Garg, S.; Rautiainen, M.; Novak, A.M.; Garrison, E.; Durbin, R.; Marschall, T. A graph-based approach to diploid genome assembly. *Bioinformatics* **2018**, *34*, i105–i114. [CrossRef]
112. He, D.; Choi, A.; Pipatsrisawat, K.; Darwiche, A.; Eskin, E. Optimal algorithms for haplotype assembly from whole-genome sequence data. *Bioinformatics* **2010**, *26*, i183–i190. [CrossRef]
113. Zhang, X.S.; Wang, R.S.; Wu, L.Y.; Zhang, W. Minimum Conflict Individual Haplotyping from SNP Fragments and Related Genotype. *Evolut. Bioinform.* **2006**, *2*, [CrossRef]
114. Hermelin, D.; Rozenberg, L. Parameterized complexity analysis for the Closest String with Wildcards problem. *Theor. Comput. Sci.* **2015**, *600*, 11–18. [CrossRef]
115. Garg, S.; Martin, M.; Marschall, T. Read-based phasing of related individuals. *Bioinformatics* **2016**, *32*, i234–i242. [CrossRef]
116. Li, Z.p.; Wu, L.y.; Zhao, Y.y.; Zhang, X.s. A Dynamic Programming Algorithm for the k-Haplotyping Problem. *Acta Mathematicae Applicatae Sinica* **2006**, *22*, 405–412. [CrossRef]
117. Bao, R.; Huang, L.; Andrade, J.; Tan, W.; Kibbe, W.A.; Jiang, H.; Feng, G. Review of Current Methods, Applications, and Data Management for the Bioinformatics Analysis of Whole Exome Sequencing. *Cancer Inform.* **2014**, *13s2*, CIN.S13779. [CrossRef]
118. Garg, S. Computational Haplotyping: Theory and Practice. Ph.D. Thesis, Universität des Saarlandes, Saarbrücken, Germany, 2018. [CrossRef]
119. Gusfield, D. Haplotype Inference by Pure Parsimony. In *Combinatorial Pattern Matching*; Springer: Berlin/Heidelberg, Germany, 2003; pp. 144–155.
120. Bonizzoni, P.; Della Vedova, G.; Dondi, R.; Pirola, Y.; Rizzi, R. Pure Parsimony Xor Haplotyping. *IEEE/ACM Trans. Comput. Biol. Bioinform.* **2010**, *7*, 598–610. [CrossRef] [PubMed]
121. Graça, A.; Lynce, I.; Marques-Silva, J.; Oliveira, A.L. Haplotype Inference by Pure Parsimony: A Survey. *J. Comput. Biol.* **2010**, *17*, 969–992. [CrossRef]
122. Hubbell, E. (GRAIL, Inc. LinkedIn, Menlo Park, CA, USA). Finding a Parsimony Solution to Haplotype Phase is NP-Hard. Personal Communication, 2002.
123. Lancia, G.; Pinotti, M.C.; Rizzi, R. Haplotyping Populations by Pure Parsimony: Complexity of Exact and Approximation Algorithms. *INFORMS J. Comput.* **2004**, *16*, 348–359. [CrossRef]
124. Huang, Y.T.; Chao, K.M.; Chen, T. An Approximation Algorithm for Haplotype Inference by Maximum Parsimony. *J. Comput. Biol.* **2005**, *12*, 1261–1274. [CrossRef]
125. Sharan, R.; Halldorsson, B.V.; Istrail, S. Islands of Tractability for Parsimony Haplotyping. *IEEE/ACM Trans. Comput. Biol. Bioinform.* **2006**, *3*, 303–311. [CrossRef]

126. van Iersel, L.; Keijsper, J.; Kelk, S.; Stougie, L. Shorelines of Islands of Tractability: Algorithms for Parsimony and Minimum Perfect Phylogeny Haplotyping Problems. *IEEE/ACM Trans. Comput. Biol. Bioinform.* **2008**, *5*, 301–312. [CrossRef]
127. Lancia, G.; Rizzi, R. A polynomial case of the parsimony haplotyping problem. *Oper. Res. Lett.* **2006**, *34*, 289–295. [CrossRef]
128. van Iersel, L.J.J. Algorithms, Haplotypes and Phylogenetic Networks. Ph.D. Thesis, Eindhoven University of Technology, Eindhoven, The Netherlands, 2009.
129. Fellows, M.R.; Hartman, T.; Hermelin, D.; Landau, G.M.; Rosamond, F.A.; Rozenberg, L. Haplotype Inference Constrained by Plausible Haplotype Data. *IEEE/ACM Trans. Comput. Biol. Bioinform.* **2011**, *8*, 1692–1699. [CrossRef] [PubMed]
130. Fleischer, R.; Guo, J.; Niedermeier, R.; Uhlmann, J.; Wang, Y.; Weller, M.; Wu, X. Extended Islands of Tractability for Parsimony Haplotyping. In *Combinatorial Pattern Matching*; Springer: Berlin, Heidelberg, Germany, 2010; pp. 214–226.
131. Gusfield, D. Haplotyping As Perfect Phylogeny: Conceptual Framework and Efficient Solutions. In Proceedings of the Sixth Annual International Conference on Computational Biology RECOMB '02, Washington, DC, USA, 18–21 April 2002; ACM: New York, NY, USA, 2002; pp. 166–175. [CrossRef]
132. Ding, Z.; Filkov, V.; Gusfield, D. A Linear-Time Algorithm for the Perfect Phylogeny Haplotyping (PPH) Problem. *J. Comput. Biol.* **2006**, *13*, 522–553. [CrossRef] [PubMed]
133. Bonizzoni, P. A Linear-Time Algorithm for the Perfect Phylogeny Haplotype Problem. *Algorithmica* **2007**, *48*, 267–285. [CrossRef]
134. Chen, Z.Z.; Ma, W.; Wang, L. The Parameterized Complexity of the Shared Center Problem. *Algorithmica* **2014**, *69*, 269–293. [CrossRef]
135. Keijsper, J.; Oosterwijk, T. Tractable Cases of (*, 2)-Bounded Parsimony Haplotyping. *IEEE/ACM Trans. Comput. Biol. Bioinform.* **2015**, *12*, 234–247. [CrossRef]
136. Cicalese, F.; Milanivc, M. *On Parsimony Haplotyping*; Technical Report; Universität Bielefeld: Bielefeld, Germany, 2008.
137. Haeckel, E. *Generelle Morphologie der Organismen. Allgemeine Grundzüge der organischen Formen-Wissenschaft, mechanisch begründet durch die von C. Darwin reformirte Descendenz-Theorie, etc.*; Verlag von Georg Reimer: Berlin, Germany; 1866; Volume 2.
138. Dobzhansky, T. Nothing in Biology Makes Sense except in the Light of Evolution. *Am. Biol. Teach.* **1973**, *35*, 125–129. [CrossRef]
139. De Bruyn, A.; Martin, D.P.; Lefeuvre, P., Phylogenetic Reconstruction Methods: An Overview. In *Molecular Plant Taxonomy: Methods and Protocols*; Humana Press: Totowa, NJ, USA, 2014; pp. 257–277. [CrossRef]
140. Huson, D.H.; Rupp, R.; Scornavacca, C. *Phylogenetic Networks—Concepts, Algorithms and Applications*; Cambridge University Press: Cambridge, UK, 2010.
141. Choy, C.; Jansson, J.; Sadakane, K.; Sung, W.K. Computing the maximum agreement of phylogenetic networks. *Theor. Comput. Sci.* **2005**, *335*, 93–107. [CrossRef]
142. Betzler, N.; Guo, J.; Komusiewicz, C.; Niedermeier, R. Average parameterization and partial kernelization for computing medians. *J. Comput. Syst. Sci.* **2011**, *77*, 774–789. [CrossRef]
143. Bryant, D. A classification of consensus methods for phylogenetics. *DIMACS Ser. Discrete Math. Theor. Comput. Sci.* **2003**, *61*, 163–184.
144. Degnan, J.H., Consensus Methods, Phylogenetic. In *Encyclopedia of Evolutionary Biology*; Elsevier: Amsterdam, The Netherlands, 2016; Volume 1; pp. 341–346.
145. Steel, M. The complexity of reconstructing trees from qualitative characters and subtrees. *J. Classif.* **1992**, *9*, 91–116. [CrossRef]
146. Böcker, S.; Bryant, D.; Dress, A.W.; Steel, M.A. Algorithmic Aspects of Tree Amalgamation. *J. Algorithms* **2000**, *37*, 522–537. [CrossRef]
147. Arnborg, S.; Lagergren, J.; Seese, D. Easy problems for tree-decomposable graphs. *J. Algorithms* **1991**, *12*, 308–340. [CrossRef]
148. Courcelle, B. The monadic second-order logic of graphs. I. Recognizable sets of finite graphs. *Inf. Comput.* **1990**, *85*, 12–75. [CrossRef]
149. Bryant, D.; Lagergren, J. Compatibility of unrooted phylogenetic trees is FPT. *Theor. Comput. Sci.* **2006**, *351*, 296–302. [CrossRef]

150. Scornavacca, C.; van Iersel, L.; Kelk, S.; Bryant, D. The agreement problem for unrooted phylogenetic trees is FPT. *J. Graph Algorithms Appl.* **2014**, *18*, 385–392. [CrossRef]
151. Baste, J.; Paul, C.; Sau, I.; Scornavacca, C. Efficient FPT Algorithms for (Strict) Compatibility of Unrooted Phylogenetic Trees. In *Algorithmic Aspects in Information and Management*; Springer International Publishing: Cham, Switzerland, 2016; pp. 53–64.
152. Aho, A.; Sagiv, Y.; Szymanski, T.; Ullman, J. Inferring a Tree from Lowest Common Ancestors with an Application to the Optimization of Relational Expressions. *SIAM J. Comput.* **1981**, *10*, 405–421. [CrossRef]
153. Ng, M.P.; Wormald, N.C. Reconstruction of rooted trees from subtrees. *Discret. Appl. Math.* **1996**, *69*, 19–31. [CrossRef]
154. Maddison, W.P. Gene Trees in Species Trees. *Syst. Biol.* **1997**, *46*, 523–536. [CrossRef]
155. Linder, C.R.; Rieseberg, L.H. Reconstructing patterns of reticulate evolution in plants. *Am. J. Bot.* **2004**, *91*, 1700–1708. [CrossRef]
156. Jansson, J. On the complexity of inferring rooted evolutionary trees. *Electron. Notes Discret. Math.* **2001**, *7*, 50–53. [CrossRef]
157. Bryant, D. Building Trees, Hunting for Trees, and Comparing Trees: Theory and Methods in Phylogenetic Analysis. Ph.D. Thesis, University of Canterbury, Canterbury, UK, 1997.
158. Wu, B.Y. Constructing the Maximum Consensus Tree from Rooted Triples. *J. Comb. Optim.* **2004**, *8*, 29–39. [CrossRef]
159. Byrka, J.; Guillemot, S.; Jansson, J. New results on optimizing rooted triplets consistency. *Discret. Appl. Math.* **2010**, *158*, 1136–1147. [CrossRef]
160. Guillemot, S.; Mnich, M. Kernel and Fast Algorithm for Dense Triplet Inconsistency. In *Theory and Applications of Models of Computation*; Springer: Berlin/Heidelberg, Germany, 2010; pp. 247–257.
161. Fomin, F.V.; Lokshtanov, D.; Saurabh, S.; Zehavi, M. *Kernelization: Theory of Parameterized Preprocessing*; Cambridge University Press: Cambridge, UK, 2019. [CrossRef]
162. Paul, C.; Perez, A.; Thomassé, S. Linear kernel for Rooted Triplet Inconsistency and other problems based on conflict packing technique. *J. Comput. Syst. Sci.* **2016**, *82*, 366–379. [CrossRef]
163. Habib, M.; To, T.H. Constructing a minimum phylogenetic network from a dense triplet set. *J. Bioinform. Comput. Biol.* **2012**, *10*, 1250013. [CrossRef] [PubMed]
164. van Iersel, L.; Kelk, S. Constructing the Simplest Possible Phylogenetic Network from Triplets. *Algorithmica* **2011**, *60*, 207–235. [CrossRef]
165. Gramm, J.; Niedermeier, R. A fixed-parameter algorithm for minimum quartet inconsistency. *J. Comput. Syst. Sci.* **2003**, *67*, 723–741. [CrossRef]
166. Jansson, J.; Ng, J.H.K.; Sadakane, K.; Sung, W.K. Rooted Maximum Agreement Supertrees. *Algorithmica* **2005**, *43*, 293–307. [CrossRef]
167. Steel, M.A.; Penny, D. Distributions of Tree Comparison Metrics—Some New Results. *Syst. Biol.* **1993**, *42*, 126–141. [CrossRef]
168. Goddard, W.; Kubicka, E.; Kubicki, G.; McMorris, F. The agreement metric for labeled binary trees. *Math. Biosci.* **1994**, *123*, 215–226. [CrossRef]
169. Berry, V.; Nicolas, F. Maximum agreement and compatible supertrees. *J. Discret. Algorithms* **2007**, *5*, 564–591. [CrossRef]
170. Guillemot, S.; Berry, V. Fixed-Parameter Tractability of the Maximum Agreement Supertree Problem. *IEEE/ACM Trans. Comput. Biol. Bioinform.* **2010**, *7*, 342–353. [CrossRef]
171. Hoang, V.T.; Sung, W.K. Improved Algorithms for Maximum Agreement andÂ Compatible Supertrees. *Algorithmica* **2011**, *59*, 195–214. [CrossRef]
172. Fernández-Baca, D.; Guillemot, S.; Shutters, B.; Vakati, S. Fixed-Parameter Algorithms for Finding Agreement Supertrees. *SIAM J. Comput.* **2015**, *44*, 384–410. [CrossRef]
173. Amir, A.; Keselman, D. Maximum Agreement Subtree in a Set of Evolutionary Trees: Metrics and Efficient Algorithms. *SIAM J. Comput.* **1997**, *26*, 1656–1669. [CrossRef]
174. Farach, M.; Przytycka, T.M.; Thorup, M. On the agreement of many trees. *Inf. Process. Lett.* **1995**, *55*, 297–301. [CrossRef]
175. Wang, B.; Swenson, K.M. A Faster Algorithm for Computing the Kernel of Maximum Agreement Subtrees. *IEEE/ACM Trans. Comput. Biol. Bioinform.* **2019**, [CrossRef]

176. Downey, R.G.; Fellows, M.R.; Stege, U. Computational Tractability: The View From Mars. *Bull. EATCS* **1999**, *69*, 73–97.
177. Alber, J.; Gramm, J.; Niedermeier, R. Faster exact algorithms for hard problems: A parameterized point of view. *Discret. Math.* **2001**, *229*, 3–27. [CrossRef]
178. Berry, V.; Nicolas, F. Improved Parameterized Complexity of the Maximum Agreement Subtree and Maximum Compatible Tree Problems. *IEEE/ACM Trans. Comput. Biol. Bioinform.* **2006**, *3*, 289–302. [CrossRef]
179. Chauve, C.; Jones, M.; Lafond, M.; Scornavacca, C.; Weller, M. Constructing a Consensus Phylogeny from a Leaf-Removal Distance. *CoRR* **2017**, arXiv:abs/1705.05295.
180. Chen, Z.Z.; Ueta, S.; Li, J.; Wang, L.; Skums, P.; Li, M. Computing a Consensus Phylogeny via Leaf Removal. In *Bioinformatics Research and Applications*; Springer International Publishing: Cham, Switzerland, 2019; pp. 3–15.
181. Lafond, M.; El-Mabrouk, N.; Huber, K.; Moulton, V. The complexity of comparing multiply-labelled trees by extending phylogenetic-tree metrics. *Theor. Comput. Sci.* **2019**, *760*, 15–34. [CrossRef]
182. Shi, F.; Feng, Q.; Chen, J.; Wang, L.; Wang, J. Distances between phylogenetic trees: A survey. *Tsinghua Sci.Technol.* **2013**, *18*, 490–499. [CrossRef]
183. Whidden, C. Efficient Computation and Application of Maximum Agreement Forests. Ph.D. Thesis, Dalhousie University, Halifax, NS, Canada, 2013.
184. Hein, J.; Jiang, T.; Wang, L.; Zhang, K. On the complexity of comparing evolutionary trees. *Discret. Appl. Math.* **1996**, *71*, 153–169. [CrossRef]
185. Allen, B.L.; Steel, M. Subtree Transfer Operations and Their Induced Metrics on Evolutionary Trees. *Ann. Comb.* **2001**, *5*, 1–15. [CrossRef]
186. Hallett, M.; McCartin, C. A Faster FPT Algorithm for the Maximum Agreement Forest Problem. *Theory Comput. Syst.* **2007**, *41*, 539–550. [CrossRef]
187. Whidden, C.; Zeh, N. A Unifying View on Approximation and FPT of Agreement Forests. In *Algorithms in Bioinformatics*; Springer: Berlin/Heidelberg, Germany, 2009; pp. 390–402.
188. Kelk, S.; Linz, S. A tight kernel for computing the tree bisection and reconnection distance between two phylogenetic trees. *CoRR* **2018**, arXiv:abs/1811.06892.
189. Kelk, S.; Linz, S. New reduction rules for the tree bisection and reconnection distance. *arXiv* **2019**, arXiv:1905.01468.
190. Shi, F.; Wang, J.; Chen, J.; Feng, Q.; Guo, J. Algorithms for parameterized maximum agreement forest problem on multiple trees. *Theor. Comput. Sci.* **2014**, *554*, 207–216. [CrossRef]
191. Chen, J.; Fan, J.H.; Sze, S.H. Parameterized and approximation algorithms for maximum agreement forest in multifurcating trees. *Theor. Comput. Sci.* **2015**, *562*, 496–512. [CrossRef]
192. Baroni, M.; Grünewald, S.; Moulton, V.; Semple, C. Bounding the Number of Hybridisation Events for a Consistent Evolutionary History. *J. Math. Biol.* **2005**, *51*, 171–182. [CrossRef]
193. Bordewich, M.; Semple, C. On the Computational Complexity of the Rooted Subtree Prune and Regraft Distance. *Ann. Comb.* **2005**, *8*, 409–423. [CrossRef]
194. Bordewich, M.; McCartin, C.; Semple, C. A 3-approximation algorithm for the subtree distance between phylogenies. *J. Discret. Algorithms* **2008**, *6*, 458–471. [CrossRef]
195. Whidden, C.; Beiko, R.; Zeh, N. Fixed-Parameter Algorithms for Maximum Agreement Forests. *SIAM J. Comput.* **2013**, *42*, 1431–1466. [CrossRef]
196. Chen, Z.Z.; Wang, L. Faster Exact Computation of rSPR Distance. In *Frontiers in Algorithmics and Algorithmic Aspects in Information and Management*; Springer: Berlin/Heidelberg, Germany, 2013; pp. 36–47.
197. Chen, Z.Z.; Wang, L. Algorithms for Reticulate Networks of Multiple Phylogenetic Trees. *IEEE/ACM Trans. Comput. Biol. Bioinform.* **2012**, *9*, 372–384. [CrossRef]
198. Collins, J.S. Rekernelisation Algorithms in Hybrid Phylogenies. Ph.D. Thesis, University of Canterbury, Christchurch, New Zealand, 2009.
199. van Iersel, L.; Kelk, S.; Lekić, N.; Stougie, L. Approximation Algorithms for Nonbinary Agreement Forests. *SIAM J. Discret. Math.* **2014**, *28*, 49–66. [CrossRef]
200. Whidden, C.; Beiko, R.G.; Zeh, N. Fixed-Parameter and Approximation Algorithms for Maximum Agreement Forests of Multifurcating Trees. *Algorithmica* **2016**, *74*, 1019–1054. [CrossRef]
201. Bordewich, M.; Semple, C. Computing the Hybridization Number of Two Phylogenetic Trees Is Fixed-Parameter Tractable. *IEEE/ACM Trans. Comput. Biol. Bioinform.* **2007**, *4*, 458–466. [CrossRef]

202. Shi, F.; Chen, J.; Feng, Q.; Wang, J. A parameterized algorithm for the Maximum Agreement Forest problem on multiple rooted multifurcating trees. *J. Comput. Syst. Sci.* **2018**, *97*, 28–44. [CrossRef]
203. Linz, S.; Semple, C. Hybridization in Nonbinary Trees. *IEEE/ACM Trans. Comput. Biol. Bioinform.* **2009**, *6*, 30–45. [CrossRef]
204. Bordewich, M.; Semple, C. Computing the minimum number of hybridization events for a consistent evolutionary history. *Discret. Appl. Math.* **2007**, *155*, 914–928. [CrossRef]
205. Albrecht, B.; Scornavacca, C.; Cenci, A.; Huson, D.H. Fast computation of minimum hybridization networks. *Bioinformatics* **2011**, *28*, 191–197. [CrossRef]
206. van Iersel, L.; Kelk, S.; Lekić, N.; Whidden, C.; Zeh, N. Hybridization Number on Three Rooted Binary Trees is EPT. *SIAM J. Discret. Math.* **2016**, *30*, 1607–1631. [CrossRef]
207. van Iersel, L.; Linz, S. A quadratic kernel for computing the hybridization number of multiple trees. *Inf. Process. Lett.* **2013**, *113*, 318–323. [CrossRef]
208. van Iersel, L.; Kelk, S.; Scornavacca, C. Kernelizations for the hybridization number problem on multiple nonbinary trees. *J. Comput. Syst. Sci.* **2016**, *82*, 1075–1089. [CrossRef]
209. Alon, N.; Gutin, G.; Kim, E.J.; Szeider, S.; Yeo, A. Solving MAX-r-SAT Above a Tight Lower Bound. *Algorithmica* **2011**, *61*, 638–655. [CrossRef]
210. Piovesan, T.; Kelk, S.M. A Simple Fixed Parameter Tractable Algorithm for Computing the Hybridization Number of Two (Not Necessarily Binary) Trees. *IEEE/ACM Trans. Comput. Biol. Bioinform.* **2013**, *10*, 18–25. [CrossRef]
211. Li, Z. Fixed-Parameter Algorithm for Hybridization Number of Two Multifurcating Trees. Master's Thesis, Dalhousie University, Halifax, NS, Canada, 2014.
212. Bordewich, M.; Scornavacca, C.; Tokac, N.; Weller, M. On the fixed parameter tractability of agreement-based phylogenetic distances. *J. Math. Biol.* **2017**, *74*, 239–257. [CrossRef]
213. Kelk, S.; van Iersel, L.; Scornavacca, C.; Weller, M. Phylogenetic incongruence through the lens of Monadic Second Order logic. *J. Graph Algorithms Appl.* **2016**, *20*, 189–215. [CrossRef]
214. Klawitter, J.; Linz, S. On the Subnet Prune and Regraft Distance. *arXiv* **2018**, arXiv:1805.07839.
215. Hickey, G.; Dehne, F.; Rau-Chaplin, A.; Blouin, C. SPR Distance Computation for Unrooted Trees. *Evolut. Bioinform.* **2008**, *4*, EBO.S419. [CrossRef]
216. Bonet, M.L.; St. John, K. On the Complexity of uSPR Distance. *IEEE/ACM Trans. Comput. Biol. Bioinform.* **2010**, *7*, 572–576. [CrossRef]
217. Whidden, C.; Matsen, F. Chain Reduction Preserves the Unrooted Subtree Prune-and-Regraft Distance. *arXiv* **2016**, arXiv:1611.02351.
218. Whidden, C.; Matsen, F. Calculating the Unrooted Subtree Prune-and-Regraft Distance. *IEEE/ACM Trans. Comput. Biol. Bioinform.* **2019**, *16*, 898–911. [CrossRef]
219. Döcker, J.; van Iersel, L.; Kelk, S.; Linz, S. Deciding the existence of a cherry-picking sequence is hard on two trees. *Discret. Appl. Math.* **2019**, *260*, 131–143. [CrossRef]
220. Humphries, P.J.; Linz, S.; Semple, C. Cherry Picking: A Characterization of the Temporal Hybridization Number for a Set of Phylogenies. *Bull. Math. Biol.* **2013**, *75*, 1879–1890. [CrossRef]
221. Fischer, M.; Kelk, S. On the Maximum Parsimony Distance Between Phylogenetic Trees. *Ann. Comb.* **2016**, *20*, 87–113. [CrossRef]
222. Kelk, S.; Fischer, M. On the Complexity of Computing MP Distance Between Binary Phylogenetic Trees. *Ann. Comb.* **2017**, *21*, 573–604. [CrossRef]
223. Kelk, S.; Fischer, M.; Moulton, V.; Wu, T. Reduction rules for the maximum parsimony distance on phylogenetic trees. *Theor. Comput. Sci.* **2016**, *646*, 1–15. [CrossRef]
224. Janssen, R.; Jones, M.; Kelk, S.; Stamoulis, G.; Wu, T. Treewidth of display graphs: bounds, brambles and applications. *J. Graph Algorithms Appl.* **2019**, *23*, 715–743. [CrossRef]
225. Ma, B.; Li, M.; Zhang, L. From Gene Trees to Species Trees. *SIAM J. Comput.* **2000**, *30*, 729–752. [CrossRef]
226. Bonizzoni, P.; Vedova, G.D.; Dondi, R. Reconciling a gene tree to a species tree under the duplication cost model. *Theor. Comput. Sci.* **2005**, *347*, 36–53. [CrossRef]
227. Doyon, J.P.; Ranwez, V.; Daubin, V.; Berry, V. Models, algorithms and programs for phylogeny reconciliation. *Brief. Bioinform.* **2011**, *12*, 392–400. [CrossRef] [PubMed]
228. Szöllősi, G.J.; Tannier, E.; Daubin, V.; Boussau, B. The Inference of Gene Trees with Species Trees. *Syst. Biol.* **2014**, *64*, e42–e62. [CrossRef] [PubMed]

229. Rusin, L.Y.; Lyubetskaya, E.; Gorbunov, K.Y.; Lyubetsky, V. Reconciliation of gene and species trees. *BioMed Res. Int.* **2014**, *2014*, 642089. [CrossRef] [PubMed]
230. Scornavacca, C. Phylogenomics among Trees and Networks: A Challenging Accrobranche. 2019. in press.
231. Górecki, P.; Tiuryn, J. DLS-trees: A model of evolutionary scenarios. *Theor. Comput. Sci.* **2006**, *359*, 378–399. [CrossRef]
232. ZHANG, L. On a Mirkin-Muchnik-Smith Conjecture for Comparing Molecular Phylogenies. *J. Comput. Biol.* **1997**, *4*, 177–187. [CrossRef]
233. Zmasek, C.M.; Eddy, S.R. A simple algorithm to infer gene duplication and speciation events on a gene tree. *Bioinformatics* **2001**, *17*, 821–828. [CrossRef]
234. Harel, D.; Tarjan, R.E. Fast Algorithms for Finding Nearest Common Ancestors. *SIAM J. Comput.* **1984**, *13*, 338–355. [CrossRef]
235. Bender, M.A.; Farach-Colton, M. The LCA Problem Revisited. In *LATIN 2000: Theoretical Informatics*; Springer: Berlin/Heidelberg, Germany, 2000; pp. 88–94.
236. Chang, W.C.; Eulenstein, O. Reconciling Gene Trees with Apparent Polytomies. In *Computing and Combinatorics*; Springer: Berlin/Heidelberg, Germany, 2006; pp. 235–244.
237. Lafond, M.; Swenson, K.M.; El-Mabrouk, N. An Optimal Reconciliation Algorithm for Gene Trees with Polytomies. In *Algorithms in Bioinformatics*; Springer: Berlin/Heidelberg, Germany, 2012; pp. 106–122.
238. Tofigh, A.; Hallett, M.; Lagergren, J. Simultaneous Identification of Duplications and Lateral Gene Transfers. *IEEE/ACM Trans. Comput. Biol. Bioinform.* **2011**, *8*, 517–535. [CrossRef]
239. Doyon, J.P.; Scornavacca, C.; Gorbunov, K.Y.; Szöllősi, G.J.; Ranwez, V.; Berry, V. An Efficient Algorithm for Gene/Species Trees Parsimonious Reconciliation with Losses, Duplications and Transfers. In *Comparative Genomics*; Springer: Berlin/Heidelberg, Germany, 2010; pp. 93–108.
240. Bansal, M.S.; Alm, E.J.; Kellis, M. Efficient algorithms for the reconciliation problem with gene duplication, horizontal transfer and loss. *Bioinformatics* **2012**, *28*, i283–i291. [CrossRef] [PubMed]
241. Ovadia, Y.; Fielder, D.; Conow, C.; Libeskind-Hadas, R. The Cophylogeny Reconstruction Problem Is NP-Complete. *J. Comput. Biol.* **2011**, *18*, 59–65. [CrossRef] [PubMed]
242. Hallett, M.T.; Lagergren, J. Efficient Algorithms for Lateral Gene Transfer Problems. In Proceedings of the Fifth Annual International Conference on Computational Biology (RECOMB '01), Montreal, QC, Canada, 22–25 April 2001; ACM: New York, NY, USA, 2001; pp. 149–156. [CrossRef]
243. Hasić, D.; Tannier, E. Gene tree species tree reconciliation with gene conversion. *J. Math. Biol.* **2019**, *78*, 1981–2014. [CrossRef] [PubMed]
244. Hasić, D.; Tannier, E. Gene tree reconciliation including transfers with replacement is NP-hard and FPT. *J. Comb. Optim.* **2019**, *38*, 502–544. [CrossRef]
245. Maddison, W.P.; Knowles, L.L. Inferring Phylogeny Despite Incomplete Lineage Sorting. *Syst. Biol.* **2006**, *55*, 21–30. [CrossRef] [PubMed]
246. Bork, D.; Cheng, R.; Wang, J.; Sung, J.; Libeskind-Hadas, R. On the computational complexity of the maximum parsimony reconciliation problem in the duplication-loss-coalescence model. *Algorithms Mol. Biol.* **2017**, *12*, 6:1–6:12. [CrossRef]
247. Stolzer, M.; Lai, H.; Xu, M.; Sathaye, D.; Vernot, B.; Durand, D. Inferring duplications, losses, transfers and incomplete lineage sorting with nonbinary species trees. *Bioinformatics* **2012**, *28*, i409–i415. [CrossRef]
248. ban Chan, Y.; Ranwez, V.; Scornavacca, C. Inferring incomplete lineage sorting, duplications, transfers and losses with reconciliations. *J. Theor. Biol.* **2017**, *432*, 1–13. [CrossRef]
249. To, T.H.; Scornavacca, C. Efficient algorithms for reconciling gene trees and species networks via duplication and loss events. *BMC Genom.* **2015**, *16*, S6. [CrossRef]
250. Bromham, L. The genome as a life-history character: why rate of molecular evolution varies between mammal species. *Philos. Trans. R. Soc. B Biol. Sci.* **2011**, *366*, 2503–2513. [CrossRef]
251. Fitch, W.M. Toward Defining the Course of Evolution: Minimum Change for a Specific Tree Topology. *Syst. Biol.* **1971**, *20*, 406–416. [CrossRef]
252. Jin, G.; Nakhleh, L.; Snir, S.; Tuller, T. Parsimony Score of Phylogenetic Networks: Hardness Results and a Linear-Time Heuristic. *IEEE/ACM Trans. Comput. Biol. Bioinform.* **2009**, *6*, 495–505. [CrossRef] [PubMed]
253. Fischer, M.; van Iersel, L.; Kelk, S.; Scornavacca, C. On Computing the Maximum Parsimony Score of a Phylogenetic Network. *SIAM J. Discret. Math.* **2015**, *29*, 559–585. [CrossRef]

254. Kanj, I.A.; Nakhleh, L.; Than, C.; Xia, G. Seeing the trees and their branches in the network is hard. *Theor. Comput. Sci.* **2008**, *401*, 153–164. [CrossRef]
255. Gambette, P.; Gunawan, A.D.; Labarre, A.; Vialette, S.; Zhang, L. Solving the tree containment problem in linear time for nearly stable phylogenetic networks. *Discret. Appl. Math.* **2018**, *246*, 62–79. [CrossRef]
256. Fakcharoenphol, J.; Kumpijit, T.; Putwattana, A. A faster algorithm for the tree containment problem for binary nearly stable phylogenetic networks. In Proceedings of the 2015 12th International Joint Conference on Computer Science and Software Engineering (JCSSE), Songkhla, Thailand, 22–24 July 2015; pp. 337–342.
257. Bordewich, M.; Semple, C. Reticulation-visible networks. *Adv. Appl. Math.* **2016**, *78*, 114–141. [CrossRef]
258. Gunawan, A.D.; DasGupta, B.; Zhang, L. A decomposition theorem and two algorithms for reticulation-visible networks. *Inf. Comput.* **2017**, *252*, 161–175. [CrossRef]
259. Van Iersel, L.; Semple, C.; Steel, M. Locating a tree in a phylogenetic network. *Inf. Process. Lett.* **2010**, *110*, 1037–1043. [CrossRef]
260. Gunawan, A.D.M. Solving the Tree Containment Problem for Reticulation-Visible Networks in Linear Time. In *Algorithms for Computational Biology*; Springer International Publishing: Cham, Switzerland, 2018; pp. 24–36.
261. Weller, M. Linear-Time Tree Containment in Phylogenetic Networks. In Proceedings of the 16th International Conference on Comparative Genomics (RECOMB-CG 2018), Magog-Orford, QC, Canada, 9–12 October 2018; pp. 309–323. [CrossRef]
262. Gunawan, A.D.; Lu, B.; Zhang, L. A program for verification of phylogenetic network models. *Bioinformatics* **2016**, *32*, i503–i510. [CrossRef]
263. Gambette, P.; van Iersel, L.; Kelk, S.; Pardi, F.; Scornavacca, C. Do Branch Lengths Help to Locate a Tree in a Phylogenetic Network? *Bull. Math. Biol.* **2016**, *78*, 1773–1795. [CrossRef]
264. Huber, K.T.; van Iersel, L.; Janssen, R.; Jones, M.; Moulton, V.; Murakami, Y.; Semple, C. Rooting for phylogenetic networks. *arXiv* **2019**, arXiv:1906.07430.

© 2019 by the authors. Licensee MDPI, Basel, Switzerland. This article is an open access article distributed under the terms and conditions of the Creative Commons Attribution (CC BY) license (http://creativecommons.org/licenses/by/4.0/).

Article

Parameterized Optimization in Uncertain Graphs—A Survey and Some Results

N. S. Narayanaswamy * and R. Vijayaragunathan *

Department of Computer Science and Engineering, Indian Institute of Technology Madras (IIT Madras), Chennai 600036, India
* Correspondence: swamy@cse.iitm.ac.in (N.S.N.); vijayr@cse.iitm.ac.in (R.V.)

Received: 11 November 2019; Accepted: 12 December 2019; Published: 19 December 2019

Abstract: We present a detailed survey of results and two new results on graphical models of uncertainty and associated optimization problems. We focus on two well-studied models, namely, the Random Failure (RF) model and the Linear Reliability Ordering (LRO) model. We present an FPT algorithm parameterized by the product of treewidth and max-degree for maximizing expected coverage in an uncertain graph under the RF model. We then consider the problem of finding the maximal core in a graph, which is known to be polynomial time solvable. We show that the PROBABILISTIC-CORE problem is polynomial time solvable in uncertain graphs under the LRO model. On the other hand, under the RF model, we show that the PROBABILISTIC-CORE problem is W[1]-hard for the parameter d, where d is the minimum degree of the core. We then design an FPT algorithm for the parameter treewidth.

Keywords: probabilistic graphs; uncertain graphs; influence maximization; cascade failure; linear reliable ordering; expected coverage; probabilistic core.

1. Introduction

Network data analytics has come to play a key role in many scientific fields. A large body of such real-world networks have an associated uncertainty and optimization problems are required to be solved taking into account the uncertainty. Some of the uncertainty are due to the data collection process, machine-learning methods employed at preprocessing, privacy-preserving reasons and due to unknown causes during the operation of the network. Throughout this work, we study the case where the uncertainty is associated with the availability or the nature of relationship between the vertices of the network. The vertices themselves are assumed to be always available, in other words, the vertices are assumed to be certain. The concepts can be naturally modified to model uncertainty by associating uncertainty with the vertices. Road networks [1,2] are a natural source of optimization problems where the uncertainty is due to the traffic. Indeed, in uncertain traffic networks [2], the travel-time on a road is inherently uncertain. One way of modeling this uncertainty is by modeling the travel-time as a random variable. Indeed, the random variable is quite complex since the probability that it takes a certain value is dependent on other parameters like the day of the week and time of the day. However, our focus is only on the fact that uncertainty is modeled by an appropriately defined random variable. The natural optimization problem on an uncertain traffic network is to compute the expected minimum-time s-t path. In biological networks [3], the *protein-protein interaction* (PPI) network is an uncertain network. In a PPI network, proteins are represented by vertices and interaction between proteins are represented by edges. The interaction between proteins are derived through noisy and error-prone experiments which cause uncertainty. Sometimes the interactions are predicted by the nature of proteins instead of experiments. In this example, the uncertainty can be of two types: it can be on the presence or the absence of a protein-protein interaction or on the strength of an interaction between two proteins. Similarly, social networks are another example of uncertain networks where the members of the network are known, and the uncertainty is on the link between two members in the network. The interaction between members of an uncertain network associated with a social network are obtained using link prediction and by evaluating peer influence [4]. In all these three examples of networks with uncertainty, the uncertainty on the edges can be modeled as random variables. A random variable can be used to indicate the presence or the absence of an edge. In this case, the random variable takes values from the set $\{0,1\}$ with each value having an associated probability. A random variable can also be used to model the strength of the interaction between two entities in the network, in which case for an edge, the corresponding random variable takes values from a set of values and each value has an associated probability.

In this work, we survey the different models of uncertainty and the associated optimization problems. We then present our results on optimization problems on uncertain networks when the networks have bounded treewidth. The fundamental motivation for this direction of study is that a typical optimization problem on an uncertain graph is an expectation computation over many graphs implicitly represented by the uncertain graph. A natural question is there are problems which have efficient algorithms on a graph in which the edges have no uncertainty but become hard on uncertain graphs. The typical optimization problems considered on uncertain graphs are shortest path [5], reliability [6], minimum spanning trees [7,8], maxflows [9,10], maximum coverage [11–13], influence maximization [14] and densest subgraph [15,16]. This article is structured partly as a survey and partly as an original research article. In Section 1.1, we formally present the concepts in uncertain graphs and then present subsequent details on uncertain graphs in Section 2. In Section 1.2, we survey the different algorithmic results on uncertain graphs and in Section 1.3 we outline our results.

1.1. Uncertain Graphs-Definition and Semantics

We consider the graphs with uncertain edges and certain vertices. An uncertain graph, denoted by $\mathcal{G}$, is a triple that consists of a vertex set V, an edge set E and a set of *outcomes* $\{\mathcal{A}_e \mid e \in E\}$. For each edge $e \in E$, the outcome $\mathcal{A}_e$ is a set of values and an associated probability distribution on $\mathcal{A}_e$. The outcome $\mathcal{A}_e$ is considered to be an interval or a finite set. For each $e \in E$, if the outcome $\mathcal{A}_e$ is an interval, the associated probability distribution is called a continuous distribution, and if it is a finite set, the associated probability distribution is called a discrete distribution. In either case, the natural distribution is the uniform distribution. For the case in which for each edge $e \in E$, $\mathcal{A}_e$ is a closed interval $[\mathcal{L}_e, \mathcal{U}_e]$, the probability that e gets specific value in $\mathcal{L}_e < \mathcal{U}_e$ is $\frac{1}{\mathcal{U}_e - \mathcal{L}_e}$. In case for each $e \in E$, $\mathcal{A}_e$ is a finite set under the uniform distribution, the probability that e gets a specific value from $\mathcal{A}_e$ is $\frac{1}{|\mathcal{A}_e|}$. Such uncertain graphs were the focus of the earliest results [5,7,9,10] and we survey these in Section 1.2. However, in general, the uncertainty could be modeled by any probability distribution $\mathcal{A}_e$. Therefore, the uncertain graph is a succinct representation of the set of all edge weighted graphs such that for each edge e its weight is a value from $\mathcal{A}_e$. This set is an uncountable set when $\mathcal{A}_e$ is an interval and it is $\prod_{e \in E} |\mathcal{A}_e|$ in case each $e \in E$, $\mathcal{A}_e$ is a finite set. In this paper, we present our results for the case when for each edge e, the outcome $\mathcal{A}_e$ is the set of values $\{0,1\}$. Naturally, 0 models the absence of an edge and 1 models the presence of an edge. An uncertain graph under this condition is a succinct representation of the set of all edge subgraphs of $G = (V, E)$. In this case, the uncertain graph $\mathcal{G}$ is represented as a triple (V, E, p) where $p : E \to [0,1]$ is a function defined on E and $p(e)$ is said to be the survival probability of the edge e. The failure probability of an edge e is $1 - p(e)$. The set of all graphs represented by an uncertain graph is well-known as the *possible world semantics* (PWS) [17,18] of the uncertain graph. For each $E' \subseteq E$, $H = (V, E')$ is called a *possible world* of $\mathcal{G}$ and this is denoted by the notation $H \sqsubseteq \mathcal{G}$. For an uncertain graph $\mathcal{G} = (V, E, p)$, there are $2^{|E|}$ possible worlds. An uncertain graph and two of its possible worlds are illustrated in Figure 1.

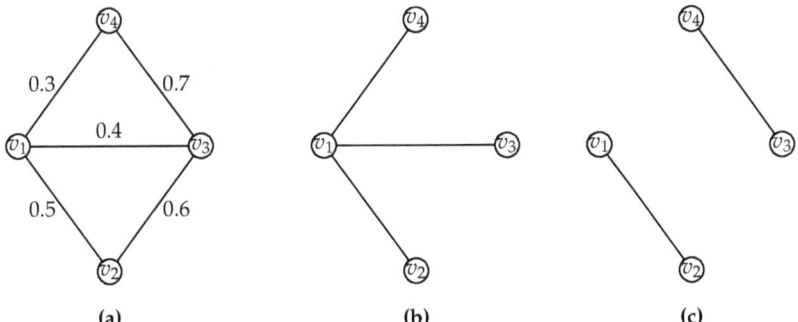

Figure 1. (a) A probabilistic graph $\mathcal{G} = (V, E, p)$; (b) A possible world H_1 of $\mathcal{G}$ with $P(H_1) = 0.0072$; (c) Another possible world H_2 of $\mathcal{G}$ with $P(H_2) = 0.0588$.

The probability associated with a possible world depends on the probability distributions on $\mathcal{A}_e$ and, most importantly, the dependence among the edge samples. A *distribution model* is a specification of the dependence among edge samples. Further, a distribution model uniquely determines a probability distribution on the possible worlds. For example, if the edge samples are all independent, the distribution model is called the Random Failure (RF) model. Under the RF model, the probability of a possible world H is given by $P(H) = \prod_{e \in E(H)} p(e) \prod_{e \in E \setminus E(H)} (1 - p(e))$.

Based on the dependence among the edge samples, the literature is rich in different distributions on the possible worlds. The distributions of interest in this paper are the Random Failure (RF) model, Independent Cascade (IC) model, Set-based Dependency (SBD) model and the Linear Reliability Ordering (LRO) model. These distribution models are all well-motivated by practical considerations on uncertain graphs in influence maximization, facility location, network reliability, to name a few. The detailed description of these distribution models and the corresponding distributions on the possible worlds are discussed in Section 2. In short, an uncertain graph along with a distribution model is a succinct description of an unique probability distribution on the PWS of the uncertain graph. Therefore, an uncertain graph along with a distribution model is equivalent to a sampling procedure to obtain a random sample from the corresponding probability distribution on the PWS of the uncertain graph. In Section 2, we describe the different distribution models by describing the corresponding sampling procedures of the edges in E.

The typical computational problem is posed for a fixed nature of the dependence among the edge samples from an input uncertain graph. The distributions on the edges and the dependence among the edge samples uniquely define the probability distribution on the possible worlds. Therefore, for a fixed dependence among the edge samples, the generic computational problem is to compute a solution that optimizes the expected value of a function over the distribution on the possible worlds. The problems that have been extensively studied are facility location to maximize coverage [19] and a selection of influentials on a social network to maximize influence [14]. Clearly, both these problems are coverage problems, studied two decades apart motivated by different considerations. In this paper, we are also motivated by understanding the parameterized complexity of problems on uncertain graphs. Historically, the earliest results considered different graph problems in which the distributions are on the set of values of the edge weights. We next present a survey of those results.

1.2. Survey of Optimization Problems in Uncertain Graphs

Graphs have been used to represent the relationships between entities. A graph is denoted by $G = (V, E)$ where V is set of vertices and $E \subseteq V \times V$ is set of edges representing relationship between pairs of vertices. PPI networks in bioinformatics, road networks and social networks are graphs with an uncertainty among the edges and are considered as uncertain graphs. The earliest ideas in graphs with uncertain edge weights were introduced by Frank and Hakimi [9] in 1965. Frank and Hakimi studied the maximum flow in a directed graph $G = (V, E)$ with uncertain capacities. On an input, consisting of an uncertain digraph $G = (V, E)$ and a continuous random variable with the uniform probability distribution on the capacity of each edge e, the probabilistic flow problem is to find the maximum flow probability and the joint distribution of the cut set values. Later in 1969, under the same setting Frank [5] studied the probabilistic shortest path problem on an undirected graph $G = (V, E)$. The probabilistic shortest path problem is to compute for each ℓ, the probability that the shortest path is at most ℓ. In 1976, J. R. Evans [10] studied the probabilistic maximum flow in a directed acyclic graph (DAG) with a discrete probability distribution on the edge capacities. A relatively recent result in 2008 is on the minimum spanning tree problem on uncertain graphs, by Erlebach et al. [7], with uncertain edge weights from a continuous distribution. Here the goal is to optimize the number of edges whose weight is sampled to obtain a spanning tree which achieves the expected MST weight over the possible worlds. This problem is different from the optimization problems of interest to us in this paper- given an uncertain graph as input, our goal is to optimize the expected value of a structural parameter over the possible worlds of the uncertain graph.

Optimization problems on uncertain graphs related to connectedness are among the most fundamental problems. Apart from their practical significance, they pose significant algorithmic challenges in different computational models. Further, the computational complexity of the problems increases significantly in the presence of uncertainty. In 1979, Valiant [6] studied the network reliability problem on uncertain graphs. The network reliability problem is a well-studied #P-hard problem [6,17,20–23]. The reliability problems have numerous applications in communication networks [24,25], biological networks [3] and social networks [26,27]. For a given network, *reliability* is defined as the ability of the network to remain operational after the failure of some of its links. The input consists of an uncertain graph $\mathcal{G} = (V, E, p)$ and a subset $S \subseteq V$ and the aim is to compute the probability that S is connected. Clearly, the reliability problem on uncertain graphs generalizes the graph connectivity problem which is polynomial time solvable. The *Canadian Traveler Problem* (CTP), formulated by Papadimitriou and Yannakakis in 1991 [28], is an online problem on uncertain graphs. Given an uncertain graph $\mathcal{G} = (V, E, p)$, a source s and a destination t, a traveler must find a walk from s to t, where an edge e is known to have survived with probability $p(e)$ only when the walk reaches one of its end points, after which it does not fail, conditioned on it having survived. The objective is to minimize the expected path length over the distribution on the possible worlds and the walker's choices.

Coverage in Uncertain Graphs. Apart from the themes of network flows and connectedness, coverage problems are very practical when the uncertainty is on the survival of the edges. In this framework, the uncertainty is on whether an edge will survive a disaster and the goal is to place facilities in the network such that the expected coverage over the possible worlds is maximized. Each possible world can be thought of as the set of edges which survive a disaster. Formally, an uncertain graph $\mathcal{G} = (V, E, p)$ is a succinct description of the set of possible worlds that can arise due to a disaster in which an edge e is known to survive with probability $p(e)$. The nature of dependence among the edge samples is used to model the nature of edge failures in the event of a disaster. As mentioned earlier, a fixed nature of dependence among the edge samples uniquely defines a probability distribution on the possible worlds. From this point, in this paper, we consider uncertain graphs where the uncertainty is on the survival of an edge (recall, the other possibility is on the uncertain edge weight). Further, all the results we present are for a fixed dependence among edge samples and such a fixed dependence among the edge samples is called a *distribution model*. Given the motivation of disasters, the distribution model is specified based on the dependence among the edge failures (recall, the failure probability of an edge e is $1 - p(e)$). In this framework, for a fixed distribution model, the function to be optimized is the coverage function.

Definition 1 (Coverage within distance r). *The input consists of an uncertain graph $\mathcal{G} = (V, E, p)$ and an integer k. The goal is to compute a k-sized vertex set S which maximizes the expected total weight of the vertices which are at distance at most r from S. The expectation is over the possible worlds represented by the uncertain graph.*

Naturally, we refer to k as the budget, r as the radius of coverage (the number of hops in the network) and S is the set of vertices where facilities have to be located. For certain graphs and $r = 1$, the facility location with unreliable edges problem is the well-studied *budgeted dominating set* problem which is known to be NP-hard [29]. The other case of natural interest is for $r = \infty$. In this case, the coverage problem is to find a set S of k vertices so as to maximize the expected number of vertices connected to S. In the case of certain graphs, this is polynomial time solvable as the problem is to find k connected components whose total vertex weight is the maximum.

Coverage in Social Networks. Coverage problems also have a natural interpretation in social networks. The dynamics of a social network based on the *word-of-mouth* effect have been of significant interest in marketing and consumer research [30], where a social network is referred to as an interpersonal network. The work by Brown and Reingen [30] state the different hypotheses for estimating the amount of uncertainty in a relationship between two persons in an interpersonal network. With the advent of social networks on the internet, the works of Domingos and Richardson [4,31] formalized the questions relating to the effective marketing of a product based on interpersonal relationships in a social network. In 1969, the work of Bass [32] had modeled the adoption of products as a diffusion process as a global phenomenon, independent of the interpersonal relationships between people in a society. Kempe, Kleinberg and Tardos (KKT) [14] brought together the earlier works on adoption of products in an interpersonal network and posed the question of selecting the most influential nodes in a social network with an aim to influence the maximum number of people to adopt a certain product or opinion. They considered the uncertainty in the social network to be the influence exerted by one individual on another individual and this is naturally modeled as an uncertain graph $\mathcal{G} = (V, E, p)$. The propagation of influence is modeled as a diffusion process which is a function of time as in Bass [32]. The influence maximization problems in KKT are considered under distribution models, which are described as diffusion phenomena, called the Independent Cascade model and the Linear Threshold Model. Among these two models, the Independent Cascade model is defined as a sampling procedure whose outcome is an edge subgraph of a given uncertain graph. Thus, this model is more relevant for our study of uncertain graphs and the IC model is defined in Section 2. In the influence maximization problems considered in KKT [14], the aim is to select k influential people S such that the expected number of people connected to S is maximized. The expected number of vertices, over the distribution model, connected to a set S is called the influence of a set S or the expected coverage of S and is denoted by $\sigma(S)$. Thus, the influence maximization problem is to find the set $\arg\max_{S \subseteq V, |S|=k} \sigma(S)$.

In Section 2, we discuss the computational complexity of $\sigma(S)$ for different distribution models. Thus, the influence maximization problem in Reference [14] is essentially a facility location problem where the distribution model is specified by a diffusion phenomenon, $r = \infty$ and each input instance consists of an uncertain network and a budget k.

Coverage and Facility Location. The survey due to Snyder [33] in 2006 collects the vast body of results on the facility location problem in uncertain graphs into a single research article. The earliest result known to us, due to Daskin [19] who formulated the maximum expected coverage problem where vertices are uncertain, is different from our case where the uncertainty is on the edges and the vertices are known. To the best of our knowledge, it was Daskin's work [19] that considered the general setting of dependence among vertex failures. In this case, the probability distribution would have been defined uniquely on the possible worlds which would have been the set of induced subgraphs. Subsequently, many variants of the facility location problem for uncertain graphs have been studied for different distribution models where the uncertainty is on the survival of edges.

Eiselt et al. [34] considered the problem with single edge failure and for $r = \infty$. In this case, exactly one edge is assumed to have failed after a disaster and the objective is to place k facilities such that the expected weight of vertices not connected to any facility is minimized. In this case, each subgraph consisting of all the edges except one is a possible world. This problem is known to be polynomial time solvable for any $k \geq 1$. When the distribution model is the Random Failure model and $r = \infty$, the *most reliable source* (MRS) is well-studied [35–38]. The input is an uncertain graph and $k = 1$. The goal is to select one vertex v such that the expected number of vertices connected to v is maximized. Melachrinoudis and Helander [38] gave a polynomial time algorithm for the MRS problem on trees, followed by linear time algorithm on uncertain trees by Ding and Xue [37]. Colbourn and Xue [35] gave an $\mathcal{O}(n^2)$-time algorithm for the MRS problem on uncertain series-parallel graphs. Wei Ding [36] gave an $\mathcal{O}(n^2)$-time algorithm for the MRS problem on uncertain ring graphs.

Apart from the RF model, the Linear Reliability Ordering (LRO) distribution model has been well-studied recently. Under this distribution model, for each integer $r \geq 1$, the facility location problem is studied as MAX-EXP-COVER-r-LRO problem [11–13]. For the case when $r = \infty$, the problem is known as the MAX-EXP-COVER-LRO problem. Hassin et al. [11] presented an algorithm to solve the MAX-EXP-COVER-LRO problem via a reduction to the MAX-WEIGHTED-k-LEAF-INDUCED-SUBTREE problem. They then showed that the MAX-WEIGHTED-k-LEAF-INDUCED-SUBTREE problem can be solved in polynomial time on trees using a greedy algorithm. Consequently, they showed that the MAX-EXP-COVER-LRO problem can be solved in polynomial time. For $r = 1$, the MAX-EXP-COVER-1-LRO problem is NP-complete on planar graphs and the hardness follows from the hardness of budgeted dominating set due to Khuller et al. [29]. The MAX-EXP-COVER-1-LRO admits a $(1 - \frac{1}{e})$-approximation algorithm [12,13]. Similarly, Kempe et al. [14] shows that the influence maximization under the IC distribution model and the LT distribution model has a $(1 - \frac{1}{e})$ approximation algorithm. Both these results naturally follow due to the submodularity of the expected neighborhood size function, the monotonicity (as r increases) and submodularity of the expected coverage function and the result of Nemhauser et al. [39] on greedy maximization of submodular functions. In the setting of parameterized algorithms, the MAX-EXP-COVER-1-LRO problem is W[1]-complete for solution size as the parameter and this follows from the hardness of budgeted dominating set problem. An FPT algorithm for the MAX-EXP-COVER-1-LRO problem with treewidth as the parameter is presented in Reference [13]. Formally stated, given an instance $\langle \mathcal{G} = (V, E, p), k\rangle$ of the MAX-EXP-COVER-1-LRO problem, an optimal solution can be computed in time $4^{\mathcal{O}(t)} n^{\mathcal{O}(1)}$ where t is the treewidth of the graph $G = (V, E)$ which is presented in the input as a tree decomposition.

Finding Communities in Uncertain Graphs. Finding communities is a significant problem in social network and in bioinformatics. A natural graph theoretic model for a community is a dense subgraph. A well-known dense subgraph is the *core*. Given an integer d, a graph $G = (V, E)$ is said to be d-core if degree of every vertex $v \in V$ is at least d. The way of obtaining the unique maximal induced subgraph of a graph G which is a d-core is to repeatedly discard vertices of degree less than d. If the procedure terminates with a non-empty graph, then the graph is a d-core of the graph G. A vertex v is said to be in a d-core if there is a d-core which contains v. This idea is generalized to the uncertain graph framework as follows: Given an uncertain graph $\mathcal{G} = (V, E, p)$ and the distribution model is the RF model, the d-core probability of a vertex $v \in V$, denoted by $q_d(v)$, is the probability that v is in the d-core of a possible world. In other words, $q_d(v) = \sum_{H \sqsubseteq \mathcal{G}} P(H) I(H, d, v)$, where $I(H, d, v)$ is an indicator function that takes value one if and only if there is a d-core of H that contains v, and $P(H)$ is probability of the possible world H. In this setting, we consider the (d, θ)-core problem defined by Peng et al. [16]. We refer to this problem as the INDIVIDUAL CORE problem and it is defined as follows.

Definition 2 (INDIVIDUAL-CORE)**.** *The input consists of an uncertain graph $\mathcal{G} = (V, E, p)$, an integer d and a probability threshold $\theta \in [0, 1]$. The objective is to compute an $S \subseteq V$ such that for each $v \in S$, $q_d(v) \geq \theta$.*

Peng et al. [16] shows that the INDIVIDUAL-CORE problem is NP-complete. Prior to the results of Peng et al. [16], Bonchi et al. [15] introduced the study of d-core problem in uncertain graphs. We refer to the d-core problem on uncertain graphs as the PROBABILISTIC-CORE problem defined the follows.

Definition 3 (PROBABILISTIC-CORE). *Given an uncertain graph $\mathcal{G} = (V, E, p)$, an integer d and a probability threshold $\theta \in (0, 1]$, then the aim of the PROBABILISTIC-CORE problem is to find a set $K \subseteq V$ such that the $\Pr(K$ is a d-core in $\mathcal{G})$ is at least θ.*

The problem of deciding on the existence of such a set K can be shown to be NP-hard using the hardness result given in Reference [16]. On the other hand, if $p(e) = 1$ for all $e \in E$, then the d-problem is polynomial time solvable as described at the beginning of this discussion.

A chronological listing of different optimization problems on uncertain graphs is presented in Table 1. The tabulation shows that there are many distribution models under which different optimization problems could be considered. The goal would be to understand the complexity of computing that expectation when the input is presented as an uncertain graph. Indeed, any NP-Complete problem on certain graphs remains NP-Complete for any distribution model when the inputs are uncertain graphs. Therefore, our natural focus is on Exact and Parameterized Computation of the expectation on uncertain graphs. The goal of the area of exact and parameterized computation is to classify problems based on their computational complexity as a function of input parameters other than the input size. The desired solution size, the treewidth of an input graph, and the input size are natural well-studied parameters. An algorithm with running time $f(k)n^{\mathcal{O}(1)}$ is said to be a Fixed Parameter Tractable algorithm with respect to the parameter k. Interestingly, there are many problems that do not have FPT algorithms with respect to some parameters, while they have FPT algorithms with respect to others. A rich complexity theory has evolved over nearly four decades of research with the W-hierarchy being the central classification. In this hierarchy problem classes are ordered in increasing order of computational complexity. In this hierarchy the problems which have FPT algorithms are considered the *simplest* in terms of computational complexity. The complete history of this line of research can be found in the most recent textbook [40].

Table 1. A chronology of studies on uncertain graphs

Work	Optimization Problem	Uncertainty Model
Frank and Hakimi, 1965 [9]	Probabilistic maximum flow	Capacities on the edges are drawn from an independent continuous distribution.
Frank, 1969 [5]	Probabilistic shortest path	Length of the edges are drawn from a continuous distribution.
Evans, 1976 [10]	Probabilistic maximum flow	Capacities on the edges are obtained from an arbitrary but unknown discrete probability distribution.
Valiant, 1979 [6]	Network reliability	The probability $p(e)$ is the same for each edge and failure of every edge is independent.
Sigal, Pritsker and Solberg, 1980 [41]	Stochastic shortest path	Edge weights are drawn from a known cumulative distribution function.
Daskin, 1983 [19]	Expected coverage	Failure probability is the same for each vertex
Papadimitriou Yannakakis, 1991 [28]	Canadian Traveler Problem	Each edge has a survival probability, edge failure is independent and algorithm knows of the failure during execution.
Guerin and Orda, 1999 [42]	Most reliable path and flows with bandwidth selection	Each edge e has a survival probability $p_e(x)$ for the availability of bandwidth x.
Hassin, Salman and Ravi, 2009 (2017) [11,12]	Expected coverage	Each edge e has a survival probability $p(e)$ and edge failure follows LRO model.
Bonchi, Gullo, Kaltenbrunner and Volkovich, 2014 [15]	PROBABILISTIC-CORE	Each edge e has a survival probability $p(e)$ and edge failure follows RF model.

1.3. Our Questions and Results

Given our focus on exact and parameterized algorithms, we think that it should be possible to classify problems based on the hardness of efficiently computing the expectation over different distribution models. One of the contributions of this paper is to collect many of the known algorithmic results on uncertain graphs and the different distribution models for which results have been obtained. Our focus is on uncertain graph optimization problems on the LRO and RF distribution models, and these models have been the focus of many previous results in the literature. Our results add to the knowledge about these two models and are among the first parameterized algorithms on uncertain graphs. They also give an increased understanding of how treewidth of uncertain graphs can be used along with the structure of the distribution models to compute the expectation in time parameterized by the treewidth. Finally, the motivation for the choice of these two models is that the number of possible worlds with non-zero probability under the LRO model is equal to the number of edges m, while the number of possible worlds under the RF model is an exponential in m, which is the maximum number of possible worlds. We define the different distribution models listed in Section 1.1 in detail and some of their properties are identified from relationships between the models in Section 2.

Our first case study on parameterized algorithms under the RF model is on maximizing expected coverage in uncertain graphs. The starting point of our work is the reduction of the MAX-EXP-COVER-LRO problem to the MAX-WEIGHTED-k-LEAF-INDUCED-SUBTREE problem, which can be solved in polynomial time [11,12]. The reduction was from maximizing the expectation coverage to maximizing the total weight of a combinatorial parameter. However, the reduction does not work for the MAX-EXP-COVER-1-LRO, as the problem is at least as hard as budgeted dominating set. In the case when the graph has bounded treewidth, we were able to show in a previous work [13] that MAX-EXP-COVER-1-LRO has an FPT algorithm parameterized by treewidth. The dynamic programming algorithm depends on properties specific to the LRO distribution model. On the other hand, the status of the question is unclear when the distribution model is the RF model. We address this by presenting a DP algorithm for the MAX-EXP-COVER-1-RF problem in Section 4, which is an FPT algorithm parameterized by the product of the treewidth and max-degree of the input graph.

The second case study on parameterized algorithms under the RF model is on finding a d-core in uncertain graphs. In Section 5, we consider the PROBABILISTIC-CORE problem, and design a polynomial time algorithm for the LRO distribution model. For the RF distribution model, we observe that the PROBABILISTIC-CORE problem is W[1]-hard for the parameter d, where d is the minimum degree of the core. Then we design a DP algorithm, which is an FPT algorithm for PROBABILISTIC-CORE with respect to treewidth as the parameter.

Essentially in both the case studies, given an uncertain graph under the RF model, the DP uses the tree decomposition to efficiently compute the expected value, which is expressed as a weighted summation over the exponentially many possible worlds.

2. Distribution Models for Uncertain Graphs

As mentioned in Section 1.1, a distribution model along with an uncertain graph $\mathcal{G} = (V, E, p)$ uniquely describes the probability distribution on the possible worlds. We describe the distribution model by describing a corresponding sampling procedure on the edges of $\mathcal{G} = (V, E, p)$. This formalism standardizes the nomenclature of optimization problems on uncertain graphs. For example, for the coverage problems on uncertain graphs, the problem names MAX-EXP-COVER-1-LRO and MAX-EXP-COVER-1-RF clearly state the distribution model and the radius of coverage relevant for the problem. An instance of each of these problems is an uncertain graph $\mathcal{G} = (V, E, p)$ and an integer $k \geq 1$. In this section, we present an edge sampling procedure from a given uncertain graph $\mathcal{G} = (V, E, p)$ corresponding to a distribution model on $\mathcal{G}$. The outcome of a sampling procedure is a possible world, which is an edge subgraph H of the graph $G = (V, E)$. An edge $e \in E(H)$ is called a *survived edge* and an edge $e \in E \setminus E(H)$ is called a *failed edge*. We also present some new observations regarding the different distribution models.

2.1. Random Failure Model

Random Failure (RF) model is the most natural distribution model. In this sampling procedure, an edge $e \in E$ is selected with probability $p(e)$ independent of the outcome of every other edge in E. Thus, each edge subgraph of G is a possible world and for an edge subgraph H, the probability that H is the outcome of the sampling procedure is denoted by $P(H)$ which is given by the equation $P(H) = \prod_{e \in E_H} p(e) \prod_{e \in E \setminus E_H} (1 - p(e))$.

2.2. Independent Cascade Model

The sampling procedure to define the Independent Cascade (IC) model is very naturally described by a diffusion process which proceeds in rounds. The process is as defined in Reference [14]. The diffusion process starts after round 0. In round 0, A_0 is a non-empty set of vertices and these are called *active* vertices. The process maintains a set of edges A_t after round $t \geq 0$. The vertices in A_t are said to be active. The sampling process is as follows. If a vertex v is in A_{t-1}, then it remains active in round t also, that is $v \in A_t$. The diffusion process in round $t+1$ is as follows: If v is in $A_t \setminus A_{t-1}$, that is v first becomes active after round $t-1$, then consider each edge $e = uv$ such that u is not in A_t. Each such edge e is sampled with probability $p(e)$ independent of the other outcomes. If $e = uv$ is selected (that is, it survives), then u is added to A_{t+1}. We reiterate that the edges incident on v will not be sampled in subsequent rounds. The sampling procedure terminates in not more than $|E|$ rounds. The outcome of this sampling procedure is the edge subgraph formed by a set of those edges which were selected when the first end point of the edge becomes active. Kempe et al. [14] showed that for any edge subgraph H of $G = (V, E)$, the probability that H is the outcome is given by $P(H) = \prod_{e \in E(H)} p(e) \prod_{e \in E \setminus E(H)} (1 - p(e))$.

Observation 1. *For an uncertain graph $\mathcal{G} = (V, E, p)$ the RF Model and the IC Model are identical.*

The next distribution model is a generalization of the RF model and was introduced by Gunnec and Salman [43].

2.3. Set-Based Dependency (SBD) Model

The uncertain graph $\mathcal{G} = (V, E, p)$ satisfies the additional properties that E is partitioned into $\{E_1, E_2, \ldots E_t\}$ of E, for some $t \geq 1$. Further, p satisfies the property that for any two edges e_1 and e_2 that belong to the same part in the partition, $p(e_1) \neq p(e_2)$. Typically, this partition is a fixed partition of E coming from the domain where the edge failures occur according to the SBD model. The sampling procedure definition of the SBD model is as follows: the edge sets are considered in order from E_1 to E_t, and for $1 \leq i < j \leq t$, the edges in the set E_i are considered before the edges in the set E_j. For each $1 \leq i \leq t$, the edges in E_i are considered in decreasing order of the value p. For each $1 \leq i \leq t$, when an edge $e \in E_i$ is considered, it is sampled with probability $p(e)$ independent of the outcome of the previous samples. If the outcome selects the edge e (that is, if e survives), then the next edge in E_i is considered. Otherwise, all the remaining edges of E_i that are to be considered after e, are considered to have failed, and the set E_{i+1} is considered. The procedure terminates after considering $E_1, \ldots, E_t$. The edge subgraph consisting of the selected edges (survived edges) is the outcome of this sampling procedure. The number of possible worlds that have a non-zero probability of being an outcome of the sampling procedure is $\prod_{i=1}^{t}(|E_i| + 1)$. To summarize, under the SBD model, for each $1 \leq i \leq t$, the survival of an edge e in a set E_i implies that every edge $e' \in E_i$ with a greater survival probability than that of e would have survived. Further, the edge samples of edges in two different sets are mutually independent. As a consequence of this, we have the following observation.

Observation 2. *The RF model on an uncertain graph $\mathcal{G} = (V, E, p)$ is identical to the SBD model on $\mathcal{G}$ and E is partitioned into m sets each containing an edge of E.*

2.4. Linear Reliable Ordering (LRO) Model

The case when the partition of E consists of only one set, the SBD model is called the Linear Reliability Ordering (LRO) model introduced by Hassin et al. [11]. Let $p(e_m) < p(e_{m-1}) < \ldots < p(e_2) < p(e_1)$. Then, under the LRO model it follows that for each $j > i$, $\Pr(e_j \text{ fails} \mid e_i \text{ fails}) = 1$. Further, the possible worlds is the set of $m+1$ graphs $\{G_0, \ldots, G_m\}$ where $E(G_0) = \emptyset$, $E(G_m) = E$, and for each $0 \leq i \leq m-1$, $E(G_{i+1}) = E(G_i) \cup \{e_{i+1}\}$. The following lemma shown by Hassin et al. [11] regarding the probability distribution on the possible worlds uniquely defined by the LRO model.

Lemma 1 (Hassin et al.[11]). *For $0 \leq i \leq m$, the probability of the possible world G_i is given by*

$$P(G_i) = \begin{cases} 1 - p(e_1) & \text{if } i = 0 \\ p(e_i) - p(e_{i+1}) & \text{if } 1 \leq i < m \\ p(e_m) & \text{if } i = m. \end{cases}$$

Coming back to the SBD model, on the uncertain graph $\mathcal{G} = (V, E, p)$ and E partitioned into $\{E_1, E_2, \ldots E_\ell\}$ for some $\ell \geq 1$, the probability of an outcome H under the SBD model naturally follows from Lemma 1. The idea is to consider an outcome of the SBD model as ℓ independent samples from the LRO model on the uncertain graphs $(V, E_1, p_1), (V, E_2, p_2), \ldots, (V, E_\ell, p_\ell)$, where for each $1 \leq i \leq \ell$, p_i is the function p restricted to E_i.

This completes our description of the distribution models known in the literature. We next present two dynamic programming algorithms on the input uncertain graph $\mathcal{G} = (V, E, p)$, when the graph $G = (V, E)$ is presented as a nice tree decomposition. The algorithms compute an expectation under the RF model and have worse running times than that of corresponding algorithms for computing the expectation under the LRO model.

3. Definitions Related to Graphs

Every graph we consider in this work are simple undirected graphs unless. A graph $G = (V, E)$ is an undirected graph with vertex set V and edge set E. We denote the number of vertices and edges by n and m, respectively. For a vertex $v \in V$, $N(v)$ denotes the set of neighbors of v and $N[v] = N(v) \cup \{v\}$ is the closed neighborhood of v. For each vertex $v \in V$, $deg_G(v)$ denote the degree of the vertex v in G. When G is clear in the context $deg(v)$ is used. The maximum degree of the graph G, denoted by $\Delta(G)$, and the minimum degree of the graph G, denoted by $\delta(G)$, are the maximum and minimum degree of its vertices. When G is clear in the context Δ and δ is used. Other than this, we follow the standard graph theoretic terminologies from Reference [44]. We define some special notations for the uncertain graphs as follows. Given a vertex $v \in V$, let $E(v) \subseteq E$ denote the set of all edges incident on v. Given a subset of vertices $C \subseteq V$, let $E(C) = \cup_{v \in C} E(v)$ denote the edge set of the vertex-induced uncertain subgraph $\mathcal{G}[C]$. Similarly, given an edge set $F \subseteq E$, let $V(F) = \cup_{e=uv \in F}\{u,v\}$ denote the vertex set of the edge-induced uncertain graph $\mathcal{G}[F]$.

We study the parameterized complexity of the coverage problems and k-core problem on uncertain graphs. We follow the standard parameterized complexity terminologies from Reference [40]. We define the parameter treewidth that will be relevant to our discussion.

Definition 4 (Tree Decomposition [45,46]). *A tree decomposition of a graph G is a pair (X, H) such that H is a tree and $X = \{X_i \subseteq V : i \in H\}$. For each node $i \in H$, X_i is referred to as bag of i. The following three conditions hold for a tree decomposition (X, H) of the graph G.*

(a) For each vertex $v \in V$, there is a node $i \in H$ such that $v \in X_i$.
(b) For each edge $uv \in E$, there is a node $i \in H$ such that $u, v \in X_i$.
(c) For each vertex $v \in V$, the induced subtree of the nodes in H that contains v is connected.

The *width* of a tree decomposition is the $\max_{i \in H}(|X_i| - 1)$. The *treewidth* of a graph, denoted by tw, is the minimum width over all possible tree decompositions of G. An example of a tree decomposition is illustrated in Figure 2. For our algorithm, we require a special kind of decomposition, called the *nice tree decomposition* which we define below.

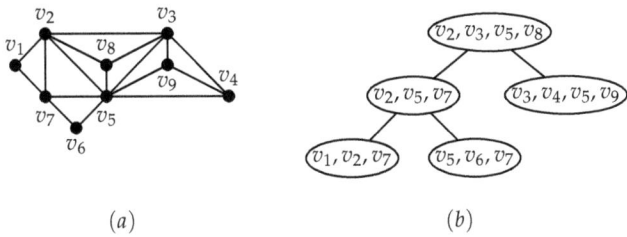

Figure 2. (a) A graph with 9 vertices; (b) An optimal tree decomposition with treewidth 3.

Definition 5 (Nice tree decomposition [46]). *A nice tree decomposition is a tree decomposition, rooted by a node r with $X_r = \emptyset$ and each node in the tree decomposition is one of the following four type of nodes.*

1. **Leaf node.** A node $i \in H$ with no child and $X_i = \emptyset$.
2. **Introduce node.** A node $i \in H$ with one child j such that $X_i = X_j \cup \{v\}$ for some $v \notin X_j$.
3. **Forget node.** A node $i \in H$ with one child j such that $X_i = X_j \setminus \{v\}$ for some $v \in X_j$.
4. **Join node.** A node $i \in H$ with two children j and g such that $X_i = X_j = X_g$.

An example of four types of nodes is illustrated in Figure 3. Given a tree decomposition (X, H) of a graph G with width k, a nice tree decomposition (X', H') with same width and $\mathcal{O}(nk)$ nodes can be computed in time $\mathcal{O}(nk^2)$ [46]. Hereafter, we will assume that the tree decompositions considered are nice. For each node $i \in H$, let H_i denote the subtree rooted at i. Let $n_i = |X_i|$. Let X_i^+ be the set of all vertices in the bag of nodes in the subtree H_i.

$$X_i^+ = \begin{cases} X_i & \text{if } i \text{ is a leaf node} \\ X_i \cup \bigcup_{j \in ch(i)} X_j^+ & \text{if } i \text{ is a non-leaf node,} \end{cases}$$

where $ch(i)$ is the set of all children of i in H. We use j and k to denote the two children of a join node i in H. Further, j denotes the left child, and g denotes the right child. In case i has only one child, as in the case of introduce and forget nodes, only the left child j is well-defined and g does not exist. In this case, X_g^+ is taken to be the empty set. We refer to Reference [40] for a thorough introduction to treewidth and its algorithmic properties.

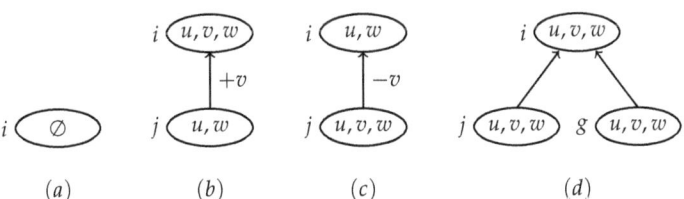

Figure 3. An example of leaf (**a**), introduce (**b**), forget (**c**) and join (**d**) nodes. Directed edges denote child to parent link.

4. Max-Exp-Cover-1-RF Problem is FPT by (Treewidth · Δ)

An instance of MAX-EXP-COVER-1-RF consists of a tuple $\langle \mathcal{G}, w, k \rangle$ where $\mathcal{G} = (V, E, p)$ is an uncertain graph, w is a function that assigns weight $w(v)$ to each vertex v, and k is the budget. The goal is to find a set $S \subseteq V$ such that $|S| \leq k$ and the expected total weight of the vertices dominated by S is maximized. The expectation is computed over the probability distribution uniquely defined on the possible worlds by the RF model. We introduce the function $\mathcal{C}$ which we refer to as the *coverage* function. The two arguments of the coverage function are subsets T and S of the vertex set V, and the value $\mathcal{C}(T,S)$ is the expected coverage of T by S, where the expectation is computed over the possible worlds. For any subsets $S, T \subseteq V$, the *Coverage* of T by S, denoted by $\mathcal{C}(T,S)$, is $\sum_{v \in T} \mathcal{C}(v,S)$. For a vertex $v \in V$ and $S \subseteq V$, the expected coverage of v by the set S, denoted by $\mathcal{C}(v,S)$, is given by

$$\mathcal{C}(v,S) = \begin{cases} w(v) & \text{if } v \in S \\ w(v)\left(1 - \prod_{u \in S \cap N(v)} (1 - p_{uv})\right) & \text{otherwise.} \end{cases} \quad (1)$$

Note that if a set is a singleton set, then we abuse notation a little and write the element instead of the set.

Our algorithm for the MAX-EXP-COVER-1-RF follows the classical bottom-up approach for dynamic programming based on the nice tree decomposition. We compute a certain number of candidate solutions for the subproblem at each node $i \in H$, using only the candidate solution values maintained in the children of i. At each node $i \in H$, we have $(5 \cdot 2^\Delta)^{n_i}$ candidate solutions which are stored in a table denoted by T_i. The final optimal solution is obtained from the solutions stored in T_r at the root node r of H. At any node i in H, the expected coverage of a vertex $u \in X_i$ by a set S in the subproblem at node i in H is computed by decomposing $N(u) \cap S$ carefully to avoid over-counting. Towards this end, we introduce notation for the coverage conditioned on the event that some edges have failed. We denote this event by a function f and call this the *surviving neighbors* (SN) function. The SN-function $f : V \to 2^V$ has the property that for each $u \in V$, $f(u) \subseteq N(u)$. In other words, we are interested in the expected coverage conditioned on the event that for each $u \in V$, all the edges to vertices in $N(u) \setminus f(u)$ have failed.

Definition 6. *Let u be a vertex and $S \subseteq V$ be a set and f be an SN-function such that $S \cap f(u) = S \cap N(u)$. The conditional coverage of u by S restricted by f is defined to be*

$$\mathcal{C}_f(u,S) = \left(\prod_{v \in N(u) \setminus f(u)} (1 - p(uv))\right) \mathcal{C}(u, S \cap f(u))$$

Extending this definition to sets, for any two vertex sets T, S, we define $\mathcal{C}_f(T,S) = \sum_{u \in T} \mathcal{C}_f(u,S)$.

For each $u \in V$, $\mathcal{C}_f(u,S)$ is the expected coverage of u by $S \cap f(u)$ conditioned on the event that u is not covered by any vertex from the set $N(u) \setminus f(u)$. Further, $\mathcal{C}_f(u,S)$ is the product of $w(u)$ and the probability of sampling a subgraph $G' \subseteq G$ in which u has a neighbor in $S \cap N(u)$ and no neighbor from $N(u) \setminus f(u)$. We refer to $\mathcal{C}_f(u,S)$ as the conditional coverage of u by S (we leave out the phrase restricted by f).

Lemma 2. *Let $u \in V$ and $X, Y \subseteq V$ such that $X \cap Y = \emptyset$, and consider an SN-function f such that $f(u) = N(u) \setminus X$. Then $\mathcal{C}(u, X \cup Y) = \mathcal{C}(u,X) + \mathcal{C}_f(u,Y)$.*

Proof.

$$\begin{aligned}
\mathcal{C}(u, X \cup Y) &= w(u)\Big(1 - \prod_{v \in (X \cup Y) \cap N(u)} (1 - p(uv))\Big) \\
&= w(u)\Big(1 - \prod_{v \in X \cap N(u)} (1 - p(uv)) \cdot \prod_{v \in Y \cap N(u)} (1 - p(uv))\Big) \\
&= w(u)\Big(\Big(1 - \prod_{v \in X \cap N(u)} (1 - p(uv))\Big) + \prod_{v \in X \cap N(u)} (1 - p(uv))\Big(1 - \prod_{v \in Y \cap N(u)} (1 - p(uv))\Big)\Big) \\
&= \mathcal{C}(u, X) + \prod_{v \in X \cap N(u)} (1 - p(uv)) \mathcal{C}(u, Y) \\
&= \mathcal{C}(u, X) + \mathcal{C}_f(u, Y)
\end{aligned}$$

Hence the proof. □

4.1. Recursive Formulation of the Value of a Solution

Let $S \subseteq V$ be a set of size k for the MAX-EXP-COVER-1-RF problem in $\mathcal{G}$. We now present a recursive formulation to compute the expected coverage of V by S. Throughout this section S denotes this set of size k. Let i be a node in H. Let j and g be the children of i. In case i is either introduce node or forget node, then g is considered to be a null node with $X_g = X_g^+ = \emptyset$. In node i, let $A = S \cap X_i$, $S_i = S \cap X_i^+$ and $Z_i = S_i \setminus A$. Let $\hat{Z} = S \setminus S_i$. For the solution S, the set X_i can be partitioned into five sets (A, C, L, R, B) as follows.

$$\begin{aligned}
A &= \{u \in X_i \cap S\} \\
C &= \{u \in X_i \setminus A \mid u \in N(A) \setminus N(Z_i)\}, \text{ in other words, } C \text{ satisfies } \mathcal{C}(C, Z_i) = 0 \\
L &= \{u \in X_i \setminus A \mid u \in N(Z_j) \setminus N(Z_g)\}, \text{ in other words, } L \text{ satisfies } \mathcal{C}(L, Z_g) = 0 \\
R &= \{u \in X_i \setminus A \mid u \in N(Z_g \setminus N(Z_j)\}, \text{ in other words, } R \text{ satisfies } \mathcal{C}(R, Z_j) = 0 \\
B &= \{u \in X_i \setminus A \mid u \in N(Z_j) \cap N(Z_g)\}.
\end{aligned}$$

Further, for the solution S, define the SN-function $f : X_i^+ \to 2^V$ as follows:

$$\begin{aligned}
f(u) &= N(u) \setminus \{v \in N(u) \mid v \in \hat{Z}\}, \text{ if } u \in X_i \\
&= N(u), \text{ if } u \in X_i^+ \setminus X_i
\end{aligned}$$

Note: f is dependent on S and i and wherever f is used, it must be used as per the definition at the corresponding node in the tree decomposition.

Let $S \subseteq V$ be a subset of size k, i be a node in H in the nice tree decomposition, f be the SN-function for X_i^+ defined using S, $P = (A, C, L, R, B)$ be a partition of X_i, and S_i, Z_i and $\hat{Z}$ be as defined above. The following two lemmas are useful in setting up a recursive definition of $\mathcal{C}(V, S)$ and a bottom-up evaluation of the recurrence.

Lemma 3. $\mathcal{C}(V, S) = \mathcal{C}(V \setminus X_i^+, \hat{Z} \cup A) + \mathcal{C}(X_i \setminus A, \hat{Z}) + \mathcal{C}_f(X_i^+, S_i)$.

Proof. The expected coverage of V by the set S is given by $\mathcal{C}(V, S) = \mathcal{C}(V \setminus X_i^+, S) + \mathcal{C}(X_i, S) + \mathcal{C}(X_i^+ \setminus X_i, S)$. Since V is partitioned into three disjoint sets X_i, $X_i^+ \setminus X_i$, and $V \setminus X_i^+$, we get

$$\mathcal{C}(V, S) = \mathcal{C}(V \setminus X_i^+, \hat{Z} \cup A) + \mathcal{C}(X_i \setminus A, \hat{Z} \cup S_i) + \mathcal{C}(A, S) + \mathcal{C}(X_i^+ \setminus X_i, S_i).$$

By applying Lemma 2 to $\mathcal{C}(X_i \setminus A, \hat{Z} \cup S_i)$, it follows that

$$\mathcal{C}(V, S) = \mathcal{C}(V \setminus X_i^+, \hat{Z} \cup A) + \mathcal{C}(X_i \setminus A, \hat{Z}) + \mathcal{C}_f(X_i \setminus A, S_i) + \mathcal{C}(A, S) + \mathcal{C}(X_i^+ \setminus X_i, S_i).$$

Further, $C(A, S_i) = C(A, S)$, since $C(A, S_i) = C(A, S) = \sum_{v \in A} w(v)$, it follows that

$$C(V, S) = C(V \setminus X_i^+, \hat{Z} \cup A) + C(X_i \setminus A, \hat{Z}) + C_f(X_i \setminus A, S_i) + C(A, S_i) + C(X_i^+ \setminus X_i, S_i).$$

By the definition of f and C_f, $C_f(X_i \setminus A, S_i) + C(A, S_i) + C(X_i^+ \setminus X_i, S_i) = C_f(X_i^+, S_i)$. Therefore we get $C(V, S) = C(V \setminus X_i^+, \hat{Z} \cup A) + C(X_i, \hat{Z}) + C_f(X_i^+, S_i)$. Hence the lemma. □

As a corollary it follows that $C(V, S) = C_f(X_r^+, S_r)$. Recall that $X_r = \emptyset$, $S_r = S$ and f is such that $f(u) = N(u)$ for all $u \in X_i^+$. We now show that for each node $i \in H$, $C_f(X_i^+, S_i)$ can be written in terms of the appropriate sub-problems in children j and g of the node i.

Lemma 4.

$$C_f(X_i^+, S_i) = w(A) + C_f(C, A) + C_f(L, A \cup Z_j) + C_f(R, A \cup Z_g) + C_f(B, S_i) + C(X_i^+ \setminus X_i, S_i).$$

Proof. By definition, the expected coverage of X_i^+ by S_i restricted by f is given by the following equations.

$$\begin{aligned} C_f(X_i^+, S_i) &= C_f(X_i, S_i) + C_f(X_i^+ \setminus X_i, S_i) \\ &= w(A) + C_f(C, S_i) + C_f(L, S_i) + C_f(R, S_i) + C_f(B, S_i) + C(X_i^+ \setminus X_i, S_i) \\ &= w(A) + C_f(C, A) + C_f(L, A \cup Z_j) + C_f(R, A \cup Z_g) + C_f(B, S_i) \\ &\quad + C(X_i^+ \setminus X_i, S_i) \end{aligned}$$

The second equation follows from the first due to the definition of the partition $P = (A, C, L, R, B)$, and the fact that for all $u \in X_i^+ \setminus X_i$, $f(u) = N(u)$. The third equation follows from the second due to the fact that $C_f(L, S_i) = C_f(L, A \cup Z_j)$ and $C_f(R, S_i) = C_f(R, A \cup Z_g)$, since L and R are sets for which $C(L, Z_g) = C(R, Z_j) = 0$. □

Lemmas 3 and 4 show that the expected coverage of a set S of size k can be computed in a bottom-up manner over the nice tree decomposition. Further, at a node $i \in H$ the expected coverage of X_i^+ by a S is $C(X_i, \hat{Z}) + C_f(X_i^+, S_i)$, where f is uniquely determined by S, i and the tree decomposition. Also, $C_f(X_i^+, S_i)$ is decomposed into six terms based on the partition P which is uniquely determined by S, i and the tree decomposition. Of these six terms, five of them are computed at the node $i \in H$, and one term comes from j and g, which are the children of i. Therefore, the search for the optimum S of size k can be performed in a bottom-up manner by enumerating all possible choices of the 5-way partition of X_i and all possible choices of the SN-function f at X_i. For each candidate partition of X_i and the SN-function f, the optimum S_i is computed by considering the *compatible* solutions at X_j and X_g. This completes the recursive formulation of the expected coverage of a set S of size k. We next present the bottom-up evaluation of the recurrence to compute the optimum set which is $\arg\max_{|S| \leq k} C(V, S)$.

4.2. Bottom-Up Computation of an Optimal Set

For each node $i \in H$, we associate a table T_i. Each row in the table T_i is a triple which consists of an integer b, a five way partitioning $P = (A, C, L, R, B)$ of X_i and an SN-function f defined on X_i^+. Throughout this section we assume that for vertices $u \in X_i^+ \setminus X_i$, $f(u) = N(u)$. The columns corresponding to a row (b, P, f) is a vertex set $T_i[b, P, f]$.Solution and a value $T_i[b, P, f]$.Value. Let S denote $T_i[b, P, f]$.Solution. To define the value $T_i[b, P, f]$.Value associated with S, consider $Z = S \setminus A$, and let $Z_j = Z \cap (X_j^+ \setminus X_j)$ and $Z_g = Z \cap (X_g^+ \setminus X_g)$. Then, $T_i[b, P, f]$.Value $= w(A) + C_f(C, A) + C_f(L, A \cup Z_j) + C_f(R, A \cup Z_g) + C_f(B, A \cup Z) + C(X_i^+ \setminus X_i, S)$. The set $T_i[b, P, f]$.Solution is a set S of size b such that $S \cap X_i = A$ and the associated value $T_i[b, P, f]$.Value is maximized.

Leaf node. Let $i \in H$ be a leaf node with bag $X_i = \emptyset$. The only possible five-way partition of an empty set is the set with five empty sets and the budget is $b = 0$. The only valid SN-function is $f : \emptyset \to \emptyset$. Therefore, the value of the corresponding row in the table is

$$T_i[b, P, f] = \{\text{Solution} = \emptyset, \text{Value} = 0\} \text{ for } b = 0, P = (\emptyset, \emptyset, \emptyset, \emptyset, \emptyset) \text{ and } f : \emptyset \to \emptyset.$$

It is clear that the empty set is the set that achieves the maximum for the MAX-EXP-COVER-1-RF problem on the empty graph with budget $b = 0$. Therefore, at the leaf nodes in H, the table T_i maintains the optimum solution for each row. The time to update an entry is $\mathcal{O}(1)$.

Introduce node. Let i be an introduce node with child j such that $X_i = X_j \cup \{v\}$ for some $v \notin X_j$. Let $0 \leq b \leq k$ be an integer, $P = (A, C, L, R, B)$ be a five-way partition of X_i and f be the SN-function defined on X_i^+. The computation of the table entry is split into two cases, depending on whether the vertex v belongs to the set A in the partition P or not.

1. **Case $v \in A$.** Define $C_v = \{u \in X_i \setminus A \mid v \in N(u) \cap f(u)\}$. Let P_j denote the partition of X_j obtained by removing vertex v from the set A of the partition P. Let $f_j : X_j \to 2^V$ be the SN-function defined as follows:

$$f_j(u) = \begin{cases} f(u) \setminus \{v\} & \text{if } u \in C_v \\ f(u) & \text{otherwise.} \end{cases}$$

Then,

$$T_i[b, P, f].\text{Solution} = T_j[b - 1, P_j, f_j].\text{Solution} \cup \{v\}$$
$$T_i[b, P, f].\text{Value} = T_j[b - 1, P_j, f_j].\text{Value} + w(v) + \mathcal{C}_f(C_v, v)$$

2. **Case $v \notin A$.** Since v is in X_i but not in X_j it follows that $N(v) \cap X_i^+ \subseteq X_i$. Therefore, the coverage of the vertex v by the vertices that occur only in $X_j^+ \setminus X_j$ is zero. Let P_j be the partition of X_j obtained by removing the vertex v from the appropriate set in the partition P. The SN-function f_j is defined as follows on the set X_j^+: For $u \in X_i \setminus \{v\}$, $f_j(u) = f(u)$.

$$T_i[b, P, f].\text{Solution} = T_j[b, P_j, f_j].\text{Solution}$$
$$T_i[b, P, f].\text{Value} = T_j[b, P_j, f_j].\text{Value} + \mathcal{C}_f(v, A)$$

Forget node. Let i be a forget node with child j such that $X_i = X_j \setminus \{v\}$ for some $v \in X_j$. Let $0 \leq b \leq k$ be a budget, $P = (A, C, L, R, B)$ be a five-way partition of X_i and f be an SN-function. We consider the following five-way partitions of X_j.

- $P_1 = (A \cup \{v\}, C, L, R, B)$
- $P_2 = (A, C \cup \{v\}, L, R, B)$
- $P_3 = (A, C, L \cup \{v\}, R, B)$
- $P_4 = (A, C, L, R \cup \{v\}, B)$
- $P_5 = (A, C, L, R, B \cup \{v\})$

Let f_j be the SN-function defined as follows: for $u \in X_i$, $f_j(u) = f(u)$ and $f_j(v) = N(v)$. Let $P_j = \arg\max\limits_{P' \in \{P_i\}_{i=1}^{5}} T_j[b, P', f_j].\text{Value}$. $T_i[b, P, f]$ is defined as follows:

$$T_i[b, P, f].\text{Solution} = T_j[b, P_j, f_j].\text{Solution}$$
$$T_i[b, P, f].\text{Value} = T_j[b, P_j, f_j].\text{Value}$$

Join node. Let i be a join node with children j and g such that $X_i = X_j = X_g$. Let $0 \leq b \leq k$ be a budget, $P = (A, C, L, R, B)$ be a five-way partition of X_i and f be an SN-function. We consider the sets $\mathcal{U}_j$ and $\mathcal{U}_g$ of all SN-functions defined on X_j^+ and X_g^+, respectively, satisfying the following conditions:

- For each $v \in L$, $f_j(v) = f(v)$ and $f_g(v) = \emptyset$.
- For each $v \in R$, $f_j(v) = \emptyset$ and $f_g(v) = f(v)$.
- For each $v \in B$, $f_j(v)$ and $f_g(v)$ are defined as follows. For each partition $f_1(v) \cup f_2(v) = f(v) \cap X_i^+$, $f_j(v) = f_1(v) \cup (f(v) \setminus X_i^+)$ and $f_g(v) = f_2(v) \cup (f(v) \setminus X_i^+)$.

We then consider all possible candidates for L_j, R_j, B_j such that $L_j \cup R_j \cup B_j = L \cup B$ and L_g, R_g, B_g such that $L_g \cup R_g \cup B_g = R \cup B$. Each such candidate defines a $P'_j = (A, C \cup R, L_j, R_j, B_j)$ of X_j and a $P'_g = (A, C \cup L, L_g, R_g, B_g)$ of X_g. Let $\mathcal{R}_j$ and $\mathcal{R}_g$ denote the set of all such candidate partitions of X_j and X_g, respectively.

Let $P_j \in \mathcal{R}_j$, $P_g \in \mathcal{R}_g$, $f_j \in \mathcal{U}_j$, $f_g \in \mathcal{U}_g$ and $0 \leq b' \leq b - |A|$ be the values at which the maximum value for Equation (2) is achieved.

$$\max_{\substack{0 \leq b_j \leq b - |A| \\ P'_j \in \mathcal{R}_j, P'_g \in \mathcal{R}_g \\ f'_j \in \mathcal{U}_j, f'_g \in \mathcal{U}_g}} T_j[b_j + |A|, P'_j, f'_j].\text{Value} + T_g[b - b_j, P'_g, f'_g].\text{Value}. \tag{2}$$

The value of the row corresponding to (b, P, f) in T_i is given as follows:

$$T_i[b, P, f].\text{Solution} = T_j[b' + |A|, P_j, f_j].\text{Solution} \cup T_g[b - b', P_g, f_g].\text{Solution}$$
$$T_i[b, P, f].\text{Value} = T_j[b' + |A|, P_j, f_j].\text{Value} + T_g[b - b', P_g, f_g].\text{Value} - C_f(A \cup C, A).$$

We now prove that the update steps presented above are correct.

Lemma 5. *For each node $i \in H$, for each row (b, P, f) in T_i, the pair $T_i[b, P, f].\text{Solution}$, $T_i[b, P, f].\text{Value}$ is such that $C(X_i^+, T_i[b, P, f].\text{Solution}) = T_i[b, P, f].\text{Value}$, and this is the maximum possible value.*

Proof. The proof is by induction on the height of a node in H. The height of a node i in a rooted tree H is the distance to the farthest leaf in the subtree rooted at i. The base case is when i is a leaf node in H, and clearly its height is 0. For a leaf node i with $X_i = \emptyset$, the row with $b = 0$, $P = (\emptyset, \emptyset, \emptyset, \emptyset, \emptyset)$ and $f : \emptyset \to \emptyset$ is the only valid row entry and its value is 0. This completes the proof of the base case. Let us assume that the claim is true for all nodes of height at most $h - 1 \geq 0$. We prove that if the claim is true at all nodes of height at most $h - 1$, then it is true for a node of height h. Let b be a budget, $P = (A, C, L, R, B)$ be a partition of X_i and f be an SN-function on X_i^+. Let $T_i[b, P, f].\text{Solution} = S = A \cup Z$. For any optimal $S' = A \cup Z'$ where $|S'| = b$, we prove that

$$T_i[b, P, f].\text{Value} \geq w(A) + C_f(C, A) + C_f(L, A \cup Z'_j) + C_f(R, A \cup Z'_g) + C_f(B, S') + C(X_i^+ \setminus X_i, S').$$

We proceed by considering the type of node i so that the induction hypothesis can be applied at nodes of height at most $h - 1$.

Case when i is an introduce node. Let j be the child of i and $X_i = X_j \cup \{v\}$ for some $v \notin X_j$. We now consider two cases depending on whether v belongs to A or not.

1. **Case $v \in A$.** Let P_j be the partition of X_j obtained from P by removing v from A. Let f_j be the SN-function on X_j^+ such that $f_j(u) = f(u) \setminus v$, if $u \in C_v$ and $f_j(u) = f(u)$, otherwise. We know from our claimed optimality of S', and that $S' \cap X_i = S \cap X_i = A$, and the value of $T_i[b, P, f]$, that

$$C_f(X_i^+, S') = C_{f_j}(X_j^+, S' \setminus \{v\}) + C_f(C_v, v)$$
$$> C_f(X_i^+, S) = C_{f_j}(X_j^+, S \setminus \{v\}) + C_f(C_v, v).$$

Therefore, it follows that $\mathcal{C}_{f_j}(X_j^+, S' \setminus \{v\}) > \mathcal{C}_{f_j}(X_j^+, S \setminus \{v\}) = T_j[b-1, P_j, f_j].\text{Value}$. In other words, we have concluded that the value for the row $(b-1, P_j, f_j)$ in T_j is not the optimum value. This contradicts our premise at node j, which is of height at most $h-1$ for which, by induction hypothesis, the table maintains the optimal values. Therefore, our assumption that $T_i[b, P, f]$ is not optimum is wrong.

2. **Case** $v \notin A$. Let P_j be the partition of X_j obtained by removing v from the appropriate set in the partition P. Let f_j be the SN-function on X_j^+ such that $f_j(u) = f(u)$ for each $u \in X_j^+$. We know from our claimed optimality of S', and that $S' \cap X_i = S \cap X_i = A$, and the value of $T_i[b, P, f]$ that $\mathcal{C}_f(X_i^+, S') = \mathcal{C}_{f_j}(X_j^+, S') + \mathcal{C}_f(v, A) > \mathcal{C}_f(X_i^+, S) = \mathcal{C}_{f_j}(X_j^+, S) + \mathcal{C}_f(v, A)$. Therefore, it follows that $\mathcal{C}_{f_j}(X_j^+, S') > \mathcal{C}_{f_j}(X_j^+, S)$. In other words, we have concluded that the value for the row (b, P_j, f_j) in T_j is not the optimum value. This contradicts our premise at node j, which is of height at most $h-1$ for which, by induction hypothesis, the table maintains the optimal values. Therefore, our assumption that $T_i[b, P, f]$ is not optimum is wrong.

Forget node. We know that $X_i^+ = X_j^+$, and v is in X_j but not in X_i, it follows that $N(v) \cap X_i = \emptyset$ and $N(v) \subseteq X_j^+$. Define f_j to be the SN-function at X_j^+ such that $f_j(u) = f(u)$ for each $u \in X_i$ and $f_j(v) = N(v)$. We have assumed $T_i[b, P, f].\text{Value} < \mathcal{C}_f(X_i^+, S')$. Further, since $X_i^+ = X_j^+$ and due to the definition of f_j, $\mathcal{C}_f(X_i^+, S') = \mathcal{C}_{f_j}(X_j^+, S')$. Since $T_i[b, P, f]$ is computed identically from some row in T_j, let us say (b, P_j, f_j), it follows that $T_j[b, P_j, f_j].\text{Value} < \mathcal{C}_{f_j}(X_j^+, S')$. This contradicts the premise that the table T_i is at the lowest height in the tree decomposition at which an entry is sub-optimal. Therefore, our premise is wrong.

Join node. We assume that S' is indeed a better solution than S for the table entry (b, P, f) of T_i. Let $S_j' = S \cap (X_j^+ \setminus X_j)$ and $S_g' = S \cap (X_g^+ \setminus X_g)$. Let $b_j' = |S_j'|$. Let $P_j' = (A, C \cup R, L_j', R_j', B_j')$ and $P_g' = (A, C \cup L, L_g', R_g', B_g')$ be the partitions of X_j and X_g defined using S_j' and S_g', respectively. Note that, $L_j' \cup R_j' \cup B_j' = L \cup B$ and $L_g' \cup R_g' \cup B_g' = R \cup B$. Let f_j' be the SN function on X_j^+ such that $f_j'(u) = f(u)$ for $u \in A \cup C \cup L \cup B$ and $f_j'(u) = \emptyset$ for $u \in R$. Let f_g' be the SN function on X_g^+ such that $f_g'(u) = f(u)$ for $u \in A \cup C \cup R$, $f_g'(u) = \emptyset$ for $u \in L$ and $f_g'(u) = f(u) \setminus S_j'$. The coverage $\mathcal{C}_f(X_i^+, S')$ can be written as $\mathcal{C}_f(X_i^+, S') = \mathcal{C}_{f_j'}(X_j^+, S_j') + \mathcal{C}_{f_g'}(X_g^+, S_g') - w(A) - \mathcal{C}_f(C, A)$, where the coverage of X_j^+ by S_j' and X_g^+ by S_g' are restricted to the partitions P_j' and P_g'.

We know that $X_i = X_j = X_g$ in case of join node. The table entry (b, P, f) is updated using the entries $(b_j + |A|, P_j, f_j)$ and $(b - b_j, P_g, f_g)$ of the table T_j and T_g, respectively. In other words, the values of variables b_j, P_j, P_g, f_j and f_g are obtained from the Equation (2) described in the recursive definition of join node. The values b_j', P_j', P_g', f_j' and f_g' are also feasible for the range given in the Equation (2). Then, we have $T_j[b_j' + |A|, P_j', f_j'].\text{Value} + T_g[b - b_j', P_g', f_g'].\text{Value} < T_j[b' + |A|, P_j, f_j].\text{Value} + T_g[b - b', P_g, f_g].\text{Value}$. Since S' is better than S for the entry (b, P, f) of T_i, we have $T_j[b' + |A|, P_j, f_j].\text{Value} + T_g[b - b', P_g, f_g].\text{Value} < \mathcal{C}_{f_j'}(X_j^+, S_j') + \mathcal{C}_{f_g'}(X_g^+, S_g')$. The above inequality shows that at least one of the table entries $T_j[b_j' + |A|, P_j', f_j']$ or $T_g[b - b_j', P_g', f_g']$ is not optimal. This would again contradict the premise that i is the node at the least height at which some table entry is sub-optimal. This completes the case analysis and our proof. Hence the lemma. □

Running Time. There are $\mathcal{O}(n\ tw)$-many nodes in the nice tree decomposition H. Each node $i \in H$ has a maximum of $(k+1)(5 \cdot 2^\Delta)^{tw}$ entries. The $2^{\Delta \cdot tw}$ comes from the fact that at each vertex in a bag, we enumerate all subsets of neighbors to come up with the SN-functions. It is clear from the description that at the leaf nodes, introduce nodes, and forget nodes, the update time is $\mathcal{O}(tw)$. At a join node, the time taken to compute an entry depends on three basic operations. The optimal partitions P_j and P_g are computed in $3^{|L|+|R|+2|B|}$ time and budget distribution can be done in $\mathcal{O}(k)$ time. The costliest operation is to enumerate the different SN-functions f_j and f_g for the given SN-function f. Since we need to consider all the 2^Δ-possible ways of distributing the $f(v)$ for each vertex v, the distribution takes $\mathcal{O}((2^\Delta)^{|B|})$ time. Therefore, the running time for an entry (b, P, f) in a join node takes $\mathcal{O}(k \cdot 3^{|L|+|R|+2|B|} \cdot (2^{\Delta})^{|B|})$ time, and this is $2^{\mathcal{O}(\Delta \cdot tw)}$. This analysis of the running time and Lemma 5 complete the proof of the following theorem which is our main result.

Theorem 1. *The* MAX-EXP-COVER-1-RF *problem can be solved in time* $2^{\mathcal{O}(\Delta \cdot tw)} n^{\mathcal{O}(1)}$.

5. Parameterized Complexity of PROBABILISTIC-CORE Problem

Let $\mathcal{G} = (V, E, p)$ be an uncertain graph and d be an integer. Given a set $K \subseteq V$, we define the probability that the set K is being a d-core in $\mathcal{G}$, denoted by $\rho_d(\mathcal{G}, K)$, to be $\sum_{H \sqsubseteq \mathcal{G}} P(H) I(H, K, d)$, where I is an indicator variable that takes value 1 if and only if the set K forms a d-core in the graph H. The decision version of the PROBABILISTIC-CORE problem is formally stated as follows: given an uncertain graph $\mathcal{G} = (V, E, p)$, an integer d and a probability $\theta \in [0, 1]$, decide if there exists a set $K \subseteq V$ such that $\rho_d(\mathcal{G}, K) \geq \theta$.

We study the PROBABILISTIC-CORE problem under the LRO and RF model. First we show that the PROBABILISTIC-CORE-LRO problem and the INDIVIDUAL-CORE problem are polynomial time solvable, due to the fact that an uncertain graph under the LRO model has only a polynomial number of worlds. Then we show that the PROBABILISTIC-CORE-RF problem is W[1]-hard for the parameter d and admits an FPT algorithm for the parameter treewidth.

5.1. An Exact Algorithm for the PROBABILISTIC-CORE-LRO *Problem*

We present a polynomial time algorithm (see as Algorithm 1) for the PROBABILISTIC-CORE-LRO problem. Let $\langle \mathcal{G} = (V, E, p), d \rangle$ be an instance of the PROBABILISTIC-CORE-LRO problem. As per the definition of the LRO model, let the set $\{G_0, \ldots, G_m\}$ be the possible worlds of $\mathcal{G}$, where $G_0 = (V, \emptyset)$ and for $0 \leq i \leq m-1$, $E(G_{i+1}) = E(G_i) \cup \{e_i\}$. For each $0 \leq i \leq m-1$, we consider the linear order in which G_i precedes G_{i+1}.

Algorithm 1: PROBABILISTIC-CORE-LRO

Data: $\mathcal{G} = (V, E, p), d$
Result: A set $C \subseteq V$ with a probability q.
for $i \leftarrow 0$ **to** m **do**
    $C \leftarrow V$;
    while $\delta(G_i[C]) < d$ **do**
        Let v be a vertex such that $deg_{G_i[C]}(v) < d$;
        $C \leftarrow C \setminus \{v\}$;
    end
    if $C \neq \emptyset$ **then**
        return $C, p(e_i)$;
    end
end
return $\emptyset, 0.0$;

Lemma 6. *The Algorithm 1 solves the* PROBABILISTIC-CORE-LRO *problem in polynomial time.*

Proof. By definition of the possible worlds, $G_m = (V, E)$. If $\delta(G_m) < d$, then G_m does not contain a d-core. Moreover, for $0 \leq i < m$, since G_i is an edge subgraph of G_m, no graph in the possible worlds has a d-core. This is indicated by the return value-pair of the $\emptyset$ and probability 0.0. On the other hand if $\delta(G_m) \geq d$, let G_i be the first graph in the linear ordering of possible worlds for which a non-empty C is computed by Algorithm 1 on exit from the While-loop. Clearly, C induces a d-core in G_i. Since G_i is the first graph which contains a d-core, for each $0 \leq j \leq i-1$, G_j does not contain a d-core. Further, for every $j > i$, C is a d-core in G_j since it is an edge-subgraph of G_i. Then,

$$\rho_d(\mathcal{G}, C) = \sum_{j=0}^{m} P(G_j) I(G_j, C, d) = \sum_{j=i}^{m} P(G_j) = p(e_i).$$

For any set $K \subseteq V$ with $\rho_d(\mathcal{G}, K) > 0$, the set K can form a d-core only in the possible worlds $\{G_i, G_{i+1}, \ldots, G_m\}$. Thus, $\rho_d(\mathcal{G}, K) \leq \rho_d(\mathcal{G}, C)$. This completes the proof. □

The following observation again uses the fact that there are only a polynomial number of possible worlds under the LRO model.

Observation 3. *The* INDIVIDUAL-CORE *problem is polynomial time solvable on uncertain graphs under the LRO model.*

Proof. Clearly, for each $0 \leq i \leq m$, the While-loop in Algorithm 1 computes the maximal d-core in G_i. Therefore, for each v, $q_d(v) = p(e_i)$, where i the smallest index such that the non-empty d-core contains v. Since the number of possible worlds are $m+1$, for each vertex $v \in V$, the probability $q_d(v)$ can be computed in polynomial time. Thus, for a given uncertain graph $\mathcal{G} = (V, E, p)$, an integer d and probability threshold θ, the set $\{v \mid q_d(v) \geq \theta\}$ is the optimum solution to the INDIVIDUAL-CORE problem. □

5.2. Parameterized Complexity of the PROBABILISTIC-CORE-RF Problem

We show that the PROBABILISTIC-CORE-RF problem is W[1]-hard. The reduction that we show for the W[1]-hardness is similar to the hardness result of the INDIVIDUAL-CORE problem shown by Peng et al. [16] (they call this the (d, θ)-CORE problem). The reduction is from the CLIQUE problem which is as follows: given a graph $G = (V, E)$ and an integer k, decide if G has a clique of size at least k. The CLIQUE problem is known to be W[1]-hard [47].

Theorem 2. *The* PROBABILISTIC-CORE-RF *problem is W[1]-hard for the parameter d.*

Proof. Let $\langle G = (V, E), k \rangle$ be an instance of the CLIQUE problem. The output of the reduction is denoted by $\langle \mathcal{G} = (V, E, p), d, \theta \rangle$, and it is an instance of PROBABILISTIC-CORE-RF problem. The vertex set and edge set of $\mathcal{G}$ are same as V and E, respectively, $d = k - 1$ and $\theta = 2^{-\binom{k}{2}}$. Further, for each edge $e \in E$, define $p(e) = \frac{1}{2}$ in $\mathcal{G}$. Now we show that the CLIQUE problem on the instance $\langle G, k \rangle$ and the PROBABILISTIC-CORE-RF problem on the instance $\langle \mathcal{G}, d, \theta \rangle$ are equivalent.

We prove the forward direction first. Let $K \subseteq V$ be a k-clique in G. The set K is indeed a d-core since every vertex $v \in K$ has $k-1$ neighbors in K. The probability that K is a d-CORE in a random sample from the possible worlds of $\mathcal{G}$ is $2^{-\binom{k}{2}}$. Thus, the set K is a feasible solution for the instance $\langle \mathcal{G}, d, \theta \rangle$ of the PROBABILISTIC-CORE-RF problem.

Now we prove the reverse direction. We claim that any feasible solution $K \subseteq V$ contains exactly $d+1$ vertices. If K has less than $d+1$ vertices then K cannot form a d-core in any possible world. Therefore, we consider the case in which that K has more than $d+1$ vertices, and each vertex in the set K has degree at least d. Consequently, the number of edges in any possible world in which K is a d-core is at least $\frac{d(d+2)}{2} = \frac{d(d+1)}{2} + \frac{d}{2} = \binom{d+1}{2} + \frac{d}{2}$. Then, the probability that the set K is a d-core in $\mathcal{G}$ is at most $2^{-\binom{d+1}{2} - \frac{d}{2}} < \theta$, and this contradicts the hypothesis that K is a feasible solution. It follows that any feasible solution $K \subseteq V$ contains exactly $d+1$ vertices and each vertex has degree d, thus K is a k clique. Hence the theorem. □

5.3. The PROBABILISTIC-CORE-RF Problem is FPT by Treewidth

We show that the PROBABILISTIC-CORE-RF problem admits an FPT algorithm with treewidth as the parameter. The input consists of an instance $\langle \mathcal{G} = (V, E, p), d \rangle$ of the PROBABILISTIC-CORE-RF problem. A nice tree decomposition $(X, \mathcal{T})$ of the graph $G = (V, E)$ is also given as part of the input.

Let i be any node in $\mathcal{T}$ and $\alpha : X_i \to \{0, 1, \ldots, d\}$ and $\beta : X_i \to \{0, 1, \ldots, d\}$ be a pair of functions on X_i. Given a set $K \subseteq V$, the set K is said to be (α, β)-constrained d-core of G if

1. for each $v \in X_i \cap K$, $|N(v) \cap K| = \alpha(v) + \beta(v)$ and $|N(v) \cap K \cap X_i| = \alpha(v)$, and
2. for each $v \in K \setminus X_i$, $|N(v) \cap K| = d$.

Let $\mathcal{G}_i^{\alpha, \beta}$ denote the uncertain graph $\mathcal{G}[X_i^+ \setminus (\alpha^{-1}(0) \cap \beta^{-1}(0))]$. We define a constrained version of the PROBABILISTIC-CORE-RF problem as follows. Given a set K, the probability that the set K is an (α, β)-constrained d-core in $\mathcal{G}_i^{\alpha, \beta}$ is given by:

$$\rho_d(\alpha, \beta, \mathcal{G}_i^{\alpha, \beta}, K) = \sum_{H \sqsubseteq \mathcal{G}_i^{\alpha, \beta}} P(H) I'(H, \alpha, \beta, K, d),$$

where $I'(H, \alpha, \beta, K, d)$ is an indicator function that takes value 1 if and only if K is an (α, β)-constrained d-core of H. The optimization version of the (α, β)-constrained PROBABILISTIC-CORE-RF problem seeks to find the set K such that $\rho_d(\alpha, \beta, \mathcal{G}_i^{\alpha, \beta}, K)$ is maximized. The solution for (α, β)-constrained PROBABILISTIC-CORE-RF problem for different values of i, α and β on $\mathcal{G}$ are partial solutions which are used to come up with a recursive specification of the optimum value. The dynamic programming (DP) formulation on the nice tree decomposition $(X, \mathcal{T})$ results in an FPT algorithm with treewidth as the parameter.

The dynamic programming formulation maintains a table T_i at every node $i \in X$. Each row in the table T_i is a pair of functions α and β such that $\alpha, \beta : X_i \to \{0, 1, \ldots, d\}$. The column corresponding to the row (α, β) is a pair (Solution, Value) which consists of a set $S_{\alpha, \beta}$ and a probability value. Further, the set $S_{\alpha, \beta}$ is an optimal solution for the (α, β)-constrained PROBABILISTIC-CORE-RF problem on the instance $\langle \mathcal{G}_i^{\alpha, \beta}, d \rangle$.

Intuitively, the functions α and β defined on X_i stand for the "in-bag-degree" and "out-bag-degree" constraints, respectively, for each vertex $v \in X_i$. We maintain the candidate solutions in the table for different values of the functions α and β. We allow the candidate solutions that are infeasible at current stage and those vertices which are not satisfied with degree d will get neighbors from nodes at a higher level in the tree decomposition. The optimal solution for the instance $\langle \mathcal{G}, d \rangle$ of the PROBABILISTIC-CORE-RF problem can be obtained from the table T_r where r is the root of the tree decomposition $\mathcal{T}$.

5.3.1. Dynamic Programming

We now present the dynamic programming formulation on the different types of nodes in $\mathcal{T}$. Let $i \in \mathcal{T}$ be a node with bag X_i. For a pair of functions $\alpha, \beta : X_i \to \{0, 1, \ldots, d\}$, we show how to compute the table entry $T_i[\alpha, \beta]$ in each type of node as follows.

Leaf node. Let i be a leaf node with bag $X_i = \emptyset$. We have one row in the table T_i. Let $\alpha, \beta : \emptyset \to \{0\}$ be the pair of functions corresponding to the row and the value of the table entry is given as:

$$T_i[\alpha, \beta] = (\emptyset, 0.0).$$

Insert node. Let i be an insert node with a child j, and let $X_i = X_j \cup \{v\}$ for some $v \notin X_j$. The row $T_i[\alpha, \beta]$ is computed based on the value of $\alpha(v)$ and $\beta(v)$. If $\beta(v) > 0$, then the row $T_i[\alpha, \beta]$ becomes infeasible since $N(v) \cap X_i^+ \subseteq X_i$. That is,

$$T_i[\alpha, \beta] = (\emptyset, 0.0).$$

In the rest of the cases we have $\beta(v) = 0$. We define $\beta' : X_j \to \{0, 1, \ldots, d\}$ to be a function such that $\beta'(u) = \beta(u)$ for all $u \in X_j$. When $\alpha(v) = 0$, then the vertex v will not be in the solution $S_{\alpha, \beta}$. Let $\alpha' : X_j \to \{0, 1, \ldots, d\}$ be a function such that $\alpha'(u) = \alpha(u)$ for all $u \in X_j$. The value of the row $T_i[\alpha, \beta]$ is same as $T_j[\alpha', \beta']$ since v is excluded. When $\alpha(v) > 0$, then v is part of $S_{\alpha, \beta}$.

Let $U = (N(v) \cap X_i) \setminus \alpha^{-1}(0)$. Then, we have $|U| \geq \alpha(v)$. Otherwise, no feasible solution exists for the row $T_i[\alpha, \beta]$. That is, degree constraint of v will not be met in any feasible solution. Assume $|U| \geq \alpha(v)$. Let $Y \subseteq U$ be a subset of U of size $\alpha(v)$, and $q(v, Y) = \prod_{u \in Y} p(uv)$. The vertices in Y that are the neighbors of v contribute degree $\alpha(v)$ to v. Then, each vertex $u \in Y$ will lose a degree from $\alpha(u)$ in the node j. Define $\alpha_Y : X_j \to \{0, 1, \ldots, d\}$ to be

$$\alpha_Y(u) = \begin{cases} \alpha(u) & \text{if } u \notin Y \\ \alpha(u) - 1 & \text{if } u \in Y. \end{cases}$$

Let

$$W = \underset{Y \subseteq U, |Y| = \alpha(v)}{\arg\max} \; q(v, Y) T_j[\alpha_Y, \beta'].\text{Value}$$

be the best $\alpha(v)$ neighbors of v such that the solution obtained from the table T_j is maximized. Then, the recursive definition of the row $T_i[\alpha, \beta]$ is given as,

$$T_i[\alpha, \beta] = (T_j[\alpha_W, \beta'].\text{Solution} \cup \{v\}, q(v, W) T_j[\alpha_W, \beta'].\text{Value}).$$

Forget node. Let i be a forget node with a child j, and let $X_i = X_j \setminus \{v\}$ for some $v \in X_j$. From the definition of the tree decomposition, it follows that $N(v) \subseteq X_j^+$. Since $N(v) \subseteq X_j^+$, either v is part of solution with constraint $\alpha(v) + \beta(v) = d$ or v is not part of solution. Let $U = (N(v) \cap X_j) \setminus \alpha^{-1}(0)$. For each $0 \leq a \leq \min(d, |U|)$ and $Y \subseteq U$ of size a, we define $\alpha_{a,Y}, \beta_{a,Y} : X_j \to \{0, 1, \ldots, d\}$ such that

$$\alpha_{a,Y}(u) = \begin{cases} \alpha(u) & \text{if } u \neq v \text{ and } u \notin Y \\ \alpha(u) + 1 & \text{if } u \neq v \text{ and } u \in Y \\ a & \text{if } u = v, \end{cases}$$

and

$$\beta_{a,Y}(u) = \begin{cases} \beta(u) & \text{if } u \neq v \text{ and } u \notin Y \\ \beta(u) - 1 & \text{if } u \neq v \text{ and } u \in Y \\ d - a & \text{if } u = v, \end{cases}$$

For $0 \leq a \leq \min(d, |U|)$, let

$$\alpha_a, \beta_a = \underset{\alpha_{a,Y}, \beta_{a,Y} : Y \subseteq U, |Y| = a}{\arg\max} \; T_j[\alpha_{a,Y}, \beta_{a,Y}].\text{Value}.$$

Let $t = \underset{0 \leq a \leq \min(d, |U|)}{\arg\max} \; T_j[\alpha_a, \beta_a].\text{Value}$. Then the value of the row $T_i[\alpha, \beta]$ is equal to $T_j[\alpha_t, \beta_t]$.

Join node. Let i be a join node with children j and g such that $X_i = X_j = X_g$. For a function α defined on X_i, we consider the function α for the child node X_j and for the other child node X_g, consider the function $\alpha' : X_i \to \{0, 1, \ldots, d\}$ such that $\alpha(u) = 0$ for all $u \in X_i$. Since β considers the neighbors from outside X_i, each vertex $v \in X_i$ with $\beta(v) > 0$ will get the d-core neighbors from the set $X_i^+ \setminus X_i$. Since $X_i^+ \setminus X_i = (X_j^+ \setminus X_j) \cup (X_g^+ \setminus X_g)$ and both the sets are disjoint, we divide $\beta(v)$ into two parts. For each vertex $v \in X_i$, we try all possible ways of dividing $\beta(v)$ into two parts. For $x : X_i \to \{0, 1, \ldots, d\}$ such that for each $v \in X_i$, $0 \leq x(v) \leq \beta(v)$, we define $\beta_x : X_j \to \{0, 1, \ldots, d\}$ and $\beta'_x : X_g \to \{0, 1, \ldots, d\}$ to be $\beta_x(v) = x(v)$ and $\beta'_x(v) = \beta(v) - x(v)$. Let

$$y = \underset{x : X_i \to \{0, 1, \ldots, d\} | \forall v \in X_i, 0 \leq x(v) \leq \beta(v)}{\arg\max} \; T_j[\alpha, \beta_x].\text{Value} \cdot T_g[\alpha', \beta'_x].\text{Value}$$

Then, the recursive definition of the row $T_i[\alpha, \beta]$ is given as,

$$T_i[\alpha, \beta] = (T_j[\alpha, \beta_y].\text{Solution} \cup T_g[\alpha', \beta'_y].\text{Solution}, T_j[\alpha, \beta_y].\text{Value} \cdot T_g[\alpha', \beta'_y].\text{Value}).$$

5.3.2. Correctness and Running Time

Lemma 7. *Let i be a node in $\mathcal{T}$. For every pair of functions $\alpha, \beta : X_i \to \{0, 1, \ldots, d\}$, the row $T_i[\alpha, \beta]$ is computed optimally.*

Proof. Let i be a node in $\mathcal{T}$. We claim that for each pair of functions $\alpha, \beta : X_i \to \{0, 1, \ldots, d\}$, the row $T_i[\alpha, \beta]$ computes the optimal solution for the instance $\mathcal{G}_i^{\alpha,\beta}$ of the (α, β)-constrained PROBABILISTIC-CORE problem. That is, the set $S_{\alpha,\beta} = T_i[\alpha, \beta].\text{Solution}$ is an optimal solution for the above instance. Let $A = X_i \setminus (\alpha^{-1}(0) \cap \beta^{-1}(0))$. We show that for any $S = A \cup Z$ for some $Z \subseteq X_i^+ \setminus X_i$,

$$\rho_d(\alpha, \beta, \mathcal{G}_i^{\alpha,\beta}, S) \leq \rho_d(\alpha, \beta, \mathcal{G}_i^{\alpha,\beta}, S_{\alpha,\beta}).$$

The proof is by induction on the height of a node in $\mathcal{T}$. The height of a node i in the rooted tree $\mathcal{T}$ is the distance to the farthest leaf in the subtree rooted at i. The base case is when i is a leaf node in $\mathcal{T}$ and height is 0. For a leaf node i with $X_i = \emptyset$, the row with $\alpha, \beta : \emptyset \to \{0, 1, \ldots d\}$ completes the proof of the base case. Let us assume that the claim is true for all nodes of height at most $h - 1 \geq 0$. We now prove that if the claim is true at all nodes of height at most $h - 1$, then it is true for a node of height h. Let i be a node of height h. Clearly i is not a leaf node.

When i is an introduce node. Let j be the child of i, and $X_i = X_j \cup \{v\}$ for some $v \notin X_j$. If $\beta(v) > 0$, then no feasible solution exists since $N(v) \cap (X_i^+ \setminus X_i) = \emptyset$. This is captured in our dynamic programming. In the further cases, we consider $\beta(v) = 0$. Let $L = N(v) \cap X_i$. Consider the case when $\alpha(v) = 0$. That is, the vertex v is not part of the solution. The recursive definition of the dynamic programming gives

$$S_{\alpha,\beta} = T_i[\alpha, \beta].\text{Solution} = T_j[\alpha', \beta'].\text{Solution}$$

where α' and β' are as defined in the dynamic programming. A feasible solution to the row $T_i[\alpha, \beta]$ should be feasible for the row $T_j[\alpha', \beta']$. Otherwise, the degree constraints are not met by the solution. For the solution $S = A \cup Z$,

$$\rho_d(\alpha', \beta', \mathcal{G}_j^{\alpha',\beta'}, S) \leq \rho_d(\alpha', \beta', \mathcal{G}_j^{\alpha',\beta'}, S_{\alpha,\beta})$$

since j is a node at height $h - 1$ and by our induction hypothesis. Also,

$$\rho_d(\alpha', \beta', \mathcal{G}_j^{\alpha',\beta'}, S) = \rho_d(\alpha, \beta, \mathcal{G}_i^{\alpha,\beta}, S)$$

since $v \notin S$. Then, we have

$$\rho_d(\alpha, \beta, \mathcal{G}_i^{\alpha,\beta}, S) = \rho_d(\alpha', \beta', \mathcal{G}_j^{\alpha',\beta'}, S) \leq \rho_d(\alpha', \beta', \mathcal{G}_j^{\alpha',\beta'}, S_{\alpha,\beta}) = \rho_d(\alpha, \beta, \mathcal{G}_i^{\alpha,\beta}, S_{\alpha,\beta}).$$

This completes the case when $\beta(v) = 0$ and $\alpha(v) = 0$.

Now we consider the case where $\alpha(v) > 0$. In the solution $S = A \cup Z$, we need $\alpha(v)$ neighbors of U in the d-core where the set $U = (N(v) \cap X_i) \setminus (\alpha^{-1}(0) \cap \beta^{-1}(0))$. Let $S_j = S \setminus \{v\}$. There exists a $Y \subseteq U$ of size $\alpha(v)$ such that the probability $\rho_d(\alpha, \beta, G_i^{\alpha,\beta}, S)$ can be written as follows:

$$\rho_d(\alpha, \beta, \mathcal{G}_i^{\alpha,\beta}, S) = \rho_d(\alpha_Y, \beta', \mathcal{G}_j^{\alpha_Y,\beta'}, S_j) \prod_{u \in Y} p(uv).$$

The solution S_j is compatible for the row $T_j[\alpha_Y, \beta']$ where the vertices in Y are neighbors of v and degree constraint of these vertices is decreased by 1 in α such that it will be satisfied by the edge from v. Since W in the dynamic programming is optimal over all possible $\alpha(v)$ sized subsets of W, we have

$$\rho_d(\alpha, \beta, \mathcal{G}_i^{\alpha,\beta}, S) = \rho_d(\alpha_Y, \beta', \mathcal{G}_j^{\alpha_Y,\beta'}, S_j) \prod_{u \in Y} p(uv) \leq \rho_d(\alpha_W, \beta', \mathcal{G}_j^{\alpha_W,\beta'}, T_j[\alpha_W, \beta'].\text{Solution}) \prod_{u \in W} p(uv)$$
$$\leq T_j[\alpha_W, \beta'].\text{Value} \prod_{u \in W} p(uv) = T_i[\alpha, \beta].\text{Value}.$$

This completes the argument of the case when i is an introduce node.

When i is a forget node. Let j be the child of i, and $X_i = X_j \setminus \{v\}$ for some $v \in X_j$. We consider two cases depending on whether v belongs to S. We first consider the case when $v \notin S$. Consider the functions $\alpha_{0,\emptyset}$ and $\beta_{0,\emptyset}$ as defined in the recursive computation of forget node. Since $v \notin S$, v will get zero degree constraint in both functions α and β, and other vertices will have same constraints as α and β values. Then, the probability $\rho_d(\alpha, \beta, \mathcal{G}_i^{\alpha,\beta}, S)$ can be written as follows:

$$\rho_d(\alpha, \beta, \mathcal{G}_i^{\alpha,\beta}, S) = \rho_d(\alpha_{0,\emptyset}, \beta_{0,\emptyset}, \mathcal{G}_j^{\alpha_{0,\emptyset}, \beta_{0,\emptyset}}, S).$$

Since j is a node at height $h-1$, we have

$$\rho_d(\alpha, \beta, \mathcal{G}_i^{\alpha,\beta}, S) = \rho_d(\alpha_{0,\emptyset}, \beta_{0,\emptyset}, \mathcal{G}_j^{\alpha_{0,\emptyset}, \beta_{0,\emptyset}}, S) \leq T_j[\alpha_{0,\emptyset}, \beta_{0,\emptyset}].\text{Value}$$
$$\leq T_j[\alpha_t, \beta_t].\text{Value} = T_i[\alpha, \beta].\text{Value}$$

Secondly, we consider the case when $v \in S$. Let $U = (N(v) \cap X_j) \setminus \alpha^{-1}(0)$. Since v is in X_j but v is not in its parent node X_i, v should have degree d in the (α, β)-constrained PROBABILISTIC-CORE problem. Then, $\alpha(v) = a$ and $\beta(v) = d - a$ for an $0 \leq a \leq d$. Let $Y \subseteq U$ of size a be the set of vertices that have an edge to v to compensate the degree constraint $\alpha(v)$ at v. Then, each vertex $u \in Y$ will gain a degree constraint $\alpha(u)$ and lose a degree constraint $\beta(u)$. Using the integer a and the set Y, the probability $\rho_d(\alpha, \beta, \mathcal{G}_i^{\alpha,\beta}, S)$ can be written as follows:

$$\rho_d(\alpha, \beta, \mathcal{G}_i^{\alpha,\beta}, S) = \rho_d(\alpha_{a,Y}, \beta_{a,Y}, \mathcal{G}_j^{\alpha_{a,Y}, \beta_{a,Y}}, S).$$

Since j is a node at height $h-1$, we know that the row $T_j[\alpha_{a,Y}, \beta_{a,Y}]$ is computed optimally. That is,

$$\rho_d(\alpha_{a,Y}, \beta_{a,Y}, \mathcal{G}_j^{\alpha_{a,Y}, \beta_{a,Y}}, S) \leq T_j[\alpha_{a,Y}, \beta_{a,Y}].\text{Value} \leq T_j[\alpha_t, \beta_t].\text{Value} = T_i[\alpha, \beta].\text{Value}$$

This completes the argument of the case when i is a forget node.

When i is a join node. Let j and g be the children of i, and $X_i = X_j = X_g$. The set $S \setminus X_i$ can be partitioned into sets S_j and S_g where $S_j = (S \setminus X_i) \cap X_j$ and $S_g = (S \setminus X_i) \cap X_g$. Let $U = \{u \in X_i \mid \alpha(v) > 0\}$ and $S_i = X_i \setminus \alpha^{-1}(0)$. For each vertex $u \in U$, the degree constraint $\alpha(u)$ should be satisfied by the edges from U to u. Since $X_i = X_j = X_g$, the degree constraint $\alpha(u)$ is either satisfied in the node j or node g and not in both. Without loss of generality we assume that the degree constraint α is satisfied in the node j and no zero degree constraint α in the node g. Then we define $\alpha' : X_g \to \{0, 1, \ldots, d\}$ to be for every vertex $v \in X_i$, $\alpha'(v) = 0$. For each vertex $v \in X_i$ with $\beta(v) > 0$, the degree constraint $\beta(v)$ can be satisfied by the sets S_j and S_g together. There exists an integer $x(v)$ such that $x(v)$ neighbors in the core are obtained from the set S_j and $\beta(v) - x(v)$ neighbors in the core are obtained from the set S_g. Then, there exists a function $x : X_i \to \{0, 1, \ldots, d\}$ such that for each vertex $v \in V$, $0 \leq x(v) \leq \beta(v)$. Using the function x, the probability $\rho_d(\alpha, \beta, \mathcal{G}_i^{\alpha,\beta}, S)$ can be given as follows:

$$\rho_d(\alpha, \beta, \mathcal{G}_i^{\alpha,\beta}, S) = \rho_d(\alpha, \beta_x, \mathcal{G}_j^{\alpha,\beta_x}, S_i \cup S_j) \cdot \rho_d(\alpha', \beta'_x, \mathcal{G}_g^{\alpha',\beta'_x}, S_i \cup S_g)$$

The functions β_x and β'_x for the given function x are defined in the description of dynamic programming. Since both j and g are nodes at height at most $h-1$, by induction hypothesis, the rows $T_j[\alpha_j, \beta_x]$ and $T_g[\alpha', \beta'_x]$ are computed optimally. Therefore, we have

$$\begin{aligned} \rho_d(\alpha, \beta, \mathcal{G}_i^{\alpha,\beta}, S) &= \rho_d(\alpha, \beta_x, \mathcal{G}_j^{\alpha,\beta_x}, S_i \cup S_j) \cdot \rho_d(\alpha', \beta'_x, \mathcal{G}_g^{\alpha',\beta'_x}, S_i \cup S_g) \\ &\leq T_j[\alpha, \beta_x].\text{Value} \cdot T_g[\alpha', \beta'_x].\text{Value} \\ &\leq T_j[\alpha, \beta_y].\text{Value} \cdot T_g[\alpha', \beta'_y].\text{Value} = T_i[\alpha, \beta].\text{Value} \end{aligned}$$

where y is the optimal distribution of the degree which results in the $T_i[\alpha, \beta]$.Value, as defined in the recursive formulation at a join node. This completes the argument for the case when i is a join node. Hence the lemma. □

For a possible world $H \sqsubseteq \mathcal{G}$, the degree of a vertex $v \in V(H)$ is at most the degree of v in $G = (V, E)$. A vertex with degree less than d in G cannot be an element of a core in any possible world. Thus, those vertices can be excluded throughout the algorithm. Such a pruning results in either the pruned graph being an empty graph or the minimum degree of the pruned graph is becoming at least d. We state the following Lemma from Koster et al. [48].

Lemma 8 (Koster et al [48]). *For a graph $G = (V, E)$ with $|V| \geq 2$, $\delta(G) \leq tw(G)$.*

In the following lemma, we analyze the running time of the dynamic programming algorithm.

Lemma 9. *The* PROBABILISTIC-CORE-RF *problem can be solved optimally in time $2^{\mathcal{O}(tw \log tw)} n^{\mathcal{O}(1)}$.*

Proof. For each node i in $\mathcal{T}$, our dynamic programming generates a table T_i with $(d+1)^{2tw} = \mathcal{O}(d^{2tw})$ rows. Each row $T_i[\alpha, \beta]$ for some $\alpha, \beta : X_i \to \{0, 1, \ldots, d\}$ is computed in our dynamic programming based on the type of node i. When i is leaf node, a single row exists in T_i and that is computed in $\mathcal{O}(1)$ time. When i is introduce node, we consider two cases that $\alpha(v) = 0$ and $\alpha(v) > 0$. If $\alpha(v) = 0$, then the functions α_j and β_j can be computed in time $\mathcal{O}(tw)$. If $\alpha(v) > 0$, the set U' can be found by enumerating all $\alpha(v)$ sized sets in U. This will take $\mathcal{O}(|Y|^{\beta(v)})$ time and using the upper bound on values we get $\mathcal{O}(tw^d)$. It follows that if i is an introduce node, then $T_i[\alpha, \beta]$ can be computed in time $\mathcal{O}(tw^d)$. When i is a forget node, for each value of $0 \leq a \leq min(d, |U|)$, we enumerate all a sized subsets of Y. This requires $\mathcal{O}(tw^k k)$ time. When i is join node, we need to compute the optimal distribution of $\beta(v)$ for each vertex $v \in X_i$. This requires $\mathcal{O}(d^{tw})$ time. From Lemma 8, we know that $d \leq tw$, we upper bound the value d by tw. Overall, a row $T_i[\alpha, \beta]$ can be computed in time $\mathcal{O}(tw^{tw})$. The entire table T_i can be computed in time $\mathcal{O}(tw^{3tw}) = 2^{\mathcal{O}(tw \log tw)}$. The nice tree decomposition $(X, \mathcal{T})$ has $\mathcal{O}(n \cdot tw)$ many nodes and a table on each node can be computed in time $\mathcal{O}(2^{\mathcal{O}(tw \log tw)}) n^{\mathcal{O}(1)}$. An optimal solution to the PROBABILISTIC-CORE-RF problem on the input uncertain graph $\mathcal{G}$ will be obtained from the table of the root node r. That is, $T_r[\alpha : \emptyset \to \{0\}, \beta : \emptyset \to \{0\}]$.Solution is the optimal solution. □

The preceding lemmas results in the following theorem.

Theorem 3. *The* PROBABILISTIC-CORE *problem admits an FPT algorithm for the parameter tw.*

6. Discussion

There are many natural questions related to the parameterized complexity of algorithms on uncertain graphs under different distribution models. The following are some open questions.

1. Are there efficient reductions between distribution models so that we can classify problems based on the efficiency of algorithms under different distribution models? This question is also

of practical significance because the distribution models are specified as sampling algorithms. Consequently, the complexity of expectation computation on uncertain graphs under different distribution models is an interesting new parameterization. Further, one concrete question is whether the LRO model is *easier* than the RF model for other optimization problems on uncertain graphs. For the two case studies considered in this paper, that is the case.

2. While our results do support the natural intuition that a tree decomposition is helpful in expectation computation, it is unclear to us how traditional techniques in parameterized algoritihms can be carried over to this setting. In particular, it is unclear to us as to whether for any distribution model, a kernelization based algorithm can give an FPT algorithm on uncertain graphs.

3. We have considered the coverage and the core problems on uncertain graphs under the LRO and RF models. However, we have not been able to get an FPT algorithm with the parameter treewidth for MAX-EXP-COVER-1-RF. Actually, any approach to avoid the exponential dependence on ($\Delta \cdot tw$) would be very interesting and would give a significant insight on other approaches to evaluate the expected coverage.

4. Even though the INDIVIDUAL-CORE-RF problem and PROBABILISTIC-CORE-RF problem are similar, we have not been able to get an FPT algorithm for the INDIVIDUAL-CORE-RF problem with treewidth as the parameter. Even for other structural parameters such as vertex-cover number and feedback vertex set number, FPT results will give a significant insight on the INDIVIDUAL-CORE problem.

Author Contributions: Investigation, N.S.N. and R.V.; Writing–original draft, N.S.N. and R.V.; All authors have read and agreed to the published version of the manuscript.

Funding: This research received no external funding

Conflicts of Interest: The authors declare no conflict of interest.

References

1. Añez, J.; Barra, T.D.L.; Pérez, B. Dual graph representation of transport networks. *Trans. Res. Part B Methodol.* **1996**, *30*, 209–216. [CrossRef]
2. Hua, M.; Pei, J. Probabilistic Path Queries in Road Networks: Traffic Uncertainty Aware Path Selection. In Proceedings of the 13th International Conference on Extending Database Technology (EDBT '10), Lausanne, Switzerland, 22–26 March 2010; pp. 347–358, doi:10.1145/1739041.1739084. [CrossRef]
3. Asthana, S.; King, O.D.; Gibbons, F.D.; Roth, F.P. Predicting protein complex membership using probabilistic network reliability. *Genome Res.* **2004**, *14 6*, 1170–5. [CrossRef]
4. Domingos, P.; Richardson, M. Mining the Network Value of Customers. In Proceedings of the Seventh ACM SIGKDD International Conference on Knowledge Discovery and Data Mining (KDD '01), San Francisco, CA, USA, 26–29 August 2001; pp. 57–66, doi:10.1145/502512.502525. [CrossRef]
5. Frank, H. Shortest Paths in Probabilistic Graphs. *Oper. Res.* **1969**, *17*, 583–599. doi:10.1287/opre.17.4.583. [CrossRef]
6. Valiant, L.G. The Complexity of Enumeration and Reliability Problems. *SIAM J. Comput.* **1979**, *8*, 410–421. doi:10.1137/0208032. [CrossRef]
7. Hoffmann, M.; Erlebach, T.; Krizanc, D.; Mihalák, M.; Raman, R. Computing Minimum Spanning Trees with Uncertainty. In Proceedings of the 25th Annual Symposium on Theoretical Aspects of Computer Science, Bordeaux, France, 21–23 February 2008; pp. 277–288, doi:10.4230/LIPIcs.STACS.2008.1358. [CrossRef]
8. Focke, J.; Megow, N.; Meißner, J. Minimum Spanning Tree under Explorable Uncertainty in Theory and Experiments. In Proceedings of the 16th International Symposium on Experimental Algorithms (SEA 2017), London, UK, 21–23 June 2017; pp. 22:1–22:14, doi:10.4230/LIPIcs.SEA.2017.22. [CrossRef]
9. Frank, H.; Hakimi, S. Probabilistic Flows Through a Communication Network. *IEEE Trans. Circuit Theory* **1965**, *12*, 413–414. doi:10.1109/TCT.1965.1082452. [CrossRef]
10. Evans, J.R. Maximum flow in probabilistic graphs-the discrete case. *Networks* **1976**, *6*, 161–183.10.1002/net.3230060208. [CrossRef]

11. Hassin, R.; Ravi, R.; Salman, F.S. Tractable Cases of Facility Location on a Network with a Linear Reliability Order of Links. In *Algorithms-ESA 2009, Proceedings of the 17th Annual European Symposium, Copenhagen, Denmark, 7–9 September 2009*; Springer: Berlin, Germany, 2009; pp. 275–276.
12. Hassin, R.; Ravi, R.; Salman, F.S. Multiple facility location on a network with linear reliability order of edges. *J. Comb. Optim.* **2017**, *34*, 1–25. [CrossRef]
13. Narayanaswamy, N.S.; Nasre, M.; Vijayaragunathan, R. Facility Location on Planar Graphs with Unreliable Links. In Proceedings of the Computer Science-Theory and Applications-13th International Computer Science Symposium in Russia, CSR 2018, Moscow, Russia, 6–10 June 2018; pp. 269–281. doi:10.1007/978-3-319-90530-3_23. [CrossRef]
14. Kempe, D.; Kleinberg, J.M.; Tardos, É. Maximizing the spread of influence through a social network. In Proceedings of the Ninth ACM SIGKDD International Conference on Knowledge Discovery and Data Mining, Washington, DC, USA, 24–27 August 2003; pp. 137–146, doi:10.1145/956750.956769. [CrossRef]
15. Bonchi, F.; Gullo, F.; Kaltenbrunner, A.; Volkovich, Y. Core decomposition of uncertain graphs. In Proceedings of the 20th ACM SIGKDD International Conference on Knowledge Discovery and Data Mining (KDD '14), New York, NY, USA, 24–27 August 2014; pp. 1316–1325, doi:10.1145/2623330.2623655. [CrossRef]
16. Peng, Y.; Zhang, Y.; Zhang, W.; Lin, X.; Qin, L. Efficient Probabilistic K-Core Computation on Uncertain Graphs. In Proceedings of the 34th IEEE International Conference on Data Engineering (ICDE), Paris, France, 16–19 April 2018; pp. 1192–1203, doi:10.1109/ICDE.2018.00110. [CrossRef]
17. Ball, M.O.; Provan, J.S. Calculating bounds on reachability and connectedness in stochastic networks. *Networks* **1983**, *13*, 253–278. doi:10.1002/net.3230130210. [CrossRef]
18. Zou, Z.; Li, J. Structural-Context Similarities for Uncertain Graphs. In Proceedings of the 2013 IEEE 13th International Conference on Data Mining, Dallas, TX, USA, 7–10 December 2013; pp. 1325–1330, doi:10.1109/ICDM.2013.22. [CrossRef]
19. Daskin, M.S. A Maximum Expected Covering Location Model: Formulation, Properties and Heuristic Solution. *Transp. Sci.* **1983**, *17*, 48–70. [CrossRef]
20. Ball, M.O. Complexity of network reliability computations. *Networks* **1980**, *10*, 153–165. [CrossRef]
21. Karp, R.M.; Luby, M. Monte-Carlo algorithms for the planar multiterminal network reliability problem. *J. Complex.* **1985**, *1*, 45–64. doi:10.1016/0885-064X(85)90021-4. [CrossRef]
22. Provan, J.S.; Ball, M.O. The Complexity of Counting Cuts and of Computing the Probability that a Graph is Connected. *SIAM J. Comput.* **1983**, *12*, 777–788. doi:10.1137/0212053. [CrossRef]
23. Guo, H.; Jerrum, M. A Polynomial-Time Approximation Algorithm for All-Terminal Network Reliability. *SIAM J. Comput.* **2019**, *48*, 964–978. doi:10.1137/18M1201846. [CrossRef]
24. Ghosh, J.; Ngo, H.Q.; Yoon, S.; Qiao, C. On a Routing Problem Within Probabilistic Graphs and its Application to Intermittently Connected Networks. In Proceedings of the 26th IEEE International Conference on Computer Communications, Joint Conference of the IEEE Computer and Communications Societies, INFOCOM, Anchorage, AK, USA, 6–12 May 2007; pp. 1721–1729, doi:10.1109/INFCOM.2007.201. [CrossRef]
25. Rubino, G. *Network Performance Modeling and Simulation*; chapter Network Reliability Evaluation; Gordon and Breach Science Publishers, Inc.: Newark, NJ, USA, 1999; pp. 275–302.
26. Swamynathan, G.; Wilson, C.; Boe, B.; Almeroth, K.C.; Zhao, B.Y. Do social networks improve e-commerce?: a study on social marketplaces. In Proceedings of the first Workshop on Online Social Networks (WOSN 2008), Seattle, WA, USA, 17–22 August 2008; pp. 1–6, doi:10.1145/1397735.1397737. [CrossRef]
27. White, D.R.; Harary, F. The Cohesiveness of Blocks In Social Networks: Node Connectivity and Conditional Density. *Soc. Methodol.* **2001**, *31*, 305–359. doi:10.1111/0081-1750.00098. [CrossRef]
28. Papadimitriou, C.H.; Yannakakis, M. Shortest paths without a map. *Theor. Comput. Sci.* **1991**, *84*, 127–150. doi:10.1016/0304-3975(91)90263-2. [CrossRef]
29. Khuller, S.; Moss, A.; Naor, J. The Budgeted Maximum Coverage Problem. *Inf. Process. Lett.* **1999**, *70*, 39–45. doi:10.1016/S0020-0190(99)00031-9. [CrossRef]
30. Brown, J.J.; Reingen, P.H. Social Ties and Word-of-Mouth Referral Behavior. *J. Consum. Res.* **1987**, *14*, 350–362. [CrossRef]
31. Richardson, M.; Domingos, P.M. Mining knowledge-sharing sites for viral marketing. In Proceedings of the Eighth ACM SIGKDD International Conference on Knowledge Discovery and Data Mining, Edmonton, AB, Canada, 23–26 July 2002; pp. 61–70, doi:10.1145/775047.775057. [CrossRef]
32. Bass, F.M. A New Product Growth for Model Consumer Durables. *Manag. Sci.* **1969**, *15*, 215–227. [CrossRef]

33. Snyder, L.V. Facility location under uncertainty: A review. *IIE Trans.* **2006**, *38*, 547–564. doi:10.1080/07408170500216480. [CrossRef]
34. Eiselt, H.A.; Gendreau, M.; Laporte, G. Location of facilities on a network subject to a single-edge failure. *Networks* **1992**, *22*, 231–246. doi:10.1002/net.3230220303. [CrossRef]
35. Colbourn, C.J.; Xue, G. A linear time algorithm for computing the most reliable source on a series-parallel graph with unreliable edges. *Theor. Comput. Sci.* **1998**, *209*, 331–345. doi:10.1016/S0304-3975(97)00124-2. [CrossRef]
36. Ding, W. Computing the Most Reliable Source on Stochastic Ring Networks. In Proceedings of the 2009 WRI World Congress on Software Engineering, Xiamen, China, 19–21 May 2009; Volume 1, pp. 345–347, doi:10.1109/WCSE.2009.31. [CrossRef]
37. Ding, W.; Xue, G. A linear time algorithm for computing a most reliable source on a tree network with faulty nodes. *Theor. Comput. Sci.* **2011**, *412*, 225–232. doi:10.1016/j.tcs.2009.08.003. [CrossRef]
38. Melachrinoudis, E.; Helander, M.E. A single facility location problem on a tree with unreliable edges. *Networks* **1996**, *27*, 219–237. doi:10.1002/(SICI)1097-0037(199605)27:3. [CrossRef]
39. Nemhauser, G.L.; Wolsey, L.A.; Fisher, M.L. An analysis of approximations for maximizing submodular set functions—I. *Math. Program.* **1978**, *14*, 265–294. doi:10.1007/BF01588971. [CrossRef]
40. Cygan, M.; Fomin, F.V.; Kowalik, L.; Lokshtanov, D.; Marx, D.; Pilipczuk, M.; Pilipczuk, M.; Saurabh, S. *Parameterized Algorithms*; Springer: Berlin, Germany, 2015. doi:10.1007/978-3-319-21275-3. [CrossRef]
41. Sigal, C.E.; Pritsker, A.A.B.; Solberg, J.J. The Stochastic Shortest Route Problem. *Oper. Res.* **1980**, *28*, 1122–1129. [CrossRef]
42. Guerin, R.A.; Orda, A. QoS routing in networks with inaccurate information: Theory and algorithms. *IEEE/ACM Trans. Netw.* **1999**, *7*, 350–364. doi:10.1109/90.779203. [CrossRef]
43. Günneç, D.; Salman, F.S. Assessing the reliability and the expected performance of a network under disaster risk. In Proceedings of the International Network Optimization Conference (INOC), Spa, Belgium, 22–25 April 2007.
44. Diestel, R. *Graph Theory*, 4th ed.; Graduate Texts in Mathematics; Springer: Berlin, Germany, 2012; Volume 173.
45. Bodlaender, H.L. A Tourist Guide through Treewidth. *Acta Cybern.* **1993**, *11*, 1–21.
46. Kloks, T. *Treewidth, Computations and Approximations*; Lecture Notes in Computer Science; Springer: Berlin, Germany, 1994; Volume 842.
47. Downey, R.G.; Fellows, M.R. Fixed-parameter intractability. In Proceedings of the Seventh Annual Structure in Complexity Theory Conference, Boston, Massachusetts, USA, 22–25 June 1992.
48. Koster, A.M.C.A.; Wolle, T.; Bodlaender, H.L. Degree-Based Treewidth Lower Bounds. In Proceedings of the 4th InternationalWorkshop, WEA 2005 Experimental and Efficient Algorithms, Santorini Island, Greece, 10–13 May 2005; pp. 101–112, doi:10.1007/11427186_11. [CrossRef]

© 2019 by the authors. Licensee MDPI, Basel, Switzerland. This article is an open access article distributed under the terms and conditions of the Creative Commons Attribution (CC BY) license (http://creativecommons.org/licenses/by/4.0/).

Article

A Survey on Approximation in Parameterized Complexity: Hardness and Algorithms

Andreas Emil Feldmann [1,*], Karthik C. S. [2,*], Euiwoong Lee [3,*] and Pasin Manurangsi [4,*]

1. Department of Applied Mathematics (KAM), Charles University, 118 00 Prague, Czech Republic
2. Department of Computer Science, Tel Aviv University, Tel Aviv 6997801, Israel
3. Courant Institute of Mathematical Sciences, New York University, New York, NY 10012, USA
4. Google Research, Mountain View, CA 94043, USA
* Correspondence: feldmann.a.e@gmail.com (A.E.F.); karthik0112358@gmail.com (K.C.S.); euiwoong@cims.nyu.edu (E.L.); pasin@google.com (P.M.)

Received: 1 October 2019; Accepted: 2 June 2020; Published: 19 June 2020

Abstract: Parameterization and approximation are two popular ways of coping with NP-hard problems. More recently, the two have also been combined to derive many interesting results. We survey developments in the area both from the algorithmic and hardness perspectives, with emphasis on new techniques and potential future research directions.

Keywords: parameterized complexity; approximation algorithms; hardness of approximation

1. Introduction

In their seminal papers of the mid 1960s, Cobham [1] and Edmonds [2] independently phrased what is now known as the Cobham–Edmonds thesis. It states that an optimization problem is feasibly solvable if it admits an algorithm with the following two properties:

1. Accuracy: the algorithm should always compute the best possible (optimum) solution.
2. Efficiency: the runtime of the algorithm should be polynomial in the input size n.

Shortly after the Cobham–Edmonds thesis was formulated, the development of the theory of NP-hardness and reducibility identified a whole plethora of problems that are seemingly intractable, i.e., for which algorithms with the above two properties do not seem to exist. Even though the reasons for this phenomenon remain elusive up to this day, this has not hindered the development of algorithms for such problems. To obtain an algorithm for an NP-hard problem, at least one of the two properties demanded by the Cobham–Edmonds thesis needs to be relaxed. Ideally, the properties are relaxed as little as possible, in order to stay close to the notion of feasible solvability suggested by the thesis.

A very common approach is to relax the accuracy condition, which means aiming for approximation algorithms [3,4]. The idea here is to use only polynomial time to compute an α-approximation, i.e., a solution that is at most a factor α times worse than the optimum solution obtainable for the given input instance. Such an algorithm may also be randomized, i.e., there is either a high probability that the output is an α-approximation, or the runtime is polynomial in expectation.

In a different direction, several relaxations of the efficiency condition have also been proposed. Popular among these is the notion of parameterized algorithms [5,6]. Here the input comes together with some parameter $k \in \mathbb{N}$, which describes some property of the input and can be expected to be small in typical applications. The idea is to isolate the seemingly necessary exponential runtime of NP-hard problems to the parameter, while the runtime dependence on the input size n remains polynomial. In particular, the algorithm should compute the optimum solution in $f(k)n^{O(1)}$ time, for some computable function $f : \mathbb{N} \to \mathbb{N}$ independent of the input size n. If such an algorithm exists for a problem it is fixed-parameter tractable (FPT), and the algorithm is correspondingly referred to as an FPT algorithm.

Approximation and FPT algorithms have been studied extensively for the past few decades, and this has lead to a rich literature on algorithmic techniques and deep links to other research fields within mathematics. However, in this process the limitations of these approaches have also become apparent. Some NP-hard problems can fairly be considered to be feasibly solvable in the respective regimes, as they admit polynomial-time algorithms with rather small approximation factors, or can be shown to be solvable optimally with only a fairly small exponential runtime overhead due to the parameter. However, many problems can also be shown not to admit any reasonable algorithms in either of these regimes, assuming some standard complexity assumptions. Thus considering only approximation and FPT algorithms, as has been mostly done in the past, we are seemingly stuck in a swamp of problems for which we have substantial evidence that they cannot be feasibly solved.

To find a way out of this dilemma, an obvious possibility is to lift both the accuracy and the efficiency requirements of the Cobham–Edmonds thesis. In this way, we obtain a parameterized α-approximation algorithm, which computes an α-approximation in $f(k)n^{O(1)}$ time for some computable function f, given an input of size n with parameter k. The study of such algorithms had been suggested dating back to the early days of parameterized complexity (cf. [5,7–9]), and we refer the readers to an excellent survey of Marx [8] for discussions on the earlier results in the area.

Recently this approach has received some increased interest, with many new results obtained in the past few years, both in terms of algorithms and hardness of approximation. The aim of this survey is to give an overview on some of these newer results. We would like to caution the readers that the goal of this survey is not to compile all known results in the field but rather to give examples that demonstrate the flavor of questions studied, techniques used to obtain them, and some potential future research directions. Finally, we remark that on the broader theme of approximation in P, there was an excellent survey recently made available by Rubinstein and Williams [10] focusing on the approximability of popular problems in P which admit simple quadratic/cubic algorithms.

Organization of the Survey

The main body of the survey is organized into two sections: one on FPT hardness of approximation (Section 3) and the other on FPT approximation algorithms (Section 4). Before these two main sections, we list some notations and preliminaries in Section 2. Finally, in Section 5, we highlight some open questions and future directions (although a large list of open problems have also been detailed throughout Sections 3 and 4).

2. Preliminaries

In this section, we review several notions that will appear regularly throughout the survey. However, we do not include definitions of basic concepts such as W-hardness, para-NP-hardness, APX-hardness, and so forth; the interested reader may refer to [3–6] for these definitions.

Parameterized approximation algorithms. We briefly specify the different types of algorithms we will consider. As already defined in the introduction, an FPT algorithm computes the optimum solution in $f(k)n^{O(1)}$ time for some parameter k and computable function $f : \mathbb{N} \to \mathbb{N}$ on inputs of size n. The common choices of parameters are the standard parameters based on solution size, structural parameters, guarantee parameters, and dual parameters.

An algorithm that computes the optimum solution in $f(k)n^{g(k)}$ time for some parameter k and computable functions $f, g : \mathbb{N} \to \mathbb{N}$, is called a slice-wise polynomial (XP) algorithm. If the parameter is the approximation factor, i.e., the algorithm computes a $(1 + \varepsilon)$-approximation in $f(\varepsilon)n^{g(\varepsilon)}$ time, then it is called a polynomial-time approximation scheme (PTAS). The latter type of algorithm has been studied avant la lettre for quite a while. This is also true for the corresponding FPT algorithm, which computes a $(1+\varepsilon)$-approximation in $f(\varepsilon)n^{O(1)}$ time, and is referred to as an efficient polynomial-time approximation scheme (EPTAS). Note that if the standard parameterization of an optimization problem is W[1]-hard, then the optimization problem does not have an EPTAS (unless FPT = W[1]) [11].

Some interesting links between these algorithms, traditionally studied from the perspective of polynomial-time approximation algorithms, and parameterized complexity have been uncovered more recently [8,11–16].

As also mentioned in the introduction, a parameterized α-approximation algorithm computes an α-approximation in $f(k)n^{O(1)}$ time for some parameter k on inputs of size n. If α can be set to $1+\varepsilon$ for any $\varepsilon > 0$ and the runtime is $f(k,\varepsilon)n^{g(\varepsilon)}$, then we obtain a parameterized approximation scheme (PAS) for parameter k. Note that this runtime is only truly FPT if we assume that ε is constant. If we forbid this and consider ε as a parameter as well, i.e., the runtime should be of the form $f(k,\varepsilon)n^{O(1)}$, then we obtain EPAS.

Kernelization. A further topic closely related to the FPT algorithms is kernelization. Here, the idea is that an instance is efficiently pre-processed by removing the "easy parts" so that only the NP-hard core of the instance remains. More concretely, a kernelization algorithm takes an instance I and a parameter k of some problem and computes a new instance I' with parameter k' of the same problem. The runtime of this algorithm is polynomial in the size of the input instance I and k, while the size of the output I' and k' is bounded as a function of the input parameter k. For optimization problems, it should also be the case that any optimum solution to I' can be converted to an optimum solution of I in polynomial time. The new instance I' is called the kernel of I (for parameter k). A fundamental result in fixed-parameter tractability is that an (optimization) problem parameterized by k is FPT if and only if it admits a kernelization algorithm for the same parameter [17]. However the size of the guaranteed kernel will in general be exponential (or worse) in the input parameter. Therefore, an interesting question is whether an NP-hard problem admits small kernels of polynomial size. This can be interpreted as meaning that the problem has a very efficient pre-processing algorithm, which can be used prior to solving the kernel. It also gives an additional dimension to the parameterized complexity landscape.

Kernelization has played a fundamental role in the development of FPT algorithms, where a pre-processing step is often used to simplify the structure of the input instance. It is therefore only natural to consider such pre-processing algorithms for parameterized approximation algorithms as well. Lokshtanov et al. [18] define an α-approximate kernelization algorithm, which computes a kernel I' such that any β-approximation in I' can be converted into an $\alpha\beta$-approximation to the input instance I in polynomial time. Again, the size of I' and k' need to be bounded as a function of the input parameter k, and the algorithm needs to run in polynomial time. The instance I' is now called an α-approximate kernel. Analogous to exact kernels, any problem has a parameterized α-approximation algorithm if and only if it admits an α-approximate kernel for the same parameter [18], which however might be of exponential size in the parameter.

An α-approximate kernelization algorithm that computes a polynomial-sized kernel, and for which we may set α to $1+\varepsilon$ for any $\varepsilon > 0$, is called a polynomial-sized approximate kernelization scheme (PSAKS). In this case ε is necessarily considered to be a constant, since any kernelization algorithms needs to run in polynomial time.

We remark here that apart from α-approximate kernels, there is another common workaround for problems with no polynomial kernels, captured using the notion of Turing kernels. There is also a lower bound framework for Turing kernels [19], and the question of approximate kernels for problems that do not even admit Turing kernels is fairly natural to ask. However we skip this discussion for the sake of brevity.

Finally, note that in literature, there is another notion of approximate kernels called α-fidelity kernelization [20] which is different from the one mentioned above. Essentially, an α-fidelity kernel is a polynomial time preprocessing procedure such that an optimal solution to the reduced instance translates to an α-approximate solution to the original. This definition allows a loss of precision in the preprocessing step, but demands that the reduced instance has to be solved to optimality. See [18] for a detailed discussion on the differences between the two approximate kernel notions.

Complexity-Theoretic Hypotheses. We assume that the readers have basic knowledge of (classic) parameterized complexity theory, including the W-hierarchy, the exponential time hypothesis (ETH), and the strong exponential time hypothesis (SETH). The reader may choose to recapitulate these definitions by referring to [6] (Sections 13 and 14).

We will additionally discuss two hypotheses that may not be standard to the community. The first is the Gap Exponential Time Hypothesis (Gap-ETH), which is a strengthening of ETH. Roughly speaking, it states that even the approximate version of 3SAT cannot be solved in subexponential time; a more formal statement of Gap-ETH can be found in Hypothesis 2. Another hypothesis we will discuss is the Parameterized Inapproximability Hypothesis (PIH), which states that the multicolored version of the DENSEST k-SUBGRAPH is hard to approximate in FPT time. Once again, we do not define PIH formally here; please refer to Hypothesis 1 for a formal statement.

3. FPT Hardness of Approximation

In this section, we focus on showing barriers against obtaining good parameterized approximation algorithms. The analogous field of study in the non-parameterized (NP-hardness) regime is the theory of hardness of approximation. The celebrated PCP Theorem [21,22] and numerous subsequent works have developed a rich set of tools that allowed researchers to show tight inapproximability results for many fundamental problems. In the context of parameterized approximation, the field is still in the nascent stage. Nonetheless, there have been quite a few tools that have already been developed, which are discussed in the subsequent subsections.

We divide this section into two parts. In Section 3.1, we discuss the results and techniques in the area of hardness of parameterized approximation under the standard assumption of $W[1] \neq FPT$. In Section 3.2, we discuss results and techniques in hardness of parameterized approximation under less standard assumptions such as the Gap Exponential Time Hypothesis, where the gap is inherent in the assumption, and the challenge is to construct gap-preserving reductions.

3.1. W[1]-Hardness of Gap Problems

In this subsection, we discuss W[1]-hardness of approximation of a few fundamental problems. In particular, we discuss the parameterized inapproximability (i.e., W[1]-hard to even approximate) of the DOMINATING SET problem, the (ONE-SIDED) BICLIQUE problem, the EVEN SET problem, the SHORTEST VECTOR problem, and the STEINER ORIENTATION problem. We emphasize here that the main difficulty that is addressed in this subsection is gap generation, i.e., we focus on how to start from a hard problem (with no gap), say k-CLIQUE (which is the canonical W[1]-complete problem), and reduce it to one of the aforementioned problems, while generating a non-trivial gap in the process.

3.1.1. Parameterized Intractability of Biclique and Applications to Parameterized Inapproximability

In this subsubsection, we will discuss the parametrized inapproximability of the one-sided biclique problem, and show how both that result and its proof technique lead to more inapproximability results.

We begin our discussion by formally stating the k-BICLIQUE problem where we are given as input a graph G and an integer k, and the goal is to determine whether G contains a complete bipartite subgraph with k vertices on each side. The complexity of k-BICLIQUE was a long standing open problem and was resolved only recently by Lin [23] where he showed that it is W[1]-hard. In fact, he showed a much stronger result and this shall be the focus of attention in this subsection.

Theorem 1 ([23]). *Given a bipartite graph $G(L \dot\cup R, E)$ and $k \in \mathbb{N}$ as input, it is W[1]-hard to distinguish between the following two cases:*

- *Completeness: There are k vertices in L with at least $n^{\Theta(\frac{1}{k})}$ common neighbors in R;*
- *Soundness: Any k vertices in L have at most $(k+6)!$ common neighbors in R.*

We shall refer to the gap problem in the above theorem as the ONE-SIDED k-BICLIQUE problem. To prove the above result, Lin introduced a technique which we shall refer to as Gadget Composition. The gadget composition technique has found more applications since [23]. We provide below a failed approach (given in [23]) to prove the above theorem; nonetheless it gives us good insight into how the gadget composition technique works.

Suppose we can construct a set family $\mathcal{T} = \{S_1, S_2, \ldots, S_n\}$ of subsets of $[n]$ for some integers k, n and $h > \ell$ (for example, $h = n^{1/k}$ and $\ell = (k+1)!$) such that:

Property 1: Any $k+1$ distinct subsets in $\mathcal{T}$ have intersection size at most ℓ;
Property 2: Any k distinct subsets in $\mathcal{T}$ have intersection size at least h.

Then we can combine $\mathcal{T}$ with an instance of k-CLIQUE to obtain a gap instance of ONE-SIDED k-BICLIQUE as follows. Given a graph G and parameter k with $V(G) \subseteq [n]$, we construct our instance of ONE-SIDED k-BICLIQUE, say $H(L \dot\cup R, E(H))$ by setting $L := E(G)$ and $R := [n]$, where for any $(v_i, v_j) \in L$ and $v \in [n]$, we have that $((v_i, v_j), v) \in E(H)$ if and only if $v \in S_i \cap S_j$. Let $s := k(k-1)/2$. It is easy to check that if G has a k-vertex clique, say $\{v_1^*, \ldots, v_k^*\}$ is a clique in G, then Property 2 implies that $|\Delta := \bigcap_{i \in [k]} S_{v_i^*}| \geq h$. It follows that the set of s vertices in L given by $\{(v_i^*, v_j^*) :$ for all $\{i,j\} \in \binom{[k]}{2}\}$ are neighbors of every vertex in $\Delta \subseteq R$. On the other hand, if G contains no k-vertex clique, then any s distinct vertices in L (i.e., s edges in G) must have at least $k+1$ vertices in G as their end points. Say V' was the set of all vertices contained the s edges. By Property 1, we know that $|\Delta' := \bigcap_{v \in V'} S_v| \leq \ell$, and thus any s distinct vertices in L have at most ℓ common neighbors in R.

It is indeed very surprising that this technique can yield non-trivial inapproximability results, as the gap is essentially produced from the gadget and is oblivious to the input! This also stands in stark contrast to the PCP theorem and hardness of approximation results in NP, where all known results were obtained by global transformations on the input. The key difference between the parameterized and NP worlds is the notion of locality. For example, consider the k-CLIQUE problem, if a graph does not have a clique of size k, then given any k vertices, a random vertex pair in these k vertices does not have an edge with probability at least $1/k^2$. It is philosophically possible to compose the input graph with a simple error correcting code to amplify this probability to a constant, as we are allowed to blowup the input size by any function of k. In contrast, when k is not fixed, like in the NP world, k is of the same magnitude as the input size, and thus we are only allowed to blow up the input size by poly(n) factor. Nonetheless, we have to point out that the gadgets typically needed to make the gadget composition technique work must be extremely rich in combinatorial structure (and are typically constructed from random objects or algebraic objects), and were previously studied extensively in the area of extremal combinatorics.

Returning to the reduction above from k-CLIQUE to ONE-SIDED k-BICLIQUE, it turns out that we do not know how to construct the set system $\mathcal{T}$, and hence the reduction does not pan out. Nonetheless Lin constructed a variant of $\mathcal{T}$, where Property 2 was more refined and the reduction from k-CLIQUE to ONE-SIDED k-BICLIQUE, went through with slightly more effort.

Before we move on to discussing some applications of Theorem 1 and the gadget composition technique, we remark about known stronger time lower bound for ONE-SIDED k-BICLIQUE under stronger running time hypotheses. Lin [23] showed a lower bound of $n^{\Omega(\sqrt{k})}$ for ONE-SIDED k-BICLIQUE assuming ETH. We wonder if this can be further improved.

Open Question 1 (Lower bound of ONE-SIDED k-BICLIQUE under ETH and SETH). *Can the running time lower bound on* ONE-SIDED k-BICLIQUE *be improved to* $n^{\Omega(k)}$ *under ETH? Can it be improved to* $n^{k-o(1)}$ *under SETH?*

We remark that a direction to address the above question was detailed in [24]. While on the topic of the k-BICLIQUE problem, it is worth noting that the lower bound of $n^{\Omega(\sqrt{k})}$ for ONE-SIDED k-BICLIQUE assuming ETH yields a running time lower bound of $n^{\Omega(\frac{\log k}{\log \log k})}$ for the k-BICLIQUE problem (due to the soundness parameters in Theorem 1). However, assuming randomized ETH, the running time lower bound for the k-BICLIQUE problem can be improved to $n^{\Omega(\sqrt{k})}$ [23]. Can this improved running time lower bound be obtained just under (deterministic) ETH? Finally, we remark that we shall discuss about the hardness of approximation of the k-BICLIQUE problem in Section 3.2.3.

Inapproximability of k-DOMINATING SET via Gadget Composition. We shall discuss about the inapproximability of k-DOMINATING SET in detail in the next subsubsection. We would like to simply highlight here how the above framework was used by Chen and Lin [25] and Lin [26] to obtain inapproximability results for k-DOMINATING SET.

In [25], the authors starting from Theorem 1, obtain the W[1]-hardness of approximating k-DOMINATING SET to a factor of almost two. Then they amplify the gap to any constant by using a specialized graph product.

We now turn our attention to a recent result of Lin [26] who provided strong inapproximability result for k-DOMINATING SET (we refer the reader to Section 3.1.2 to obtain the context for this result). Lin's proof of inapproximability of k-DOMINATING SET is a one-step reduction from an instance of k-SET COVER on a universe of size $O(\log n)$ (where n is the number of subsets given in the collection) to an instance of k-SET COVER on a universe of size $\text{poly}(n)$ with a gap of $\left(\frac{\log n}{\log \log n}\right)^{1/k}$. Lin then uses this gap-producing self-reduction to provide running time lower bounds (under different time hypotheses) for approximating k-set cover to a factor of $(1-o(1)) \cdot \left(\frac{\log n}{\log \log n}\right)^{1/k}$. Recall that k-DOMINATING SET is essentially [27,28] equivalent to k-SET COVER.

Elaborating, Lin designs a gadget by combining the hypercube partition gadget of Feige [29] with a derandomizing combinatorial object called universal set, to obtain a gap gadget, and then combines the gap gadget with the input k-SET COVER instance (on small universe but with no gap) to obtain a gap k-SET COVER instance. This is another success story of the gadget composition technique.

Finally, we remark that Lai [30] recently extended Lin's inapproximability results for dominating set (using the same proof framework) to rule out constant-depth circuits of size $f(k)n^{o(\sqrt{k})}$ for any computable function f.

Even Set. A recent success story of Theorem 1 is its application to resolve a long standing open problem called k-MINIMUM DISTANCE PROBLEM (also referred to as k-EVEN SET), where we are given as input a generator matrix $\mathbf{A} \in \mathbb{F}_2^{n \times m}$ of a binary linear code and an integer k, and the goal is to determine whether the code has distance at most k. Recall that the distance of a linear code is $\min_{\vec{0} \neq \mathbf{x} \in \mathbb{F}_2^m} \|\mathbf{A}\mathbf{x}\|_0$ where $\|\cdot\|_0$ denote the 0-norm (aka the Hamming norm).

In [31], the authors showed that k-EVEN SET is W[1]-hard under randomized reductions. The result was obtained by starting from the inapproximability result stated in Theorem 1 followed by a series of intricate reductions. In fact they proved the following stronger inapproximability result.

Theorem 2 ([31]). *For any $\gamma \geq 1$, given input $(\mathbf{A},k) \in \mathbb{F}^{n \times m} \times \mathbb{N}$, it is W[1]-hard (under randomized reductions) to distinguish between*

- *Completeness: Distance of the code generated by $\mathbf{A}$ is at most k, and,*
- *Soundness: Distance of the code generated by $\mathbf{A}$ is more than $\gamma \cdot k$.*

We emphasize that even to obtain the W[1]-hardness of k-EVEN SET (with no gap), they needed to start from the gap problem given in Theorem 1.

The proof of the above theorem proceeds by first showing FPT hardness of approximation of the non-homogeneous variant of k-MINIMUM DISTANCE PROBLEM called the k-NEAREST CODEWORD PROBLEM. In k-NEAREST CODEWORD PROBLEM, we are given a target vector $\mathbf{y}$ (in $\mathbb{F}^n$) in addition to $(\mathbf{A}, k)$, and the goal is to find whether there is any $\mathbf{x}$ (in $\mathbb{F}^m$) such that the Hamming norm of $\mathbf{A}\mathbf{x} - \mathbf{y}$ is at most k. As an intermediate step of the proof of Theorem 2, they showed that k-NEAREST CODEWORD PROBLEM is W[1]-hard to approximate to any constant factor.

An important intermediate problem which was studied by [31] to prove the inapproximability of k-NEAREST CODEWORD PROBLEM, was the k-LINEAR DEPENDENT SET problem where given a set $\mathbf{A}$ of n vectors over a finite field $\mathbb{F}_q$ and an integer k, the goal is to decide if there are k vectors in $\mathbf{A}$ that are linearly dependent. They ruled out constant factor approximation algorithms for this problem running in FPT time. Summarizing, the high level proof overview of Theorem 2 follows by reducing ONE-SIDED k-BICLIQUE to k-LINEAR DEPENDENT SET, which is then reduced to k-NEAREST CODEWORD PROBLEM, followed by a final randomized reduction to k-MINIMUM DISTANCE PROBLEM.

Finally, we note that there is no reason to define k-MINIMUM DISTANCE PROBLEM only for binary code, but can instead be defined over larger fields as well. It turns out that [31] cannot rule out FPT algorithms for k-MINIMUM DISTANCE PROBLEM over $\mathbb{F}_p$ with $p > 2$, when p is fixed and is not part of the input. Thus we have the open problem.

Open Question 2. *Is it W[1]-hard to decide k-MINIMUM DISTANCE PROBLEM over $\mathbb{F}_p$ with $p > 2$, when p is fixed and is not part on the input?*

Shortest Vector Problem. Theorem 1 (or more precisely the constant inapproximability of k-LINEAR DEPENDENT SET stated above) was also used to resolve the complexity of the parameterized k-SHORTEST VECTOR PROBLEM in lattices, where the input (in the ℓ_p norm) is an integer $k \in \mathbb{N}$ and a matrix $\mathbf{A} \in \mathbb{Z}^{n \times m}$ representing the basis of a lattice, and we want to determine whether the shortest (non-zero) vector in the lattice has length at most k, i.e., whether $\min_{\vec{0} \neq \mathbf{x} \in \mathbb{Z}^m} \|\mathbf{A}\mathbf{x}\|_p \leq k$. Again, k is the parameter of the problem. It should also be noted here that (as in [32]), we require the basis of the lattice to be integer valued, which is sometimes not enforced in literature (e.g., [33,34]). This is because, if $\mathbf{A}$ is allowed to be any matrix in $\mathbb{R}^{n \times m}$, then parameterization is meaningless because we can simply scale $\mathbf{A}$ down by a large multiplicative factor.

In [31], the authors showed that k-SHORTEST VECTOR PROBLEM is W[1]-hard under randomized reductions. In fact they proved the following stronger inapproximability result.

Theorem 3 ([31]). *For any $p > 1$, there exists a constant $\gamma_p > 1$ such that given input $(\mathbf{A}, k) \in \mathbb{Z}^{n \times m} \times \mathbb{N}$, it is W[1]-hard (under randomized reductions) to distinguish between*

- *Completeness: The ℓ_p norm of the shortest vector of the lattice generated by $\mathbf{A}$ is $\leq k$, and,*
- *Soundness: The ℓ_p norm of the shortest vector of the lattice generated by $\mathbf{A}$ is $> \gamma_p \cdot k$.*

Notice that Theorem 2 rules out FPT approximation algorithms with any constant approximation ratio for k-EVEN SET. In contrast, the above result only prove FPT inapproximability with some constant ratio for k-SHORTEST VECTOR PROBLEM in ℓ_p norm for $p > 1$. As with k-EVEN SET, even to prove the W[1]-hardness of k-SHORTEST VECTOR PROBLEM (with no gap), they needed to start from the gap problem given in Theorem 1.

The proof of the above theorem proceeds by first showing FPT hardness of approximation of the non-homogeneous variant of k-SHORTEST VECTOR PROBLEM called the k-NEAREST VECTOR PROBLEM. In k-NEAREST VECTOR PROBLEM, we are given a target vector $\mathbf{y}$ (in $\mathbb{Z}^n$) in addition to $(\mathbf{A}, k)$, and the goal is to find whether there is any $\mathbf{x}$ (in $\mathbb{Z}^m$) such that the ℓ_p norm of $\mathbf{A}\mathbf{x} - \mathbf{y}$ is at most k. As an intermediate step of the proof of Theorem 2, they showed that k-NEAREST VECTOR PROBLEM is W[1]-hard to approximate to any constant factor. Summarizing, the high level proof overview of Theorem 3 follows by reducing ONE-SIDED k-BICLIQUE to k-LINEAR DEPENDENT SET, which is then reduced to k-NEAREST VECTOR PROBLEM, followed by a final randomized reduction to k-SHORTEST VECTOR PROBLEM.

An immediate open question left open from their work is whether Theorem 3 can be extended to k-SHORTEST VECTOR PROBLEM in the ℓ_1 norm. In other words,

Open Question 3 (Approximation of k-SHORTEST VECTOR PROBLEM in ℓ_1 norm). *Is k-SHORTEST VECTOR PROBLEM in the ℓ_1 norm in FPT?*

3.1.2. Parameterized Inapproximability of Dominating Set

In the k-DOMINATING SET problem we are given an integer k and a graph G on n vertices as input, and the goal is to determine if there is a dominating set of size at most k. It was a long standing open question to design an algorithm which runs in time $T(k) \cdot \text{poly}(n)$ (i.e., FPT-time), that would find a dominating set of size at most $F(k) \cdot k$ whenever the graph G has a dominating set of size k, for any computable functions T and F.

The first non-trivial progress on this problem was by Chen and Lin [25] who ruled out the existence of such algorithms (under W[1] $\neq$ FPT) for all constant functions F (i.e., $F(n) = c$, where c is any universal constant). We discussed their proof technique in the previous subsubsection. A couple of years later, Karthik C. S. et al. [35] completely settled the question, by ruling out the existence of such an algorithm (under W[1] $\neq$ FPT) for any computable function F. Thus, k-DOMINATING SET was shown to be totally inapproximable. We elaborate on their proof below.

Theorem 4 ([35]). *Let $F : \mathbb{N} \to \mathbb{N}$ be any computable function. Given an instance (G, k) of k-DOMINATING SET as input, it is W[1]-hard to distinguish between the following two cases:*

- *Completeness: G has a dominating set of size k.*
- *Soundness: Every dominating set of G is of size at least $F(k) \cdot k$.*

The overall proof follows by reducing k-MULTICOLOR CLIQUE to the gap k-DOMINATING SET with parameters as given in the theorem statement. In the k-MULTICOLOR CLIQUE problem, we are given an integer k and a graph G on vertex set $V := V_1 \dot\cup V_2 \dot\cup \cdots \dot\cup V_k$ as input, where each V_i is an independent set of cardinality n, and the goal is to determine if there is a clique of size k in G. Following a straightforward reduction from the k-CLIQUE problem, it is fairly easy to see that k-MULTICOLOR CLIQUE is W[1]-hard.

The reduction from k-MULTICOLOR CLIQUE to the gap k-DOMINATING SET proceeds in two steps. In the first step we reduce k-MULTICOLOR CLIQUE to k-GAP CSP. This is the step where we generate the gap. In the second step, we reduce k-GAP CSP to gap k-DOMINATING SET. This step is fairly standard and mimics ideas from Feige's proof of the NP-hardness of approximating the MAX COVERAGE problem [29].

Before we proceed with the details of the above two steps, let us introduce a small technical tool from coding theory that we would need. We need codes known in literature as good codes, these are binary error correcting codes whose rate and relative distances are both constants bounded away from 0 (see [36] (Appendix E.1.2.5) for definitions). The reader may think of them as follows: for every $\ell \in \mathbb{N}$, we say that $C_\ell \subseteq \{0,1\}^\ell$ is a good code if (i) $|C_\ell| = 2^{\rho\ell}$, for some universal constant $\rho > 0$, (ii) for any distinct $c, c' \in C_\ell$ we have that c and c' have different values on at least $\delta\ell$ fraction of coordinates, for some universal constant $\delta > 0$. An encoding of C_ℓ is an injective function $\mathcal{E}_{C_\ell} : \{0,1\}^{\rho\ell} \to C_\ell$. The encoding is said to be efficient if $\mathcal{E}_{C_\ell}(x)$ can be computed in $\text{poly}(\ell)$ time for any $x \in \{0,1\}^{\rho\ell}$.

Let us fix $k \in \mathbb{N}$ and $F : \mathbb{N} \to \mathbb{N}$ as in the theorem statement. We further define

$$\alpha := 1 - \frac{1}{\left(\binom{k}{2} \cdot F(\binom{k}{2})\right)^{\binom{k}{2}}}.$$

From k-MULTICOLOR CLIQUE to k-GAP CSP. Starting from an instance of k-MULTICOLOR CLIQUE, say G on vertex set $V := V_1 \dot\cup V_2 \dot\cup \cdots \dot\cup V_k$, we write down a set of constraints $\mathcal{P}$ on a variable set $X := \{x_{i,j} \mid i, j \in [k], i \neq j\}$ as follows. For every $i, j \in [k]$, such that $i \neq j$, define $E_{i,j}$ to be the set of all edges in G whose end points are in V_i and V_j. An assignment to variable $x_{i,j}$ is an element of $E_{i,j}$, i.e., a pair of vertices, one from V_i and the other from V_j. Suppose that $x_{i,j}$ was assigned the edge $\{v_i, v_j\}$, where $v_i \in V_i$ and $v_j \in V_j$. Then we define the assignment of $x_{i,j}^i$ to be v_i and the assignment of $x_{i,j}^j$ to be v_j. We define $\mathcal{P} := \{P_1, \ldots, P_k\}$, where the constraint P_i is defined to be satisfied if the assignment to all of $x_{1,i}^i, x_{2,i}^i, \ldots, x_{i-1,i}^i, x_{i+1,i}^i, \ldots, x_{k,i}^i$ are the same. We refer to the problem of determining if there is an assignment to the variables in X such that all the constraints are satisfied as the k-CSP problem. Notice that while this is a natural way to write k-MULTICOLOR CLIQUE as a CSP, where we have tried to check if all variables having a vertex in common, agree on its assignment, there is no gap yet in the k-CSP problem. In particular, if there was a clique of size k in G then there is an assignment to the variables of X (by assigning the edges of the clique in G to the corresponding variable in X) such that all the constraints in $\mathcal{P}$ are satisfied; however, if every clique in G is of size less than k then there every assignment to the variables of X may violate only one constraint in $\mathcal{P}$ (and not more).

In order to amplify the gap, we rewrite the set of constraints $\mathcal{P}$ in a different way to obtain the set of constraints $\mathcal{P}'$, on the same variable set X, as follows. Suppose that $x_{i,j}$ was assigned the edge $\{v_i, v_j\}$, where $v_i \in V_i$ and $v_j \in V_j$, then for $\beta \in [\log n]$, we define the assignment of $x_{i,j}^{i,\beta}$ to be the βth coordinate of v_i. Recall that $|V_i| = n$ and therefore we can label all vertices in V_i by vectors in $\{0,1\}^{\log n}$. We define $\mathcal{P}' := \{P'_1, \ldots, P'_{\log n}\}$, where the constraint P'_β is defined to be satisfied if and only if the following holds for all $i \in [k]$: the assignment to all of $x_{1,i}^{i,\beta}, x_{2,i}^{i,\beta}, \ldots, x_{i-1,i}^{i,\beta}, x_{i+1,i}^{i,\beta}, \ldots, x_{k,i}^{i,\beta}$ are the same. Again notice that there is an assignment to the variables of X such that all the constraints in $\mathcal{P}$ are satisfied if and only if the same assignment also satisfies all the constraints in $\mathcal{P}'$.

However, rewriting $\mathcal{P}$ as $\mathcal{P}'$ allows us to simply apply the error correcting code C_ℓ (with parameters ρ and δ, and encoding function $\mathcal{E}_{C_\ell}$) to the constraints in $\mathcal{P}'$, to obtain a gap! In particular, we choose ℓ to be such that $\rho\ell = \log n$. Consider a new set of constraints $\mathcal{P}''$, on the same variable set X, as follows. For any $z \in \{0,1\}^{\log n}$ and $\beta \in [\ell]$, we denote by $\mathcal{E}_{C_\ell}(z)_\beta$, the βth coordinate of $\mathcal{E}_{C_\ell}(z)$. We define $\mathcal{P}'' := \{P''_1, \ldots, P''_\ell\}$, where the constraint P''_β is defined to be satisfied if and only if the following holds for all $i \in [k]$: the assignment to all of $\mathcal{E}_{C_\ell}(x_{1,i}^i)_\beta, \mathcal{E}_{C_\ell}(x_{2,i}^i)_\beta, \ldots, \mathcal{E}_{C_\ell}(x_{i-1,i}^i)_\beta, \mathcal{E}_{C_\ell}(x_{i+1,i}^i)_\beta, \ldots, \mathcal{E}_{C_\ell}(x_{k,i}^i)_\beta$ are the same.

Notice, as before, that there is an assignment to the variables of X such that all the constraints in $\mathcal{P}'$ are satisfied if and only if the same assignment also satisfies all the constraints in $\mathcal{P}''$. However, for every assignment to X that violates at least one constraint in $\mathcal{P}'$, we have that the same assignment violates at least δ fraction of the constraints in $\mathcal{P}''$. To see this, consider an assignment that violates the constraint P_1 in $\mathcal{P}'$. This implies that there is some $i \in [k]$ such that the assignment to all of $x_{1,i}^{i,1}, x_{2,i}^{i,1}, \ldots, x_{i-1,i}^{i,1}, x_{i+1,i}^{i,1}, \ldots, x_{k,i}^{i,1}$ are not the same. Let us suppose, without loss of generality, that the assignment to $x_{1,i}^{i,1}$ and $x_{2,i}^{i,1}$ are different. In other words, we have that $x_{1,i}^i \neq x_{2,i}^i$, where we think of $x_{1,i}^i, x_{2,i}^i$ as $\log n$ bit vectors. Let $\Delta \subseteq [\ell]$ such that $\beta \in \Delta$ if and only if $\mathcal{E}_{C_\ell}(x_{1,i}^i)_\beta \neq \mathcal{E}_{C_\ell}(x_{2,i}^i)_\beta$. By the distance of the code C_ℓ we have that $|\Delta| \geq \delta\ell$. Finally, notice that for all $\beta \in \Delta$, we have that the assignment does not satisfy constraint P_β in $\mathcal{P}''$. We refer to the problem of distinguishing if there is an assignment to X such that all the constraints are satisfied or if every assignment to X does not satisfy a constant fraction of the constraints, as the k-GAP CSP problem.

In order to rule out $F(k)$ approximation FPT algorithms for k-DOMINATING SET, we will need that for every assignment to X that violates at least one constraint in $\mathcal{P}'$, we have that the same assignment violates at least α fraction of the constraints in $\mathcal{P}''$ (instead of just δ; note that α is very close to 1, whereas δ can be at most half). To boost the gap [37] we apply a simple repetition/direct-product trick to our constraint system. Starting from $\mathcal{P}''$, we construct a new set of constraints $\mathcal{P}^*$, on the same variable set X, as follows.

$$\mathcal{P}^* := \{P_S \mid S \in [\ell]^t\},$$

where $t = \frac{\log(1-\alpha)}{\log(1-\delta)}$. For every $S \in [\ell]^t$, we define P_S to be satisfied if and only if for all $\beta \in S$, the constraint P_β is satisfied.

It is easy to see that $\mathcal{P}''$ and $\mathcal{P}^*$ have the same set of completely satisfying assignments. However, for every assignment to X that violates δ fraction of constraints in $\mathcal{P}''$, we have that the same assignment violates at least α fraction of the constraints in $\mathcal{P}^*$. To see this, consider an assignment that violates δ fraction of constraints in $\mathcal{P}''$, say it violates all constraints $P_\beta \in \mathcal{P}''$, for every $\beta \in \Delta \subseteq [\ell]$. This implies that the assignment satisfies constraint P_S if and only if $S \in ([\ell] \setminus \Delta)^t$. This implies that the fraction of constraints in $\mathcal{P}^*$ that the assignment can satisfy is upper bounded by $(1-\delta)^t = 1 - \alpha$.

From k-GAP CSP to gap k-DOMINATING SET. In the second part, starting from the aforementioned instance of k-GAP CSP (after boosting the gap), we construct an instance H of k-DOMINATING SET. The construction is due to Feige [29,38] and it proceeds as follows. Let $\mathcal{F}$ be the set of all functions from $\{0,1\}^{tk}$ to $\binom{k}{2}$, i.e., $\mathcal{F} := \{f : \{0,1\}^{tk} \to \binom{k}{2}\}$. The graph H is on vertex set $U = A \dot\cup B$, where $A = \mathcal{P}^* \times \mathcal{F}$ and $B = E(G)$, i.e., B is simply the edge set of G. We introduce an edge between all pairs of vertices in B. We introduce an edge between $a := (S := (s_1, \ldots, s_t) \in [\ell]^t, f : \{0,1\}^{tk} \to \binom{k}{2}) \in A$ and $e := (v_i, v_j) \in E$ if and only if the following holds.

$$\exists \tau := (\tau_1, \ldots, \tau_t) \in \{0,1\}^{kt}, \text{ such that } f(\tau) = \{i,j\} \text{ and}$$
$$\forall r \in [t], \text{ we have } \mathcal{E}_{C_\ell}(v_i)_{s_r} = (\tau_r)_i \text{ and } \mathcal{E}_{C_\ell}(v_j)_{s_r} = (\tau_r)_j.$$

Notice that the number of vertices in H is $|A| + |B| \le (\frac{\log n}{\rho})^t \cdot \binom{k}{2}^{2^{tk}} + n^2 < \eta(k) \cdot n^{2.01}$, for some computable function η. It is not hard to check that the following hold:

- (Completeness) If there is an assignment to X that satisfies all constraints in $\mathcal{P}^*$, then the corresponding $\binom{k}{2}$ vertices in B dominate all vertices in the graph H.
- (Soundness) If each assignment can only satisfy $(1-\alpha)$ fraction of constraints in $\mathcal{P}^*$, then any dominating set of H has size at least $F\left(\binom{k}{2}\right) \cdot \binom{k}{2}$.

We skip presenting details of this part of the proof here. The proofs have been derived many times in literature; if needed, the readers may refer to Appendix A of [35]. This completes our sketch of the proof of Theorem 4.

A few remarks are in order. First, the k-GAP CSP problem described in the proof above, is formalized as the k-MAXCOVER problem in [35] (and was originally introduced in [39]). In particular, the formalism of k-MAXCOVER (which may be thought of as the parameterized label cover problem) is generic enough to be used as an intermediate gap problem to reduce to both k-DOMINATING SET (as in [35]) and k-CLIQUE (as in [39]). Moreover, it was robust enough to capture stronger running time lower bounds (under stronger hypotheses); this will elaborated below. However, in order to keep the above proof succinct, we skipped introducing the k-MAXCOVER problem, and worked with k-GAP CSP, which was sufficient for the above proof.

Second, Karthik C. S. et al. [35] additionally showed that for every computable functions $T, F : \mathbb{N} \to \mathbb{N}$ and every constant $\varepsilon > 0$:

- Assuming the Exponential Time Hypothesis (ETH), there is no $F(k)$-approximation algorithm for k-DOMINATING SET that runs in $T(k) \cdot n^{o(k)}$ time.
- Assuming the Strong Exponential Time Hypothesis (SETH), for every integer $k \ge 2$, there is no $F(k)$-approximation algorithm for k-DOMINATING SET that runs in $T(k) \cdot n^{k-\varepsilon}$ time.

In order to establish Theorem 4 and the above two results, Karthik C. S. et al. [35] introduced a framework to prove parameterized hardness of approximation results. In this framework, the objective was to start from either the $W[1] \neq$ FPT hypothesis, ETH, or SETH, and end up with the gap k-DOMINATING SET, i.e., they design reductions from instances of k-CLIQUE, 3-CNF-SAT, and ℓ-CNF-SAT, to an instance of gap k-DOMINATING SET. A prototype reduction in this framework has two modular parts. In the first part, which is specific to the problem they start from, they generate a gap and obtain hardness of gap k-MAXCOVER. In the second part, they show a gap preserving reduction from gap k-MAXCOVER to gap k-DOMINATING SET, which is essentially the same as the reduction from k-GAP CSP to k-DOMINATING SET in the proof of Theorem 4.

The first part of a prototype reduction from the computational problem underlying a hypothesis of interest to gap k-MAXCOVER follows by the design of an appropriate communication protocol. In particular, the computational problem is first reduced to a constraint satisfaction problem (CSP) over k (or some function of k) variables over an alphabet of size n. The predicate of this CSP would depend on the computational problem underlying the hypothesis from which we started. Generalizing ideas from [40], they then show how a protocol for computing this predicate in the multiparty (number of players is the number of variables of the CSP) communication model, can be combined with the CSP to obtain an instance of gap k-MAXCOVER. For example, for the $W[1] \neq$ FPT hypothesis and ETH, the predicate is a variant of the equality function, and for SETH, the predicate is the well studied disjointness function. The completeness and soundness of the protocols computing these functions translate directly to the completeness and soundness of k-MAXCOVER.

Third, we recall that Lin [26] recently provided alternate proofs of Theorem 4 and the above mentioned stronger running time lower bounds. While we discussed about his proof technique in Section 3.1.1, we would like to discuss about his result here. Following the right setting of parameters in the proof of Theorem 4 (for example set $\alpha = 1 - \frac{1}{(\log n)^{\Omega(1/k)}}$), we can obtain that approximating k-DOMINATING SET to a factor of $(\log n)^{1/k^3}$ is $W[1]$-hard. Lin improved the exponent of $1/k^3$ in the approximation factor to $h(k)$ for any computable function h. Can this inapproximability be further improved? On the other hand, can we do better than the simple polynomial time greedy algorithm which provides a $(1+\ln n)$ factor approximation? This leads us to the following question:

Open Question 4 (Tight inapproximability of k-DOMINATING SET). *Is there a $(\log n)^{1-o(1)}$ factor approximation algorithm for k-DOMINATING SET running in time $n^{k-0.1}$?*

We conclude the discussion on k-DOMINATING SET with an open question on $W[2]$-hardness of approximation. As noted earlier, k-DOMINATING SET is a $W[2]$-complete problem, and Theorem 4 shows that the problem is $W[1]$-hard to approximate to any $F(k)$ factor. However, is there some computable function F for which approximating k-DOMINATING SET is in $W[1]$? In other words we have:

Open Question 5 ($W[2]$-completeness of approximating k-DOMINATING SET). *Can we base total inapproximability of k-DOMINATING SET on $W[2] \neq$ FPT?*

3.1.3. Parameterized Inapproximability of Steiner Orientation by Gap Amplification

Gap amplification is a widely used technique in the classic literatures on (NP-)hardness of approximation (e.g., [41–43]). In fact, the arguably simplest proof of the PCP theorem, due to Dinur [43], is indeed via repeated gap amplification. The overall idea here is simple: we start with a hardness of approximation for a problem with small factor (e.g., $1 + 1/n$). At each step, we perform an operation that transforms an instance of our problem to another instance, in such a way that the gap becomes bigger; usually this new instance will also be bigger than our instance. By repeatedly applying this operation, one can finally arrive at a constant, or even super constant, factor hardness of approximation.

There are two main parameters that determine the success/failure of such an approach: how large the new instance is compared to the old instance (i.e., size blow-up) and how large the new gap is compared to the old gap, in each operation. To see how these two come into the picture, let us first consider a case study where a (straightforward) gap amplification procedure does not work: k-CLIQUE. The standard way to amplify the gap for k-CLIQUE is through graph product. Recall that the (tensor) graph product of a graph $G = (V, E)$ with itself, denoted by $G^{\otimes 2}$, is a graph whose vertex set is V^2 and there is an edge between (u_1, u_2) and (v_1, v_2) if and only if $(u_1, v_1) \in E$ and $(u_2, v_2) \in E$. It is not hard to check that, if we can find a clique of size t in $G^{\otimes 2}$, then we can find one of size $\sqrt{t}$ in G (and vice versa). This implies that, if we have an instance of clique that is hard to approximate to within a factor of $(1 + \varepsilon)$, then we may take the graph product with itself which yields an instance of CLIQUE that is hard to approximate to within a factor of $(1 + \varepsilon)^2$.

Now, let us imagine that we start with the hard instance of an exact version of k-CLIQUE. We may think of this as being hard to approximate to within a factor of $(1 - 1/k)$. Hence, we may apply the above gap amplification procedure $\log k$ times, resulting in an instance of CLIQUE that is hard to approximate to within a factor of $(1 - 1/k)^{2^{\log k}}$, which is a constant bounded away from one (i.e., $\approx 1/e$). The bad news here is that the number of the vertices of the final graph is $n^{2^{\log k}} = n^k$, where n is the number of vertices of the initial graph. This does not give any lower bound, because we can solve k-CLIQUE in the original graph in $n^{O(k)}$ time trivially! In the next subsection, we will see a simple way to prove hardness of approximating k-CLIQUE, assuming stronger assumptions. However, it remains an interesting and important open question how to prove such hardness from a non-gap assumption:

Open Question 6. *Is it W[1]-hard or ETH-hard to approximate k-CLIQUE to within a constant factor in FPT time?*

Having seen a failed attempt, we may now move on to a success story. Remarkably, Wlodarczyk [44] recently managed to use gap amplification to prove hardness of approximation for connectivity problems, including the k-STEINER ORIENTATION problem. Here we are given a mixed graph G, whose edges are either directed or undirected, and a set of k terminal pairs $\{(s_i, t_i)\}_{i \in [k]}$. The goal is to orient all the undirected edges in such a way that maximizes the number of t_i that can be reached from s_i. The problem is known to be in XP [45] but is W[1]-hard even when all terminal pairs can be connected [46]. Starting from this W[1]-hardness, Wlodarczyk [44] devises a gap amplification step that implies a hardness of approximation with factor $(\log k)^{o(1)}$ for the problem. Due to the technicality of the gap amplification step, we will not go into the specifics in this survey. However, let us point out the differences between this gap amplification and the (failed) one for CLIQUE above. The key point here is that the new instance of Wlodarczyk's gap amplification has size of the form $f(k) \cdot n$ instead of n^2 as in the graph product. This means that, even if we are applying Wlodarczyk's gap amplification step $\log(k)$ times, or, more generally, $g(k)$ times, it only results in an instance of size $\underbrace{f(f(\cdots(f(k))))}_{g(k)\text{times}} \cdot n$, which is still FPT! Since the technique is still quite new, it is an exciting frontier to examine whether other parameterized problems allow such similar gap amplification steps.

3.2. Hardness from Gap Hypotheses

In the previous subsection, we have seen that several hardness of approximation results can be proved based on standard assumptions. However, as alluded to briefly, some basic problems, including k-CLIQUE, still evades attempts at proving such results. This motivates several researchers in the community to come up with new assumptions that allow more power and flexibility in proving inapproximability results. We will take a look at two of these hypotheses in this subsection; we note that there have also been other assumptions formulated, but we only focus on these two since they arguably have been used most often.

The first assumption, called the Parameterized Inapproximability Hypothesis (PIH) for short, can be viewed as a gap analogue of the W[1] $\ne$ FPT assumption. There are many (equivalent) ways to state PIH. We choose to state it in terms of an inapproximability of the colored version of DENSEST k-SUBGRAPH. In MULTICOLORED DENSEST k-SUBGRAPH, we are given a graph $G = (V, E)$ where the vertex set V is partition in to k parts $V_1, \ldots, V_k$. The goal is to select k vertices $v_1 \in V_1, v_2 \in V_2, \ldots, v_k \in V_k$ such that $\{v_1, \ldots, v_k\}$ induces as many edges as possible.

It is easy to see that the exact version of this problem is W[1]-hard, via a straightforward reduction from k-CLIQUE. PIH postulates that even the approximate version of this problem is hard:

Hypothesis 1 (Parameterized Inapproximability Hypothesis (PIH) [47,48]). *For some constant $\varepsilon > 0$, there is no $(1 + \varepsilon)$ factor FPT approximation algorithm for* MULTICOLORED DENSEST k-SUBGRAPH.

There are two important remarks about PIH. First, the factor $(1 + \varepsilon)$ is not important, and the conjecture remains equivalent even if we state it for a factor C for any arbitrarily large constant C; this is due to gap amplification via parallel repetition [42]. Second, PIH implies that k-CLIQUE is hard to approximate to within any constant factor:

Lemma 1. *Assuming PIH, there is no constant factor FPT approximation algorithm for k-CLIQUE.*

The above result can be shown via a classic reduction of Feige, Goldwasser, Lovász, Safra and Szegedy (henceforth FGLSS) [49], which was one of the first works connecting proof systems and hardness of approximation. Specifically, the FGLSS reduction transforms G to another graph G' by viewing the edges of G as vertices of G'. Then, we connect $\{u_1, v_1\}$ and $\{u_2, v_2\}$ except when the union $\{u_1, v_1\} \cup \{u_2, v_2\}$ contains two distinct vertices from the same partition. One can argue that the size of the largest clique in G' is exactly equal to the number of edges in the optimal solution of MULTICOLORED DENSEST k-SUBGRAPH on G. As a result, PIH implies hardness of approximation of the former. Interestingly, however, it is not known if the inverse is true and this remains an interesting open question:

Open Question 7. *Does PIH hold if we assume that k-CLIQUE is FPT inapproximable to within any constant factor?*

As demonstrated by the FGLSS reduction, once we have a gap, it is much easier to give a reduction to another hardness of approximation result, because we do not have to create the initial gap ourselves (as in the previous subsection) but only need to preserve or amplify the gap. Indeed, PIH turns out to be a pretty robust hypothesis that gives FPT inapproximability for many problems, including k-CLIQUE, DIRECTED ODD CYCLE TRAVERSAL [48] and STRONGLY CONNECTED STEINER SUBGRAPH [50]. We remark that the current situation here is quite similar to that of the landscape of the classic theory of hardness of approximation before the PCP Theorem [21,22] was proved. There, Papadimitriou and Yannakakis introduced a complexity class MAX-SNP and show that many optimization problems are hard (or complete) for this class [51]. Later, the PCP Theorem confirms that these problems are NP-hard. In our case of FPT inapproximability, PIH seems to be a good analogy of MAX-SNP for problems in W[1] and, as mentioned before, PIH has been used as a starting point of many hardness of approximation results. However, there has not yet been many reverse reductions to PIH, and this is one of the motivation behind Question 7 above.

Despite the aforementioned applications of PIH, there are still quite a few questions that seem out of reach of PIH, such as whether there is an $o(k)$ factor FPT approximation for k-CLIQUE or questions related to running time lower bounds of approximation algorithms. On this front, another stronger conjecture called the Gap Exponential Time Hypothesis (Gap-ETH) is often used instead:

Hypothesis 2 (Gap Exponential Time Hypothesis (Gap-ETH) [52,53]). *For some constants $\varepsilon, \delta > 0$, there is no $O(2^{\delta n})$-time algorithm that can, given a 3CNF formula, distinguish between the following two cases:*

- *(Completeness) the formula is satisfiable.*
- *(Soundness) any assignment violates more than ε fraction of the clauses.*

Here n denotes the number of clauses [54].

Clearly, Gap-ETH is a strengthening of ETH, which can be thought of in the above form but with $\varepsilon = 1/n$. Another interesting fact is that Gap-ETH is stronger than PIH. This can be shown via the standard reduction from 3SAT to k-CLIQUE that establishes $N^{\Omega(k)}$ lower bound for the latter. The reduction, due to Chen et al. [55,56], proceed as follows. First, we partition the set of clauses C into $C_1, \ldots, C_k$ each of size n/k. For each C_i, we create a partition V_i in the new graph where each vertex corresponds to all partial assignments (to variables that appear in at least one clause of C_k) that satisfy all the clauses in C_k. Two vertices are connected if the corresponding partial assignments are consistent, i.e., they do not assign a variable to different values.

If there is an assignment that satisfies all the clauses, then clearly the restrictions of this assignment to each clause corresponds to k vertices from different partitions that form a clique. On the other hand, it is also not hard to argue that, in the soundness case, the number of edges induced by any k vertices from different partitions is at most $1 - \Theta(\varepsilon)$. Thus, Gap-ETH implies PIH as claimed.

Now that we have demonstrated that Gap-ETH is at least as strong as PIH, we may go further and ask how much more can we achieve from Gap-ETH, compared to PIH. The obvious consequences of Gap-ETH is that it can give explicit running time lower bounds for FPT hardness of approximation results. Perhaps more surprising, however, is that it can be used to improve the inapproximability ratio as well. The rest of this subsection is devoted to present some of these examples, together with brief overviews of how the proofs of these results work.

3.2.1. Strong Inapproximability of k-Clique

Our first example is the k-CLIQUE problem. Obviously, we can approximate k-CLIQUE to within a factor of k, by just outputting any single vertex. It had long been asked whether an $o(k)$-approximation is achievable in FPT time. As we saw above, PIH implies that a constant factor FPT approximation does not exist, but does not yet resolve this question. Nonetheless, assuming Gap-ETH, this question can be resolved in the negative:

Theorem 5 ([39]). *Assuming Gap-ETH, there is no $o(k)$-FPT-approximation for k-CLIQUE.*

The reduction used in [39] to prove the above inapproximability is just a simple modification of the above reduction [55,56] that we saw for k-CLIQUE. Suppose that we would like to rule out a $\frac{k}{g}$-approximation, where $g = g(k)$ is a function such that $\lim_{k \to \infty} g(k) = \infty$. The only change in the reduction is that, instead of letting $C_1, \ldots, C_k$ be the partition of the set of clauses C, we let each C_i be a set of $\frac{Dn}{g}$ clauses for some sufficiently large constant $D > 0$. The rest of the reduction works similar to before: for each C_i, we create a vertex corresponding to each partial assignment that satisfies all the clauses in C_i. Two vertices are joined by an edge if and only if they are consistent. This completes the description of the reduction.

To see that the reduction yields Theorem 5, first note that, if there is an assignment that satisfies the CNF formula, then we can again pick the restrictions on this formula onto $C_1, \ldots, C_k$; these gives k vertices that induces a clique in the graph.

On the other hand, suppose that every assignment violates more than an ε fraction of clauses. We will argue that there is no clique of size g in the constructed graph. The only property we need from the subsets $C_1, \ldots, C_k$ is that the union of any g such subsets contain at least $(1 - \varepsilon)$ fraction of the clauses. It is not hard to show that this is true with high probability, when we choose D to be sufficiently large. Now, suppose for the sake of contradiction that there exists a clique of size g in the graph. Since the vertices corresponding to the same subset C_i form an independent set, it must be that these g vertices are from different subsets. Let us call these subsets $C_{i_1}, \cdots, C_{i_g}$. Because these vertices induce a clique, we can find a global assignment that is consistent with each vertex. This global assignment satisfies all the clauses in $C_{i_1} \cup \cdots \cup C_{i_g}$. However, $C_{i_1} \cup \cdots \cup C_{i_g}$ contains at least $1 - \varepsilon$ fraction of all clauses, which contradicts to our assumption that every assignment violates more than ε fraction of the clauses.

Now, if we can $o(k)$-approximate k-CLIQUE in $T(k) \cdot N^{O(1)}$ time. Then, we may run this algorithm to distinguish the two cases in Gap-ETH in $f(k) \cdot (2^{Dn/g})^{O(1)} = 2^{o(n)}$ time, which violates Gap-ETH. This concludes our proof sketch. We end by remarking that the reduction may also be viewed as an instantiation of the randomized graph product [41,57,58], and it can also be derandomized. We omit the details of the latter here. Interested readers may refer to [39] for more detail.

3.2.2. Strong Inapproximability of Multicolored Densest k-Subgraph and Label Cover

For our second example, we go back to the MULTICOLORED DENSEST k-SUBGRAPH once again. Recall that PIH asserts that this problem is hard to approximate to some constant factor, and we have seen above that Gap-ETH also implies this. On the approximation front, however, only the trivial k-approximation algorithm is known: just pick a vertex that has edges to as many partitions as possible. Then, output that vertex and one of its neighbors from each partition. It is hence a natural question to ask whether it is possible to beat this approximation ratio. This question has been, up to lower order terms, answered in the negative, assuming Gap-ETH:

Theorem 6 ([59]). *Assuming Gap-ETH, there is no $k^{1-o(1)}$-approximation for* MULTICOLORED DENSEST k-SUBGRAPH.

An interesting aspect of the above result is that, even in the NP-hardness regime, no NP-hardness of factor k^γ for some constant $\gamma > 0$ is known. In fact, the problem is closely related to (and is a special case of) a well-known conjecture in the hardness of approximation community called the Sliding Scale Conjecture (SSC) [60–62]. (See [59] for more discussion on the relation between the two.) Thus, this is yet another instance where taking a parameterized complexity perspective helps us advance knowledge even in the classical settings.

To prove Theorem 6, arguably the most natural reduction here is the above reduction for Clique! Note that we now view the vertices corresponding to each subset C_i as forming a partition V_i. The argument in the YES case is exactly the same as before: if the formula is satisfiable, then there is a (multicolored) k-clique. However, as the readers might have noticed, the argument in the NO case does not go through anymore. In particular, even when the graph is quite dense (e.g., having half of the edges present), it may not contain any large clique at all and hence it is unclear how to recover back an assignment that satisfies a large fraction of constraints.

This obstacle was overcomed in [59] by proving an agreement testing theorem (i.e., direct product theorem), which is of the following form. Given k local functions $f_1, \ldots, f_k$, where $f_i : S_i \to \{0,1\}$ is a boolean function whose domain S_i is a subset of a universe $\mathcal{U}$. If some (small) ζ fraction of the pairs agree [63] with each other, then we can find (i.e., "decode") a global function $h : [n] \to \{0,1\}$ that "approximately agrees" with roughly ζ fraction of the local functions. The theorem in [59] works when $S_1, \ldots, S_k$ are sets of size $\Omega(n)$.

Due to the technical nature of the definitions, we will not fully formalize the notions in the previous paragraph. Nonetheless, let us sketch how to apply the agreement testing theorem to prove the NO case for our reduction. Suppose for the sake of contradiction that the formula is not $(1-\delta)$-satisfiable and that there exists a k-subgraph with density $\zeta \geq \frac{1}{k^{1-o(1)}}$. Recall that each selected vertex is simply a partial assignment onto the subset of clauses C_i for some i; we may view this as a function $f_i : S_i \to \{0,1\}$ where S_i denote the set of variables that appear in C_i. Here the universe $\mathcal{U}$ is the set of all variables. With this perspective, we can apply the agreement testing theorem to recover a global function $h : \mathcal{U} \to \{0,1\}$ that "approximately agrees" with roughly ζk of the local functions. Notice that, in this context, h is simply a global assignment for the CNF formula. Previously in the proof for inapproximability of Clique, we had a global assignment that (perfectly) agrees with g local functions, from which we can conclude that this assignment satisfies all but δ fraction of the clauses. It turns out that relaxing "perfect agreement" to "approximate agreement" does not affect the proof too much, and the latter still implies that h satisfies all but δ fraction of clauses as desired.

As for the proof of the agreement testing theorem itself, we will not delve too much into detail here. However, we note that the proof is based on looking at different "agreement levels" and the graph associated with them. It turns out that such a graph has a certain transitivity property, which allows one to "decode" back the global function h. This general approach of looking at different agreement levels and their transitivity properties is standard in the direct product/agreement testing literature [64–66]. The main challenge in [59] is to make the proof works for ζ as small as $1/k$, which requires a new notion of transitivity.

To end this subsection, we remark that the MULTICOLORED DENSEST k-SUBGRAPH is known as the 2-ary Constraint Satisfaction Problem (2-CSP) in the classical hardness of approximation community. The problem, and in particular its special case called Label Cover, serves as the starting point of almost all known NP-hardness of approximation (see e.g., [67–69]). The technique in [59] can also be used to show inapproximability for Label Cover with strong running time lower bound of the form $f(k) \cdot N^{\Omega(k)}$ [70]. Due to known reductions, this has numerous consequences. For example, it implies, assuming Gap-ETH, that approximating k-EVEN SET to within any factor less than two cannot be done in $f(k) \cdot N^{o(k)}$ time, considerably improving the lower bound mentioned in the previous subsection.

3.2.3. Inapproximability of k-Biclique and Densest k-Subgraph

While PIH (or equivalently the MULTICOLORED DENSEST k-SUBGRAPH problem) can serve as a starting point for hardness of approximation of many problems, there are some problems for which not even a constant factor hardness is known under PIH, but strong inapproximability results can be obtained via Gap-ETH. We will see two examples of this here.

First is the k-BICLIQUE problem. Recall that in this problem, we are given a bipartite graph and we would like to determine whether there is a complete bipartite subgraph of size k. As stated earlier in the previous subsection, despite its close relationship to k-CLIQUE, k-BICLIQUE turned out to be a much more challenging problem to prove intractibility and even its W[1]-hardness was only shown recently [23]. This difficulty is corroborated by its approximability status in the classical (non-parameterized) regime; while CLIQUE is long known to be NP-hard to approximate to within $N^{1-o(1)}$ factor [71], BICLIQUE is not even known to be NP-hard to approximate to within say 1.01 factor [72–75]. With this in mind, it is perhaps not a surprise that k-BICLIQUE is not known to be hard to approximate under PIH. Nonetheless, when we assume Gap-ETH, we can in fact prove a very strong hardness of approximation for the problem:

Theorem 7 ([39]). *Assuming Gap-ETH, there is no $o(k)$-FPT-approximation for k-BICLIQUE.*

Note that, similar to k-CLIQUE, a k-approximation for BICLIQUE can be easily achieved by outputting a single edge. Hence, in terms of the inapproximability ratio, the above result is tight.

Due to its technicality, we only sketch an outline of the proof of Theorem 7 here. Firstly, the reduction starts by constructing a graph that is similar (but not the same) to that of k-Clique that we describe above. The main properties of this graph is that (i) in the YES case where the formula is satisfiable, the graph contains many copies of k-BICLIQUE, and (ii) in the NO case where the formula is not even $(1 - \delta)$-satisfiable, the graph contains few copies of g-BICLIQUE. The construction and these properties were in fact shown in [76]. In [39], it was observed that, if we subsample the graph by keeping each vertex independently with probability p for an appropriate value of p, then (i) ensures that at least one of the k-BICLIQUE survives the subsampling, whereas (ii) ensures that no g-BICLIQUE survives. This indeed gives the claimed result in the above theorem.

We remark that, while Theorem 7 seems to resolve the approximability of k-BICLIQUE, there is still one aspect that is not yet completely understood: the running time lower bound. To demonstrate this, recall that, for k-CLIQUE, the reduction that gives hardness of k-VS-g CLIQUE has size $2^{O(n/g)}$; this means that we have a running time lower bound of $f(k) \cdot N^{\Omega(g)}$ on the problem. This is of course tight, because we can determine whether a graph has a g-clique in $N^{O(g)}$ time. However, for k-BICLIQUE, the known reduction that gives hardness for k-VS-g BICLIQUE has size $2^{O(n/\sqrt{g})}$. This results in a running time lower bound of only $f(k) \cdot N^{\Omega(\sqrt{g})}$. Specifically, for the most basic setting of constant factor approximation, Theorem 7 only rules out algorithms with running time $f(k) \cdot N^{o(\sqrt{k})}$. Hence, an immediate question here is:

Open Question 8. *Is there an $f(k) \cdot N^{o(k)}$-time algorithm that approximates k-BICLIQUE to within a constant factor?*

To put things into perspective, we note that, even for exact algorithms for k-BICLIQUE, the best running time lower bound is still $f(k) \cdot N^{\Omega(\sqrt{k})}$ [23] (under any reasonable complexity assumption). This means that, to answer Question 8, one has to first settle the best known running time lower bound for exact algorithms, which would already be a valuable contribution to the understanding of the problem.

Let us now point out an interesting consequence of Theorem 7 for the DENEST k-SUBGRAPH problem. This is the "uncolored" version of the MULTICOLORED DENSEST k-SUBGRAPH problem as defined above, where there are no partitions $V_1, \ldots, V_k$ and we can pick any k vertices in the input graph G with the objective of maximizing the number of induced edges. The approximability status of DENEST k-SUBGRAPH very much mirrors that of k-BICLIQUE. Namely, in the parameterized setting, PIH is not known to imply hardness of approximation for DENSEST k-SUBGRAPH. Furthermore, in the classic (non-parameterized) setting, DENSEST k-SUBGRAPH is not known [72,76–79] to be NP-hard to approximate even to within a factor of say 1.01. Despite these, Gap-ETH does give a strong inapproximability for DENSEST k-SUBGRAPH, as stated below:

Theorem 8 ([39]). *Assuming Gap-ETH, there is no $k^{o(1)}$-FPT-approximation for DENSEST k-SUBGRAPH.*

In fact, the above result is a simple consequence of Theorem 7. To see this, recall the following classic result in extremal graph theory commonly referred to as the Kővári-Sós-Turán (KST) Theorem [80]: any k-vertex graph that does not contain a g-biclique as a subgraph has density at most $O(k^{-1/g})$. Now, the hardness for k-BICLIQUE from Theorem 7 tells us that there is no FPT time algorithm that can distinguish between the graph containing k-biclique from one that does not contain g-biclique for any $g = \omega(1)$. When the graph contains a k-biclique, we have a k-vertex subgraph with density (at least) $1/2$. On the other hand, when the graph does not even contain a g-biclique, the KST Theorem ensures us that any k-vertex subgraph has density at most $O(k^{1/g})$. This indeed gives a gap of $O(k^{1/g})$ in terms of approximation Densest k-Subgraph and finishes the proof sketch for Theorem 8.

Unfortunately, Theorem 8 does not yet resolve the FPT approximability of DENSEST k-SUBGRAPH. In particular, while the hardness is only of the form $k^{o(1)}$, the best known algorithm (which is the same as that of the multicolored version discussed above) only gives an approximation ratio of k. Hence, we may ask whether this can be improved:

Open Question 9. *Is there an $o(k)$-FPT-approximation algorithm for DENSEST k-SUBGRAPH?*

This should be contrasted with Theorem 6, for which the FPT approximability of MULTICOLORED DENSEST k-SUBGRAPH is essentially resolved (up to lower order terms).

4. Algorithms

In this section we survey some of the developments on the algorithmic side in recent years. The organization of this section is according to problem types. We begin with basic packing and covering problems in Sections 4.1 and 4.2. We then move on to clustering in Section 4.3, network design in Section 4.4, and cut problems in Section 4.5. In Section 4.6 we present width reduction problems.

The algorithms in the above mentioned subsections compute approximate solutions to problems that are W[1]-hard. Therefore it is necessary to approximate, even when using parameterization. However, one may also aim to obtain faster parameterized runtimes than the known FPT algorithms, by sacrificing in the solution quality. We present some results of this type in Section 4.7.

4.1. Packing Problems

For a packing problem the task is to select as many combinatorial objects of some mathematical structure (such as a graph or a set system) as possible under some constraint, which restricts some objects to be picked if others are. A basic example is the INDEPENDENT SET problem, for which a maximum sized set of vertices of a graph needs to be found, such that none of them are adjacent to each other.

4.1.1. Independent Set

The INDEPENDENT SET problem is notoriously hard in general. Not only is there no polynomial time $n^{1-\varepsilon}$-approximation algorithm [81] for any constant $\varepsilon > 0$, unless P = NP, but also, under Gap-ETH, no $g(k)$-approximation can be computed in $f(k)n^{O(1)}$ time [39] for any computable functions f and g, where k is the solution size. On the other hand, for planar graphs a PTAS exists [82]. Hence a natural question is how the problem behaves for graphs that are "close" to being planar.

One way to generalize planar graphs is to consider minor-free graphs, because planar graphs are exactly those excluding K_5 and $K_{3,3}$ as minors. When parameterizing by the size of an excluded minor, the INDEPENDENT SET problem is paraNP-hard, since the problem is NP-hard on planar graphs [83]. Nevertheless a PAS can be obtained for this parameter [84].

Theorem 9 ([84,85]). *Let H be a fixed graph. For H-minor-free graphs, INDEPENDENT SET admits an $(1 + \varepsilon)$-approximation algorithm that runs in $f(H, \varepsilon)n^{O(1)}$ time for some function f.*

This result is part of the large framework of "bidimensionality theory" where any graph in an appropriate minor-closed class has treewidth bounded above in terms of the problem's solution value, typically by the square root of that value. These properties lead to efficient, often subexponential, fixed-parameter algorithms, as well as polynomial-time approximation schemes, for bidimensional problems in many minor-closed graph classes. The bidimensionality theory is based on algorithmic and combinatorial extensions to parts of the Robertson–Seymour Graph Minor Theory, in particular initiating a parallel theory of graph contractions. The foundation of this work is the topological theory of drawings of graphs on surfaces. We refer the reader to the survey of [86] and more recent papers [85,87,88].

A different way to generalize planar graphs is to consider a planar deletion set, i.e., a set of vertices in the input graph whose removal leaves a planar graph. Taking the size of such a set as a parameter, INDEPENDENT SET is again paraNP-hard [83]. However, by first finding a minimum sized planar deletion set, then guessing the intersection of this set with the optimum solution to INDEPENDENT SET, and finally using the PTAS for planar graphs [82], a PAS can be obtained parameterized by the size of a planar deletion set [8].

Theorem 10 ([8]). *For the INDEPENDENT SET problem a $(1 + \varepsilon)$-approximation can be computed in $2^k n^{O(1/\varepsilon)}$ time for any $\varepsilon > 0$, where k is the size of a minimum planar deletion set.*

Ideas using linear programming allow us to generalize and handle larger noise at the expense of worse dependence on ε. Bansal et al. [89] showed that given a graph obtained by adding δn edges to some planar graph, one can compute a $(1 + O(\varepsilon + \delta))$-approximate independent set in time $n^{O(1/\varepsilon^4)}$, which is faster than the $2^k n^{O(1/\varepsilon)}$ running time of Theorem 10 for large $k = \delta n$. Magen and Moharrami [90] showed that for every graph H and $\varepsilon > 0$, given a graph $G = (V, E)$ that can be made H-minor-free after at most δn deletions and additions of vertices or edges, the size of the maximum independent set can be approximately computed within a factor $(1 + \varepsilon + O(\delta |H| \sqrt{\log |H|}))$ in time $n^{f(\varepsilon, H)}$. Note that this algorithm does not find an independent set. Recently, Demaine et al. [91] presented a general framework to obtain better approximation algorithms for various problems including INDEPENDENT SET and CHROMATIC NUMBER, when the input graph is close to well-structured graphs (e.g., bounded degeneracy, degree, or treewidth).

It is also worth noting here that INDEPENDENT SET problem can be generalized to the d-SCATTERED SET problem where we are given an (edge-weighted) graph and are asked to select at least k vertices, so that the distance between any pair is at least d [92]. Recently in [93] some lower and upper bounds on the approximation of the d-SCATTERED SET problem have been provided.

A special case of INDEPENDENT SET is the INDEPENDENT SET OF RECTANGLES problem, where a set of axis-parallel rectangles is given in the two-dimensional plane, and the task is to find a maximum sized subset of non-intersecting rectangles. This is a special case, since pairwise intersections of rectangles can be encoded by edges in a graph for which the vertices are the rectangles. Parameterized by the solution size, the problem is W[1]-hard [94], and while a QPTAS is known [95], it is a challenging open question whether a PTAS exists. It was shown [96] however that both a PAS and a PSAKS exist for INDEPENDENT SET OF RECTANGLES parameterized by the solution size, even for the weighted version.

The runtime of this PAS is $f(k, \varepsilon) n^{g(\varepsilon)}$ for some functions f and g, where k is the solution size. Note that the dependence on ε in the degree of the polynomial factor of this algorithm cannot be removed, unless FPT=W[1], since any efficient PAS with runtime $f(k, \varepsilon) n^{O(1)}$ could be used to compute the optimum solution in FPT time by setting ε to $\frac{1}{k+1}$ in the W[1]-hard unweighted version of the problem [94]. However, in the so-called shrinking model an efficient PAS can be obtained [97] for INDEPENDENT SET OF RECTANGLES. The parameter in this case is a factor $0 < \delta < 1$ by which every rectangle is shrunk before computing an approximate solution, which is compared to the optimum solution without shrinking.

Theorem 11 ([96,97]). *For the INDEPENDENT SET OF RECTANGLES problem a $(1 + \varepsilon)$-approximation can be computed in $k^{O(k/\varepsilon^8)} n^{O(1/\varepsilon^8)}$ time for any $\varepsilon > 0$, where k is the size of the optimum solution, or in $f(\delta, \varepsilon) n^{O(1)}$ time for some computable function f and any $\varepsilon > 0$ and $0 < \delta < 1$, where δ is the shrinking factor. Moreover, a $(1 + \varepsilon)$-approximate kernel with $k^{O(1/\varepsilon^8)}$ rectangles can be computed in polynomial time.*

Another special case of INDEPENDENT SET is the INDEPENDENT SET ON UNIT DISK GRAPH problem, where given set of n unit disks in the Euclidean plane, the task is to determine if there exists a set of k non-intersecting disks. The problem is NP-hard [98] but admits a PTAS [99]. Marx [94] showed that, when parameterized by the solution size, the problem is W[1]-hard; this also rules out EPTAS (and even efficient PAS) for the problem, assuming FPT $\neq$ W[1]. On the other hand, in [100] the authors give an FPT algorithm for a special case of INDEPENDENT SET ON UNIT DISK GRAPH when there is a lower bound on the distance between any pair of centers.

4.1.2. Vertex Coloring

A problem related to INDEPENDENT SET is the VERTEX COLORING problem, for which the vertices need to be colored with integer values, such that no two adjacent vertices have the same color (which means that each color class forms an independent set in the graph). The task is to minimize the number of used colors. For planar graphs the problem has a polynomial time 4/3-approximation algorithm [8] via the celebrated Four Color Theorem, and a better approximation is not possible in polynomial time [101]. Using this algorithm, a 7/3-approximation can be computed in FPT time when parameterizing by the size of a planar deletion set [8]. When generalizing planar graphs by excluding any fixed minor, and taking its size as the parameter, a 2-approximation can be computed in FPT time [102]. Due to the NP-hardness for planar graphs [101], neither of these two parameterizations admits a PAS, unless P = NP.

Theorem 12 ([8,102]). *For the* VERTEX COLORING *problem*

- *a 7/3-approximation can be computed in $k^k n^{O(1)}$ time, where k is the size of a minimum planar deletion set, and*
- *a 2-approximation can be computed in $f(k)n^{O(1)}$ time for some function f, where k is the size of an excluded minor of the input graph.*

One way to generalize VERTEX COLORING is to see each color class as an induced graph of degree 0. The DEFECTIVE COLORING problem [103] correspondingly asks for a coloring of the vertices, such that each color class induces a graph of maximum degree Δ, for some given Δ. The aim again is to minimize the number of used colors. In contrast to VERTEX COLORING, the DEFECTIVE COLORING problem is W[1]-hard [104] parameterized by the treewidth. This parameter measures how "tree-like" a graph is, and is defined as follows.

Definition 1. *A tree decomposition of a graph $G = (V, E)$ is a tree T for which every node is associated with a bag $X \subseteq V$, such that the following properties hold:*

1. *the union of all bags is the vertex set V of G,*
2. *for every edge (u, v) of G, there is a node of T for which the associated bag contains u and v, and*
3. *for every vertex u of G, all nodes of T for which the associated bags contain u, induce a connected subtree of T.*

The width of a tree decomposition is the size of the largest bag minus 1 (which implies that a tree has a decomposition of width 1 where each bag contains the endpoints of one edge). The treewidth of a graph is the smallest width of any of its tree decompositions.

Treewidth is fundamental parameter of a graph and will be discussed more elaborately in Section 4.6.1. However, it is worth mentioning here that VERTEX COLORING is in FPT when parameterized by treewidth.

The strong polynomial-time approximation lower bound of $n^{1-\varepsilon}$ for VERTEX COLORING [81] naturally carries over to the more general DEFECTIVE COLORING problem. A much improved approximation factor of 2 is possible though in FPT time if the parameter is the treewidth [104]. It can be shown however, that a PAS is not possible in this case, as there is no $(3/2 - \varepsilon)$-approximation algorithm for any $\varepsilon > 0$ parameterized by the treewidth [104], unless FPT = W[1]. A natural question is whether the bound Δ of DEFECTIVE COLORING can be approximated instead of the number of colors. For this setting, a bicriteria PAS parameterized by the treewidth exists [104], which computes a solution with the optimum number of colors where each color class induces a graph of maximum degree at most $(1 + \varepsilon)\Delta$.

Theorem 13 ([104]). *For the* DEFECTIVE COLORING *problem, given a tree decomposition of width k of the input graph,*

- *a solution with the optimum number of colors where each color class induces a graph of maximum degree $(1 + \varepsilon)\Delta$ can be computed in $(k/\varepsilon)^{O(k)} n^{O(1)}$ time for any $\varepsilon > 0$,*
- *a 2-approximation (of the optimum number of colors) can be computed in $k^{O(k)} n^{O(1)}$ time, but*
- *no $(3/2 - \varepsilon)$-approximation (of the optimum number of colors) can be computed in $f(k)n^{O(1)}$ time for any $\varepsilon > 0$ and computable function f, unless FPT = W[1].*

The algorithms of the previous theorem build on the techniques of [105] using approximate addition trees in combination with dynamic programs that yield XP algorithms for these problems. This technique can be applied to various problems (cf. Section 4.2), including a different generalization of VERTEX COLORING called EQUITABLE COLORING. Here the aim is to color the vertices of a graph with as few colors as possible, such that every two adjacent vertices receive different colors, and all color classes contain the same number of vertices. It is a generalization of VERTEX COLORING, since one may add a sufficiently large independent set (i.e., a set of isolated vertices) to a graph such that the number of colors needed for an optimum VERTEX COLORING solution is the same as for an optimum EQUITABLE COLORING solution.

The EQUITABLE COLORING problem is W[1]-hard even when combining the number of colors needed and the treewidth of the graph as parameters [106]. On the other hand, a PAS exists [105] if the parameter is the cliquewidth of the input graph. This is a weaker parameter than treewidth, as the cliquewidth of a graph is bounded as a function of its treewidth. However, while bounded treewidth graphs are sparse, cliquewidth also allows for dense graphs (such as complete graphs). Formally, a graph of cliquewidth ℓ can be constructed using the following recursive operations using ℓ labels on the vertices:

1. Introduce(x): create a graph containing a singleton vertex labelled $x \in \{1, \ldots, \ell\}$.
2. Union(G_1, G_2): return the disjoint union of two vertex-labelled graphs G_1 and G_2.
3. Join(G, x, y): add all edges connecting a vertex of label x with a vertex of label y to the vertex-labelled graph G.
4. Rename(G, x, y): change the label of every vertex of G with label x to $y \in \{1, \ldots, \ell\}$.

A cliquewidth expression with ℓ labels is a recursion tree describing how to construct a graph using the above four operations using only labels from the set $\{1, \ldots, \ell\}$. Notice that the cliquewidth of a complete graph is two and therefore we have graphs of bounded clique-width but unbounded treewidth. As stated earlier the cliquewidth of a graph is bounded above exponentially in its treewidth and this dependence is tight for some graph families [107].

The PAS for EQUITABLE COLORING will compute a coloring using at most k colors such that the ratio between the sizes of any two color classes is at most $1 + \varepsilon$. In this sense it is a bicriteria approximation algorithm.

Theorem 14 ([105]). *For the EQUITABLE COLORING problem, given a cliquewidth expression with ℓ labels for the input graph, a solution with optimum number of colors where the ratio between the sizes of any two color classes is at most $1 + \varepsilon$, can be computed in $(k/\varepsilon)^{O(k\ell)} n^{O(1)}$ time [108,109] for any $\varepsilon > 0$, where k is the optimum number of colors.*

A variant of VERTEX COLORING is the MIN SUM COLORING problem, where, instead of minimizing the number of colors, the aim is to minimize the sum of (integer) colors, where the sum is taken over all vertices. This problem is FPT parameterized by the treewidth [110], but the related MIN SUM EDGE COLORING problem is NP-hard [111] on graphs of treewidth 2 (while being polynomial time solvable on trees [112]). For this problem the edges need to be colored with integer values, so that no two edges sharing a vertex have the same color, and the aim again is to minimize the total sum of colors. Despite being APX-hard [111] and also paraNP-hard for parameter treewidth, MIN SUM EDGE COLORING admits a PAS for this parameter [113].

Theorem 15 ([113]). *For the MIN SUM EDGE COLORING problem a $(1 + \varepsilon)$-approximation can be computed in $f(k, \varepsilon)n$ time for any $\varepsilon > 0$, where k is the treewidth of the input graph.*

4.1.3. Subgraph Packing

A special family of packing problems can be obtained by subgraph packing. Let H be a fixed "pattern" graph. The H-PACKING problem, given the "host" graph G, asks to find the maximum number of vertex-disjoint copies of H. One can also let H be a family of graphs and ask the analogous problem. There is another choice whether each copy of H is required to be an induced subgraph or a regular subgraph. We focus on the regular subgraph case here.

When H is a single graph with k vertices, a simple greedy algorithm that finds an arbitrary copy of H and adds it to the packing, guarantees a k-approximation in time $f(H,n) \cdot n$. Here $f(H,n)$ denotes the time to find a copy of H in an n-vertex graph. Following a general result for k-SET PACKING, a $(k+1+\varepsilon)/3$-approximation algorithm that runs in polynomial time for fixed k, ε exists [114]. When H is 2-vertex-connected or a star graph, even for fixed k, it is NP-hard to approximate the problem better than a factor $\Omega(k/\text{polylog}(k))$ [115]. There is no known connected H that admits an FPT (or even XP) algorithm achieving a $k^{1-\delta}$-approximation for some $\delta > 0$; in particular, the parameterized approximability of k-PATH PACKING is wide open. It is conceivable that k-PATH PACKING admits a parameterized $o(k)$-approximation algorithm, given an $O(\log k)$-approximation algorithm for k-PATH DELETION [116] and an improved kernel for INDUCED P_3 PACKING [117].

When H is the family of all cycles, the problem becomes the VERTEX CYCLE PACKING problem, for which the largest number of vertex-disjoint cycles of a graph needs to be found. No polynomial time $O(\log^{1/2-\varepsilon} n)$-approximation is possible for this problem [118] for any $\varepsilon > 0$, unless every problem in NP can be solved in randomized quasi-polynomial time. Furthermore, despite being FPT [119] parameterized by the solution size, VERTEX CYCLE PACKING does not admit any polynomial-sized exact kernel for this parameter [120], unless NP$\subseteq$coNP/poly. Nevertheless, a PSAKS can be found [18].

Theorem 16 ([18]). *For the* VERTEX CYCLE PACKING *problem, a* $(1 + \varepsilon)$-*approximate kernel of size* $k^{O(1/(\varepsilon \log \varepsilon))}$ *can be computed in polynomial time, where k is the solution size.*

4.1.4. Scheduling

Yet another packing problem on graphs, which, however, has applications in scheduling and bandwidth allocation, is the UNSPLITTABLE FLOW ON A PATH problem. Here a path with edge capacities is given together with a set of tasks, each of which specifies a start and an end vertex on the path and a demand value. The goal is to find the largest number of tasks such that for each edge on the path the total demand of selected tasks for which the edge lies between its start and end vertex, does not exceed the capacity of the edge. This problem admits a QPTAS [121], but it remains a challenging open question whether a PTAS exists. When parameterizing by the solution size, UNSPLITTABLE FLOW ON A PATH is W[1]-hard [122]. However a PAS exists [122] for this parameter.

Theorem 17 ([122]). *For the* UNSPLITTABLE FLOW ON A PATH *problem a* $(1 + \varepsilon)$-*approximation can be computed in* $2^{O(k \log k)} n^{g(\varepsilon)}$ *time for some computable function g and any $\varepsilon > 0$, where k is the solution size.*

Another scheduling problem is FLOW TIME SCHEDULING, for which a set of jobs is given, each of which is specified by a processing time, a release date, and a weight. The jobs need to be scheduled on a given number of machines, such that no job is processed before its release date and a job only runs on one machine at a time. Given a schedule, the flow time of a job is the weighted difference between its completion time and release date, and the task for the FLOW TIME SCHEDULING problem is to minimize the sum of all flow times. Two types of schedules are distinguished: in a preemtive schedule a job may be interrupted on one machine and then resumed on another, while in a non-preemtive schedule every job runs on one machine until its completion once it was started. If pre-emptive schedules are allowed, FLOW TIME SCHEDULING has no polynomial time $O(\log^{1-\varepsilon} p)$-approximation algorithm [123], unless P = NP, where p is the maximum processing time. For the more restrictive non-preemtive setting, no $O(n^{1/2-\varepsilon})$-approximation can be computed in polynomial time [124], unless P = NP, where n is the number of jobs. The latter lower bound is in fact even valid for only one machine, and thus parameterizing FLOW TIME SCHEDULING by the number of machines will not yield any better approximation ratio in this setting. A natural parameter for FLOW TIME SCHEDULING is the maximum over all processing times and weights of the given jobs. It is not known whether the problem is FPT or W[1]-hard for this parameter. However, when combining this parameter with the number of machines, a PAS can be obtained [125] despite the strong polynomial time approximation lower bounds.

Theorem 18 ([125]). *For the* FLOW TIME SCHEDULING *problem a* $(1+\varepsilon)$-*approximation can be computed in* $(mk)^{O(mk^3/\varepsilon)} n^{O(1)}$ *time in the preemtive setting, and in* $(mk/\varepsilon)^{O(mk^5)} n^{O(1)}$ *time in the non-preemtive setting, for any* $\varepsilon > 0$, *where m is the number of machines and k is an upper bound on every processing time and weight.*

4.2. Covering Problems

For a covering problem the task is to select a set of k combinatorial objects in a mathematical structure, such as a graph or set system (i.e., hypergraph), under some constraints that demands certain other objects to be intersected/covered. A basic example is the SET COVER problem where we are given a set system, which is simply a collection of subsets of a universe. The goal is to determine whether there are k subsets whose union cover the whole universe.

There are two ways define optimization based on covering problems. First, we may view the covering demands as strict constraints and aim to find a solution that minimize the constraint/cost while covering all objects (i.e., relaxing the size-k constraint); this results in a minimization problem. Second, we may view the size constraint as a strict constraint and aim to find a solution that covers as many objects as possible; this results in a maximization problem. We divide our discussion mainly into two parts, based on these two types of optimization problems. In Section 4.2.3, we discuss problems related to covering that fall into neither category.

4.2.1. Minimization Variants

We start out discussion with the minimization variants. For brevity, we overload the problem name and use the same name for the minimization variant (e.g., we use SET COVER instead of the more cumbersome MIN SET COVER). Later on, we will use different names for the maximization versions; hence, there will be no confusion.

Set Cover, Dominating Set and Vertex Cover. As discussed in detail in Section 3.1.2, SET COVER and equivalently DOMINATING SET are very hard to approximate in the general case. Hence, special cases where some constraints are placed on the set system are often considered. Arguably the most well-studied special case of SET COVER is the VERTEX COVER problem, in which the set system is a graph. That is, we would like to find the smallest set of vertices such that every edge has at least one endpoint in the selected set (i.e., the edge is "covered"). VERTEX COVER is well known to be FPT [126] and admit a linear-size kernel [127]. A generalization of VERTEX COVER on d-uniform hypergraph, where the input is now a hypergraph and the goal is to find the smallest set of vertices such that every hyperedge contain at least one vertex from the set, is also often referred to as d-HITTING SET in the parameterized complexity community. However, we will mostly use the nomenclature VERTEX COVER on d-uniform hypergraph because many algorithms generalizes well from VERTEX COVER in graphs to hypergraphs. Indeed, branching algorithms for VERTEX COVER on graphs can be easily generalized to hypergraphs, and hence the latter is also FPT. Polynomial-size kernels are also known for VERTEX COVER on d-uniform hypergraphs [128].

While VERTEX COVER both on graphs and d-uniform hypergraphs are already tractable, approximation can still help make algorithms even faster and kernels even smaller. We defer this discussion to Section 4.7.

Connected Vertex Cover. A popular variant of VERTEX COVER that is the CONNECTED VERTEX COVER problem, for which the computed solution is required to induce a connected subgraph of the input. Just as VERTEX COVER, the problem is FPT [129]. However, unlike VERTEX COVER, CONNECTED VERTEX COVER does not admit a polynomial-time kernel [130], unless NP$\subseteq$coNP/poly. In spite of this, a PSAKS for CONNECTED VERTEX COVER exists:

Theorem 19 ([18]). *For any $\varepsilon > 0$, an $(1+\varepsilon)$-approximate kernel with $k^{O(1/\varepsilon)}$ vertices can be computed in polynomial time.*

The ideas behind [18] is quite neat and we sketch it here. There are two reduction rules: (i) if there exists a vertex with degree more than $\Delta := 1/\varepsilon$ just "select" the vertex and (ii) if we see a vertex with more than k false twins, i.e., vertices with the same set of neighbors, then we simply remove it from the graph. An important observation for (i) is that, since we have to either pick the vertex or all $\geq \Delta$ neighbors anyway, we might as well just select it even in the second case because it affects the size of the solution by a factor of at most $\frac{1+\Delta}{\Delta} = 1 + \varepsilon$. For (ii), it is not hard to see that we either select one of the false twins or all of them; hence, if a vertex has more than k false twins, then it surely cannot be in the optimal solution. Roughly speaking, these two observations show that this is an $(1+\varepsilon)$-kernel. Of course, in the actual proof, "selecting" a vertex needs to be defined more carefully, but we will not do it here. Nonetheless, imagine the end step when we cannot apply these two reduction rules anymore. Essentially speaking, we end up with a graph where some (less than $(1+\varepsilon)k$) vertices are marked as "selected" and the remaining vertices have degree at most Δ. Now, every vertex is either inside the solution, or all of its neighbors must lie in the solution. There are only (at most) k vertices in the first case. For the second case, note that these vertices have degree at most Δ and they have at most k false twins, meaning that there are at most $k^{1+\Delta} = k^{1+1/\varepsilon}$ such vertices. In other words, the kernel is of size $k^{O(1/\varepsilon)}$ as desired. This constitutes the main ideas in the proof; let us stress again that the actual proof is of course more complicated than this since we did not define rule (i) formally.

Recently, Krithika et al. [131] considered the following structural parameters beyond the solution size: split deletion set, clique cover and cluster deletion set. In each case, the authors provide a PSAKS for the problem. We will not fully define these parameters here, but we note that the first parameter (split deletion set) is always no larger than the size of the minimum vertex cover of the graph. In another very recent work, Majumdar et al. [132] give a PSAKS for each of the following parameters, each of which is always no larger than the solution size: the deletion distance of the input graph to the class of cographs, the class of bounded treewidth graphs, and the class of all chordal graphs. Hence, these results may be viewed as a generalization of the aforementioned PSAKS from [18].

Connected Dominating Set. Similar to CONNECTED VERTEX COVER, the CONNECTED DOMINATING SET problem is the variant of DOMINATING SET for which the solution additionally needs to induce a connected subgraph of the input graph. When placing no restriction on the input graph, the problem is as hard to approximate as DOMINATING SET. However, for some special classes of graphs, PSAKS or bi-PSAKS [133] are known; these include graphs with bounded expansion, nowhere dense graphs, and d-biclique-free graphs [134].

Covering Problems parameterized by Graph Width Parameters. Several works in literature also study the approximability of variants of VERTEX COVER and DOMINATING SET parameterized by graph widths [105,135]. These variants include:

- POWER VERTEX COVER (PVC). Here, along with the input graph, each edge has an integer demand and we have to assign (power) values to vertices, such that each edge has at least one endpoint with a value at least its demand. The goal is to minimize the total assigned power. Note that this is generalizes of VERTEX COVER, where edges have unit demands.
- CAPACITATED VERTEX COVER (CVC). The problem is similar to VERTEX COVER, except that each vertex has a capacity which limits the number of edges that it can cover. Once again, VERTEX COVER is a special case of CVC where each vertex's capacity is ∞.
- CAPACITATED DOMINATING SET (CDS). Analogous to CVC, this is a generalization of DOMINATING SET where each vertex has a capacity and it can only cover/dominate at most that many other vertices.

All problems above are FPT under standard parameter (i.e., the optimum) [135,136]. However, when parameterizing by the treewidth [137], all three problems become W[1]-hard [135]. (This is in contrast to VERTEX COVER and DOMINATING SET, both of which admit straightforward dynamic programming FPT algorithms parameterized by treewidth.) Despite this, good FPT approximation algorithms are known for the problem. In particular, a PAS is known for PVC [135]. For CVC and CDS, a bicriteria PAS exists for the problem [105], which in this case computes a solution of size at most the optimum, so that no vertex capacity is violated by more than a factor of $1 + \varepsilon$.

The approximation algorithms for CVC and CDS are results of a more general approach of Lampis [105]. The idea is to execute an "approximate" version of dynamic programming in tree decomposition instead of the exact version; this helps reduce the running time from $n^{O(w)}$ to $(\log n/\varepsilon)^{O(w)}$, which is FPT. The approach is quite flexible: several approximation for graphs problems including covering problems can be achieved via this method and it also applies to clique-width. Please refer to [105] for more details.

Packing-Covering Duality and Erdős-Pósa Property. Given a set system $(V, \mathcal{C})$ where V is the universe and $\mathcal{C} = \{C_1, \ldots, C_m\}$ is a collection of subsets of V, HITTING SET is the problem of computing the smallest $S \subseteq V$ that intersects every C_i, and SET PACKING is the problem of computing the largest subcollection $\mathcal{C}' \subseteq \mathcal{C}$ such that no two sets in $\mathcal{C}'$ intersect. It can also be observed that the optimal value for HITTING SET is at least the optimal value for SET PACKING, while the standard LP relaxations for them (covering LP and packing LP) have the same optimal value by strong duality. Studying the other direction of the inequality (often called the packing-covering duality) for natural families of set systems has been a central theme in combinatorial optimization. The gap between the covering optimum and packing optimum is large in general (e.g., DOMINATING SET/INDEPENDENT SET), but can be small for some families of set systems (e.g., s-t CUT/s-t DISJOINT PATHS and VERTEX COVER/MATCHING especially in bipartite graphs).

One notion that has been important for both parameterized and approximation algorithms is the Erdős-Pósa property [138]. A family of set systems is said to have the Erdős-Pósa property when there is a function $f : \mathbb{N} \to \mathbb{N}$ such that for any set system in the family, if the packing optimum is k, the covering optimum is at most $f(k)$. This immediately implies that the multiplicative gap between these two optima is at most $f(k)/k$, and constructive proofs for the property for various set systems have led to $(f(k)/k)$-approximation algorithms. Furthermore, for some problems including CYCLE PACKING, the Erdős-Pósa property gives an immediate parameterized algorithm. We refer the reader to a recent survey [139] and papers [119,140,141].

The original paper of Erdős and Pósa [138] proved the property for set systems $(V, \mathcal{C})$ when there is an underlying graph $G = (V, E)$ and $\mathcal{C}$ is the set of cycles, which corresponds to the pair CYCLE PACKING/FEEDBACK VERTEX SET; every graph either has at least k vertex-disjoint cycles or there is a feedback vertex set of size at most $O(k \log k)$. Many subsequent papers also studied natural set systems arising from graphs where V is the set of vertices or edges and $\mathcal{C}$ denotes a collection of subgraphs of interest. For those set systems, Erdős-Pósa Properties are closely related to Set Packing introduced in Section 4.1.3 and $\mathcal{F}$-DELETION problems introduced in Section 4.6.

4.2.2. Maximization Variants

We now move on to the maximization variants of covering problems. To our knowledge, these covering problems are much less studied in the context of parameterized approximability compared to their minimization counterparts. In particular, we are only aware of works on the maximization variants of SET COVER and VERTEX COVER, which are typically called MAX k-COVERAGE and MAX k-VERTEX COVER respectively.

Max k-Coverage. Recall that here we are given a set system and the goal is to select k subsets whose union is maximized. It is well known that the simple greedy algorithm yields an $\left(\frac{e}{e-1}\right)$-approximation [142]. Furthermore, Fiege shows, in his seminal work [29], show that this is tight: $\left(\frac{e}{e-1} - \varepsilon\right)$-approximation is NP-hard for any constant $\varepsilon > 0$. In fact, recently it has been shown that this inapproximability applies also to the parameterized setting. Specifically, under Gap-ETH, $\left(\frac{e}{e-1} - \varepsilon\right)$-approximation cannot be achieved in FPT time [143] or even $f(k) \cdot n^{o(k)}$ time [70]. In other words, the trivial algorithm is tight in terms of running time, the greedy algorithm is tight in terms of approximation ratio, and there is essentially no trade-off possible between these two extremes. We remark here that this hardness of approximation is also the basis of hardness for k-MEDIAN and k-MEANS [143] (see Section 4.3).

Due to the strong inapproximability result for the general case of MAX k-COVERAGE, different parameters have to be considered in order to obtain a PAS for MAX k-COVERAGE. An interesting positive result here is when the parameters are k and the VC dimension of the set system, for which a PAS exists while the exact version of the problem is W[1]-hard [144].

Max k-Vertex Cover. Another special case of MAX k-COVERAGE is the restriction when each element belongs to at most d subsets in the system. This corresponds exactly to the maximization variant of the VERTEX COVER problem on d-uniform hypergraph, which will refer to as MAX k-VERTEX COVER. Note here that, for such set systems, their VC-dimensions are also bounded by $\log d + 1$ and hence the aforementioned PAS of [144] applies here as well. Nonetheless, MAX k-VERTEX COVER admits a much simpler PAS (and even PSAKS) compared to MAX k-COVERAGE parameterized by k and VC-dimension, as we will discuss more below.

MAX k-VERTEX COVER was first studied in the context of parameterized complexity by Guo et al. [145] who showed that the problem is W[1]-hard. Marx, in his survey on parameterized approximation algorithms [8], gave a PAS for the problem with running time $2^{\tilde{O}(k^3/\varepsilon)}$. Later, Lokshtanov et al. [18] shows that Marx's approach can be used to give a PSAKS of size $O(k^5/\varepsilon^2)$. Both of these results mainly focus on graphs. Later, Skowron and Faliszewski [146–148] gave a more general argument that both works generally for any d-uniform hypergraph and improves the running time and kernel size:

Theorem 20 ([146]). *For the* MAX k-VERTEX COVER *problem in d-uniform hypergraphs, a $(1 + \varepsilon)$-approximation can be computed in $O^*\left((d/\varepsilon)^k\right)$ time for any $\varepsilon > 0$. Moreover, an $(1 + \varepsilon)$-approximate kernel with $O(dk/\varepsilon)$ vertices can be computed in polynomial time.*

The main idea of the above proof is simple and elegant, and hence we will include it here. For convenience, we will only discuss the graph case, i.e., $d = 2$. It suffices to just give the $O(k/\varepsilon)$-vertex kernel; the PAS immediately follows by running the brute force algorithm on the output instance from the kernel. The kernel is as simple as it gets: just keep $2k/\varepsilon$ vertices with highest degrees and throw the remaining vertices away! Note that there is a subtle point here, which is that we do not want to throw away the edges linking from the kept vertices to the remaining vertices. If self-loops are allowed in a graph, this is not an issue since we may just add a self-loop to each vertex for each edge adjacent to it with the other endpoint being discarded. When self-loops are not allows, it is still possible to overcome this issue but with slightly larger kernel; we refer the readers to Section 3.2 of [147] for more detail.

Having defined the kernel, let us briefly discuss the intuition as to why it works. Let $V_{2k/\varepsilon}$ denote the set of $2k/\varepsilon$ highest-degree vertices. The main argument of the proof is that, if there is an optimal solution S, then we may modify it to be entirely contained in $V_{2k/\varepsilon}$ while preserving the number of covered edges to within $(1 + \varepsilon)$ factor. The modification is simple: for every vertex that is outside of $V_{2k/\varepsilon}$, we replace it with a random vertex from $V_{2k/\varepsilon}$. Notice here that we always replace a vertex with a higher-degree vertex. Naturally, this should be good in terms of covering more edges, but there is a subtle point here: it is possible that the high degree vertices are "double counted" if a particular edge is covered by both endpoints. The size $2k/\varepsilon$ is selected exactly to combat this issue; since the set is large enough, "double counting" is rare for random vertices. This finishes our outline for the intuition.

We end by remarking that MAX k-VERTEX COVER on graphs is already APX-hard [149], and hence the PASes mentioned above once again demonstrate additional power of FPT approximation algorithms over polynomial-time approximation algorithms.

4.2.3. Other Related Problems

There are several other covering-related problems that do not fall into the two categories we discussed so far. We discuss a couple such problems below.

Min k-Uncovered. The first is the MIN k-UNCOVERED problem, where the input is a set system and we would like to select k sets as to minimize the number of uncovered elements. When we are concerned with exact solutions, this is of course the SET COVER. However, the optimization version becomes quite different from MAX k-COVERAGE. In particular, since it is hard to determine whether we can find k subsets that cover the whole universe, the problem is not approximable at all in the general case. However, if restrict ourselves to graphs and hypergraphs (for which we refer to the problem as MIN k-VERTEX COVER), it is possible to get a (randomized) PAS for the problem [146]:

Theorem 21 ([146]). *For the* MIN k-VERTEX UNCOVERED *problem in d-uniform hypergraphs, a $(1+\varepsilon)$-approximation can be computed in $O^*\left((d/\varepsilon)^k\right)$ time for any $\varepsilon > 0$.*

The algorithm is based on the following simple randomized branching: pick a random uncovered element and branch on all possibilities of selecting a subset that contains it. Notice that since an element belongs to only d subsets, the branching factor is at most d. The key intuition in the approximation proof is that, when the number of elements we have covered so far is still much less than that in the optimal solution, there is a relatively large probability (i.e., ε) that the random element is covered in the optimal solution. If we always pick such a "good" element in most branching steps, then we would end up with a solution close to the optimum. Skowron and Faliszewski [146] formalizes this intuition by showing that the algorithm outputs an $(1+\varepsilon)$-approximate solution with probability roughly ε^k. Hence, by repeating the algorithm $(1/\varepsilon)^k$ time, one arrives at the claimed PAS. To the best of our knowledge, it is unknown whether a PSAKS exists for the problem.

Min k-Coverage. Another variant of the SET COVER problem studied is MIN k-COVERAGE [150–152], where we would like to select k subsets that minimizes the number of covered elements. We stress here that this problem is not a relaxation of SET COVER but rather is much more closely related to graph expansion problems (see [151]).

It is known that, when there is no restriction on the input set system, the problem is (up to a polynomial factor) as hard to approximate as the DENEST k-SUBGRAPH problem [150]. Hence, by the inapproximability of the latter discussed earlier in the survey (Theorem 8), we also have that there is no $k^{o(1)}$-approximation algorithm for the problem that runs in FPT time.

Once again, the special case that has been studied in literature is when the input set system is a graph, in which case we refer to the problem as MIN k-VERTEX COVER. Gupta, Lee, and Li [153,154] used the technique of Marx [8] to give a PAS for the problem with running time $O^*((k/\varepsilon)^{O(k)})$. The running time was later improved in [147] to $O^*((1/\varepsilon)^{O(k)})$. The algorithm there is again based on branching, but the rules are more delicate and we will not discuss them here. An interesting aspect to note here is that, while both MAX k-VERTEX COVER and MIN k-VERTEX COVER have PSAKS of the (asymptotically) same running time, the former admits a PSAKS whereas the latter does not (assuming a variant of the Small Set Expansion Conjecture) [147].

To the best of our knowledge, MIN k-VERTEX COVER has not been explicitly studied on d-uniform hypergraphs before, but we suspect that the above results should carry over from graphs to hypergraphs as well.

4.3. Clustering

Clustering is a representative task in unsupervised machine learning that has been studied in many fields. In combinatorial optimization communities, it is often formulated as the following: Given a set P of points and a set F of candidate centers (also known as facilities), and a metric on $X \supseteq P \cup F$ given by the distance $\rho : X \times X \to \mathbb{R}^+ \cup \{0\}$, choose k centers $C \subseteq F$ to minimize some objective function cost := $\text{cost}(P, C)$. To fully specify the problem, the choices to make are the following. Let $\rho(C, p) := \min_{c \in C} \rho(c, p)$.

- Objective function: Three well-studied objective functions are

    - k-MEDIAN ($\text{cost}(P, C) := \sum_{p \in P} \rho(C, p)$).
    - k-MEANS ($\text{cost}(P, C) := \sum_{p \in P} \rho(C, p)^2$).
    - k-CENTER ($\text{cost}(P, C) := \max_{p \in P} \rho(C, p)$).

- Metric space: The ambient metric space X can be

    - A general metric space explicitly given by the distance $\rho : X \times X \to \mathbb{R}^+ \cup \{0\}$.
    - The Euclidean space $\mathbb{R}^d$ equipped with the ℓ_2 distance.
    - Other structured metric spaces including metrics with bounded doubling dimension or bounded highway dimension.

While many previous results on clustering focused on non-parameterized polynomial time, there are at least three natural parameters one can parameterize: The number of clusters k, the dimension d (if defined), and the approximation accuracy parameter ε. In general metric spaces, parameterized approximation algorithms (mainly with parameter k) were considered very recently, but in Euclidean spaces, many previous results already give parameterized approximation algorithms with parameters $k, d,$ and ε.

4.3.1. General Metric Space

We can assume $X = P \cup F$ without loss of generality. Let $n := |X|$ and note that the distance $\rho : X \times X \to \mathbb{R}^+ \cup \{0\}$ is explicitly specified by $\Theta(n^2)$ numbers. A simple exact algorithm running in time $O(n^{k+1})$ can be achieved by enumerating all k centers $c_1, \ldots, c_k \in F$ and assign each point p to the closest center. In this setting, the best approximation ratios achieved by polynomial time algorithms are $2.611 + \varepsilon$ for k-MEDIAN [155], $9 + \varepsilon$ for k-MEANS [156], and 3 for k-CENTER [157,158]. From the hardness side, it is NP-hard to approximate k-MEDIAN within a factor $1 + 2/e - \varepsilon \approx 1.73 - \varepsilon$, k-MEDIAN within a factor $1 + 8/e - \varepsilon \approx 3.94 - \varepsilon$, k-CENTER within a factor $3 - \varepsilon$ [159].

While there are some gaps between the best algorithms and the best hardness results for k-MEDIAN and k-MEANS, it is an interesting question to ask how parameterization by k changes the approximation ratios for both problems. Cohen-Addad et al. [143] studied this question and gave exact answers.

Theorem 22 ([143]). *For any $\varepsilon > 0$, there is an $(1 + 2/e + \varepsilon)$-approximation algorithm for k-MEDIAN, and an $(1 + 8/e + \varepsilon)$-approximation algorithm for k-MEANS, both running in time $(O(k \log k / \varepsilon^2))^k n^{O(1)}$.*

There exists a function $g : \mathbb{R}^+ \to \mathbb{R}^+$ such that assuming the Gap-ETH, for any $\varepsilon > 0$, any $(1 + 2/e - \varepsilon)$-approximation algorithm for k-MEDIAN, and any $(1 + 8/e - \varepsilon)$-approximation algorithm for k-MEANS, must run in time at least $n^{kg(\varepsilon)}$.

These results show that if we parameterize by k, $1 + 2/e$ (for k-MEDIAN) and $1 + 8/e$ (for k-MEANS) are the exact limits of approximation for parameterized approximation algorithms. Similar reductions also show that no parameterized approximation algorithm can achieve $(3 - \varepsilon)$-approximation for k-CENTER for any $\varepsilon > 0$ (only assuming W[2] $\neq$ FPT), so the power of parameterized approximation is exactly revealed for all three objective functions.

Algorithm for k-MEDIAN. We briefly describe ideas for the algorithm for k-MEDIAN in Theorem 22. The main technical tool that the algorithm uses is a coreset, which will be also frequently used for Euclidean subspaces in the next subsection.

When S is a set of points with weight functions $w : S \to \mathbb{R}^+$, let us extend the definition of the objective function $\text{cost}(S, C)$ such that

$$\text{cost}(S, C) := \sum_{p \in S} w(p) \cdot \rho(C, s).$$

Given a clustering instance (P, F, ρ, k) and $\varepsilon > 0$, a subset $S \subseteq P$ with weight functions $w : S \to \mathbb{R}^+$ is called a (strong) corset if for any k centers $C = \{c_1, \ldots, c_k\} \subseteq F$,

$$|\text{cost}(S, C) - \text{cost}(P, C)| \leq \varepsilon \cdot \text{cost}(P, C).$$

For a general metric space, Chen [160] gave a coreset of cardinality $\tilde{O}(k^2 \log^2 n / \varepsilon^2)$. (In this subsection, $\tilde{O}(\cdot)$ hides $\text{poly}(\log \log n, \log k, \log(1/\varepsilon))$.) This was improved by Feldman and Langeberg [161] to $O(k \log n / \varepsilon^2)$. We introduce high-level ideas of [160] later.

Given the coreset, it remains to give a good parameterized approximation algorithm for the problem for a much smaller (albeit weighted) point set $|P| = O(\text{poly}(\log n, k))$. Note that $|F|$ can be still as large as n, so naively choosing k centers from F will take n^k time and exhaustively partitioning P into k sets will take $k^{|P|} = n^{\text{poly}(k)}$ time. (Indeed, exactly solving this small case will give an EPAS, which will contradict the Gap-ETH.)

Fix an optimal solution, and let $C^* = \{c_1^*, \ldots, c_k^*\}$ are the optimal centers and $P_i^* \subseteq P$ is the cluster assigned to c_i^*. One information we can guess is, for each $i \in [k]$, the point $p_i \in P_i^*$ closest to c_i^* and (approximate) $\rho(c_i^*, p_i)$. Since $|P| = \text{poly}(k \log n)$, guessing them only takes time $(k \log n)^{O(k)}$, which can be made FPT by separately considering the case $(\log n)^k \leq n$ and the case $(\log n)^k \geq n$, in which $k = \Omega(\log n / \log \log n)$ and $(\log n)^k = (k \log k)^{O(k)}$.

Let $F_i \subseteq F$ be the set of candidate centers that are at distance approximately r_i from p_i, so that $c_i^* \in F_i$ for each i. The algorithm chooses k centers $C \subseteq F$ such that $|C \cap F_i| \geq 1$ for each $i \in [k]$. Let $c_i \in C \cap F_i$. For any point $p \in P$ (say $p \in P_j^*$, though the algorithm doesn't need to know j), then

$$\rho(C, p) \leq \rho(c_j, p) \leq \rho(c_j, p_j) + \rho(p_j, c_j^*) + \rho(c_j^*, p) \leq 3\rho(c_j^*, p),$$

where $\rho(c_j, p_j) \approx \rho(p_j, c_j^*) \leq \rho(c_j^*, p)$ by the choice of p_j.

This immediately gives a 3-approximation algorithm in FPT time, which is worse than the best polynomial time approximation algorithm. To get the optimal $(1 + 2/e)$-approximation algorithm, we further reduce the job of finding $c_i \in F_i$ to maximizing a monotone submodular function with a partition matroid constraint, which is known to admit an optimal $(1 - 1/e)$-approximation algorithm [162]. Then we can ensure that for $(1 - 1/e)$ fraction of points, the distance to the chosen centers is shorter than in the optimal solution, and for the remaining $1/e$ fraction, the distance is at most three times the distance in the optimal solution. We refer the reader to [143] for further details.

Constructing a coreset. As discussed above, a coreset is a fundamental building block for optimal parameterized approximation algorithms for k-MEDIAN and k-MEANS for general metrics. We briefly describe the construction of Chen [160] that gives a coreset of cardinality $\tilde{O}(k^2 \log^2 n / \varepsilon^2)$ for k-MEDIAN. Similar ideas can be also used to obtain an EPAS for Euclidean spaces parameterized by k, though better specific constructions are known in Euclidean spaces.

We first partition P into $P_1, \ldots P_\ell$ such that $\ell = O(k \log n)$ and

$$\sum_{i=1}^{\ell} |P_i| \, \text{diam}(P_i) = O(\text{OPT}).$$

Such a partition can be obtained by using a known (bicriteria) constant factor approximation algorithm for k-MEDIAN. Next, let $t = \widetilde{O}(k \log n)$, and for each $i = 1, \ldots, \ell$, we let $S_i = \{s_1, \ldots, s_t\}$ be a random subset of t points of P_i where each s_j is an independent and uniform sample from P_i and is given weight $|P_i|/t$. (If $|P_i| \leq t$, we simply let $S_i = P_i$ with weights 1.) The final coreset S is the union of all S_i's.

To prove that it works, we simply need to show that for any set of k centers $C \subseteq F$ with $|C| = k$,

$$\Pr[|\operatorname{cost}(S,C) - \operatorname{cost}(P,C)| > \varepsilon \cdot \operatorname{cost}(P,C)] \leq o(1/n^k),$$

so that the union bound over $\binom{n}{k}$ choices of C works. Indeed, we show that for each $i = 1, \ldots, \ell$,

$$\Pr[|\operatorname{cost}(S_i,C) - \operatorname{cost}(P_i,C)| > \varepsilon \cdot |P_i| \operatorname{diam}(P_i)] \leq o(1/(\ell \cdot n^k)), \tag{1}$$

so that we can also union bound and sum over $i \in [\ell]$, using the fact that $\sum_i |P_i| \operatorname{diam}(P_i) = O(\text{OPT}) \leq O(\operatorname{cost}(P,C))$.

It is left to prove (1). Fix C and i (let $P_i = \{p_1, \ldots, p_{|P_i|}\}$), and recall that

$$\operatorname{cost}(P_i, C) = \sum_{j=1}^{|P_i|} \rho(C, p_j).$$

When $|P_i| \leq t$, $S_i = P_i$, so (1) holds. Otherwise, recall that $S_i = \{s_1, \ldots, s_t\}$ where each s_j is an independent and uniform sample from P_i with weight $w := |P_i|/t$. For $j = 1, \ldots, t$, let $X_j := w \cdot \rho(C, s_j)$. Note that $\operatorname{cost}(S_i, C) = \sum_j X_j$ and $\operatorname{cost}(P_i, C) = t \cdot \mathbb{E}[X_j] = \mathbb{E}[\operatorname{cost}(S_i, C)]$. A crucial observation is that that $|\rho(C, p_j) - \rho(C, p_{j'})| \leq \operatorname{diam}(P_i)$ for any $j, j' \in [|P_i|]$, so that $|X_j - X_{j'}| \leq w \cdot \operatorname{diam}(P_i)$ for any $j, j' \in [t]$. If we let $X_{\min} := \min_{j \in [|P_i|]}(w\rho(C, p_j))$ and $Y_j := (X_j - X_{\min})/(w \cdot \operatorname{diam}(P_i))$, Y_j's are t i.i.d. random variables that are supported in $[0, 1]$. The standard Chernoff–Hoeffding inequality gives

$$\Pr\left[\left|\sum_j X_j - t\mathbb{E}[X_j]\right| > \varepsilon|P_i|\operatorname{diam}(P_i)\right] = \Pr\left[\left|\sum_j Y_j - t\mathbb{E}[Y_j]\right| > \varepsilon t\right] \leq \exp(O(\varepsilon^2 t)) \leq o(1/(\ell \cdot n^k)),$$

proving (1) for $t = \widetilde{O}(k \log n)$ and finishing the proof. A precise version of this argument was stated in Haussler [163].

4.3.2. Euclidean Space

For Euclidean spaces, we assume that $X = F = \mathbb{R}^d$ for some $d \in \mathbb{N}$, endowed with the standard ℓ_2 metric. Let $n = |P|$ in this subsection. Now we have k and d as natural structural parameters of clustering tasks. Many previous approximation algorithms in EUCLIDEAN k-MEDIAN and EUCLIDEAN k-MEANS in Euclidean spaces, without explicit mention to parameterized complexity, are parameterized approximation algorithms parameterized by k or d (or both). The highlight of this subsection is that for both EUCLIDEAN k-MEDIAN and EUCLIDEAN k-MEANS, an EPAS exists with only one of k and d as a parameter. Without any parameterization, both EUCLIDEAN k-MEDIAN and k-MEANS are known to be APX-hard [164,165]. We introduce these results in the chronological order, highlighting important ideas.

EUCLIDEAN k-MEDIAN with parameter d. The first PTAS for EUCLIDEAN k-MEDIAN in Euclidean spaces with fixed d appears in Arora et al. [166]. The techniques extend Arora's previous PTAS for the EUCLIDEAN TRAVELING SALESMAN problem in Euclidean spaces [167], first proving that there exists a near-optimal solution that interacts with a quadtree (a geometric division of $\mathbb{R}^d$ into a hierarchy of square regions) in a restricted sense, and finally finding such a tour using dynamic programming. The running time is $n^{O(1/\varepsilon)}$ for $d = 2$ and $n^{(\log n/\varepsilon)^{d-2}}$ for $d > 2$. Kolliopoulous and Rao [168] improved the running time to $2^{O((\log(1/\varepsilon)/\varepsilon)^{d-1})} n \log^{d+6} n$, which is an EPAS with parameter d.

EUCLIDEAN k-MEDIAN and EUCLIDEAN k-MEANS with parameter k.

An EPAS for EUCLIDEAN k-MEANS even with parameters both k and d took longer to be discovered, and first appeared when Matoušek [169] gave an approximation scheme that runs in time $O(n\varepsilon^{2k^2d}\log^k n)$. After this, several improvements on the running time followed Bādoiu et al. [170], De La Vega et al. [171], Har-Peled and Mazumdar [172].

Kumar et al. [173,174] gave approximation schemes for both k-MEDIAN and k-MEANS, running in time $2^{(k/\varepsilon)^{O(1)}}dn$. This shows that an EPAS can be obtained by using only k as a parameter. Using this result and and improved coresets, more improvements followed [160,161,175]. The current best runtime to get $(1+\varepsilon)$-approximation is $O(ndk + d \cdot \text{poly}(k/\varepsilon) + 2^{\tilde{O}(k/\varepsilon)})$ for k-MEANS [175], and $O(ndk + 2^{\text{poly}(1/\varepsilon,k)})$ time for k-MEDIAN [161].

A crucial property of the Euclidean space that allows an EPAS with parameter k (which is ruled out for general metrics by Theorem 22) is the sampling property, which says that for any set $Q \subseteq \mathbb{R}^d$ as one cluster, there is an algorithm that is given only $g(1/\varepsilon)$ samples from Q and outputs $h(1/\varepsilon)$ candidate centers such that one of them is ε-close to the optimal center for the entire cluster Q for some functions g, h. (For example, for k-MEANS, the mean of $O(1/\varepsilon)$ random samples ε-approximates the actual mean with constant probability.) This idea leads to an $(1+\varepsilon)$-approximation algorithm running in time $|P|^{f(\varepsilon,k)}$. Together even with a general coreset construction of size $\text{poly}(k,\log n,1/\varepsilon)$, one already gets an EPAS with parameter k. Better coresets construction are also given in Euclidean spaces. Recent developments [176–178] construct core-sets of size $\text{poly}(k,1/\varepsilon)$ (no dependence on n or d), which is further extended to the shortest-path metric of an excluded-minor graph [179].

EUCLIDEAN k-MEANS with parameter d.

Cohen-Addad et al. [180] and Friggstad et al. [181] recently gave approximation schemes running in time $n^{f(d,\varepsilon)}$ using local search techniques. These results were improved to an EPAS in [182], and also extended to doubling metrics [183].

Other metrics and k-CENTER. For the k-CENTER problem an EPAS exists when parametrizing by both k and the doubling dimension [184], and also for planar graphs there is an EPAS for parameter k, which is implied by the EPTAS of Fox-Epstein et al. [185] (cf. [184]).

There are also parameterized approximation schemes for metric spaces with bounded highway dimension [184,186,187] and various graph width parameters [108].

Capacitated clustering and other variants. Another example where the parameterization by k helps is CAPACITATED k-MEDIAN, where each possible center $c \in F$ has a capacity $u_c \in \mathbb{N}$ and can be assigned at most u_c points. It is not known whether there exists a constant-factor approximation algorithm, and known constant factor approximation algorithms either open $(1+\varepsilon)k$ centers [188] or violate capacity constraints by an $(1+\varepsilon)$ factor [189]. Adamczyk et al. [190] gave a $(7+\varepsilon)$-approximation algorithm in $f(k,\varepsilon)n^{O(1)}$ time, showing that a constant factor parameterized approximation algorithm is possible. The approximation ratio was soon improved to $(3+\varepsilon)$ [191]. For CAPACITATED EUCLIDEAN k-MEANS, [192] also gave a $(69+\varepsilon)$-approximation algorithm for in $f(k,\varepsilon)n^{O(1)}$ time.

While the capacitated versions of clustering look much harder than their uncapacitated counterparts, there is no known theoretical separation between the capacitated version and the uncapacitated version in any clustering task. Since the power of parameterized algorithms for uncapacitated clustering is well understood, it is a natural question to understand the "capacitated VS uncapacitated question" in the FPT setting.

Open Question 10. *Does* CAPACITATED k-MEDIAN *admit an* $(1+2/e)$*-approximation algorithm in FPT time with parameter k? Do* CAPACITATED EUCLIDEAN k-MEANS/k-MEDIAN *admit an EPAS with parameter k or d?*

Since clustering is a universal task, like capacitated versions, many variants of clustering tasks have been studied including k-MEDIAN/k-MEANS WITH OUTLIERS [193] and MATROID/KNAPSACK MEDIAN [194]. While no variant is proved to harder than the basic versions, it would be interesting to see whether they all have the same parameterized approximability with the basic versions.

4.4. Network Design

In network design, the task is to connect some set of vertices in a metric, which is often given by the shortest-path metric of an edge-weighted graph. Two very prominent problems of this type are the TRAVELLING SALESPERSON (TSP) and STEINER TREE problems. For TSP all vertices need to be connected in a closed walk (called a route), and the length of the route needs to be minimized [195]. For STEINER TREE a subset of the vertices (called terminals) is given as part of the input, and the objective is to connect all terminals by a tree of minimum weight in the metric (or graph). Both of these are fundamental problems that have been widely studied in the past, both on undirected and directed input graphs.

Undirected graphs. A well-studied parameter for STEINER TREE is the number of terminals, for which the problem has been known to be FPT since the early 1970s due to the work of Dreyfus and Wagner [196]. Their algorithm is based on dynamic programming and runs in $3^k n^{O(1)}$ time if k is the number of terminals. Faster algorithms based on the same ideas with runtime $(2+\delta)^k n^{O(1)}$ for any constant $\delta > 0$ exist [197] (here the degree of the polynomial depends on δ). The unweighted STEINER TREE problem also admits a $2^k n^{O(1)}$ time algorithm [198] using a different technique based on subset convolution. Given any of these exact algorithms as a subroutine, a faster PAS can also be found [20] (cf. Section 4.7). On the other hand, no exact polynomial-sized kernel exists [130] for the STEINER TREE problem, unless NP$\subseteq$coNP/poly. Interestingly though, a PSAKS can be obtained [18].

This kernel is based on a well-known fact proved by Borchers and Du [199], which is very useful to obtain approximation algorithms for the STEINER TREE problem. It states that any Steiner tree can be covered by smaller trees containing few terminals, such that these trees do not overlap much. More formally, a full-component is a subtree of a Steiner tree, for which the leaves coincide with its terminals. For the optimum Steiner tree T and any $\varepsilon > 0$, there exist full-components $C_1, \ldots, C_\ell$ of T such that

1. each full-component C_i contains at most $2^{\lceil 1/\varepsilon \rceil}$ terminals (leaves),
2. the sum of the weights of the full-components is at most $1 + \varepsilon$ times the cost of T, and
3. taking any collection of Steiner trees $T_1, \ldots, T_\ell$, such that each tree T_i connects the subset of terminals that forms the leaves of full-component C_i, the union $\bigcup_{i=1}^{\ell} T_i$ is a feasible solution to the input instance.

Not knowing the optimum Steiner tree, it is not possible to know the subsets of terminals of the full-components corresponding to the optimum. However, it is possible to compute the optimum Steiner tree for every subset of terminals of size at most $2^{\lceil 1/\varepsilon \rceil}$ using an FPT algorithm for STEINER TREE. The time to compute all these solutions is $k^{O(2^{1/\varepsilon})} n^{O(1)}$, using for instance the Dreyfus and Wagner [196] algorithm. Now the above three properties guarantee that the graph given by the union of all the computed Steiner trees, contains a $(1+\varepsilon)$-approximation for the input instance. In fact, the best polynomial time approximation algorithm known to date [200] uses an iterative rounding procedure to find a $\ln(4)$-approximation of the optimum solution in the union of these Steiner trees. To obtain a kernel, the union needs to be sparsified, since it may contain many Steiner vertices and also the edge weights might be very large. However, Lokshtanov et al. [18] show that the number of Steiner vertices can be reduced using standard techniques, while the edge weights can be encoded so that their space requirement is bounded in the parameter and the cost of any solution is distorted by at most a $1 + \varepsilon$ factor.

Theorem 23 ([18,20]). *For the* STEINER TREE *problem a* $(1 + \varepsilon)$-*approximation can be computed in* $(2 + \delta)^{(1-\varepsilon/2)} n^{O(1)}$ *time for any constant* $\delta > 0$ *(and in* $2^{(1-\varepsilon/2)} n^{O(1)}$ *time in the unweighted case) for any* $\varepsilon > 0$, *where k is the number of terminals. Moreover, a* $(1 + \varepsilon)$-*approximate kernel of size* $(k/\varepsilon)^{O(2^{1/\varepsilon})}$ *can be computed in polynomial time.*

A natural alternative to the number of terminals is to consider the vertices remaining in the optimum tree after removing the terminals: a folklore result states that STEINER TREE is W[2]-hard parameterized by the number of non-terminals (called Steiner vertices) in the optimum solution. At the same time, unless P = NP there is no PTAS for the problem, as it is APX-hard [201]. However an approximation scheme is obtainable when parametrizing by the number of Steiner vertices k in the optimum, and also a PSAKS is obtainable under this parameterization.

To obtain both of these results, Dvořák et al. [202] devise a reduction rule that is based on the following observation: if the optimum tree contains few Steiner vertices but many terminals, then the tree must contain (1) a large component containing only terminals, or (2) a Steiner vertex that has many terminal neighbours. Intuitively, in case (2) we would like to identify a large star with terminal leaves and small cost in the current graph, while in case (1) we would like to find a cheap edge between two terminals. Note that such a single edge also is a star with terminal leaves. The reduction rule will therefore find the star with minimum weight per contained terminal, which can be done in polynomial time. This rule is applied until the number of terminals, which decreases after each use, falls below a threshold depending on the input parameter k and the desired approximation ratio $1 + \varepsilon$. Once the number of terminals is bounded by a function of k and ε, the Dreyfus and Wagner [196] algorithm can be applied on the remaining instance, or a kernel can be computed using the PSAKS of Theorem 23. It can be shown that the reduction rule does not distort the optimum solution by much as long as the threshold is large enough, which implies the following theorem.

Theorem 24 ([202]). *For the* STEINER TREE *problem a* $(1 + \varepsilon)$-*approximation can be computed in* $2^{O(k^2/\varepsilon^4)} n^{O(1)}$ *time for any* $\varepsilon > 0$, *where k is the number of non-terminals in the optimum solution. Moreover, a* $(1 + \varepsilon)$-*approximate kernel of size* $(k/\varepsilon)^{2^{O(1/\varepsilon)}}$ *can be computed in polynomial time.*

This theorem is also generalizable to the STEINER FOREST problem, where a list of terminal pairs is given and the task is to find a minimum weight forest in the input graph connecting each pair. In this case though, the parameter has to be combined with the number of connected components of the optimum forest [202].

A variation of the STEINER FOREST problem is the SHALLOW-LIGHT STEINER NETWORK (SLSN) problem. Here a graph with both edge costs and edge lengths is given, together with a set of terminal pairs and a length threshold L. The task is to compute a minimum cost subgraph, which connects each terminal pair with a path of length at most L. For this problem a dichotomy result was shown [203] in terms of the pattern given by the terminal pairs. More precisely, the terminal pairs are interpreted as edges in a graph for which the vertices are the terminals: if $\mathcal{C}$ is some class of graphs, then SLSN$_\mathcal{C}$ is the SHALLOW-LIGHT STEINER NETWORK problem restricted to sets of terminal pairs that span some graph in $\mathcal{C}$. Let $\mathcal{C}^*$ denote the class of all stars, and $\mathcal{C}_\lambda$ the class of graphs with at most λ edges. The SLSN$_{\mathcal{C}^*}$ problem is APX-hard [201], as it is a generalization of STEINER TREE (where $L = \infty$). At the same time, both the SLSN$_{\mathcal{C}^*}$ and SLSN$_{\mathcal{C}_\lambda}$ problems parameterized by the number of terminals are paraNP-hard [204], since they are generalizations of the RESTRICTED SHORTEST PATH problem (where there is exactly one terminal pair). A PAS can however be obtained for both of these problems (whenever λ is a constant), but for no other class $\mathcal{C}$ of demand patterns [203].

Theorem 25 ([203]). *For any constant* $\lambda > 0$, *there is an FPTAS for the* SLSN$_{\mathcal{C}_\lambda}$ *problem. For the* SLSN$_{\mathcal{C}^*}$ *problem a* $(1 + \varepsilon)$-*approximation can be computed in* $4^k (n/\varepsilon)^{O(1)}$ *time for any* $\varepsilon > 0$, *where k is the number of terminal pairs. Moreover, under Gap-ETH no* $(5/3 - \varepsilon)$-*approximation for* SLSN$_\mathcal{C}$ *can be computed in*

$f(k)n^{O(1)}$ time for any $\varepsilon > 0$ and computable function f, whenever $\mathcal{C}$ is a recursively enumerable class for which $\mathcal{C} \not\subseteq \mathcal{C}^* \cup \mathcal{C}_\lambda$ for every constant λ.

A notable special case is when all edge lengths are 1 but edge costs are arbitrary. Then $\text{SLSN}_{\mathcal{C}_\lambda}$ is polynomial time solvable for any constant λ, while $\text{SLSN}_{\mathcal{C}^*}$ is FPT parameterized by the number of terminals [203]. At the same time the parameterized approximation lower-bound of Theorem 25 is still valid for this case. It is not known however, whether constant approximation factors can be obtained for $\text{SLSN}_\mathcal{C}$ when $\mathcal{C}$ is a class different from $\mathcal{C}_\lambda$ and $\mathcal{C}^*$. More generally we may ask the following question.

Open Question 11. *Given some class of graphs $\mathcal{C} \not\subseteq \mathcal{C}^* \cup \mathcal{C}_\lambda$, which approximation factor $\alpha_\mathcal{C}$ can be obtained in FPT time for $\text{SLSN}_\mathcal{C}$ parameterized by the number of terminals?*

Turning to the TSP problem, a generalization of TSP introduces deadlines until which vertices need to be visited by the computed tour. A natural parameterization in this setting is the number of vertices that have deadlines. It can be shown [205] that no approximation better than 2 can be computed when using this parameter. Nevertheless, a 2.5-approximation can be computed in FPT time [205]. The algorithm will guess the order in which the vertices with deadlines are visited by the optimum solution. It then computes a 3/2-approximation for the remaining vertices using Christofides algorithm [3]. The approximation ratio follows, since the optimum tour can be thought of as two tours, of which one visits only the deadline vertices, while the other contains all remaining vertices. The approximation algorithm incurs a cost of OPT for the former, and a cost of $\frac{3}{2} \cdot$ OPT for the latter part of the optimum tour.

Theorem 26 ([205]). *For the* DLTSP *problem a 2.5-approximation can be computed in $O(k! \cdot k) + n^{O(1)}$ time, if the number of vertices with deadlines is k. Moreover, no $(2 - \varepsilon)$-approximation can be computed in $f(k)n^{O(1)}$ time for any $\varepsilon > 0$ and computable function f, unless P = NP.*

Low dimensional metrics. Just as for clustering problems, another well-studied parameter in network design is the dimension of the underlying geometric space. A typical setting is when the input is assumed to be a set of points in some k-dimensional ℓ_p-metric, where distances between points x and y are given by a function $\text{dist}(x,y) = (\sum_{i=1}^{k} |x_i - y_i|^p)^{1/p}$. Two prominent examples are Euclidean metrics (where $p = 2$) and Manhattan metrics (where $p = 1$). The dimension k of the metric space has been studied as a parameter from the parameterized approximation point-of-view avant la lettre for quite a while. It was shown [206,207] that both STEINER TREE and TSP are paraNP-hard for this parameter (since they are NP-hard even if $k = 2$), and that they are APX-hard in general metrics [201,208]. However, a PAS for Euclidean metrics both for the STEINER TREE and the TSP problems were shown to exist in the seminal work of Arora [167,209]. The techniques are similar to those used for clustering, and we refer to Section 4.3.2 for an overview.

Theorem 27 ([167]). *For the* STEINER TREE *and* TSP *problems a $(1 + \varepsilon)$-approximation can be computed in $k^{O(\sqrt{k}/\varepsilon)^{k-1}} n^2$ time for any $\varepsilon > 0$, if the input consists of n points in k-dimensional Euclidean space.*

This result also holds for the t-MST and t-TSP problems [167], where the cheapest tree or tour, respectively, on at least t nodes needs to be found. In this case the runtime has to be multiplied by t however.

A related setting is the parameterization by the doubling dimension of the underlying metric. That is, when the parameter k is the smallest integer such that any ball in the metric can be covered by 2^k balls of half the radius. Any point set in a k dimensional ℓ_p-metric has doubling dimension $O(k)$, and thus the latter parameter generalizes the former. For the TSP problem the above theorem can be generalized [210] to a PAS parameterized by the doubling dimension.

Theorem 28 ([210]). *For the* STSP *problem a* $(1+\varepsilon)$-*approximation can be computed in* $2^{(k/\varepsilon)^{O(k^2)}} n \log^2 n$ *time for any* $\varepsilon > 0$, *if the input consists of n points with doubling dimension k.*

Given that a PAS exists for STEINER TREE in the Euclidean case, it is only natural to ask whether this is also possible for low doubling metrics. Only a QPTAS is known so far [211]. Moreover, a related parameter is the highway dimension, which is used to model transportation networks. As shown by Feldmann et al. [212] the techniques of Talwar [211] for low doubling metrics can be generalized to the highway dimension to obtain a QPTAS as well. Again, it is quite plausible to assume that a PAS exists.

Open Question 12. *Is there a PAS for* STEINER TREE *parameterized by the doubling dimension? Is there a PAS for either* STEINER TREE *or* TSP *parameterized by the highway dimension?*

Directed Graphs. When considering directed input graphs (asymmetric metrics), the DIRECTED STEINER TREE problem takes as input a terminal set and a special terminal called the root. The task is to compute a directed tree of minimum weight that contains a path from each terminal to the root. In general no $f(k)$-approximation can be computed in FPT time for any computable function f, when the parameter k is the number of Steiner vertices in the optimum solution [202]. A notable special case is the unweighted DIRECTED STEINER TREE problem, which for this parameter admits a PAS. The techniques here are the same as those used to obtain Theorem 24 for the undirected case. However, in contrast to the undirected case which admits a PSAKS, no polynomial-sized $(2-\varepsilon)$-approximate kernelization exists for DIRECTED STEINER TREE [202], unless NP⊆coNP/poly. It is an intriguing question whether a 2-approximate kernel exists.

Open Question 13. *Is there a polynomial-sized 2-approximate kernel for the unweighted* DIRECTED STEINER TREE *problem parameterized by the number of Steiner vertices in the optimum solution?*

If the parameter is the number of terminals, the (weighted) DIRECTED STEINER TREE problem is FPT, using the same algorithm as for the undirected version [196,197]. A different variant of STEINER TREE in directed graphs is the STRONGLY CONNECTED STEINER SUBGRAPH problem, where a terminal set needs to be strongly connected in the cheapest possible way. This problem is W[1]-hard parameterized by the number of terminals [213], and no $O(\log^{2-\varepsilon} n)$-approximation can be computed in polynomial time [214], unless NP $\subseteq$ ZTIME($n^{\text{polylog}(n)}$). However, a 2-approximation can be computed in FPT time [215].

The crucial observation for this algorithm is that in any strongly connected solution, fixing some terminal as the root, every terminal can be reached from the root, while at the same time the root can be reached from each terminal. Thus the optimum solution is the union of two directed trees, of which one is directed towards the root and the other is directed away from the root, and the leaves of both trees are terminals. Hence it suffices to compute two solutions to the DIRECTED STEINER TREE problem, which can be done in FPT time, to obtain a 2-approximation for STRONGLY CONNECTED STEINER SUBGRAPH. Interestingly, no better approximation is possible with this runtime [50].

Theorem 29 ([50,215]). *For the* STRONGLY CONNECTED STEINER SUBGRAPH *problem a 2-approximation can be computed in* $(2+\delta)^k n^{O(1)}$ *time for any constant* $\delta > 0$, *where k is the number of terminals. Moreover, under* Gap-ETH *no* $(2-\varepsilon)$-*approximation can be computed in* $f(k)n^{O(1)}$ *time for any* $\varepsilon > 0$ *and computable function* f.

A generalization of both DIRECTED STEINER TREE and STRONGLY CONNECTED STEINER SUBGRAPH is the DIRECTED STEINER NETWORK problem [216], for which an edge-weighted directed graph is given together with a list of ordered terminal pairs. The aim is to compute the cheapest subgraph that contains a path from s to t for every terminal pair (s,t). If k is the number of terminals, then for this problem no $k^{1/4-o(1)}$-approximation can be computed in $f(k)n^{O(1)}$ time [59] for any computable function f, under Gap-ETH. Both a PAS and a PSAKS exist [50] for the special case when the input graph is planar and bidirected, i.e., for every directed edge uv the reverse edge vu exists and has the same cost.

Similar to the PSAKS for the STEINER TREE problem, these two algorithms are based on a generalization of Borchers and Du [199]. That is, Chitnis et al. [50] show that a planar solution in a bidirected graph can be covered by planar graphs with at most $2^{O(1/\varepsilon)}$ terminals each, such that the sum of their costs is at most $1+\varepsilon$ times the cost of the solution. These covering graphs may need to contain edges that are reverse to those in the solution, but are themselves not part of the solution. For this the underlying graph needs to be bidirected. Analogous to STEINER TREE, to obtain a kernel it then suffices to compute solutions for every possible list of ordered pairs of at most $2^{O(1/\varepsilon)}$ terminals. In contrast to STEINER TREE however, there is no FPT algorithm for this. Instead, an XP algorithm with runtime $2^{O(k^{3/2}\log k)}n^{O(\sqrt{k})}$ needs to be used, which runs in polynomial time for $k \leq 2^{O(1/\varepsilon)}$ terminals with ε being a constant. After taking the union of all computed solutions, the number of Steiner vertices and the encoding length of the edge weights can be reduced in a similar way as for the STEINER TREE problem. To obtain a PAS, the algorithm will guess how the planar optimum can be covered by solutions involving only small numbers of terminals. It will then compute solutions on these subsets of at most $2^{O(1/\varepsilon)}$ terminals using the same XP algorithm.

Theorem 30 ([50]). *For the* DIRECTED STEINER NETWORK *problem on planar bidirected graphs a* $(1+\varepsilon)$-*approximation can be computed in* $\max\{2^{k^{2^{O(1/\varepsilon)}}}, n^{2^{O(1/\varepsilon)}}\}$ *time for any* $\varepsilon > 0$, *where k is the number of terminals. Moreover, a* $(1+\varepsilon)$-*approximate kernel of size* $(k/\varepsilon)^{2^{O(1/\varepsilon)}}$ *can be computed in polynomial time.*

4.5. Cut Problems

Starting from Menger's theorem and the corresponding algorithm for s-t CUT, graph cut problems have always been at the heart of combinatorial optimization. While many natural generalizations of s-t CUT are NP-hard, further study of these cut problems yielded beautiful techniques such as flow-cut gaps and metric embeddings in approximation algorithms [217,218], and also important separators and randomized contractions in parameterized algorithms [219–222].

4.5.1. Multicut

An instance of UNDIRECTED MULTICUT (resp. DIRECTED MULTICUT) is an undirected (resp. directed) graph $G = (V, E)$ with k pairs of vertices $(s_1, t_1), \ldots, (s_k, t_k)$. The goal is to remove the minimum number of edges such that there is no path from s_i to t_i for every $i \in [k]$. UNDIRECTED MULTIWAY CUT (resp. DIRECTED MULTIWAY CUT) is a special case of UNDIRECTED MULTICUT (resp. DIRECTED MULTIWAY CUT) where k vertices are given as terminals and the goal is to make sure there is no path between any pair of terminals. They have been actively studied from both approximation and parameterized algorithms perspectives. We survey parameterized approximation algorithms for these problems with parameters k and the solution size OPT.

Undirected Multicut. UNDIRECTED MULTICUT admits an $O(\log k)$-approximation algorithm [223] in polynomial time, and is NP-hard to approximate within any constant factor assuming the Unique Games Conjecture [224]. UNDIRECTED MULTIWAY CUT admits an 1.2965-approximation algorithm [225] in polynomial time, and is NP-hard to approximate within a factor 1.20016 [226]. UNDIRECTED MULTICUT (and thus UNDIRECTED MULTIWAY CUT) admits an exact algorithm parametrized by OPT [219,220].

With k as a parameter, we cannot hope for an exact algorithm or an approximation scheme, since even UNDIRECTED MULTIWAY CUT with 3 terminals is NP-hard to approximate within a factor $12/11 - \varepsilon$ for any $\varepsilon > 0$ under the Unique Games Conjecture. However, for UNDIRECTED MULTICUT with k pairs $(s_1, t_1), \ldots, (s_k, t_k)$, one can reduce it to $k^{O(k)}$ instances of UNDIRECTED MULTIWAY CUT with at most $2k$ terminals, by guessing a partition of these $s_1, t_1, \ldots, s_k, t_k$ according to the connected components containing them in the optimal solution (e.g., s_i and t_i should be always in different groups), merging the vertices in the same group into one vertex, and solving UNDIRECTED MULTIWAY CUT with the merged vertices as terminals. This shows an 1.2965-approximation algorithm for UNDIRECTED MULTICUT that runs in time $k^{O(k)} n^{O(1)}$.

Some recent results improve or generalize this observation. For graphs with bounded genus g, Cohen-Addad et al. [227] gave an EPAS running in time $f(g, k, \varepsilon) \cdot n \log n$. Chekuri and Madan [228] considered the demand graph H, which is the graph formed by k edges $(s_1, t_1), \ldots, (s_k, t_k)$. When t is the smallest integer such that H does not contain t disjoint edges as an induced subgraph, they presented a 2-approximation algorithm that runs in time $k^{O(t)} n^{O(1)}$.

Directed Multicut. Generally, DIRECTED MULTICUT is a much harder computational task than UNDIRECTED MULTICUT in terms of both approximation and parameterized algorithms. DIRECTED MULTICUT admits a $\min(k, \widetilde{O}(n^{11/23}))$-approximation algorithm [229]. It is NP-hard to approximate within a factor $k - \varepsilon$ for any $\varepsilon > 0$ for fixed k [230] under the Unique Games Conjecture, or $2^{\Omega(\log^{1-\varepsilon} n)}$ for any $\varepsilon > 0$ [231] for general k. DIRECTED MULTIWAY CUT admits an 2-approximation algorithm [232], which is tight even when $k = 2$ [230]. Parameterizing by OPT, DIRECTED MULTICUT is FPT for $k = 2$, but DIRECTED MULTICUT is W[1]-hard even when $k = 4$ [46]. DIRECTED MULTIWAY CUT on the other hand is in FPT [221].

Since it is hard to improve the trivial k-approximation algorithm even for fixed k [230], parameterizing by k does not yield a better approximation algorithm. Chitnis and Feldmann [233] gave a $k/2$-approximation algorithm that runs in time $2^{O(\text{OPT}^2)} n^{O(1)}$, and also proved that the problem under the same parameterization is still hard to approximate within a factor $59/58$ with $k = 4$.

Open Question 14. *What is the best approximation ratio (as a function of k) achieved by a parameterized algorithm (with parameter* OPT*)? Will it be close to $O(1)$ or $\Omega(k)$?*

4.5.2. Minimum Bisection and Balanced Separator

Given a graph $G = (V, E)$, MINIMUM EDGE BISECTION (resp. MINIMUM VERTEX BISECTION) asks to remove the fewest number of edges such that the graph is partitioned into two parts A and B with $||A| - |B|| \leq 1$. BALANCED EDGE SEPARATOR (resp. BALANCED VERTEX SEPARATOR) is a more relaxed version of the problem where the goal is to bound the size of the largest component by αn for some $1/2 < \alpha < 1$. It has been actively studied from approximation algorithms, culminating in $O(\sqrt{\log n})$-approximation algorithms for both BALANCED EDGE SEPARATOR and BALANCED VERTEX SEPARATOR [218,234], and an $O(\log n)$-approximation algorithm for MINIMUM EDGE BISECTION [235].

If we parameterize by the size of optimal separator k, MINIMUM EDGE BISECTION admits an exact parameterized algorithm [222]. While MINIMUM VERTEX BISECTION is W[1]-hard [219], Feige and Mahdian [236] gave an algorithm that given $2/3 \leq \alpha < 1$ and $\varepsilon > 0$, in time $2^{O(k)} n^{O(1)}$ returns an $(\alpha + \varepsilon)$ separator of size at most k.

4.5.3. k-Cut

Given an undirected graph $G = (V, E)$ and an integer $k \in \mathbb{N}$, the k-CUT problem asks to remove the smallest number of edges such that G is partitioned into at least k non-empty connected components. The edge contraction algorithm by Karger and Stein [237] yields a randomized exact XP algorithm running in time $O(n^{2k})$, which was made deterministic by Thorup [238]. There were recent improvements to the running time [154,239]. There is an exact parameterized algorithm with parameter

OPT [221,240]. For general k, it admits a $(2-2/k)$-approximation algorithm [241], and is NP-hard to approximate within a factor $(2-\varepsilon)$ for any $\varepsilon > 0$ under the Small Set Expansion Hypothesis [74].

A simple reduction shows that k-CUT captures $(k-1)$-CLIQUE, so an exact FPT algorithm with parameter k is unlikely to exist. Gupta et al. [153] gave an $(2-\delta)$-approximation algorithm for a small universal constant $\delta > 0$ that runs in time $f(k) \cdot n^{O(1)}$. The approximation ratio was improved to 1.81 in [154], and further to 1.66 [242]. Very recently, Lokshtanov et al. [243] gave a PAS that runs in time $(k/\varepsilon)^{O(k)} n^{O(1)}$, thereby (essentially) resolving the parameterized approximability of k-CUT.

4.6. $\mathcal{F}$-DELETION Problems

Let $\mathcal{F}$ be a vertex-hereditary family of undirected graphs, which means that if $G \in \mathcal{F}$ and H is a vertex-induced subgraph of G, then $H \in \mathcal{F}$ as well. $\mathcal{F}$-DELETION is the problem where given a graph $G = (V, E)$, we are supposed to find $S \subseteq V$ such that the subgraph induced by $V \setminus S$ (denoted by $G \setminus S$) belongs to $\mathcal{F}$. The goal is to minimize $|S|$. The natural weighted version, where there is a non-negative weight $w(v)$ for each vertex v and the goal is to minimize the sum of the weights of the vertices in S, is called WEIGHTED $\mathcal{F}$-DELETION.

$\mathcal{F}$-DELETION captures numerous combinatorial optimization problems, including VERTEX COVER (when $\mathcal{F}$ includes all graphs with no edges), FEEDBACK VERTEX SET (when $\mathcal{F}$ is the set of all forests), and ODD CYCLE TRANSVERSAL (when $\mathcal{F}$ is the set of all bipartite graphs). There are a lot more interesting graph classes $\mathcal{F}$ studied in structural and algorithmic graph theory. Some famous examples include planar graphs, perfect graphs, chordal graphs, and graphs with bounded treewidth.

In addition to beautiful structural results that give multiple equivalent characterizations, these graph classes often admit very efficient algorithms for some tasks that are believed to be hard in general graphs. Therefore, a systematic study of $\mathcal{F}$-DELETION for more graph classes is not only an interesting algorithmic task by itself, but also a way to obtain better algorithms for other optimization problems when the given graph G is close to a nice class $\mathcal{F}$ (i.e., deleting few vertices from G makes it belong to $\mathcal{F}$.) Indeed, some algorithms for INDEPENDENT SET for noisy planar/minor-free graphs discussed in Section 4.1 use an algorithm for $\mathcal{F}$-DELETION as a subroutine [89].

For the maximization version where the goal is to maximize $|V \setminus S|$, a powerful but pessimistic characterization is known. Lund and Yannakakis [244] showed that whenever $\mathcal{F}$ is vertex-hereditary and nontrivial (i.e., there are infinitely many graphs in $\mathcal{F}$ and out of $\mathcal{F}$), the maximization version is hard to approximate within a factor $2^{\log^{1/2-\varepsilon} n}$ for any $\varepsilon > 0$. So no nontrivial $\mathcal{F}$ is likely to admit even a polylogarithmic approximation algorithm. However, the situation is different for the minimization problem, since VERTEX COVER admits a 2-approximation algorithm, while ODD CYCLE TRANSVERSAL [245] and PERFECT DELETION [246] are NP-hard to approximate within any constant factor approximation algorithm. (The first result assumes the Unique Games Conjecture.) It indicates that a characterization of approximabilities for the minimization versions will be more complex and challenging.

There are two (closely related) frameworks to capture large graph classes.

- Choose a graph width parameter (e.g., treewidth, pathwidth, cliquewidth, rankwidth, etc.) and $k \in \mathbb{N}$. Let $\mathcal{F}$ be the set of graphs G with the chosen width parameter at most k. The parameter of $\mathcal{F}$-DELETION is k.
- Choose a notion of subgraph (e.g., subgraph, induced subgraph, minor, etc.) and a finite family of forbidden graphs $\mathcal{H}$. Let $\mathcal{F}$ be the set of graphs G that do not have any graph in $\mathcal{H}$ as the chosen notion of subgraph. The parameter of $\mathcal{F}$-DELETION is $|\mathcal{H}| := \sum_{H \in \mathcal{H}} |V(H)|$.

Many interesting classes are capture by the above frameworks. For example, to express FEEDBACK VERTEX SET, we can take $\mathcal{F}$ to be the set of graphs with treewidth at most 1, or equivalently, the set of graphs that does not have the triangle graph K_3 as a minor. In the rest of the subsection, we introduce known results of $\mathcal{F}$-DELETION under the above two parameterization. Note that under these two parameterizations, the need for approximation is inherent since the simplest problem in both frameworks, VERTEX COVER, already does not admit a polynomial-time $(2-\varepsilon)$-approximation algorithm under the Unique Games Conjecture.

Finally, we mention that the parameterization by the size of the optimal solution has been studied more actively from the parameterized complexity community, where many important problems are shown to be in FPT [247–249].

4.6.1. Treewidth and Planar Minor Deletion

The treewidth of a graph (see Definition 1) is arguably the most well-studied graph width parameter with numerous structural and algorithmic applications. It is one of the most important concepts in the graph minor project of Robertson and Seymour. Algorithmically, Courcelle's theorem [250] states that every problem expressible in the monadic second-order logic of graphs can be solved in FPT time parameterized by treewidth. We refer the reader to the survey of Bodlaender [251]. Computing treewidth is NP-hard in general [252], but if we parameterize by treewidth, it can be done in FPT time [253], and there is a faster constant-factor parameterized approximation algorithm [254].

Let $k \in \mathbb{N}$ be the parameter. TREEWIDTH k-DELETION (also known as TREEWIDTH k-MODULATOR in the literature) is a special case of $\mathcal{F}$-DELETION where $\mathcal{F}$ is the set of all graphs with treewidth at most k. Note the case $k = 0$ yields VERTEX COVER and $k = 1$ yields FEEDBACK VERTEX SET.

Fomin et al. [247] gave a randomized $f(k)$-approximation algorithm that runs in $g(k) \cdot nm$ for some computable functions f and g. The approximation ratio was improved by Gupta et al. [255] that gave a deterministic $O(\log k)$-approximation algorithm that runs in $f(k) \cdot n^{O(1)}$ some f.

This result has immediate applications to minor deletion problems. Let $\mathcal{H}$ be a finite set of graphs, and consider $\mathcal{H}$-MINOR DELETION, which is a special case of $\mathcal{F}$-DELETION when $\mathcal{F}$ is the set of all graphs that do not have any graph in $\mathcal{H}$ as a minor. Its parameterized and kernelization complexity (with parameter OPT) for family $\mathcal{H}$ has been actively studied [247,256,257].

When $\mathcal{H}$ contains a planar graph H (also known as PLANAR $\mathcal{H}$-DELETION in the literature), by the polynomial grid-minor theorem [258], any graph $G \in \mathcal{F}$ has treewidth at most $k := \text{poly}(|V(H)|)$. Therefore, in order to solve $\mathcal{H}$-MINOR DELETION, one can first solve TREEWIDTH k-DELETION to reduce the treewidth to k and then solve $\mathcal{H}$-MINOR DELETION optimally using Courcelle's theorem [250]. Combined with the above algorithm for TREEWIDTH k-DELETION [255], this strategy yields an $O(\log k)$-approximation algorithm that runs in $f(|\mathcal{H}|) \cdot n^{O(1)}$ time.

Beyond PLANAR $\mathcal{H}$-DELETION, there are not many results known for $\mathcal{H}$-MINOR DELETION. The case $\mathcal{H} = \{K_5, K_{3,3}\}$ is called MINIMUM PLANARIZATION and was recently shown to admit an $O(\log^{O(1)} n)$-approximation algorithm in $n^{O(\log n / \log \log n)}$ time [259].

While the unweighted versions of TREEWIDTH k-DELETION and PLANAR $\mathcal{H}$-DELETION admit an approximation algorithm whose approximation ratio only depends on k not n, such an algorithm is not known for WEIGHTED TREEWIDTH k-DELETION or WEIGHTED PLANAR $\mathcal{H}$-DELETION. Agrawal et al. [260] gave a randomized $O(\log^{1.5} n)$-approximation algorithm and a deterministic $O(\log^2 n)$-approximation algorithm that run in polynomial time for fixed k, i.e., the degree of the polynomial depends on k. Bansal et al. [89] gave an $O(\log n \log \log n)$-approximation algorithm for the edge deletion version. The only graphs H whose weighted minor deletion problem is known to admit a constant factor approximation algorithm are single edge (WEIGHTED VERTEX COVER), triangle (WEIGHTED FEEDBACK VERTEX SET), and diamond [261]. For the weighted versions, no hardness beyond VERTEX COVER is known.

Open Question 15. *Does* WEIGHTED TREEWIDTH k-DELETION *admit an $f(k)$-approximation algorithm with parameter k for some function f? Does* TREEWIDTH k-DELETION *admit a c-approximation algorithm with parameter k for some universal constant c?*

Algorithms for TREEWIDTH k-DELETION. Here we present high-level ideals of [255,260] for TREEWIDTH k-DELETION and WEIGHTED TREEWIDTH k-DELETION respectively. These two algorithms share the following two important ingredients:

1. Graphs with bounded treewidth admit good separators.
2. There are good approximation algorithms to find such separators.

Given an undirected and vertex-weighted graph $G = (V, E)$ and an integer $k \in \mathbb{N}$, let (WEIGHTED) k-VERTEX SEPARATOR be the problem whose goal is to remove the vertices of minimum total weight so that each connected component has at most k vertices. An algorithm is called an α-bicriteria approximation algorithm if it returns a solution whose total weight is at most $\alpha \cdot \text{OPT}$ and each connected component has at most $1.1k$ vertices [262]. The case $k = 2n/3$ is called BALANCED SEPRATOR and has been actively studied in the approximation algorithms community, and the best approximation algorithm achieves $O(\sqrt{\log n})$-bicriteria approximation [234]. When k is small, $O(\log k)$-bicriteria approximation is also possible [116].

WEIGHTED TREEWIDTH k-DELETION. Agrawal et al. [260] achieves an $O(\log^{1.5} n)$-approximation for WEIGHTED TREEWIDTH k-DELETION in time $n^{O(k)}$. It would be interesting to see whether the running time can be made FPT with parameter k.

The main structure of their algorithm is top-down recursive. Deleting the optimal solution S^* from G reduces the treewidth of $G \setminus S^*$ to k, so from the forest decomposition of $G \setminus S^*$, there exists a set $M^* \subseteq G \setminus S^*$ with at most $k+1$ vertices such that each connected component of $G \setminus (M^* \cup S^*)$ has at most $2n/3$ vertices. While we do not know S^*, we can exhaustively try every possible $M \subseteq V$ with $|M| \leq k+1$ and use the bicriteria approximation algorithm for BALANCED SEPRATOR to find M and S such that (1) $|M| \leq k+1$, (2) $w(S) \leq O(\sqrt{\log n})\text{OPT}$, and (3) $G \setminus (M \setminus S)$ has at most $1.1 \cdot (2n/3) \leq 3n/4$ vertices.

Let $G_1, \ldots, G_t$ be the resulting connected components of $G \setminus (S \cap M)$. We solve each G_i recursively to compute S_i such that each $G_i \setminus S_i$ has treewidth at most k. The weight of S was already bounded in terms of OPT, but the weight of M was not, so we finally need to consider the graph induced by $M \cup V(G_1) \cup \cdots \cup V(G_t)$ and delete vertices of small weight to ensure small treewidth. However, this task is easy since since the treewidth of each G_i is bounded by k and $|M| \leq k+1$, which bounds the treewidth of the considered graph by $2k+1$. So we can fetch the algorithm for small treewidth graphs to solve the problem optimally. Note that the total weight of removed vetices in this recursive call is at most $(O(\sqrt{\log n}) + 1)\text{OPT}$. Since $\sum_i \text{OPT}(G_i) \leq \text{OPT}(G)$ and the recursion depth is at most $O(\log n)$, the total approximation ratio is $O(\log^{1.5} n)$.

TREEWIDTH k-DELETION. Gupta et al. [255] give an $O(\log k)$-approximation algorithm that runs in time $f(k) \cdot n^{O(1)}$ for the unweighted version of TREEWIDTH k-DELETION. The main structure of this algorithm is bottom-up iterative refinement. The algorithm maintains a feasible solution $S \subseteq V$ (we can start with $S = V$), and iteratively uses S to obtain another feasible solution S'. If the new solution is not smaller (i.e., $|S'| \geq |S|$), then $|S| \leq O(\log k) \cdot \text{OPT}$.

Let us focus on one refinement step with the current feasible solution S. Let S^* be the optimal solution, so that $G \setminus S^*$ has treewidth at most k. We use the following simple lemma showing the existence of a good separator of G in a finer scale than before.

Lemma 2 ([87,255]). *Let H be a graph with treewidth at most k, $T \subseteq V(H)$ be any subset of vertices, and $\varepsilon > 0$. There exists $R \subseteq V(H)$ such that (1) $|R| \leq \varepsilon|T|$ and (2) every connected component of $H \setminus R$ has at most $O(k/\varepsilon)$ vertices from T.*

Plugging $H \leftarrow G \setminus S^*$, $T \leftarrow S$ in the above lemma and letting $S' = R \cup S^*$, we can conclude that there exists $S' \subseteq V$ such that $|S'| \leq |R| + |S^*| \leq \varepsilon|S| + \text{OPT}$ and each connected component of $G \setminus S'$ has at most $O(k/\varepsilon)$ vertices from S.

How can we find such a set S' efficiently? Note that if $S = V$, then S' is an $O(k/\varepsilon)$-vertex separator of G. Lee [116] defined a generalization of k-VERTEX SEPARATOR called k-SUBSET VERTEX SEPARATOR, where the input consists of $G = (V, E)$, $S \subseteq V$, $k \in \mathbb{N}$, and the goal is to remove the smallest number of vertices so that each connected component has at most k vertices from S, and gave an $O(\log k)$-bicriteria approximation algorithm.

Since the above lemma guarantees that OPT for $O(k/\varepsilon)$-SUBSET VERTEX SEPARATOR is at most OPT for TREEWIDTH k-DELETION plus $\varepsilon|S|$, applying this bicriteria approximation algorithm yields S' such that $|S'| \leq O(\log k)(\text{OPT} + \varepsilon|S|)$ and each connected component of $G \setminus S'$ has at most $O(k/\varepsilon)$ vertices from S. Since S is a feasible solution, it implies that the treewidth of each connected component is bounded by $O(k/\varepsilon)$, so we can solve each component optimally in time $f(k/\varepsilon) \cdot n^{O(1)}$. By setting $\varepsilon = 0.5$, we can see the size of new solution is strictly decreased unless $|S| = O(\log k) \cdot \text{OPT}$, finishing the proof.

4.6.2. Subgraph Deletion

Let H be a fixed pattern graph with k vertices. Given a host graph G, deciding whether H is a subgraph of G (in the usual sense) is known as SUBGRAPH ISOMORPHISM, whose parameterized complexity with various parameters (e.g., k, $\mathbf{tw}(H)$, $\mathbf{genus}(G)$, etc.) was studied by Marx and Pilipczuk [263].

Guruswami and Lee [115] studied the corresponding vertex deletion problem H-SUBGRAPH DELETION (called H-TRANSVERSAL in the paper), which is a special case of $\mathcal{F}$-DELETION where $\mathcal{F}$ is the set of graphs that do not have H as a subgraph. Note that the problem admits a simple k-approximation algorithm that runs in time $O(n \cdot f(n, H))$, where $f(n, H)$ denotes time to solve SUBGRAPH ISOMORPHISM with the pattern graph H and a host graph with n vertices. Their main hardness result states that assuming the Unique Games Conjecture, whenever H is 2-vertex connected, for any $\varepsilon > 0$, no polynomial time algorithm (including algorithms running in time $n^{f(k)}$ for any f) can achieve a $(k - \varepsilon)$-approximation. (Without the UGC, they still ruled out a $(k - 1 - \varepsilon)$-approximation.)

Among H that are not 2-vertex-connected, there is an $O(\log k)$-approximation algorithm when H is a star (in time $n^{O(1)}$) or a path (in time $f(k)n^{O(1)}$) [115,116,264]. The algorithm for k-path follows from the result for TREEWIDTH k-DELETION, because any graph without a k-path has treewidth at most k. Whenever H is a tree with k vertices, detecting a copy of H in G with n vertices can be done in $2^{O(k)}n^{O(1)}$ time [265], and it is open whether there is an $O(\log k)$-approximation algorithm for H-SUBGRAPH DELETION in time $f(k) \cdot n^{O(1)}$.

4.6.3. Other Deletion Problems

Chordal graphs. A graph is chordal if it does not have an induced cycle of length ≥ 4. Chordal graphs form a subclass of perfect graphs that have been actively studied. Initially motivated by efficient kernels, approximation algorithms for CHORDAL DELETION have been developed recently. The current best results are a $\text{poly}(\text{OPT})$-approximation [140,266] and a $O(\log^2 n)$-approximation [260].

Edge versions. While this subsection focused on the vertex deletion problem, there are some results on the edge deletion, edge addition, and edge modification versions. (Edge modification allows both addition and deletion.) Cao and Sandeep [267] studied MINIMUM FILL-IN, whose goal is to add the minimum number of edges to make a graph chordal. They gave new inapproximability results implying improved time lower bounds for parameterized algorithms. Giannopoulou et al. [268] gave $O(1)$-approximation algorithms for PLANAR $\mathcal{H}$-IMMERSION DELETION parameterized by $\mathcal{H}$. Bliznets et al. [269] considered H-free edge modification for a forbidden induced subgraph H and give an almost complete characterization on its approximability depending on H.

Directed graphs. There is also a large body of work on parameterized algorithms for vertex deletion problems in directed graphs. While many of the known problems (including DIRECTED FEEDBACK

VERTEX SET [270]) admit an exact FPT algorithm, Lokshtanov et al. [48] studied DIRECTED ODD CYCLE TRANSVERSAL, and proved that it is W[1]-hard and is unlikely to admit an PAS under the Parameterized Inapproximability Hypothesis (or Gap-ETH). They complemented the result by showing a 2-approximation algorithm running in time $f(\text{OPT})n^{O(1)}$.

4.7. Faster Algorithms and Smaller Kernels via Approximation

The focus of this section so far has been on problems for which its exact version is intractable (i.e., W[1]/W[2]-hard) and the goal is to obtain good approximations in FPT time. In this subsection, we shift our focus slightly by asking: does approximation allow us to find faster algorithms for problems already known to be in FPT?

To illustrate this, let us consider VERTEX COVER. It is of course well-known that the exact version of the problem can be solved in FPT time, with the current best running time being $O^*(1.2738^k)$ [271]. The question here would be: if we are allowed to output an $(1-\varepsilon)$-approximate solution, instead of just an exact one, can we speed up the algorithm?

To the best of our knowledge, such a question was tackled for the first time by Bourgeois et al. [272] and revisited quite a few times in the literature [20,273–276]. As one might have suspected, the answer to this question is a YES, as stated below.

Theorem 31 ([20]). *Let $\delta > 0$ be such that there exists an $O^*(\delta^k)$-time algorithm for* VERTEX COVER *(e.g., $\delta = 1.2738$). Then, for any $\varepsilon > 0$, there is an $(1+\varepsilon)$-approximation for* VERTEX COVER *that runs in $O^*(\delta^{(1-\varepsilon)k})$ time.*

The main idea of the algorithm is inspired by the "local ratio" method in the approximation algorithms literature (see e.g., [277]) and we sketch it here. The algorithm works in two stages. In the first stage, we run the greedy algorithm: as long as we have picked less than $2\varepsilon k$ vertices so far and not all edges are covered, pick an uncovered edge and add both endpoints to our solution. In the second stage, we run the exact algorithm on the remaining part of the graph to find a VERTEX COVER of size $(1-\varepsilon)k$. Since the first stage runs in polynomial time, the running time of the entire algorithm is dominated by the second stage, whose running time is $O^*(\delta^{(1-\varepsilon)k})$ as desired. The correctness of the algorithm follows from the fact that, for each selected edge in the first step, the optimal solution still needs to pick at least one endpoint. As a result, the optimal solution must pick at least εk vertices with respect to the first stage (compared to $2\varepsilon k$ picked by the algorithm). Thus, when the optimal solution is of size at most k, there must be a solution in the second stage of size at most $(1-\varepsilon)k$, meaning that the algorithm finds such a solution and outputs a vertex cover of size $(1+\varepsilon)k$ as claimed.

The above "approximate a small fraction and brute force the rest" approach of Fellows et al. [20] generalizes naturally to problems beyond VERTEX COVER. Fellows et al. [20] formalized the method in terms of α-fidelity kernelization and apply it to several problems, including CONNECTED VERTEX COVER, d-HITTING SET and STEINER TREE. For these problems, the method gives an $(1+\varepsilon)$-approximation algorithm that runs in time $O^*(\delta^{(1-\Omega(\varepsilon))k})$, where $\delta > 0$ denotes a constant for which a $O^*(\delta^k)$-time algorithm is known for the exact version of the corresponding problem. The approach, in some form or another, is also applicable both to other parameterized problems [278,279] and to non-parameterized problems (e.g., [272]); since the latter is out-of-scope for the survey, we will not discuss the specifics here.

An intriguing question related to this line of work is whether it must be the case that the running time of $(1+\varepsilon)$-approximation algorithms is of the form $O^*(\delta^{(1-\Omega(\varepsilon))k})$. That is, can we get a $(1+o(1))$-approximation for these problems in time $O^*(\lambda^k)$ where λ is a constant strictly smaller than δ? More specifically, we may ask the following:

Open Question 16. Let $\delta > 0$ be the smallest (known) constant such that an $O^*(\delta^k)$-time exact algorithm exists for VERTEX COVER. Is there an algorithm that, for any $\varepsilon > 0$, runs in time $f(1/\varepsilon) \cdot O^*(\lambda^k)$ for some constant $\lambda < \delta$?

Of course, the question applies not only for VERTEX COVER but other problems in the list as well. The informal crux of this question is whether, in the regime of very good approximation factors (i.e. $1 + o(1)$), approximation can still be exploited in such a way that the algorithm works significantly better than the approach "approximate a $o(1)$ fraction and then brute force".

Turning back once again to our running example of VERTEX COVER, it turns out that algorithms faster than "approximate a small fraction and then brute force" are known [273,275,276] but only for the regime of large approximation ratios. In particular, Brankovic and Fernau [273] give faster algorithms than in Theorem 31 already for approximation ratios as small as 3/2. The algorithms in [275,276] focus on the case of "barely non-trivial" $(2 - \rho)$-approximation factors. (Recall the greedy algorithm yields a 2-approximation and, under the Unique Games Conjecture, the problem is NP-hard to approximate to within any constant factor less than 2.) The algorithm in [275] has a running time of $O^*(2^{k/2^{\Omega(1/\rho)}})$, which was later improved in [276] to $O^*(2^{k/2^{\Omega(1/\rho^2)}})$. These running times should be contrasted with that of "approximate a small fraction and then brute force" (i.e., applying Theorem 31 directly with $\varepsilon = 1 - \rho$) which gives an algorithm with running time $O^*(2^{k\rho})$. In other words, Refs. [275,276] improve the "saving factor" from $1/\rho$ to $2^{\Omega(1/\rho)}$ and $2^{\Omega(1/\rho^2)}$ respectively. It should be noted however that, since the known $(2 - o(1))$-factor hardness of approximation is shown via the Unique Games Conjecture and unique games admit subexponential time algorithms [280,281], it is still entirely possible that this regime of approximating VERTEX COVER admits subexponential time algorithms as well. This is perhaps the biggest open question in the "barely non-trivial" approximation range:

Open Question 17. Is there an algorithm that runs in $2^{o(k)} n^{O(1)}$ time and achieves an approximation ratio of $(2 - \rho)$ for some absolute constant $\rho > 0$?

Let us now briefly discuss the techiques used in some of the aforementioned works. The algorithms in [273,275] are based on branching in conjunction with certain approximation techniques. (See also [282] where a similar technique is used for a related problem TOTAL VERTEX COVER.) A key idea in [273,275] is that (i) if the (average or maximum) degree of the graph is small, then good polynomial-time approximation algorithms are known [283] and (ii) if the degree is large, then branching algorithms are naturally already fast. The second part of [273] involves a delicate branching rule. However, for [275], it is quite simple: for some threshold d (to be specified), as long as there exists a vertex with degree at least d, then (1) with some probability, simply add the vertex to the vertex cover, or (2) branch on both possibilities of it being inside the cover and outside. After this branching finishes and we are left with low-degree graphs, just run the known polynomial-time approximation algorithms [283] on these graphs. The point here is that the "error" incurred if option (1) is chosen will be absorbed by the approximation. By carefully selecting d and the probability, one can arrive at the desired running time and approximation guarantee. This algorithm is randomized, but can be derandomized using the sparsification lemma [284].

To the best of our knowledge, this "barely non-trivial approximation" regime has not been studied beyond VERTEX COVER. In particular, while Bansal et al. [275] apply their techniques on several problems, these are not parameterized problems and we are not aware of any other parameterized study related to the regime discussed here.

Parallel to the running time questions we have discussed so far, one may ask an analogous question in the kernelization regime: does approximation allow us to find smaller kernels for problems that already admit polynomial-size kernels? As is the case with exact algorithms, parameterized approximation algorithms go hand in hand with approximate kernels. Indeed, many algorithmic improvements mentioned can also be viewed as improvements in terms of the size of the kernels.

In particular, recall the proof sketch of Theorem 31 for VERTEX COVER. If we stop and do not proceed with brute force in the second step, then we are left with an $(1+\varepsilon)$-approximate kernel. It is also not hard to argue that, by for instance applying the standard $2k$-size kernelization at the end, we are left with at most $2(1-\varepsilon)k$ vertices. This improves upon the best known $2k - \Theta(\log k)$ bound for the exact kernel [285]. A similar improvement is known also for d-HITTING SET [20].

5. Future Directions

Although we have provided open questions along the way, we end this survey by zooming out and discussing some general future directions or meta-questions, which we find to be interesting and could be the basis for future work.

5.1. Approximation Factors

The quality of a polynomial-time approximation algorithm is mainly measured by the obtainable approximation factor α: the smaller it is the more feasibly solvable the problem is. Therefore, a lot of work has been invested into determining the smallest obtainable approximation factor α for all kinds of computationally hard problems. In the non-parameterized (i.e., NP-hardness) world, a whole spectrum of approximability has been discovered (cf. [3,4]): the most feasibly solvable NP-hard problems (e.g., the KNAPSACK problem) admit a so-called polynomial-time approximation scheme (PTAS), which is an algorithm computing a $(1+\varepsilon)$-approximation for any given constant $\varepsilon > 0$. Some problems can be shown not to admit a PTAS (under reasonable complexity assumptions), but still allow constant approximation factors (e.g., the STEINER TREE problem). Yet others can only be approximated within a polylogarithmic factor (e.g., the SET COVER problem), while some are even harder than this, as the best approximation factor obtainable is polynomial in the input size (e.g., the CLIQUE problem).

In contrast to polynomial-time approximation algorithms, a full spectrum of obtainable approximation ratios is still missing when allowing parameterized runtimes. Instead, only some scattered basic results are known. In particular, most of parameterized approximation problems belongs to one of the following categories:

- A parameterized approximation scheme (PAS) exists, i.e., for any constant $\varepsilon > 0$ a $(1+\varepsilon)$-approximation can be computed in $f(k)n^{O(1)}$ time for some parameter k. These are currently the most prevalent types of results in the literature. To just mention one example, the STEINER TREE problem is APX-hard, but admits a PAS [202] when parameterized by the number of non-terminals (so-called Steiner vertices) in the optimum solution (cf. Section 4.4).
- A lower bound excluding any non-trivial approximation factor exists. For example, under ETH the DOMINATING SET problem has no $g(k)$-approximation in $f(k)n^{o(k)}$ time [35] for any functions g and f, where k is the size of the largest dominating set.
- A polynomial-time approximation algorithm can achieve a similar approximation ratio, i.e., the parameterization is not very helpful. For instance, for the k-CENTER problem [286] a 2-approximation can be computed in polynomial time [287], but even when parameterizing by k no $(2-\varepsilon)$-approximation is possible [187] for any $\varepsilon > 0$, under standard complexity assumptions. A similar situation holds for MAX k-COVERAGE, which we discussed in Section 4.2.2.
- Constant or logarithmic approximation ratios can be shown, and which beat any approximation ratio obtainable in polynomial time. For instance, STRONGLY CONNECTED STEINER SUBGRAPH problem : under standard complexity assumptions, for this problem no polynomial-time $O(\log^{2-\varepsilon} n)$-approximation algorithm exists [214], and there is no FPT algorithm parameterized by the number k of terminals [213]. However it is not hard to compute a 2-approximation in $2^{O(k)}n^{O(1)}$ time [215], and no $(2-\varepsilon)$-approximation algorithm with runtime $f(k)n^{O(1)}$ exists [50] under Gap-ETH, for any function f and any $\varepsilon > 0$ (cf. Section 4.4).

For many problems discussed in this survey, including DENSEST k-SUBGRAPH, STEINER TREE with bounded doubling/highway dimension, it has not been determined which category they belong.

There are also a lot of problems in the final category for which asymptotically tight approximation ratios have not been found, including DIRECTED MULTICUT, TREEWIDTH k-DELETION (both weighted and unweighted). The parameterized approximability of $\mathcal{H}$-MINOR DELETION for non-planar $\mathcal{H}$ is also widen open except MINIMUM PLANARIZATION ($\mathcal{H} = \{K_5, K_{3,3}\}$) [259]. It is an immediate but still interesting direction to prove tight parameterized approximation ratios for these (and more) problems.

Digressing, we remark that this survey does not include FPT-approximation of counting problems, such as approximately counting number of k-paths in a graph. The best $(1 + \varepsilon)$-multiplicative factor algorithm known [288,289] for counting number of k-paths runs in time $4^k f(\varepsilon) poly(n)$ for some subexponential function f (cf. [290]). So a natural question is: can we count k-paths approximately in time c^k, where c is as close to the base of running time of the algorithm of deciding existence of k-Path in a graph (the best currently known c is roughly 1.657 [291,292])?

5.2. Parameterized Running Times

The quality of FPT algorithms is mainly measured in the obtainable runtime. Given a parameter k, for some problems the optimum solution can be computed in $f(k)n^{g(k)}$ time, for some functions f and g independent of the input size n (i.e., the degree of the polynomial also depends on the parameter). If such an algorithm exists the problem is slice-wise polynomial (XP), and the algorithm is called an XP algorithm. A typical example is if a solution of size k is to be found within a data set of size n, in which case often an $n^{O(k)}$ time exhaustive search algorithm exists. However, an FPT algorithm with runtime, say, $O(2^k n)$ is a lot more efficient than an XP algorithm with runtime $n^{O(k)}$, and therefore the aim is usually to find FPT algorithms, while XP algorithms are counted as prohibitively slow. The discovery of the W-hierarchy in complexity theory has paved the way to providing evidence when an FPT algorithm is unlikely to exist. Assuming ETH, it is even possible to provide lower bounds on the runtimes obtainable by any FPT or XP algorithm. Similar to approximation algorithms, this has lead to the discovery of a spectrum of tractability (cf. [6]): starting from slightly sub-exponential $2^{O(\sqrt{k})}n^{O(1)}$ time, through single exponential $2^{O(k)}n^{O(1)}$ time, to double exponential $2^{2^{O(k)}}n^{O(1)}$ time for FPT algorithms with matching asymptotic lower bounds under ETH (e.g., for the PLANAR VERTEX COVER, VERTEX COVER, and EDGE CLIQUE COVER problems, respectively, each parameterized by the solution size). For XP algorithms, asymptotically tight runtime bounds of the form $n^{O(\sqrt{k})}$ and $n^{O(k)}$ can be obtained under ETH (e.g., for the CLIQUE problem parameterized by the solution size, and the PLANAR BIDIRECTED STEINER NETWORK problem parameterized by the number of terminals [50], respectively). Finally, problems that are NP-hard when the given parameter is constant do not even allow XP algorithms unless P = NP (e.g., the GRAPH COLOURING problem where the parameter is the number of colours).

In terms of tight runtime bounds, existing results on parameterized approximation algorithms are few and far between. In particular, most of them show that for a given parameter k one of the following cases applies.

- An approximation is possible in $f(k)n^{O(1)}$ time for some function f. Most current results are only concerned with the existence of an algorithm with this type of runtime, i.e., they do not provide any evidence that the obtained runtime is best possible, or try to optimize it. The only lower bounds known exclude certain types of approximation schemes when a hardness result for the parameterization by the solution size exists. For instance, it is known that if some problem does not admit a $2^{o(k)}n^{O(1)}$ time algorithm for this parameter k then it also does not admit an EPTAS with runtime $2^{o(1/\varepsilon)}n^{O(1)}$ (cf. [5,8]).

- A certain approximation ratio cannot be obtained in $f(k)n^{O(1)}$ time for any function f. For example, it is known that while a 2-approximation for the STRONGLY CONNECTED STEINER SUBGRAPH problem can be computed in $2^{O(k)}n^{O(1)}$ time [215], where k is the number of terminals, no $(2 - \varepsilon)$-approximation can be computed in $f(k)n^{O(1)}$ time [50] for any function f, under Gap-ETH (cf. Section 4.4).

Hence, matching lower bounds on the time needed to compute an approximation are missing. For example, is the runtime of $2^{O(k)}n^{O(1)}$ best possible to compute a 2-approximation for the STRONGLY CONNECTED STEINER SUBGRAPH problem? Could there be a $2^{O(\sqrt{k})}n^{O(1)}$ time algorithm to compute a 2-approximation as well? For PASs the exact obtainable runtime is often elusive, even if certain types of approximation schemes can be excluded. For instance, for the STEINER TREE problem parameterized by the number of Steiner vertices in the optimum solution a $(1+\varepsilon)$-approximation can be computed in $2^{O(k^2/\varepsilon^4)}n^{O(1)}$ time [202]. Is the dependence on k and ε best possible? Could there be a $2^{O(k/\varepsilon^4)}n^{O(1)}$ or $2^{O(k^2/\varepsilon)}n^{O(1)}$ time algorithm as well?

We remark that, for problems for which straightforward algorithms are known to be (essentially) the best possible in FPT time, or for which an improvement over polynomial time approximation is not possible, sometimes tight running time lower bounds are known in conjunction with tight inapproximability ratios. This includes k-DOMINATING SET (Section 3.1.2), k-CLIQUE (Section 3.2.1) and MAX k-COVERAGE (Section 4.2.2).

5.3. Kernel Sizes

The development of compositionality has lead to a theory from which lower bounds on the size of the smallest possible kernel of a problem can be derived (under reasonable complexity assumptions). The spectrum (cf. [6]) here reaches from polynomial-sized kernels (e.g., for any $q \geq 3$ and $\varepsilon > 0$ the q-SAT problem parameterized by the number of variables n has no $O(n^{q-\varepsilon})$-sized kernel) to exponential-sized kernels (e.g., the STEINER TREE problem parameterized by the number of terminals does not admit any polynomial-sized kernel despite being FPT).

For approximate kernels, only a small number of publications exist, and the few known results fall into two categories:

- A polynomial-sized approximate kernelization scheme (PSAKS) exists, i.e., for any $\varepsilon > 0$ there is a $(1+\varepsilon)$-approximate kernelization algorithm that computes a $(1+\varepsilon)$-approximate kernel of size polynomial in the parameter k. For example, the STEINER TREE problem admits a PSAKS for both the parameterization in the number of terminals [18] and in the number of Steiner vertices in the optimum [202], even though neither of these two parameters admits a polynomial-sized (exact) kernel.
- A lower bound excluding any approximation factor for polynomial-sized kernels exists. For example, the LONGEST PATH problem parameterized by the maximum path length has no α-approximate polynomial-sized kernel for any α [18], despite being FPT for this parameter [6].

Hence again the intermediate cases, for which tight constant or logarithmic approximation factors can be proved for polynomial-sized kernels, are missing. Studying approximate kernelization algorithms however is of undeniable importance to the field of parameterized approximation algorithms, as witnessed by the importance of exact kernelization to fixed-parameter tractability.

5.4. Completeness in Hardness of Approximation

A final direction we would like to highlight is to obtain more completeness in inapproximability results. Most of the results so far for FPT hardness of approximation either (i) rely on gap hypothesis or (ii) yield a hardness in terms of the W-hierarchy but the exact version of the problem is known to be complete on an even higher level (e.g., DOMINATING SET is known to be W[1]-hard to approximate but its exact version is W[2]-complete). We have discussed (i) extensively in Section 3.2 and some examples of (ii) in Section 3.1. There are also some examples of (ii) that are not covered here; for instance, Marx [293] showed W[t]-hardness for certain monotone/anti-monotone circuit satisfiability problems and the exact versions of these problems are known to be complete for higher levels of the W-hierarchy. The situation here is unlike that in the theory of NP-hardness of approximation; there the PCP Theorem [21,22] implies NP-completeness of optimization problems [294].

Thus, in the parameterized inapproximability arena, the main question here is whether we can prove completeness results for hardness of approximation for the aforementioned problems. The two important examples here are: is k-CLIQUE W[1]-hard to approximate, and is k-DOMINATING SET W[2]-hard to approximate? As discussed in Section 3.2, the former is also closely related to resolving PIH.

Finally, we note that, while completeness results are somewhat rare in FPT hardness of approximation, some are known. We give two such examples here. First is the k-STEINER ORIENTATION problem, discussed in Section 3.1.3; it is W[1]-complete to approximate [44]. Second is the MONOTONE CIRCUIT SATISFIABILITY problem (without depth bound), which was proved to be W[P]-complete by Marx [293]. However, it does not seem clear to us whether these techniques can be applied elsewhere, e.g., for k-CLIQUE.

Author Contributions: Writing—original draft and writing—review and editing, A.E.F., K.C.S., E.L. and P.M. All authors have read and agreed to the published version of the manuscript.

Funding: Andreas Emil Feldmann is supported by the Czech Science Foundation GACR (grant #19-27871X), and by the Center for Foundations of Modern Computer Science (Charles Univ. project UNCE/SCI/004). Karthik C. S. is supported by ERC-CoG grant 772839, the Israel Science Foundation (grant number 552/16), and from the Len Blavatnik and the Blavatnik Family foundation. Euiwoong Lee is supported by the Simons Collaboration on Algorithms and Geometry.

Conflicts of Interest: The authors declare no conflict of interest.

References and Notes

1. Cobham, A. The intrinsic computational difficulty of functions. In Proceedings of the 1964 Congress for Logic, Methodology, and the Philosophy of Science, Paris, France, 23–25 July 1964; pp. 24–30.
2. Edmonds, J. Paths, Trees, and Flowers. *Can. J. Math.* **1965**, *17*, 449–467. [CrossRef]
3. Vazirani, V.V. *Approximation Algorithms*; Springer: Berlin, Germany, 2001.
4. Williamson, D.P.; Shmoys, D.B. *The Design of Approximation Algorithms*; Cambridge University Press: Cambridge, UK, 2011.
5. Downey, R.G.; Fellows, M.R. *Fundamentals of Parameterized Complexity*; Springer: Berlin, Germany, 2013; Volume 4.
6. Cygan, M.; Fomin, F.V.; Kowalik, L.; Lokshtanov, D.; Marx, D.; Pilipczuk, M.; Pilipczuk, M.; Saurabh, S. *Parameterized Algorithms*; Springer: Berlin, Germany, 2015.
7. Cai, L.; Chen, J. On fixed-parameter tractability and approximability of NP optimization problems. *J. Comput. Syst. Sci.* **1997**, *54*, 465–474. [CrossRef]
8. Marx, D. Parameterized complexity and approximation algorithms. *Comput. J.* **2008**, *51*, 60–78. [CrossRef]
9. Flum, J.; Grohe, M. *Parameterized Complexity Theory*; Springer: Berlin, Germany, 2006.
10. Rubinstein, A.; Williams, V.V. SETH vs. Approximation. *SIGACT News* **2019**, *50*, 57–76. [CrossRef]
11. Cesati, M.; Trevisan, L. On the efficiency of polynomial time approximation schemes. *Inf. Process. Lett.* **1997**, *64*, 165–171. [CrossRef]
12. Cygan, M.; Lokshtanov, D.; Pilipczuk, M.; Pilipczuk, M.; Saurabh, S. Lower Bounds for Approximation Schemes for Closest String. In Proceedings of the 15th Scandinavian Symposium and Workshops on Algorithm Theory, Reykjavik, Iceland, 22–24 June 2016.
13. Cai, L.; Chen, J. On fixed-parameter tractability and approximability of NP-hard optimization problems. In Proceedings of the IEEE 2nd Israel Symposium on Theory and Computing Systems, Natanya, Israel, 7–9 June 1993; pp. 118–126.
14. Chen, J.; Huang, X.; Kanj, I.A.; Xia, G. Polynomial time approximation schemes and parameterized complexity. *Discret. Appl. Math.* **2007**, *155*, 180–193. [CrossRef]
15. Kratsch, S. Polynomial kernelizations for MIN $F^+\Pi_1$ and MAX NP. *Algorithmica* **2012**, *63*, 532–550. [CrossRef]
16. Guo, J.; Kanj, I.; Kratsch, S. Safe approximation and its relation to kernelization. In *International Symposium on Parameterized and Exact Computation*; Springer: Berlin, Germany, 2011; pp. 169–180.
17. Cai, L.; Chen, J.; Downey, R.G.; Fellows, M.R. Advice Classes of Parameterized Tractability. *Ann. Pure Appl. Log.* **1997**, *84*, 119–138. [CrossRef]

18. Lokshtanov, D.; Panolan, F.; Ramanujan, M.; Saurabh, S. Lossy Kernelization. In Proceedings of the 49th Annual ACM SIGACT Symposium on Theory of Computing, Montreal, QC, Canada, 19–23 June 2017; pp. 224–237.
19. Hermelin, D.; Kratsch, S.; Soltys, K.; Wahlström, M.; Wu, X. A Completeness Theory for Polynomial (Turing) Kernelization. *Algorithmica* **2015**, *71*, 702–730. [CrossRef]
20. Fellows, M.R.; Kulik, A.; Rosamond, F.; Shachnai, H. Parameterized approximation via fidelity preserving transformations. In *International Colloquium on Automata, Languages, and Programming*; Springer: Berlin/Heidelberg, Germany, 2012; pp. 351–362.
21. Arora, S.; Lund, C.; Motwani, R.; Sudan, M.; Szegedy, M. Proof Verification and the Hardness of Approximation Problems. *J. ACM* **1998**, *45*, 501–555. [CrossRef]
22. Arora, S.; Safra, S. Probabilistic Checking of Proofs; A New Characterization of NP. In Proceedings of the 33rd Annual Symposium on Foundations of Computer Science, Pittsburgh, PA, USA, 24–27 October 1992; pp. 2–13. [CrossRef]
23. Lin, B. The Parameterized Complexity of the K-Biclique Probl. *J. ACM* **2018**, *65*, 34:1–34:23. [CrossRef]
24. Karthik C. S.; Manurangsi, P. On Closest Pair in Euclidean Metric: Monochromatic is as Hard as Bichromatic. In Proceedings of the 10th Innovations in Theoretical Computer Science Conference ITCS, San Diego, CA, USA, 10–12 January 2019; pp. 17:1–17:16. [CrossRef]
25. Chen, Y.; Lin, B. The Constant Inapproximability of the Parameterized Dominating Set Problem. *SIAM J. Comput.* **2019**, *48*, 513–533. [CrossRef]
26. Lin, B. A Simple Gap-Producing Reduction for the Parameterized Set Cover Problem. In Proceedings of the 46th International Colloquium on Automata, Languages, and Programming ICALP, Patras, Greece, 9–12 July 2019; pp. 81:1–81:15. [CrossRef]
27. Kann, V. On the Approximability of NP-complete Optimization Problems. Ph.D. Thesis, Royal Institute of Technology, Stockholm, Sweden, 1992.
28. Recall that there is a pair of polynomial-time L-reductions between the minimum dominating set problem and the set cover problem [27].
29. Feige, U. A threshold of $\ln n$ for approximating set cover. *J. ACM (JACM)* **1998**, *45*, 634–652. [CrossRef]
30. Lai, W. The Inapproximability of k-DominatingSet for Parameterized AC 0 Circuits. *Algorithms* **2019**, *12*, 230. [CrossRef]
31. Bhattacharyya, A.; Bonnet, É.; Egri, L.; Ghoshal, S.; Karthik C. S.; Lin, B.; Manurangsi, P.; Marx, D. Parameterized Intractability of Even Set and Shortest Vector Problem. *Electron. Colloq. Comput. Complex. (ECCC)* **2019**, *26*, 115.
32. Downey, R.G.; Fellows, M.R.; Vardy, A.; Whittle, G. The Parametrized Complexity of Some Fundamental Problems in Coding Theory. *SIAM J. Comput.* **1999**, *29*, 545–570. [CrossRef]
33. van Emde-Boas, P. *Another NP-Complete Partition Problem and the Complexity of Computing Short Vectors in a Lattice*; Report Department of Mathematics; University of Amsterdam: Amsterdam, The Netherlands, 1981.
34. Ajtai, M. The Shortest Vector Problem in ℓ_2 is NP-hard for Randomized Reductions (Extended Abstract). In Proceedings of the Thirtieth Annual ACM Symposium on Theory of Computing, Dallas, TX, USA, 23–26 May 1998; pp. 10–19. [CrossRef]
35. Karthik C. S.; Laekhanukit, B.; Manurangsi, P. On the parameterized complexity of approximating dominating set. *J. ACM* **2019**, *66*, 33.
36. Goldreich, O. *Computational Complexity: A Conceptual Perspective*, 1st ed.; Cambridge University Press: New York, NY, USA, 2008.
37. We could have skipped this boosting step, had we chosen a different good code with distance α but over a larger alphabet. For example, taking the Reed Solomon code over alphabet $\frac{\log n}{1-\alpha}$ would have sufficed. We chose not to do so, to keep the proof as elementary as possible.
38. This reduction (which employs the hypercube set system) is used in [29] for proving hardness of approximating MAX k-COVERAGE; for SET COVER, Feige used a more efficient set system which is not needed in our context.
39. Chalermsook, P.; Cygan, M.; Kortsarz, G.; Laekhanukit, B.; Manurangsi, P.; Nanongkai, D.; Trevisan, L. From Gap-ETH to FPT-Inapproximability: Clique, Dominating Set, and More. In Proceedings of the 58th IEEE Annual Symposium on Foundations of Computer Science (FOCS), Berkeley, CA, USA, 15–17 October 2017; pp. 743–754.

40. Abboud, A.; Rubinstein, A.; Williams, R.R. Distributed PCP Theorems for Hardness of Approximation in P. In Proceedings of the 58th IEEE Annual Symposium on Foundations of Computer Science, FOCS, Berkeley, CA, USA, 15–17 October 2017; pp. 25–36.
41. Berman, P.; Schnitger, G. On the Complexity of Approximating the Independent Set Problem. *Inf. Comput.* **1992**, *96*, 77–94. [CrossRef]
42. Raz, R. A Parallel Repetition Theorem. *SIAM J. Comput.* **1998**, *27*, 763–803. [CrossRef]
43. Dinur, I. The PCP theorem by gap amplification. *J. ACM* **2007**, *54*, 12. [CrossRef]
44. Wlodarczyk, M. Inapproximability within W[1]: The case of Steiner Orientation. *arXiv* **2019**, arXiv:1907.06529.
45. Cygan, M.; Kortsarz, G.; Nutov, Z. Steiner Forest Orientation Problems. *SIAM J. Discret. Math.* **2013**, *27*, 1503–1513. [CrossRef]
46. Pilipczuk, M.; Wahlström, M. Directed Multicut is W[1]-hard, Even for Four Terminal Pairs. *TOCT* **2018**, *10*, 13:1–13:18. [CrossRef]
47. We remark that the original conjecture in [48] says that the problem is W[1]-hard to approximate. However, we choose to state the more relaxed form here.
48. Lokshtanov, D.; Ramanujan, M.S.; Saurabh, S.; Zehavi, M. Parameterized Complexity and Approximability of Directed Odd Cycle Transversal. In Proceedings of the 2020 ACM-SIAM Symposium on Discrete Algorithms, SODA 2020, Salt Lake City, UT, USA, 5–8 January 2020; pp. 2181–2200. [CrossRef]
49. Feige, U.; Goldwasser, S.; Lovász, L.; Safra, S.; Szegedy, M. Interactive Proofs and the Hardness of Approximating Cliques. *J. ACM* **1996**, *43*, 268–292. [CrossRef]
50. Chitnis, R.; Feldmann, A.E.; Manurangsi, P. Parameterized Approximation Algorithms for Bidirected Steiner Network Problems. In Proceedings of the 26th Annual European Symposium on Algorithms (ESA), Helsinki, Finland, 20–22 August 2018; pp. 20:1–20:16. [CrossRef]
51. Papadimitriou, C.H.; Yannakakis, M. Optimization, Approximation, and Complexity Classes. *J. Comput. Syst. Sci.* **1991**, *43*, 425–440. [CrossRef]
52. Dinur, I. Mildly exponential reduction from gap 3SAT to polynomial-gap label-cover. *Electron. Colloq. Comput. Complex. (ECCC)* **2016**, *23*, 128.
53. Manurangsi, P.; Raghavendra, P. A Birthday Repetition Theorem and Complexity of Approximating Dense CSPs. In Proceedings of the 44th International Colloquium on Automata, Languages, and Programming ICALP, Warsaw, Poland, 10–14 July 2017; pp. 78:1–78:15. [CrossRef]
54. The version where n denotes the number of variables is equivalent to the current formulation, because we can always assume without loss of generality that $m = O(n)$ (see [52,53]).
55. Chen, J.; Huang, X.; Kanj, I.A.; Xia, G. Linear FPT reductions and computational lower bounds. In Proceedings of the 36th Annual ACM Symposium on Theory of Computing (STOC), Chicago, IL, USA, 13–16 June 2004; pp. 212–221. [CrossRef]
56. Chen, J.; Huang, X.; Kanj, I.A.; Xia, G. Strong computational lower bounds via parameterized complexity. *J. Comput. Syst. Sci.* **2006**, *72*, 1346–1367. [CrossRef]
57. Bellare, M.; Goldreich, O.; Sudan, M. Free Bits, PCPs, and Nonapproximability-Towards Tight Results. *SIAM J. Comput.* **1998**, *27*, 804–915. [CrossRef]
58. Zuckerman, D. Simulating BPP Using a General Weak Random Source. *Algorithmica* **1996**, *16*, 367–391. [CrossRef]
59. Dinur, I.; Manurangsi, P. ETH-Hardness of Approximating 2-CSPs and Directed Steiner Network. In Proceedings of the 9th Innovations in Theoretical Computer Science Conference (ITCS), Cambridge, MA, USA, 11–14 January 2018; pp. 36:1–36:20.
60. Bellare, M.; Goldwasser, S.; Lund, C.; Russeli, A. Efficient probabilistically checkable proofs and applications to approximations. In Proceedings of the Twenty-Fifth Annual ACM Symposium on Theory of Computing, San Diego, CA, USA, 16–18 May 1993; pp. 294–304. [CrossRef]
61. Moshkovitz, D. The Projection Games Conjecture and the NP-Hardness of ln n-Approximating Set-Cover. *Theory Comput.* **2015**, *11*, 221–235. [CrossRef]
62. See also the related Projection Game Conjecture (PGC) [61].
63. Naturally, we say that two functions f_i and f_j agree iff $f_i(x) = f_j(x)$ for all $x \in S_i \cap S_j$.

64. Raz, R.; Safra, S. A Sub-Constant Error-Probability Low-Degree Test, and a Sub-Constant Error-Probability PCP Characterization of NP. In Proceedings of the Twenty-Ninth Annual ACM Symposium on the Theory of Computing, El Paso, TX, USA, 4–6 May 1997; pp. 475–484. [CrossRef]
65. Impagliazzo, R.; Kabanets, V.; Wigderson, A. New Direct-Product Testers and 2-Query PCPs. *SIAM J. Comput.* **2012**, *41*, 1722–1768. [CrossRef]
66. Dinur, I.; Navon, I.L. Exponentially Small Soundness for the Direct Product Z-Test. In Proceedings of the 32nd Computational Complexity Conference, CCC, Riga, Latvia, 6–9 July 2017; pp. 29:1–29:50. [CrossRef]
67. Arora, S.; Babai, L.; Stern, J.; Sweedyk, Z. The Hardness of Approximate Optimia in Lattices, Codes, and Systems of Linear Equations. In Proceedings of the 34th Annual Symposium on Foundations of Computer Science, Palo Alto, CA, USA, 3–5 November 1993; pp. 724–733. [CrossRef]
68. Håstad, J. Some optimal inapproximability results. *J. ACM* **2001**, *48*, 798–859. [CrossRef]
69. Chan, S.O. Approximation Resistance from Pairwise-Independent Subgroups. *J. ACM* **2016**, *63*, 27. [CrossRef]
70. Manurangsi, P. Tight Running Time Lower Bounds for Strong Inapproximability of Maximum k-Coverage, Unique Set Cover and Related Problems (via t-Wise Agreement Testing Theorem). In Proceedings of the 2020 ACM-SIAM Symposium on Discrete Algorithms (SODA), Salt Lake City, UT, USA, 5–8 January 2020; pp. 62–81. [CrossRef]
71. Håstad, J. Clique is Hard to Approximate Within $n^{1-\varepsilon}$. In Proceedings of the 37th Annual Symposium on Foundations of Computer Science, FOCS '96, Burlington, VT, USA, 14–16 October 1996; pp. 627–636. [CrossRef]
72. Khot, S. Ruling Out PTAS for Graph Min-Bisection, Dense k-Subgraph, and Bipartite Clique. *SIAM J. Comput.* **2006**, *36*, 1025–1071. [CrossRef]
73. Bhangale, A.; Gandhi, R.; Hajiaghayi, M.T.; Khandekar, R.; Kortsarz, G. Bi-Covering: Covering Edges with Two Small Subsets of Vertices. *SIAM J. Discret. Math.* **2017**, *31*, 2626–2646. [CrossRef]
74. Manurangsi, P. Inapproximability of maximum biclique problems, minimum k-cut and densest at-least-k-subgraph from the small set expansion hypothesis. *Algorithms* **2018**, *11*, 10. [CrossRef]
75. We note, however, that strong inapproximability of BICLIQUE is known under stronger assumptions [72–74].
76. Manurangsi, P. Almost-polynomial ratio ETH-hardness of approximating densest k-subgraph. In Proceedings of the 49th Annual ACM SIGACT Symposium on Theory of Computing STOC, Montreal, QC, Canada, 19–23 June 2017; pp. 954–961. [CrossRef]
77. Raghavendra, P.; Steurer, D. Graph expansion and the unique games conjecture. In Proceedings of the ACM Forty-Second ACM Symposium on Theory of Computing, Cambridge, MA, USA, 6–8 June 2010; pp. 755–764.
78. Alon, N.; Arora, S.; Manokaran, R.; Moshkovitz, D.; Weinstein, O. Inapproximabilty of Densest k-Subgraph from Average Case Hardness. Unpublished Manuscript.
79. Again, similar to BICLIQUE, DENSEST k-SUBGRAPH is known to be hard to approximate under stronger assumptions [72,76–78].
80. Kővári, T.; Sós, V.T.; Turán, P. On a problem of K. Zarankiewicz. *Colloq. Math.* **1954**, *3*, 50–57. [CrossRef]
81. Zuckerman, D. Linear degree extractors and the inapproximability of max clique and chromatic number. In Proceedings of the ACM Thirty-Eighth Annual ACM Symposium on Theory of Computing, Seattle, WA, USA, 21–23 May 2006; pp. 681–690.
82. Baker, B.S. Approximation algorithms for NP-complete problems on planar graphs. *J. ACM (JACM)* **1994**, *41*, 153–180. [CrossRef]
83. Johnson, D.S.; Garey, M.R. *Computers and Intractability: A Guide to the Theory of NP-Completeness*; WH Freeman: San Francisco, CA, USA; New York, NY, USA, 1979; Volume 1.
84. Demaine, E.D.; Hajiaghayi, M. Equivalence of local treewidth and linear local treewidth and its algorithmic applications. In Proceedings of the Fifteenth Annual ACM-SIAM Symposium on Discrete Algorithms, New Orleans, LA, USA, 11–13 January 2004; pp. 840–849.
85. Grohe, M.; Kawarabayashi, K.I.; Reed, B. A simple algorithm for the graph minor decomposition: Logic meets structural graph theory. In Proceedings of the Twenty-Fourth Annual ACM-SIAM Symposium on Discrete Algorithms, New Orleans, LA, USA, 6–8 January 2013; pp. 414–431.
86. Demaine, E.D.; Hajiaghayi, M. The bidimensionality theory and its algorithmic applications. *Comput. J.* **2008**, *51*, 292–302. [CrossRef]

87. Fomin, F.V.; Lokshtanov, D.; Raman, V.; Saurabh, S. Bidimensionality and EPTAS. In Proceedings of the Twenty-Second Annual ACM-SIAM Symposium on Discrete Algorithms, San Francisco, CA, USA, 23–25 January 2011; pp. 748–759.
88. Demaine, E.D.; Hajiaghayi, M.; Kawarabayashi, K.i. Contraction decomposition in H-minor-free graphs and algorithmic applications. In Proceedings of the Forty-Third Annual ACM Symposium on Theory of Computing, San Jose, CA, USA, 6–8 June 2011; pp. 441–450.
89. Bansal, N.; Reichman, D.; Umboh, S.W. LP-based robust algorithms for noisy minor-free and bounded treewidth graphs. In Proceedings of the Twenty-Eighth Annual ACM-SIAM Symposium on Discrete Algorithms, Barcelona, Spain, 16–19 January 2017; pp. 1964–1979.
90. Magen, A.; Moharrami, M. Robust algorithms for on minor-free graphs based on the Sherali-Adams hierarchy. In *Approximation, Randomization, and Combinatorial Optimization. Algorithms and Techniques*; Springer: Berlin, Germany, 2009; pp. 258–271.
91. Demaine, E.D.; Goodrich, T.D.; Kloster, K.; Lavallee, B.; Liu, Q.C.; Sullivan, B.D.; Vakilian, A.; van der Poel, A. Structural Rounding: Approximation Algorithms for Graphs Near an Algorithmically Tractable Class. In Proceedings of the 27th Annual European Symposium on Algorithms (ESA), Dagstuhl, Germany, 9–11 September 2019.
92. Katsikarelis, I.; Lampis, M.; Paschos, V.T. Structurally Parameterized d-Scattered Set. In Proceedings of the Graph-Theoretic Concepts in Computer Science—44th International Workshop WG, Cottbus, Germany, 27–29 June 2018; pp. 292–305._24. [CrossRef]
93. Katsikarelis, I.; Lampis, M.; Paschos, V.T. Improved (In-)Approximability Bounds for d-Scattered Set. In Proceedings of the Approximation and Online Algorithms—17th International Workshop, WAOA, Munich, Germany, 12–13 September 2019; Revised Selected Papers; pp. 202–216._14. [CrossRef]
94. Marx, D. Efficient approximation schemes for geometric problems? In *European Symposium on Algorithms*; Springer: Berlin, Germany, 2005; pp. 448–459.
95. Adamaszek, A.; Wiese, A. Approximation schemes for maximum weight independent set of rectangles. In Proceedings of the 2013 IEEE 54th Annual Symposium on Foundations of Computer Science, Berkeley, CA, USA, 27–29 October 2013; pp. 400–409.
96. Grandoni, F.; Kratsch, S.; Wiese, A. Parameterized Approximation Schemes for Independent Set of Rectangles and Geometric Knapsack. In Proceedings of the 27th Annual European Symposium on Algorithms (ESA), Munich/Garching, Germany, 9–11 September 2019; pp. 53:1–53:16.
97. Pilipczuk, M.; van Leeuwen, E.J.; Wiese, A. Approximation and Parameterized Algorithms for Geometric Independent Set with Shrinking. In Proceedings of the 42nd International Symposium on Mathematical Foundations of Computer Science (MFCS), Aalborg, Denmark, 21–25 August 2017; 42:1–42:13.
98. Clark, B.N.; Colbourn, C.J.; Johnson, D.S. Unit disk graphs. *Discret. Math.* **1990**, *86*, 165–177. [CrossRef]
99. III, H.B.H.; Marathe, M.V.; Radhakrishnan, V.; Ravi, S.S.; Rosenkrantz, D.J.; Stearns, R.E. NC-Approximation Schemes for NP- and PSPACE-Hard Problems for Geometric Graphs. *J. Algorithms* **1998**, *26*, 238–274. [CrossRef]
100. Alber, J.; Fiala, J. Geometric separation and exact solutions for the parameterized independent set problem on disk graphs. *J. Algorithms* **2004**, *52*, 134–151. [CrossRef]
101. Stockmeyer, L. Planar 3-colorability is NP-complete. *ACM Sigact News* **1973**, *5*, 19–25. [CrossRef]
102. Demaine, E.D.; Hajiaghayi, M.T.; Kawarabayashi, K.i. Algorithmic graph minor theory: Decomposition, approximation, and coloring. In Proceedings of the IEEE 46th Annual IEEE Symposium on Foundations of Computer Science (FOCS'05), Pittsburgh, PA, USA, 23–25 October 2005; pp. 637–646.
103. Sometimes called IMPROPER COLORING.
104. Belmonte, R.; Lampis, M.; Mitsou, V. Parameterized (Approximate) Defective Coloring. In Proceedings of the 35th Symposium on Theoretical Aspects of Computer Science (STACS), Caen, France, 28 February–3 March 2018; pp. 10:1–10:15.
105. Lampis, M. Parameterized Approximation Schemes Using Graph Widths. In Proceedings of the Automata, Languages, and Programming—41st International Colloquium (ICALP), Copenhagen, Denmark, 8–11 July 2014; pp. 775–786.
106. Fellows, M.R.; Fomin, F.V.; Lokshtanov, D.; Rosamond, F.; Saurabh, S.; Szeider, S.; Thomassen, C. On the complexity of some colorful problems parameterized by treewidth. *Inf. Comput.* **2011**, *209*, 143–153. [CrossRef]

107. Corneil, D.G.; Rotics, U. On the Relationship Between Clique-Width and Treewidth. *SIAM J. Comput.* **2005**, *34*, 825–847. [CrossRef]
108. Katsikarelis, I.; Lampis, M.; Paschos, V.T. Structural parameters, tight bounds, and approximation for (k, r)-center. *Discret. Appl. Math.* **2019**, *264*, 90–117. [CrossRef]
109. In [105] the runtime of these algorithms is stated as $(\log n/\varepsilon)^{O(k)} 2^{k\ell} n^{O(1)}$, which can be shown to be upper bounded by $(k/\varepsilon)^{O(k\ell)} n^{O(1)}$ (see e.g., ([108] Lemma 1)).
110. Salavatipour, M.R. On sum coloring of graphs. *Discret. Appl. Math.* **2003**, *127*, 477–488. [CrossRef]
111. Marx, D. Complexity results for minimum sum edge coloring. *Discret. Appl. Math.* **2009**, *157*, 1034–1045. [CrossRef]
112. Giaro, K.; Kubale, M. Edge-chromatic sum of trees and bounded cyclicity graphs. *Inf. Process. Lett.* **2000**, *75*, 65–69. [CrossRef]
113. Marx, D. Minimum sum multicoloring on the edges of planar graphs and partial k-trees. In *International Workshop on Approximation and Online Algorithms*; Springer: Berlin, Germany, 2004; pp. 9–22.
114. Cygan, M. Improved approximation for 3-dimensional matching via bounded pathwidth local search. In Proceedings of the 2013 IEEE 54th Annual Symposium on Foundations of Computer Science, Berkeley, CA, USA, 27–29 October 2013; pp. 509–518.
115. Guruswami, V.; Lee, E. Inapproximability of H-Transversal/Packing. In Proceedings of the Approximation, Randomization, and Combinatorial Optimization, Algorithms and Techniques (APPROX/RANDOM), Princeton, NJ, USA, 24–26 August 2015; pp. 284–304.
116. Lee, E. Partitioning a graph into small pieces with applications to path transversal. In Proceedings of the Twenty-Eighth Annual ACM-SIAM Symposium on Discrete Algorithms, Barcelona, Spain, 16–19 January 2017; pp. 1546–1558.
117. Fomin, F.V.; Le, T.N.; Lokshtanov, D.; Saurabh, S.; Thomassé, S.; Zehavi, M. Subquadratic kernels for implicit 3-hitting set and 3-set packing problems. *ACM Trans. Algorithms (TALG)* **2019**, *15*, 1–44.
118. Friggstad, Z.; Salavatipour, M.R. Approximability of packing disjoint cycles. In *International Symposium on Algorithms and Computation*; Springer: Berlin, Germany, 2007; pp. 304–315.
119. Lokshtanov, D.; Mouawad, A.E.; Saurabh, S.; Zehavi, M. Packing Cycles Faster Than Erdos–Posa. *SIAM J. Discret. Math.* **2019**, *33*, 1194–1215. [CrossRef]
120. Bodlaender, H.L.; Thomassé, S.; Yeo, A. Kernel bounds for disjoint cycles and disjoint paths. *Theor. Comput. Sci.* **2011**, *412*, 4570–4578. [CrossRef]
121. Batra, J.; Garg, N.; Kumar, A.; Mömke, T.; Wiese, A. New approximation schemes for unsplittable flow on a path. In Proceedings of the Twenty-Sixth Annual ACM-SIAM Symposium on Discrete Algorithms, San Diego, CA, USA, 4–6 January 2015; pp. 47–58.
122. Wiese, A. A $(1+\varepsilon)$-approximation for Unsplittable Flow on a Path in fixed-parameter running time. In Proceedings of the 44th International Colloquium on Automata, Languages, and Programming (ICALP), Warsaw, Poland, 10–14 July 2017; pp. 67:1–67:13.
123. Garg, N.; Kumar, A.; Muralidhara, V. Minimizing Total Flow-Time: The Unrelated Case. In *International Symposium on Algorithms and Computation*; Springer: Berlin/Heidelberg, Germany, 2008; pp. 424–435.
124. Kellerer, H.; Tautenhahn, T.; Woeginger, G. Approximability and nonapproximability results for minimizing total flow time on a single machine. *SIAM J. Comput.* **1999**, *28*, 1155–1166. [CrossRef]
125. Wiese, A. Fixed-Parameter approximation schemes for weighted flowtime. In Proceedings of the Approximation, Randomization, and Combinatorial Optimization, Algorithms and Techniques (APPROX/RANDOM 2018), Princeton, NJ, USA, 20–22 August 2018; pp. 28:1–28:19.
126. Buss, J.F.; Goldsmith, J. Nondeterminism Within P. *SIAM J. Comput.* **1993**, *22*, 560–572. [CrossRef]
127. Nemhauser, G.L.; Trotter, L.E., Jr. Vertex packings: Structural properties and algorithms. *Math. Program.* **1975**, *8*, 232–248. [CrossRef]
128. Abu-Khzam, F.N. A kernelization algorithm for d-Hitting Set. *J. Comput. Syst. Sci.* **2010**, *76*, 524–531. [CrossRef]
129. Cygan, M. Deterministic parameterized connected vertex cover. In *Scandinavian Workshop on Algorithm Theory*; Springer: Berlin, Germany, 2012; pp. 95–106.
130. Dom, M.; Lokshtanov, D.; Saurabh, S. Kernelization Lower Bounds Through Colors and IDs. *ACM Trans. Algorithms* **2014**, *11*, 1–20. [CrossRef]

131. Krithika, R.; Majumdar, D.; Raman, V. Revisiting connected vertex cover: FPT algorithms and lossy kernels. *Theory Comput. Syst.* **2018**, *62*, 1690–1714. [CrossRef]
132. Majumdar, D.; Ramanujan, M.S.; Saurabh, S. On the Approximate Compressibility of Connected Vertex Cover. *arXiv* **2019**, arXiv:1905.03379.
133. Recall that a bi-kernel is similar to a kernel except that its the output need not be an instance of the original problem. Bi-PSAKS can be defined analogously to PSAKS, but with bi-kernel instead of kernel. In the case of CONNECTED DOMINATING SET, the bi-kernel outputs an instance of an annotated variant of CONNECTED DOMINATING SET, where some vertices are marked and do not need to be covered by the solution.
134. Eiben, E.; Kumar, M.; Mouawad, A.E.; Panolan, F.; Siebertz, S. Lossy Kernels for Connected Dominating Set on Sparse Graphs. In Proceedings of the STACS, Caen, France, 28 February–3 March 2018; Volume 96, pp. 29:1–29:15.
135. Angel, E.; Bampis, E.; Escoffier, B.; Lampis, M. Parameterized power vertex cover. In *International Workshop on Graph-Theoretic Concepts in Computer Science*; Springer: Berlin, Germany, 2016; pp. 97–108.
136. Dom, M.; Lokshtanov, D.; Saurabh, S.; Villanger, Y. Capacitated domination and covering: A parameterized perspective. In *International Workshop on Parameterized and Exact Computation*; Springer: Berlin, Germany, 2008; pp. 78–90.
137. See Definition 1 for the definition of the treewidth.
138. Erdős, P.; Pósa, L. On Independent Circuits Contained in a Graph. *Can. J. Math.* **1965**, *17*, 347–352. [CrossRef]
139. Raymond, J.F.; Thilikos, D.M. Recent techniques and results on the Erdős–Pósa property. *Discret. Appl. Math.* **2017**, *231*, 25–43. [CrossRef]
140. Kim, E.J.; Kwon, O.j. Erdős-Pósa property of chordless cycles and its applications. In Proceedings of the Twenty-Ninth Annual ACM-SIAM Symposium on Discrete Algorithms, New Orleans, LA, USA, 7–10 January 2018; pp. 1665–1684.
141. Van Batenburg, W.C.; Huynh, T.; Joret, G.; Raymond, J.F. A tight Erdős-Pósa function for planar minors. In Proceedings of the Thirtieth Annual ACM-SIAM Symposium on Discrete Algorithms, San Diego, CA, USA, 6–9 January 2019; pp. 1485–1500.
142. Cornuejols, G.; Nemhauser, G.L.; Wolsey, L.A. Worst-case and probabilistic analysis of algorithms for a location problem. *Oper. Res.* **1980**, *28*, 847–858. [CrossRef]
143. Cohen-Addad, V.; Gupta, A.; Kumar, A.; Lee, E.; Li, J. Tight FPT Approximations for k-Median and k-Means. In *46th International Colloquium on Automata, Languages, and Programming (ICALP 2019)*; Baier, C., Chatzigiannakis, I., Flocchini, P., Leonardi, S., Eds.; Volume 132, Leibniz International Proceedings in Informatics (LIPIcs); Schloss Dagstuhl–Leibniz-Zentrum fuer Informatik: Dagstuhl, Germany, 2019; pp. 42:1–42:14. [CrossRef]
144. Badanidiyuru, A.; Kleinberg, R.; Lee, H. Approximating low-dimensional coverage problems. In Proceedings of the Twenty-Eighth Annual Symposium on Computational Geometry, Chapel Hill, NC, USA, 17–20 June 2012; pp. 161–170.
145. Guo, J.; Niedermeier, R.; Wernicke, S. Parameterized complexity of vertex cover variants. *Theory Comput. Syst.* **2007**, *41*, 501–520. [CrossRef]
146. Skowron, P.; Faliszewski, P. Chamberlin–Courant Rule with Approval Ballots: Approximating the MaxCover Problem with Bounded Frequencies in FPT Time. *J. Artif. Intell. Res.* **2017**, *60*, 687–716. [CrossRef]
147. Manurangsi, P. A Note on Max k-Vertex Cover: Faster FPT-AS, Smaller Approximate Kernel and Improved Approximation. In *2nd Symposium on Simplicity in Algorithms (SOSA 2019)*; Schloss Dagstuhl-Leibniz-Zentrum fuer Informatik: Wadern, Germany, 2018.
148. The argument of [146] was later independently rediscovered in [147] as well.
149. Petrank, E. The hardness of approximation: Gap location. *Comput. Complex.* **1994**, *4*, 133–157. [CrossRef]
150. Chlamtáč, E.; Dinitz, M.; Konrad, C.; Kortsarz, G.; Rabanca, G. The Densest k-Subhypergraph Problem. *SIAM J. Discret. Math.* **2018**, *32*, 1458–1477. [CrossRef]
151. Chlamtáč, E.; Dinitz, M.; Makarychev, Y. Minimizing the Union: Tight Approximations for Small Set Bipartite Vertex Expansion. In Proceedings of the Twenty-Eighth Annual ACM-SIAM Symposium on Discrete Algorithms SODA, Barcelona, Spain, 16–19 January 2017; pp. 881–899. [CrossRef]
152. The problem has also been referred to as MIN k-UNION and SMALL SET BIPARTITE VERTEX EXPANSION in the literature [150,151].

153. Gupta, A.; Lee, E.; Li, J. An FPT algorithm beating 2-approximation for k-cut. In Proceedings of the Twenty-Ninth Annual ACM-SIAM Symposium on Discrete Algorithms, New Orleans, LA, USA, 7–10 January 2018; pp. 2821–2837.
154. Gupta, A.; Lee, E.; Li, J. Faster exact and approximate algorithms for k-cut. In Proceedings of the 2018 IEEE 59th Annual Symposium on Foundations of Computer Science (FOCS), Philadelphia, PA, USA, 18–21 October 2018; pp. 113–123.
155. Byrka, J.; Pensyl, T.; Rybicki, B.; Srinivasan, A.; Trinh, K. An improved approximation for k-median, and positive correlation in budgeted optimization. In Proceedings of the Twenty-Sixth Annual ACM-SIAM Symposium on Discrete Algorithms, Budapest, Hungary, 26–29 August 2014; pp. 737–756.
156. Kanungo, T.; Mount, D.M.; Netanyahu, N.S.; Piatko, C.D.; Silverman, R.; Wu, A.Y. A local search approximation algorithm for k-means clustering. *Comput. Geom.* **2004**, *28*, 89–112. [CrossRef]
157. Gonzalez, T.F. Clustering to minimize the maximum intercluster distance. *Theor. Comput. Sci.* **1985**, *38*, 293–306. [CrossRef]
158. A special case that has received significant attention assumes $P = F$. In this case, the best approximation ratio for k-CENTER becomes 2.
159. Guha, S.; Khuller, S. Greedy strikes back: Improved facility location algorithms. *J. Algorithms* **1999**, *31*, 228–248. [CrossRef]
160. Chen, K. On k-median clustering in high dimensions. In Proceedings of the Seventeenth Annual ACM-SIAM Symposium on Discrete Algorithm, Miami, FL, USA, 22–26 January 2006; pp. 1177–1185.
161. Feldman, D.; Langberg, M. A unified framework for approximating and clustering data. In Proceedings of the Forty-Third Annual ACM Symposium on Theory of Computing, San Jose, CA, USA, 6–8 June 2011; pp. 569–578.
162. Calinescu, G.; Chekuri, C.; Pál, M.; Vondrák, J. Maximizing a monotone submodular function subject to a matroid constraint. *SIAM J. Comput.* **2011**, *40*, 1740–1766. [CrossRef]
163. Haussler, D. Decision theoretic generalizations of the PAC model for neural net and other learning applications. *Inf. Comput.* **1992**, *100*, 78–150. [CrossRef]
164. Lee, E.; Schmidt, M.; Wright, J. Improved and simplified inapproximability for k-means. *Inf. Process. Lett.* **2017**, *120*, 40–43. [CrossRef]
165. Cohen-Addad, V.; Karthik, C.S. Inapproximability of Clustering in L_p-metrics. In Proceedings of the 2019 IEEE 60th Annual Symposium on Foundations of Computer Science, Baltimore, MD, USA, 9–12 November 2019.
166. Arora, S.; Raghavan, P.; Rao, S. Approximation Schemes for Euclidean k-Medians and Related Problems. In Proceedings of the Thirtieth Annual ACM Symposium on the Theory of Computing, Dallas, TX, USA, 23–26 May 1998; Volume 98, pp. 106–113.
167. Arora, S. Polynomial time approximation schemes for Euclidean traveling salesman and other geometric problems. *J. ACM (JACM)* **1998**, *45*, 753–782. [CrossRef]
168. Kolliopoulos, S.G.; Rao, S. A nearly linear-time approximation scheme for the Euclidean k-median problem. In Proceedings of the European Symposium on Algorithms, Prague, Czech Republic, 16–18 July 1999; Springer: Berlin, Germany, 1999; pp. 378–389.
169. Matoušek, J. On approximate geometric k-clustering. *Discret. Comput. Geom.* **2000**, *24*, 61–84. [CrossRef]
170. Bādoiu, M.; Har-Peled, S.; Indyk, P. Approximate clustering via core-sets. In Proceedings of the Thiry-Fourth Annual ACM Symposium on Theory of Computing, Montreal, QC, Canada, 19–21 May 2002; pp. 250–257.
171. De La Vega, W.F.; Karpinski, M.; Kenyon, C.; Rabani, Y. Approximation schemes for clustering problems. In Proceedings of the Thirty-Fifth Annual ACM Symposium on Theory of Computing, San Diego, CA, USA, 9–11 June 2003; pp. 50–58.
172. Har-Peled, S.; Mazumdar, S. On coresets for k-means and k-median clustering. In Proceedings of the Thirty-Sixth Annual ACM Symposium on Theory of Computing, Chicago, IL, USA, 13–15 June 2004; pp. 291–300.
173. Kumar, A.; Sabharwal, Y.; Sen, S. A simple linear time $(1 + \varepsilon)$-approximation algorithm for k-means clustering in any dimensions. In Proceedings of the 45th Annual IEEE Symposium on Foundations of Computer Science, Rome, Italy, 17–19 October 2004; pp. 454–462.

174. Kumar, A.; Sabharwal, Y.; Sen, S. Linear time algorithms for clustering problems in any dimensions. In *International Colloquium on Automata, Languages, and Programming*; Springer: Berlin, Germany, 2005; pp. 1374–1385.
175. Feldman, D.; Monemizadeh, M.; Sohler, C. A PTAS for k-means clustering based on weak coresets. In Proceedings of the Twenty-Third Annual Symposium on Computational Geometry, Gyeongju, Korea, 6–8 June 2007; pp. 11–18.
176. Sohler, C.; Woodruff, D.P. Strong coresets for k-median and subspace approximation: Goodbye dimension. In Proceedings of the 2018 IEEE 59th Annual Symposium on Foundations of Computer Science (FOCS), Paris, France, 7–9 October 2018; pp. 802–813.
177. Becchetti, L.; Bury, M.; Cohen-Addad, V.; Grandoni, F.; Schwiegelshohn, C. Oblivious dimension reduction for k-means: Beyond subspaces and the Johnson-Lindenstrauss lemma. In Proceedings of the 51st Annual ACM SIGACT Symposium on Theory of Computing, Phoenix, AZ, USA, 23–26 June 2019; pp. 1039–1050.
178. Huang, L.; Vishnoi, N.K. Coresets for Clustering in Euclidean Spaces: Importance Sampling is Nearly Optimal. *arXiv* **2020**, arXiv:2004.06263.
179. Braverman, V.; Jiang, S.H.C.; Krauthgamer, R.; Wu, X. Coresets for Clustering in Excluded-minor Graphs and Beyond. *arXiv* **2020**, arXiv:2004.07718.
180. Cohen-Addad, V.; Klein, P.N.; Mathieu, C. Local search yields approximation schemes for k-means and k-median in euclidean and minor-free metrics. *SIAM J. Comput.* **2019**, *48*, 644–667. [CrossRef]
181. Friggstad, Z.; Rezapour, M.; Salavatipour, M.R. Local search yields a PTAS for k-means in doubling metrics. *SIAM J. Comput.* **2019**, *48*, 452–480. [CrossRef]
182. Cohen-Addad, V. A fast approximation scheme for low-dimensional k-means. In *Twenty-Ninth Annual ACM-SIAM Symposium on Discrete Algorithms*; Society for Industrial and Applied Mathematics: Philadelphia, PA, USA, 2018; pp. 430–440.
183. Cohen-Addad, V.; Feldmann, A.E.; Saulpic, D. Near-Linear Time Approximation Schemes for Clustering in Doubling Metrics. *arXiv* **2019**, arXiv:1812.08664.
184. Feldmann, A.E.; Marx, D. The Parameterized Hardness of the k-Center Problem in Transportation Networks. In Proceedings of the 16th Scandinavian Symposium and Workshops on Algorithm Theory (SWAT), Malmö, Sweden, 18–20 June 2018; pp. 19:1–19:13. [CrossRef]
185. Fox-Epstein, E.; Klein, P.N.; Schild, A. Embedding Planar Graphs into Low-Treewidth Graphs with Applications to Efficient Approximation Schemes for Metric Problems. In Proceedings of the Thirtieth Annual ACM-SIAM Symposium on Discrete Algorithms (SODA), San Diego, CA, USA, 6–9 January 2019; pp. 1069–1088.
186. Becker, A.; Klein, P.N.; Saulpic, D. Polynomial-time approximation schemes for k-center, k-median, and capacitated vehicle routing in bounded highway dimension. In Proceedings of the 26th Annual European Symposium on Algorithms (ESA), Helsinki, Finland, 20–22 August 2018; pp. 8:1–8:15.
187. Feldmann, A.E. Fixed Parameter Approximations for k-Center Problems in Low Highway Dimension Graphs. In *42nd International Colloquium on Automata, Languages, and Programming (ICALP)*; Springer: Berlin/Heidelberg, Germany, 2015; pp. 588–600._47. [CrossRef]
188. Li, S. Approximating capacitated k-median with $(1+\varepsilon)k$ open facilities. In Proceedings of the Twenty-Seventh Annual ACM-SIAM Symposium on Discrete Algorithms, Arlington, VA, USA, 10–12 January 2016; pp. 786–796.
189. Demirci, G.; Li, S. Constant Approximation for Capacitated k-Median with $(1+\varepsilon)$-Capacity Violation. *arXiv* **2016**, arXiv:1603.02324.
190. Adamczyk, M.; Byrka, J.; Marcinkowski, J.; Meesum, S.M.; Włodarczyk, M. Constant factor FPT approximation for capacitated k-median. *arXiv* **2018**, arXiv:1809.05791.
191. Cohen-Addad, V.; Li, J. On the Fixed-Parameter Tractability of Capacitated Clustering. In Proceedings of the 46th International Colloquium on Automata, Languages, and Programming (ICALP), Patras, Greece, 9–12 July 2019; pp. 41:1–41:14.
192. Xu, Y.; Zhang, Y.; Zou, Y. A constant parameterized approximation for hard-capacitated k-means. *arXiv* **2019**, arXiv:1901.04628.
193. Krishnaswamy, R.; Li, S.; Sandeep, S. Constant approximation for k-median and k-means with outliers via iterative rounding. In Proceedings of the 50th Annual ACM SIGACT Symposium on Theory of Computing, Los Angeles, CA, USA, 25–29 June 2018; pp. 646–659.

194. Swamy, C. Improved approximation algorithms for matroid and knapsack median problems and applications. *ACM Trans. Algorithms (TALG)* **2016**, *12*, 49. [CrossRef]
195. Sometimes also the non-metric version of TSP is considered, which however is much harder than the metric one. We only consider the metric version here.
196. Dreyfus, S.E.; Wagner, R.A. The Steiner problem in graphs. *Networks* **1971**, *1*, 195–207. [CrossRef]
197. Fuchs, B.; Kern, W.; Molle, D.; Richter, S.; Rossmanith, P.; Wang, X. Dynamic programming for minimum Steiner trees. *Theory Comput. Syst.* **2007**, *41*, 493–500. [CrossRef]
198. Nederlof, J. Fast Polynomial-Space Algorithms Using Möbius Inversion: Improving on Steiner Tree and Related Problems. In Proceedings of the Automata, Languages and Programming, 36th International Colloquium, ICALP, Rhodes, Greece, 5–12 July 2009; pp. 713–725.
199. Borchers, A.; Du, D.Z. The k-Steiner Ratio in Graphs. *SIAM J. Comput.* **1997**, *26*, 857–869. [CrossRef]
200. Byrka, J.; Grandoni, F.; Rothvoss, T.; Sanità, L. Steiner Tree Approximation via Iterative Randomized Rounding. *J. ACM* **2013**, *60*, 1–33. [CrossRef]
201. Chlebík, M.; Chlebíková, J. The Steiner tree problem on graphs: Inapproximability results. *Theor. Comput. Sci.* **2008**, *406*, 207–214. [CrossRef]
202. Dvořák, P.; Feldmann, A.E.; Knop, D.; Masařík, T.; Toufar, T.; Veselý, P. Parameterized Approximation Schemes for Steiner Trees with Small Number of Steiner Vertices. In Proceedings of the 35th Symposium on Theoretical Aspects of Computer Science (STACS), Caen, France, 28 February–3 March 2018; pp. 26:1–26:15. [CrossRef]
203. Babay, A.; Dinitz, M.; Zhang, Z. Characterizing Demand Graphs for (Fixed-Parameter) Shallow-Light Steiner Network. In Proceedings of the 38th IARCS Annual Conference on Foundations of Software Technology and Theoretical Computer Science (FSTTCS), Ahmedabad, India, 11–13 December 2018; pp. 33:1–33:22.
204. Hassin, R. Approximation schemes for the restricted shortest path problem. *Math. Oper. Res.* **1992**, *17*, 36–42. [CrossRef]
205. Bockenhauer, H.J.; Hromkovic, J.; Kneis, J.; Kupke, J. The parameterized approximability of TSP with deadlines. *Theory Comput. Syst.* **2007**, *41*, 431–444. [CrossRef]
206. Papadimitriou, C.H. The Euclidean travelling salesman problem is NP-complete. *Theor. Comput. Sci.* **1977**, *4*, 237–244. [CrossRef]
207. Garey, M.R.; Graham, R.L.; Johnson, D.S. The complexity of computing Steiner minimal trees. *SIAM J. Appl. Math.* **1977**, *32*, 835–859. [CrossRef]
208. Karpinski, M.; Lampis, M.; Schmied, R. New inapproximability bounds for TSP. *JCSS* **2015**, *81*, 1665–1677. [CrossRef]
209. In [167] the runtime of these algorithms is stated as $O(n(\log n)^{O(\sqrt{k}/\varepsilon)^{k-1}})$, which can be shown to be upper bounded by $k^{O(\sqrt{k}/\varepsilon)^{k-1}} n^2$ (see e.g., ([108] Lemma 1)).
210. Gottlieb, L. A Light Metric Spanner. In Proceedings of the 56th Annual Symposium on Foundations of Computer Science, FOCS, Berkeley, CA, USA, 17–20 October 2015; pp. 759–772.
211. Talwar, K. Bypassing the embedding: Algorithms for low dimensional metrics. In Proceedings of the 36th Annual ACM Symposium on Theory of Computing, Chicago, IL, USA, 13–16 June 2004; pp. 281–290.
212. Feldmann, A.E.; Fung, W.S.; Könemann, J.; Post, I. A (1+ε)-Embedding of Low Highway Dimension Graphs into Bounded Treewidth Graphs. *SIAM J. Comput.* **2018**, *47*, 1667–1704. [CrossRef]
213. Guo, J.; Niedermeier, R.; Suchý, O. Parameterized Complexity of Arc-Weighted Directed Steiner Problems. *SIAM J. Discret. Math.* **2011**, *25*, 583–599. [CrossRef]
214. Halperin, E.; Krauthgamer, R. Polylogarithmic inapproximability. In Proceedings of the 35th Annual ACM Symposium on Theory of Computing, San Diego, CA, USA, 9–11 June 2003; pp. 585–594.
215. Chitnis, R.; Hajiaghayi, M.; Kortsarz, G. Fixed-Parameter and Approximation Algorithms: A New Look. In Proceedings of the Parameterized and Exact Computation—8th International Symposium, IPEC, Sophia Antipolis, France, 4–6 September 2013; pp. 110–122.
216. Sometimes also called DIRECTED STEINER FOREST; note however that the optimum is not necessarily a forest.
217. Leighton, T.; Rao, S. Multicommodity max-flow min-cut theorems and their use in designing approximation algorithms. *J. ACM (JACM)* **1999**, *46*, 787–832. [CrossRef]
218. Arora, S.; Rao, S.; Vazirani, U. Expander flows, geometric embeddings and graph partitioning. *J. ACM (JACM)* **2009**, *56*, 5. [CrossRef]
219. Marx, D. Parameterized graph separation problems. *Theor. Comput. Sci.* **2006**, *351*, 394–406. [CrossRef]

220. Marx, D.; Razgon, I. Fixed-parameter tractability of multicut parameterized by the size of the cutset. *SIAM J. Comput.* **2014**, *43*, 355–388. [CrossRef]
221. Chitnis, R.; Cygan, M.; Hajiaghayi, M.; Pilipczuk, M.; Pilipczuk, M. Designing FPT algorithms for cut problems using randomized contractions. *SIAM J. Comput.* **2016**, *45*, 1171–1229. [CrossRef]
222. Cygan, M.; Lokshtanov, D.; Pilipczuk, M.; Pilipczuk, M.; Saurabh, S. Minimum Bisection is fixed-parameter tractable. *SIAM J. Comput.* **2019**, *48*, 417–450. [CrossRef]
223. Garg, N.; Vazirani, V.V.; Yannakakis, M. Approximate max-flow min-(multi) cut theorems and their applications. *SIAM J. Comput.* **1996**, *25*, 235–251. [CrossRef]
224. Chawla, S.; Krauthgamer, R.; Kumar, R.; Rabani, Y.; Sivakumar, D. On the hardness of approximating multicut and sparsest-cut. *Comput. Complex.* **2006**, *15*, 94–114. [CrossRef]
225. Sharma, A.; Vondrák, J. Multiway cut, pairwise realizable distributions, and descending thresholds. *arXiv* **2013**, arXiv:1309.2729.
226. Bérczi, K.; Chandrasekaran, K.; Király, T.; Madan, V. Improving the Integrality Gap for Multiway Cut. In *International Conference on Integer Programming and Combinatorial Optimization*; Springer: Berlin, Germany, 2019; pp. 115–127.
227. Cohen-Addad, V.; De Verdière, É.C.; De Mesmay, A. A near-linear approximation scheme for multicuts of embedded graphs with a fixed number of terminals. In Proceedings of the Twenty-Ninth Annual ACM-SIAM Symposium on Discrete Algorithms, New Orleans, LA, USA, 7–10 January 2018; pp. 1439–1458.
228. Chekuri, C.; Madan, V. Approximating multicut and the demand graph. In Proceedings of the Twenty-Eighth Annual ACM-SIAM Symposium on Discrete Algorithms, Barcelona, Spain, 16–19 January 2017; pp. 855–874
229. Agarwal, A.; Alon, N.; Charikar, M.S. Improved approximation for directed cut problems. In Proceedings of the Thirty-Ninth Annual ACM Symposium on Theory of Computing, San Diego, CA, USA, 11–13 June 2007; pp. 671–680.
230. Lee, E. Improved Hardness for Cut, Interdiction, and Firefighter Problems. In *44th International Colloquium on Automata, Languages, and Programming (ICALP 2017)*; Schloss Dagstuhl-Leibniz-Zentrum fuer Informatik: Wadern, Germany, 2017.
231. Chuzhoy, J.; Khanna, S. Polynomial flow-cut gaps and hardness of directed cut problems. *J. ACM (JACM)* **2009**, *56*, 6. [CrossRef]
232. Naor, J.; Zosin, L. A 2-approximation algorithm for the directed multiway cut problem. In Proceedings of the 38th Annual Symposium on Foundations of Computer Science, Miami Beach, FL, USA, 19–22 October 1997; pp. 548–553.
233. Chitnis, R.; Feldmann, A.E. FPT Inapproximability of Directed Cut and Connectivity Problems. *arXiv* **2019**, arXiv:1910.01934.
234. Feige, U.; Hajiaghayi, M.; Lee, J.R. Improved approximation algorithms for minimum weight vertex separators. *SIAM J. Comput.* **2008**, *38*, 629–657. [CrossRef]
235. Räcke, H. Optimal hierarchical decompositions for congestion minimization in networks. In Proceedings of the Fortieth Annual ACM Symposium on Theory of Computing, Budapest, Hungary, 26–29 August 2008; pp. 255–264.
236. Feige, U.; Mahdian, M. Finding small balanced separators. In Proceedings of the Thirty-Eighth Annual ACM Symposium on Theory of Computing, Seattle, WA, USA, 21–23 May 2006; pp. 375–384.
237. Karger, D.R.; Stein, C. A new approach to the minimum cut problem. *J. ACM (JACM)* **1996**, *43*, 601–640. [CrossRef]
238. Thorup, M. Minimum k-way cuts via deterministic greedy tree packing. In Proceedings of the Fortieth Annual ACM Symposium on Theory of Computing, Budapest, Hungary, 26–29 August 2008; pp. 159–166.
239. Gupta, A.; Lee, E.; Li, J. The number of minimum k-cuts: Improving the Karger-Stein bound. In Proceedings of the 51st Annual ACM SIGACT Symposium on Theory of Computing, Phoenix, AZ, USA, 23–26 June 2019; pp. 229–240.
240. Kawarabayashi, K.i.; Thorup, M. The minimum k-way cut of bounded size is fixed-parameter tractable. In Proceedings of the 2011 IEEE 52nd Annual Symposium on Foundations of Computer Science, Palm Springs, CA, USA, 22–25 October 2011; pp. 160–169.
241. Saran, H.; Vazirani, V.V. Finding k cuts within twice the optimal. *SIAM J. Comput.* **1995**, *24*, 101–108. [CrossRef]

242. Kawarabayashi, K.I.; Lin, B. A nearly 5/3-approximation FPT Algorithm for Min-k-Cut. In Proceedings of the 2020 ACM-SIAM Symposium on Discrete Algorithms, SODA 2020, Salt Lake City, UT, USA, 5–8 January 2020; pp. 990–999.
243. Lokshtanov, D.; Saurabh, S.; Surianarayanan, V. A Parameterized Approximation Scheme for Min k-Cut. *arXiv* **2020**, arXiv:2005.00134.
244. Lund, C.; Yannakakis, M. The approximation of maximum subgraph problems. In *International Colloquium on Automata, Languages, and Programming*; Springer: Berlin, Germany, 1993; pp. 40–51.
245. Khot, S. On the power of unique 2-prover 1-round games. In Proceedings of the Thiry-Fourth Annual ACM Symposium on Theory of Computing, Montreal, QC, Canada, 19–21 May 2002; pp. 767–775.
246. Heggernes, P.; Van't Hof, P.; Jansen, B.M.; Kratsch, S.; Villanger, Y. Parameterized complexity of vertex deletion into perfect graph classes. In *International Symposium on Fundamentals of Computation Theory*; Springer: Berlin, Germany, 2011; pp. 240–251.
247. Fomin, F.V.; Lokshtanov, D.; Misra, N.; Saurabh, S. Planar F-deletion: Approximation, kernelization and optimal FPT algorithms. In Proceedings of the 2012 IEEE 53rd Annual Symposium on Foundations of Computer Science, Brunswick, NJ, USA, 20–23 October 2012; pp. 470–479.
248. Marx, D. Chordal deletion is fixed-parameter tractable. *Algorithmica* **2010**, *57*, 747–768. [CrossRef]
249. Cao, Y.; Marx, D. Interval deletion is fixed-parameter tractable. In Proceedings of the Twenty-Fifth Annual ACM-SIAM Symposium on Discrete Algorithms, Portland, OR, USA, 5–7 January 2014; pp. 122–141.
250. Courcelle, B. The monadic second-order logic of graphs. I. Recognizable sets of finite graphs. *Inf. Comput.* **1990**, *85*, 12–75. [CrossRef]
251. Bodlaender, H.L. Treewidth: Structure and algorithms. In *International Colloquium on Structural Information and Communication Complexity*; Springer: Berlin, Germany, 2007; pp. 11–25.
252. Arnborg, S.; Corneil, D.G.; Proskurowski, A. Complexity of finding embeddings in ak-tree. *SIAM J. Algebr. Discret. Methods* **1987**, *8*, 277–284. [CrossRef]
253. Bodlaender, H.L. A linear-time algorithm for finding tree-decompositions of small treewidth. *SIAM J. Comput.* **1996**, *25*, 1305–1317. [CrossRef]
254. Bodlaender, H.L.; Drange, P.G.; Dregi, M.S.; Fomin, F.V.; Lokshtanov, D.; Pilipczuk, M. A $c^k n$ 5-Approximation Algorithm for Treewidth. *SIAM J. Comput.* **2016**, *45*, 317–378. [CrossRef]
255. Gupta, A.; Lee, E.; Li, J.; Manurangsi, P.; Włodarczyk, M. Losing treewidth by separating subsets. In Proceedings of the Thirtieth Annual ACM-SIAM Symposium on Discrete Algorithms, SODA, San Diego, CA, USA, 6–9 January 2019; pp. 1731–1749.
256. Jansen, B.M.; Pieterse, A. Polynomial Kernels for Hitting Forbidden Minors under Structural Parameterizations. In Proceedings of the 26th Annual European Symposium on Algorithms (ESA), Helsinki, Finland, 20–22 August 2018; pp. 48:1–48:15.
257. Donkers, H.; Jansen, B.M. A Turing Kernelization Dichotomy for Structural Parameterizations of F-Minor-Free Deletion. In *International Workshop on Graph-Theoretic Concepts in Computer Science*; Springer: Berlin, Germany, 2019; pp. 106–119.
258. Chekuri, C.; Chuzhoy, J. Polynomial bounds for the grid-minor theorem. *J. ACM (JACM)* **2016**, *63*, 40. [CrossRef]
259. Kawarabayashi, K.i.; Sidiropoulos, A. Polylogarithmic approximation for minimum planarization (almost). In Proceedings of the 2017 IEEE 58th Annual Symposium on Foundations of Computer Science (FOCS), Berkeley, CA, USA, 15–17 October 2017; pp. 779–788.
260. Agrawal, A.; Lokshtanov, D.; Misra, P.; Saurabh, S.; Zehavi, M. Polylogarithmic Approximation Algorithms for Weighted-F-Deletion Problems. In Proceedings of the Approximation, Randomization, and Combinatorial Optimization, Algorithms and Techniques (APPROX/RANDOM 2018), Princeton, NJ, USA, 20–22 August 2018; pp. 1:1–1:15.
261. Fiorini, S.; Joret, G.; Pietropaoli, U. Hitting diamonds and growing cacti. In *International Conference on Integer Programming and Combinatorial Optimization*; Springer: Berlin, Germany, 2010; pp. 191–204.
262. Here 1.1 can be replaced by $1 + \varepsilon$ for any constant $\varepsilon > 0$.
263. Marx, D.; Pilipczuk, M. Everything you always wanted to know about the parameterized complexity of Subgraph Isomorphism (but were afraid to ask). In Proceedings of the 31st International Symposium on Theoretical Aspects of Computer Science (STACS), Lyon, France, 5–8 March 2014; pp. 542–553.
264. Ebenlendr, T.; Kolman, P.; Sgall, J. An Approximation Algorithm for Bounded Degree Deletion. *Preprint* **2009**.

265. Alon, N.; Yuster, R.; Zwick, U. Color-coding. *J. ACM* **1995**, *42*, 844–856. [CrossRef]
266. Jansen, B.M.; Pilipczuk, M. Approximation and kernelization for chordal vertex deletion. *SIAM J. Discret. Math.* **2018**, *32*, 2258–2301. [CrossRef]
267. Cao, Y.; Sandeep, R. Minimum fill-in: Inapproximability and almost tight lower bounds. In Proceedings of the Twenty-Eighth Annual ACM-SIAM Symposium on Discrete Algorithms, Barcelona, Spain, 16–19 January 2017; pp. 875–880.
268. Giannopoulou, A.C.; Pilipczuk, M.; Raymond, J.F.; Thilikos, D.M.; Wrochna, M. Linear Kernels for Edge Deletion Problems to Immersion-Closed Graph Classes. In Proceedings of the 44th International Colloquium on Automata, Languages, and Programming ICALP, Warsaw, Poland, 10–14 July 2017; pp. 57:1–57:15.
269. Bliznets, I.; Cygan, M.; Komosa, P.; Pilipczuk, M. Hardness of approximation for H-free edge modification problems. *ACM Trans. Comput. Theory (TOCT)* **2018**, *10*, 9. [CrossRef]
270. Chen, J.; Liu, Y.; Lu, S. Directed feedback vertex set problem is FPT. In Proceedings of the Structure Theory and FPT Algorithmics for Graphs, Digraphs and Hypergraphs, Dagstuhl, Germany, 8–13 July 2007.
271. Chen, J.; Kanj, I.A.; Xia, G. Improved parameterized upper bounds for vertex cover. In *International Symposium on Mathematical Foundations of Computer Science*; Springer: Berlin, Germany, 2006; pp. 238–249.
272. Bourgeois, N.; Escoffier, B.; Paschos, V.T. Efficient Approximation of Combinatorial Problems by Moderately Exponential Algorithms. In Proceedings of the Algorithms and Data Structures, 11th International Symposium, WADS, Banff, AB, Canada, 21–23 August 2009; pp. 507–518._44. [CrossRef]
273. Brankovic, L.; Fernau, H. Combining Two Worlds: Parameterised Approximation for Vertex Cover. In *International Symposium on Algorithms and Computation*; Springer: Berlin/Heidelberg, Germany, 2010; pp. 390–402.
274. Brankovic, L.; Fernau, H. A novel parameterised approximation algorithm for minimum vertex cover. *Theor. Comput. Sci.* **2013**, *511*, 85–108. [CrossRef]
275. Bansal, N.; Chalermsook, P.; Laekhanukit, B.; Nanongkai, D.; Nederlof, J. New Tools and Connections for Exponential-Time Approximation. *Algorithmica* **2019**, *81*, 3993–4009. [CrossRef]
276. Manurangsi, P.; Trevisan, L. Mildly Exponential Time Approximation Algorithms for Vertex Cover, Balanced Separator and Uniform Sparsest Cut. In Proceedings of the Approximation, Randomization, and Combinatorial Optimization. Algorithms and Techniques, APPROX/RANDOM, Princeton, NJ, USA, 20–22 August 2018; pp. 20:1–20:17. [CrossRef]
277. Bar-Yehuda, R.; Bendel, K.; Freund, A.; Rawitz, D. Local ratio: A unified framework for approxmation algrthms. *ACM Comput. Surv.* **2004**, *36*, 422–463. [CrossRef]
278. Escoffier, B.; Monnot, J.; Paschos, V.T.; Xiao, M. New results on polynomial inapproximability and fixed parameter approximability of edge dominating set. *Theory Comput. Syst.* **2015**, *56*, 330–346. [CrossRef]
279. Bonnet, É.; Paschos, V.T.; Sikora, F. Parameterized exact and approximation algorithms for maximum k-set cover and related satisfiability problems. *RAIRO-Theor. Inform. Appl.* **2016**, *50*, 227–240. [CrossRef]
280. Arora, S.; Barak, B.; Steurer, D. Subexponential Algorithms for Unique Games and Related Problems. *J. ACM* **2015**, *62*, pp. 42:1–42:25. [CrossRef]
281. Barak, B.; Raghavendra, P.; Steurer, D. Rounding Semidefinite Programming Hierarchies via Global Correlation. In Proceedings of the IEEE 52nd Annual Symposium on Foundations of Computer Science, FOCS, Palm Springs, CA, USA, 22–25 October 2011; pp. 472–481. [CrossRef]
282. Fernau, H. Saving on Phases: Parameterized Approximation for Total Vertex Cover. In Proceedings of the Combinatorial Algorithms, 23rd International Workshop, IWOCA, Tamil Nadu, India, 19–21 July 2012; Revised Selected Papers; pp. 20–31._3. [CrossRef]
283. Halperin, E. Improved Approximation Algorithms for the Vertex Cover Problem in Graphs and Hypergraphs. *SIAM J. Comput.* **2002**, *31*, 1608–1623. [CrossRef]
284. Impagliazzo, R.; Paturi, R.; Zane, F. Which Problems Have Strongly Exponential Complexity? *J. Comput. Syst. Sci.* **2001**, *63*, 512–530. [CrossRef]
285. Lampis, M. A kernel of order 2 k-c log k for vertex cover. *Inf. Process. Lett.* **2011**, *111*, 1089–1091. [CrossRef]
286. Here we consider the version where the set of candidate centers is not separately given.
287. Hochbaum, D.S.; Shmoys, D.B. A unified approach to approximation algorithms for bottleneck problems. *J. ACM* **1986**, *33*, 533–550. [CrossRef]

288. Brand, C.; Dell, H.; Husfeldt, T. Extensor-coding. In Proceedings of the 50th Annual ACM SIGACT Symposium on Theory of Computing, STOC 2018, Los Angeles, CA, USA, 25–29 June 2018; pp. 151–164. [CrossRef]
289. Björklund, A.; Lokshtanov, D.; Saurabh, S.; Zehavi, M. Approximate Counting of k-Paths: Deterministic and in Polynomial Space. In Proceedings of the 46th International Colloquium on Automata, Languages, and Programming, ICALP, Patras, Greece, 9–12 July 2019; pp. 24:1–24:15. [CrossRef]
290. Pratt, K. Waring Rank, Parameterized and Exact Algorithms. In Proceedings of the 2019 IEEE 60th Annual Symposium on Foundations of Computer Science (FOCS), Baltimore, MD, USA, 9–12 November 2019; pp. 806–823.
291. Björklund, A. Determinant Sums for Undirected Hamiltonicity. *SIAM J. Comput.* **2014**, *43*, 280–299. [CrossRef]
292. Björklund, A.; Husfeldt, T.; Kaski, P.; Koivisto, M. Narrow sieves for parameterized paths and packings. *J. Comput. Syst. Sci.* **2017**, *87*, 119–139. [CrossRef]
293. Marx, D. Completely inapproximable monotone and antimonotone parameterized problems. *J. Comput. Syst. Sci.* **2013**, *79*, 144–151. [CrossRef]
294. To be more precise, these problems need to be phrased as promise problems and NP-hardness is with respect to these. We will not go into details here.

© 2020 by the authors. Licensee MDPI, Basel, Switzerland. This article is an open access article distributed under the terms and conditions of the Creative Commons Attribution (CC BY) license (http://creativecommons.org/licenses/by/4.0/).

Editorial

Special Issue "New Frontiers in Parameterized Complexity and Algorithms": Foreward by the Guest Editors

Neeldhara Misra [1], Frances Rosamond [2,*] and Meirav Zehavi [3]

[1] Department of Computer Science and Engineering, Indian Institute of Technology Gandhinagar, Palaj, Gandhinagar 382355, India; neeldhara.m@iitgn.ac.in
[2] Department of Informatics, University of Bergen, Postboks 7803, 5020 Bergen, Norway
[3] Department of Computer Science, Ben-Gurion University, Beer-Sheva 8410501, Israel; meiravze@bgu.ac.il
* Correspondence: Frances.Rosamond@uib.no

Received: 11 September 2020; Accepted: 15 September 2020; Published: 18 September 2020

Abstract: This Special Issue contains eleven articles—surveys and research papers—that represent fresh and ambitious new directions in the area of Parameterized Complexity. They provide ground-breaking research at the frontiers of knowledge, and they contribute to bridging the gap between theory and practice. The scope and impact of the field continues to increase. Promising avenues and new research challenges are highlighted in this Special Issue.

Keywords: FPT; parameterized complexity; kernelization; lower bounds; fine-grained; Heuristics; turbo-charged; ETH/SETH

1. Introductory Remarks

Parameterized Complexity has developed strongly and there is a long way to go, as evidenced by the directions of the papers in this Special Issue.

The high-quality research in the field has been recognized by funding agencies (ERC, NSF, many national research councils), by research fellowships (EATCS, ACM, Marie Curie, Lise Meitner and others), by memberships in Royal Societies, Academy Europaea, and Academy of Science in various countries, and by prizes and awards (Humboldt Research Prize, Witold Lipski Prize, Krill Prize, NSF Career and others). Foundational papers are evinced in a broad sweep of areas including AI, Access Control, Business, Bioinformatics, Computational Geometry, Computational Social Choice, Cognitive Science, Computational Medicine, Machine Learning, Phylogeny, Psychology, Operations Research and Scheduling, with many Best Paper Awards. About 8% of papers in top theoretical computer science conferences are related to Parameterized Complexity.

Since 2015, the annual Parameterized Algorithms and Computational Experiments Challenge (PACE) has served to deepen the relationship between parameterized algorithms and practice, and help bridge the divide between the theory of algorithm design and analysis, and the practice of algorithm engineering, as well as inspire new theoretical developments.

Books in the field include *Parameterized Complexity in the Polynomial Hierarchy: Extending Parameterized Complexity Theory to Higher Levels of the Hierarchy* (de Haan 2020). Two new books were published in 2019: *Cognition and Intractability: A Guide to Classical and Parameterized Complexity Analysis* (Iris van Rooij, et al. 2019), and *Kernelization: Theory of Parameterized Preprocessing* (Fomin et al. 2019). Many open problems remain that were proposed in *Parameterized Complexity*, 1999 and *Foundations of Parameterized Complexity*, 2013 (Downey and Fellows), *Invitation to Fixed-Parameter Algorithms* (Niedermeier 2002), *Parameterized Complexity Theory* (Flum, Grohe 2006), and *Parameterized Algorithms*

(Cygan et al. 2016). The wiki helps to keep the community informed, and archives the *Parameterized Complexity Newsletter* (www.wikidot.com).

The articles in this Special Issue represent fresh and ambitious research, in new directions in this area.

We would like to thank all the authors for the excellent surveys and papers in this Special Issue. We thank all the reviewers. Experts take reviewing very seriously, and the detailed help in this is valued by the authors. We thank Jones Zhang and the staff at the Algorithms Editorial Office, who worked closely with us to make this Special Issue a smooth and enjoyable experience. As an Open Access journal, the *Algorithms* Journal publishes quickly, which is a boon to the rapid dissemination of ground-breaking research, which you will find in this Special Issue.

2. New Frontiers

Several frontiers in Parameterized Complexity, either new or with well-established roots, but still in their infancy, are represented in this collection. All of these frontiers go "hand in hand" with the overarching goal of increasing the scope and impact of Parameterized Complexity. To this end, first and foremost, parameterized algorithms should be made practical.

This, of course, means that specific methods to enhance performance in practice are to be developed. Novel approaches to do so, with an emphasis on treewidth, the most well-studied width measure in the field, are given by Salvchev et al., and by Bannach and Brendt. Further, basic machinery in Parameterized Complexity should be exported to work when the input is not a graph. Indeed, most problems in practice are not about graphs, yet some of them can be modeled by graphs. For example, Integer Linear Programming (ILP) is a ubiquitous language to formulate problems, which is used in practice, and Ganian and Ordyniak survey how it can be modeled by parameterized graph problems. Additionally, Lin et al. do so for the kidney exchange problem. Taking a broader perspective, Bulteau and Weller survey new challenges for parameterized complexity in bioinformatics, where problems often concern entities that are not "just" graphs or are not graphs at all.

To design practical parameterized algorithms, additional considerations are often required. Specifically, in practice, it may not be clear how to determine what the "best" solution is. Therefore, we might wish to compute several solutions. Creignou et al. present results for the enumeration of solutions with FPT delay, and Baste et al. highlight diversity as a measure of the quality of a set of solutions. In addition to uncertainty with respect to the best solution, the input itself might be uncertain (in a different sense). This issue is discussed by Narayanaswamy and Vijayaragunathan. Further, the paradigm of parameterized complexity by itself might not be enough, for example, when no efficient fixed-parameter algorithm, or no fixed-parameter algorithm at all, is likely to exist for some problems. In this regard, a research direction that has received growing attention in recent years is that of parameterized approximation, surveyed by Feldmann et al., and a specific result is given by Li. Another framework to handle this is offered by SAT-solvers, which are often employed in practice, as surveyed by de Haan and Szeider.

Beyond Fixed-Parameter Tractability: New Paradigms. In recent years, the paradigm of Prameterized Complexity has been "combined" with other algorithm design paradigms, including approximation algorithms, streaming algorithms, dynamic algorithms and distributed algorithms. The last three focus on the design of parameterized algorithms that can be used under additional constraints/features of the input and the computation entities involved, such as a lack of space or the necessity to efficiently deal with small changes in the input that occur over time. Approximation algorithms, like Parameterized Complexity, is a framework to deal with computationally hard problems. Here, we compromise the "quality" of the solution, but still demand that the algorithm run in polynomial time. By compromising both quality and running time, we can compromise each of them by just a little bit rather than one of them by a lot. For example, we can achieve a $(1+\epsilon)$-approximate solution in FPT time for a problem that is unlikely to admit any constant-factor approximation algorithm in polynomial time and which is W[1]-hard, and hence enjoy both. In this collection, Feldmann et al. survey both positive algorithmic and hardness results in Parameterized Approximation. Further, Li presents a result in the context of circuits, showing negative results regarding the approximability of the Dominating Set by para-AC^0 circuits.

Another framework to deal with problems that are computationally hard even when FPT time is allowed is to employ SAT solvers. In particular, SAT solvers are often used in practice, and are very effective for this purpose. Set in the framework of Parameterized Complexity, this can mean that the objective is to design FPT-time reductions in SAT. Of course, this makes sense only for problems that are supposed not to be NP-complete, but belong to higher levels of the Polynomial Hierarchy. In this way, we harness the power of both Parameterized Complexity and SAT solvers, aiming to develop practical algorithms for very hard problems. In this collection, a comprehensive compendium by de Haan and Szeider discusses this framework and presents a long list of relevant problems that are known to be reducible in FPT time to SAT.

The "Best" Solution and the "Correct" Input. Motivated by both theoretical and practical considerations, the study of enumeration and counting problems is of broad interest. Indeed, in many scenarios that arise in practice, we might not be interested in computing a single solution, but we may be interested in several—and possibly even all—solutions to a given problem, or just to count how many solutions there are. For example, in many problems that arise in bioinformatics, it is not clear what the "best" solution is, since different solutions can represent different biological explanations (e.g., describe different evolutionary events), and it is not clear which is the correct one. In some cases, but not all, after offering these solutions, further experiments can be used to further sieve them. Therefore, perhaps the t "best" (or some t) solutions are sought. This yields different questions in enumeration, where we wish to list all solutions, some t best solutions, or each of the solutions one by one with modest delay. Clearly, the last option is the least restrictive if we enumerate the solution from "best" to "worse". The study of enumeration in the framework of Parameterized Complexity is relatively young. Here, when interested in the last option, we want the delay time to be FPT. Problems that admit enumeration algorithm with an FPT delay time are said to belong to the class DelayFPT. Creignou et al. study this setting, and present corresponding algorithms for modification problems on graphs and beyond.

When seeking some collection, say, of size t, of solutions, we may wish them to be "diverse", in some sense. Indeed, we may not have a reliable measure for what the t "best" solutions are, or perhaps we want to represent the entire solution space. In both cases, we may want the t output solutions to be, in some sense, as "diverse" as possible. There are various measures for the diversity of a collection. In particular, the work of Baste et al. makes use of the sum of the "distances" between every pair of output solutions. Specifically, they present parameterized algorithms to output a collection of t solutions for the d-HITTING SET problem (which are hitting sets whose size is, at most, some input parameter k), whose diversity measure is, at least, some input parameter p, where the distance between two solutions is the size of their symmetric difference.

Besides being uncertain about which solution we seek, we might be uncertain about the input itself. In particular, for a graph problem, we might be uncertain about which edges should exist—for example, in protein–protein-interaction networks, where vertices represent proteins and edges represent interactions between them, noisy and error-prone experiments to determine the interactions lead to uncertainty. Narayanaswamy and Vijayaragunathan survey different models of uncertainty as well as problems corresponding to them, some of which are motivated by biological and social networks. In addition, they zoom into the problem of finding the maximal core of a graph under uncertainty in two models, and present polynomial-time and FPT algorithms as well as a W[1]-hardness result.

Width Measures: New Perspectives. This collection also contains two articles that address practical issues, both focused on treewidth and tree decompositions, but from two different perspectives. One concerns the computation of treewidth itself, and the other concerns the solution of graph problems parameterized by treewidth (where a corresponding tree decomposition is given as input). Roughly speaking, treewidth is a measure of how close a graph is to a tree, where a tree decomposition is a structure that witnesses it, and this measure is among the most well-studied structural parameters in Parameterized Complexity. Its computation was also posed several times as a challenge in Parameterized Algorithms and Computational Experiments (PACE), an annual competition to design the fastest (in practice) algorithms for the parameterized problems posed as challenges in that year. On the one hand, the article by Slavchev, Masliankova and Kelk presents a new approach based on Machine Learning to obtain a faster (in practice) algorithm for the computation of the treewidth of a given graph. Specifically, it considers known algorithms to compute treewidth, and presents an approach to automatically learn which input characteristics make which algorithm among them better. On the other hand, the article by Bannach and Berndt considers algorithms based on dynamic programming on tree decompositions. It is noteworthy that almost all existing algorithms based on treewidth make use of dynamic programming, though this is sometimes coupled with other modern techniques. Specifically, Bannach and Berndt focus on the practicality of these algorithms, and also present a software for this purpose.

While many problems that need to be solved in practice do not concern graphs, some of them can be represented by using graphs, and thereby give rise to the usage of structural graph parameters such as treewidth. Ganian and Ordyniak present a survey that has this view with respect to Integer Linear Programming (ILP), a well-known, very general problem that encompasses many basic problems in computer science. Specifically, they consider a standard graphical representation from the study of constraint satisfaction problems that describe the interactions between the variables and constraints (namely, which variable is present, having a nonnegative coefficient, in which constraint). They survey recent development in this regard, considering width parameters such as treewidth, treedepth and cliquewidth.

Exporting Fixed-Parameter Tractability to Various Application Domains. Many computational problems that arise in Bioinformatics are NP-hard and require dealing with inputs of a very large size which are, nevertheless, structured, and hence naturally fit the framework of Parameterized Complexity. Indeed, Bioinformatics has been a main application domain for this framework for two decades. The survey by Bulteau and Weller presents an up-to-date survey of several topics in Bioinformatics from the viewpoint of Parameterized Complexity, with an emphasis on problems solved in practice, and addressing current questions, aiming, also, by providing concrete open problems, to advance further research in this regard. The survey presents an overview on various central topics: genome comparison and completion, genome assembly and sequence analysis, haplotyping and phylogenetics. For example, one direction for research in the first topic concerns distances between genomes, particularly rearrangement distances. Here, given two genomes, the objective is to decide whether one can be transformed to the other using, at most, k operations of a specific kind, motivated by biological considerations. This aims to reflect the evolutionary events that have occurred since their last common ancestor.

An article by Lin et al. considers a specific problem, called the Kidney Exchange problem, which aims to effectively utilize living-donor kidneys. This is modelled as a (di)graph problem, where the objective is to find a maximum weight packing of vertex-disjoint cycles and chains, where the length of cycles is upper-bounded by a given constant. This represents, in some sense, barter exchange. For example, when the cycle is of length 2, this corresponds to the scenario where a patient–donor pair can donate a kidney to some other pair and receive a compatible kidney from that pair in exchange.

Funding: The Norwegian NFR Toppforsk PCPC Project, The Bergen (BFS) Toppforsk; Israel Science Foundation grant no. 1176/18; United States-Israel Binational Science Foundation grant no. 2018302.

Conflicts of Interest: The authors declare no conflict of interest.

© 2020 by the authors. Licensee MDPI, Basel, Switzerland. This article is an open access article distributed under the terms and conditions of the Creative Commons Attribution (CC BY) license (http://creativecommons.org/licenses/by/4.0/).

MDPI
St. Alban-Anlage 66
4052 Basel
Switzerland
www.mdpi.com

Algorithms Editorial Office
E-mail: algorithms@mdpi.com
www.mdpi.com/journal/algorithms

Disclaimer/Publisher's Note: The statements, opinions and data contained in all publications are solely those of the individual author(s) and contributor(s) and not of MDPI and/or the editor(s). MDPI and/or the editor(s) disclaim responsibility for any injury to people or property resulting from any ideas, methods, instructions or products referred to in the content.

www.ingramcontent.com/pod-product-compliance
Lightning Source LLC
LaVergne TN
LVHW070505100526
838202LV00014B/1791